W9-AFB-428

Real Variables

Claude W. Burrill
IBM Corporation

John R. Knudsen
New York University

HOLT, RINEHART AND WINSTON, INC.
New York Chicago San Francisco Atlanta Dallas Montreal Toronto London Sydney

Library of Congress Catalog Card Number: 69-14521

SBN: 03-073110-0

Printed in the United States of America

9 8 7 6 5 4 3 2 1

To Kathleen and Ruth

Preface

The genesis of the present text lies in classroom notes prepared by the authors for students in mathematical analysis at New York University. Over a period of some ten years such notes were developed, revised, rewritten, and expanded more times than one can recall. The effort, a joint one between teachers and students, adjusted itself to the level of rigor demanded by the teacher, and the capacity to absorb demonstrated by the student. A reasonable course, substantially meeting both requirements, seems to have evolved. The pace of the book tends to be slow and deliberate at first, yielding to more rapid development only as the student acquires the capacity to accelerate.

The level of instruction is designed to meet the requirements of the mathematics major at the senior and possibly even the junior level, as well as those of the student of engineering and science at the senior or graduate level who is interested in a rigorous basis for applied mathematics. The text has been employed very successfully in these various categories. As far as prerequisites are concerned, essentially the needed mathematical background is a degree of maturity such as might be acquired in a course in advanced calculus or differential equations.

What ingredients constitute a reasonable basic course in real variable theory is a matter for debate. Many formulas are in vogue in current textbook literature. No claim is made that the recipe employed here is a new one, but perhaps its presentation is new since it is dictated solely by the needs of the student.

The authors have chosen to begin a course in real variable theory with a rigorous development of the real number system. They do not agree with those authors who begin by listing properties of the real numbers and assume that this suffices to provide the student with a knowledge of, and a feeling for, the real number system.

The first chapter deals with an axiomatic foundation for the system of integers and introduces as well the nuances of epsilon and delta, induction, the nature of proof, direct and indirect, and the elements of necessity and sufficiency. Chapter 2 expounds the ingredients of set theory necessary for subsequent usage. A careful analysis of infinite sets is presented, and at this point, too, the concept of function is introduced.

Chapter 3 develops the rational number system on the basis of equivalence classes and introduces the definition of a real number as a Dedekind cut. The authors firmly believe that the time spent on the details of Dedekind cuts has a value beyond that of portraying the nature of number; it provides the student with an opportunity to use the tools of mathematical reasoning, and it establishes a frame of mind desirable for the analysis to follow.

Chapter 4 is an introduction to line topology. The role of neighborhood and the nature of closed and open sets is treated in detail sufficient to the needs of the measure concept in later chapters.

Chapters 5 and 6 deal with sequences and series of real numbers, 7 and 8 with limits and continuity of functions, 9 with differentiability, and 10 with the Riemann integral. In the latter, three equivalent definitions of integral are given to provide a basis for generalization to Stieltjes and Lebesgue integrals in later chapters. In Chapters 11 and 12 sequences and series of functions are studied in some detail. Much is made of uniform convergence and its role in the determination of properties of limit functions. Power series are investigated fully, as is the relationship between a function and its Taylor expansion.

Chapters 13 and 14 introduce the notions of monotone function, bounded variation, and the Stieltjes integral. Measure of a set of real numbers and the concept of measurable function are introduced in Chapter 15, followed by the Lebesgue integral in Chapter 16. Finally, in Chapter 17 the notion of general measure is discussed, based on distribution and outer measure functions, as well as the ultimate generalization of integration.

The book is intended for a full year course, but is such that a one-semester course, not including measure and advanced integration, is possible. The authors have given one-semester courses to applied science students, taking them through the first ten chapters or, in the case of better students, through twelve chapters. The text is so written that a cursory study of Chapter 3 is possible. Details of proofs of Dedekind cut material may be omitted and replaced by the summary of properties of real numbers at the end of the chapter.

The text abounds in illustrative examples and counterexamples. In all there are 773 exercises for the student, some routine and some of moderate or greater difficulty.

To express thanks to those individuals who have made a contribution to the spirit, presentation, or contents of this book would be to name the thousand and more students who saw the authors through the various stages of development. A blanket expression of gratitude must suffice. To the Bell Telephone Laboratories for every consideration and encouragement and for supplying students of top calibre, also our sincere appreciation.

Special thanks go also to C. W. Langley, M. Meisner, and I. F. Ritter for constructive criticism.

NEW YORK, NEW YORK *C.W.B. J.R.K.*

JANUARY 1969

Contents

1

The System of Integers

"Where shall I begin, please your Majesty?" he asked.
"Begin at the beginning," the King said, very gravely,
"and go on till you come to the end: then stop."
LEWIS CARROLL (1832–1898), *Alice in Wonderland*

1-1 Introduction

When we speak of the *set of all integers*, denote it by the symbol I, and say of each integer that it is a *member of* or *belongs to* I, we employ terminology which we assume to be meaningful to the reader. Actually, the concepts of *set* and *member* and the notion of *belonging to a set* will remain undefined in terms of more basic concepts. This is characteristic of the early stages in the study of any branch of mathematics. Some point must serve as a beginning; accordingly, an understanding of certain terminology and denotation must be assumed.

We shall employ the term *system of integers* when we think of the members of I not only as distinguished by membership in a common set but also as subject to a variety of interrelations based on such notions as addition, multiplication, and order. The symbol I will be used to denote the *system* as well as the *set* of integers. It is convenient also to refer to the system of integers as a particular instance of a *system of numbers* and, as such, to call each member of I a *number*. The development and analysis of *number systems*—in particular those of *integers rational numbers*, and *real numbers*—will be our first concern. The last will serve as a basis for the study of *functions of a real variable*.

1-2 The Axiomatic Foundation

While we assume that the reader is conversant with the system of integers and its properties, we propose here to discuss its structure as though it were being encountered by the reader for the first time. This we do not only to emphasize the role of the integers as a natural beginning or basis for the study of the real number system but also to set the stage for the level of rigorous discussion which will characterize the chapters to come.

We assert that we have for consideration a set I whose members are called integers (or numbers) and are denoted by symbols such as a, b, c, m, and n. Concerning the members of I, we postulate certain principles or *axioms* which we neither prove nor defend but which, in effect, formally define the integers as a system of numbers. Certain of the axioms serve to declare the existence of integers playing special roles. Others particularize the properties of the operations of addition and multiplication as well as the order relation. Still others spell out characteristics of the set I considered as a single entity.

For simplicity we omit here discussion of the consistency of the axioms set forth, assuming such to be evident. Also, we make no attempt to minimize the number of axioms; our concern is not with compactness of presentation but rather with clarity of exposition.

It will be important in what follows that the notion of equality of integers be clear. Properly speaking, to say of two integers a and b that they are *equal* and to write $a = b$ signifies nothing more than that a and b are two symbols denoting one and the same integer. Thus, equal integers are identical integers. It is trivial that $a = a$ for any integer a. This feature is described as the *reflexive* property of equality. It is also true that equality is *symmetric*; that is, if $a = b$, then $b = a$. Further, it satisfies the *transitive law*, which means that if $a = b$ and $b = c$, then $a = c$.

The negation of the relation of equality between two integers will be denoted by the symbol $\neq$. Thus $a \neq b$ means that a and b are different members of the set I.[1]

1-3 Addition and Multiplication

As a first step in the construction of the number system of integers, we assert concerning the set I that corresponding to each pair of members a and b there exist unique integers, one to be known as their *sum* and one their *product*. The sum of a and b we denote by $a + b$ and their product by ab (or $a \cdot b$). The operation associated with the determination of the number $a + b$ when a and b are given is called *addition* and that with ab *multiplication*. The character of each

[1] Quite generally, the mark / drawn through a symbol for a relationship serves to denote the negation or denial of that relationship.

of these operations and the interrelations between them will follow from their properties as set forth in the axioms below. At the outset we simply declare that the operations of addition and multiplication are such that integers $a + b$ and ab both exist for each pair of integers a and b.

Two obvious rules follow directly from the stated meaning of equality and the stipulated uniqueness of sums and products. If a and b are symbols for one integer (that is, $a = b$) and if c and d are likewise (that is, $c = d$), then it is true that $a + c = b + d$, since $a + c$ and $b + d$ are by definition one and the same integer. Similarly, $ac = bd$.

The sum and product of two integers a and b are not to depend on the order in which the integers a and b are presented or written. Although we speak of $a + b$ as the sum obtained by *adding b to a* and of $b + a$ as the result of *adding a to b*, we mean the two sums to be one and the same integer. Similarly, this is to be true of the integers ab and ba obtained, respectively, from the *multiplication of a by b* and that of *b by a*. Formally, we express this independence in our first two axioms.

Axiom 1 For any integers a and b, $a + b = b + a$.

Axiom 2 For any integers a and b, $ab = ba$.

These we refer to, respectively, as the *commutative laws for addition and multiplication.*[2]

If to the sum $a + b$ of the numbers a and b is added any integer c, the resulting sum is reasonably denoted by $(a + b) + c$. Three or even more integers may be added by performing a succession of additions of two at a time. Naturally this can be done in several ways, and, conceptually at least, the outcomes may differ. Thus $a + (b + c)$, where the sum of b and c is added to a, is not equal to $(a + b) + c$ simply by virtue of the existence of both sums. For multiplication this is likewise the case; it is not a consequence of definition that the symbols $(ab)c$ and $a(bc)$, with obvious significance, represent one and the same integer.

It is scarcely necessary to point out the many complications that must arise if the sum or product of three or more numbers depends on the manner in which these integers are arranged or grouped. Before going further we take care to avoid such complication by making it axiomatic that the arrangement of integers in an addition or multiplication does not influence the sum or product. Thus we have two further axioms.

Axiom 3 For any integers a, b, and c, $(a + b) + c = a + (b + c)$.

Axiom 4 For any integers a, b, and c, $(ab)c = a(bc)$.

[2] The reader is possibly aware of systems of entities in which operations are defined that are not commutative. An obvious example is that of the *vector* or *cross-product* of two vectors **A** and **B**. Thus $\mathbf{A} \times \mathbf{B} = -\mathbf{B} \times \mathbf{A} \neq \mathbf{B} \times \mathbf{A}$.

In matrix algebra, too, an operation called multiplication is so defined that if A and B denote any two matrices, then their products AB and BA may be such that one or both do not exist, or if both exist, they need not be equal. Thus, matrix multiplication is not commutative.

These are referred to as the *associative laws for addition and multiplication*, respectively. Because of such laws it is clear that without ambiguity we may denote the sum and product of a, b, and c by the respective symbols $a + b + c$ and abc.

1-4 The Law of Cancellation for Addition

The concept of addition assures the existence of a unique sum corresponding to two given integers. However, it does not follow, either from the definition or from the above axioms, that if a and b are given integers, then an integer x exists with the property that $a + x = b$. To the end that such be the case, we postulate in Axioms 5 and 6 two particular instances of this property.

Axiom 5 There exists an integer 0 (*zero*) such that $a + 0 = a$ for every integer a.

Actually there can exist only one integer with this property. For if $0'$ and $0''$ are such that $a + 0' = a$ and $a + 0'' = a$ for any a, then with $a = 0''$ in the first of these and with $a = 0'$ in the second we obtain $0'' + 0' = 0''$ and $0' + 0'' = 0'$. Because of Axiom 1 we have $0'' + 0' = 0' + 0''$, and it follows at once that $0'' = 0'$. This proves the uniqueness of the integer 0. We refer to the number 0 as the *identity with respect to addition*.

Axiom 6 Corresponding to each integer a there exists in I an integer $-a$ (*minus* a) with the property that $a + (-a) = 0$.

We refer to $-a$ as the *negative of* a or as the *inverse of* a *with respect to addition*. The negative of a given integer a is easily seen to be unique. For if b and c are such that $a + b = 0$ and $a + c = 0$, then $b = b + 0 = b + (a + c) = (b + a) + c = 0 + c = c$.

Following Axiom 5 we proved that there is only one member of I having the special property described in that axiom. Axiom 6 also is followed by a uniqueness proof, but the statement in this instance is quite different. Here we merely say that *after* an integer a has been prescribed, it follows that there is only one element of I, conveniently denoted by $-a$, for which $a + (-a) = 0$. It does not follow that $b + (-a) = 0$ for any integer b. Rather, corresponding to b is a unique integer $-b$ such that $b + (-b) = 0$. We would expect that $-b$ and $-a$ are different if b and a differ, and this is readily verified. For if a and b are two integers and if $-a = -b$, then $a = a + (-b) + b = a + (-a) + b = b$ or $a = b$. Thus if a and b differ, so must $-a$ and $-b$.

An immediate consequence of the definition of the negative $-a$ of an integer a is that a is the negative of $-a$. Thus, from $(-a) + a = 0$ follows that a is that integer otherwise denoted by $-(-a)$; that is, $-(-a) = a$. Also note that $-0 = 0$.

A property of the system of integers, which we refer to as the *law of cancellation for addition*, follows directly from axioms.

Theorem 1-1 If a, b, and c are any integers with the property that $a + c = b + c$, then $a = b$.

Proof This conclusion is immediate since $a + c = b + c$ implies that $a + c + (-c) = b + c + (-c)$ or $a + 0 = b + 0$. Hence $a = b$▲

Corollary If $a + b = a$, then $b = 0$.

A particularly important consequence alluded to earlier of Axioms 5 and 6 we now state in the form of a theorem.

Theorem 1-2 Corresponding to any integers a and b there exists a unique integer x with the property that $a + x = b$.

Proof Assuming x to exist, it follows from $a + x = b$ that $a + x + (-a) = b + (-a)$ or $x = b + (-a)$. We conclude that if x exists, then it must be the (unique) integer $b + (-a)$. That the latter actually possesses the desired property is seen by direct computation. Thus $a + [b + (-a)] = [a + (-a)] + b = 0 + b = b$▲

The law expressed in Theorem 1-2 we refer to as the *principle of solvability for addition*. Observe that Axioms 5 and 6 are special cases of Theorem 1-2 but that these two axioms alone do not imply the theorem.

The number $b + (-a)$ encountered above is conveniently denoted by the symbol $b - a$ and is referred to as the *difference of b and a*. We speak of it as having been obtained as a result of the operation of *subtraction*; that is, a is *subtracted* from b. Observe that the operation of subtraction, unlike that of addition, does not exhibit the properties of commutativity or associativity.

1-5 The Law of Cancellation for Multiplication

With respect to the operation of multiplication, we postulate the existence of an identity integer by means of the following axiom.

Axiom 7 There exists an integer 1 (*one*) such that $a1 = a$ for any integer a.

The number 1 is the *identity with respect to multiplication*. Its uniqueness follows readily, employing an argument like that used to establish the uniqueness of 0.

The parallelism between the properties of addition and multiplication evident in the first four axioms and in Axioms 5 and 7 will not be pursued further. In particular, we shall not insist on the existence of an inverse with respect to multiplication for each integer a. The effect of this is to forego a principle of solvability for multiplication. Indeed, we shall see later that the equation $ax = b$ will not admit of a solution for all pairs of integers a and b.

In the case of addition, a law of cancellation was consequent upon the existence of an inverse. For multiplication we cannot deduce such a law since no inverse is assumed; hence the following axiom.

Axiom 8 If a, b, and c are integers such that $c \neq 0$ and $ac = bc$, then $a = b$.

This is the *law of cancellation for multiplication.*

1-6 The Distributive Law

Each of the axioms so far encountered has dealt with one operation exclusively. A property of the system of integers which relates the operations of addition and multiplication is postulated in the next axiom.

Axiom 9 For any integers a, b, and c, $a(b + c) = ab + ac$.

This is the so-called *distributive law for multiplication with respect to addition.* However, we shall not include among our axioms, nor will it be derivable from those set forth, that addition is distributive with respect to multiplication. Thus, in general, $a + bc$ is not equal to $(a + b)(a + c)$.

1-7 Some Implications of the Axioms

Certain properties of the system of integers follow directly from the axioms set forth so far. Examples of such are demonstrated in the following theorems.

Theorem 1-3 $a0 = 0$ for any integer a.

Proof We have, from the defining relation for the integer 0, the simple fact that $1 + 0 = 1$. Hence $a(1 + 0) = a1$. Axiom 9 may then be applied to yield $a1 + a0 = a1$, whence $a0 = 0$ by the corollary to Theorem 1-1▲

Theorem 1-4 If a and b are integers such that $ab = 0$, then $a = 0$ or $b = 0$ (or both).[3]

Proof From the hypothesis we have $ab = 0$, and from Theorem 1-3 we have $a0 = 0$. It follows that $ab = a0$. If $a = 0$, there is nothing further to prove. On the other hand, if $a \neq 0$, it then follows from Axiom 8 that $b = 0$▲

Theorem 1-5 $(-a)b = a(-b) = -(ab)$ for any integers a and b.

[3] It is conventional in mathematics that the word *or* is used in the inclusive sense and signifies *both* as well as *either*. Thus the statement "$a=0$ or $b=0$" implies one of three possibilities: (1) $a=0$, $b\neq 0$; (2) $a\neq 0$, $b=0$; (3) $a=0$, $b=0$. We shall adhere to this convention throughout, employing the term "or both" now and then for emphasis only.

Proof We have $ab + (-a)b = [a + (-a)]b = 0b = 0$ employing the distributive law and the definition of a negative. We conclude that $(-a)b = -(ab)$. From this and from Axiom 2 it follows further that $a(-b) = (-b)a = -(ba) = -(ab)$ ▲

For simplicity we write the negative of ab as $-ab$; thus $-(ab) = -ab$. Because of Theorem 1-5 this will lead to no ambiguity.

Theorem 1-6 $(-a)(-b) = ab$ for any integers a and b.

Proof We have $(-a)(-b) = (-a)(-b) + (-a)b + ab = (-a)[(-b) + b] + ab = 0 + ab = ab$ ▲

Exercises

1–1 Prove that the integer 1 of Axiom 7 is unique.

1–2 Prove that if a, b, and c are any integers, then (a) $(-1)a = -a$, (b) $-(a+b) = -a-b$, (c) $-(a-b) = -a+b$, and (d) $a(b-c) = ab - ac$.

1–3 Consider Axiom 8′: If a and b are any integers such that $ab = 0$, then $a = 0$ or $b = 0$. Prove that Axiom 8′, together with Axioms 1 through 7 and 9, implies Axiom 8.

1–4 Let F denote a system in which operations of addition and multiplication are defined satisfying Axioms 1 through 8 together with Axiom 9′: If a, b, and c are any members of F, then $(a+b)(a+c) = a + bc$. Prove that F consists of a single member only; this implies, in particular, that $0 = 1$ in F.

1-8 The Order Relation

We now postulate for the members of the set I an *order relation*, which we describe by the term *less than* and denote by the symbol $<$. The relation $a < b$, which we read "a is less than b," we shall also write $b > a$ and read "b is greater than a." We employ the notation $a \leq b$ to signify that either $a < b$ or $a = b$. Similarly for $a \geq b$. Concerning the order relation, we specify four axioms.

Axiom 10 If a, b, and c are integers such that $a < b$ and $b < c$, then $a < c$.

This is the so-called *transitive law* for the order relation. From it follows the transitive law for the relation $\leq$.

If a, b, and c are integers such that $a < b$ and $b < c$, we may write compositely $a < b < c$ and say that b is *between* a and c. The symbols $a \leq b \leq c$, $a < b \leq c$ and $a \leq b < c$ are self-explanatory.

Axiom 11 If a, b, and c are integers such that $a < b$, then $a + c < b + c$.

This axiom implies its own converse; that is, from $a + c < b + c$ follows $a + c + (-c) < b + c + (-c)$, whence $a + 0 < b + 0$ or $a < b$.

Axiom 12 Any integers a and b are related by exactly one of the following: $a < b$, $a = b$, $a > b$.

This is sometimes referred to as the *trichotomy* for the order relation. It implies that each of the relations $a < b$ and $a > b$ implies $a \neq b$, and, conversely, $a \neq b$ implies either $a < b$ or $a > b$.

The notions of positive and negative integers are defined through the order relation. Thus, an integer a is said to be *positive* if $0 < a$ and *negative* if $a < 0$. The set of all positive integers will be denoted by I_+. It is sometimes referred to as the set of *natural numbers*. The set of negative integers we symbolize by I_-. The members of I_+ and I_- together with the integer 0 (which is neither positive nor negative) constitute all the members of the set I. It is clear that every negative integer is less than every positive integer. An axiom involving the concept of positive integers follows.

Axiom 13 If a, b, and c are integers such that $a < b$ and $0 < c$, then $ac < bc$.

1-9 Further Consequences of the Axioms

Theorem 1-7 If a and b are positive integers, then ab and $a + b$ are positive.

Proof From $0 < a$ and $0 < b$ it follows by Axiom 13 that $0b < ab$, or, what is the same, $0 < ab$. Also, from Axiom 11 we have $0 + b < a + b$ or $b < a + b$. Since $0 < b$, it follows by Axiom 10 that $0 < a + b$▲

Theorem 1-8 If a is a positive integer, then $-a$ is a negative integer. Also, if a is negative, then $-a$ is positive.

Proof From $0 < a$ it follows by Axiom 11 that $0 + (-a) < a + (-a)$ or $-a < 0$. Similarly for the second part of the theorem▲

The reader is asked to supply proofs for Theorems 1-9 and 1-10.

Theorem 1-9 (a) If a and b are negative integers, then ab is positive. (b) If $a \neq 0$, then a^2 is positive.[4] (c) If a is positive and b is negative, then ab is negative.

Theorem 1-10 (a) If $0 < a < b$ and $0 < c < d$, then $ac < bd$. (b) $a < b$ if and only if $a - b < 0$.

1-10 Elimination of a Trivial System

The totality of different integers that makes up the set I has not been referred to so far. It has been tacitly assumed that I is sufficiently endowed with members that it is meaningful to speak of two different members so that, among other

[4] a^2 is used to denote the product aa.

things, the order relation actually has application. Further, the identities for addition and multiplication have been represented by different symbols 0 and 1 with the suggestion that they denote different integers.

Actually the axioms postulated thus far do not warrant such conclusion because all the axioms are satisfied by a system consisting of the single member 0. This is clearly the case in Axioms 1 through 6 and in Axiom 9 if the four symbols a, b, c, and $-a$ are all identical with 0. In Axiom 7 it is not inconsistent to identify the integer 1 with 0 since the only value a may have is 0. Axiom 8 is satisfied vacuously, as are Axioms 10 through 13, since all concern members that do not exist. Thus, in Axiom 8, for example, anything at all may be said to be a consequence of $c \neq 0$ if no integer c exists to deny it.

It is apparent that if the system of integers under construction is to be other than the trivial system consisting of the single member 0, then an axiom must be postulated to that effect. To this end we include the following:

Axiom 14 There exists in the set I at least one integer that is different from 0.

As a consequence we have the next theorem.

Theorem 1-11 The integers 0 and 1 are not identical.

Proof Let a be a member of I which is different from 0. Then, on the supposition that $1 = 0$, it follows that $a1 = a0$; and this in turn implies that $a = 0$. The obvious contradiction points to the absurdity of the supposition $1 = 0$. Therefore $1 \neq 0$▲

We can now observe that not every equation $ax = b$ has a solution. In particular, there can be no integer x such that $0x = 1$; for if there were, it would follow that $0 = 0x = 1$, contrary to Theorem 1-11.

Theorem 1-12 The integer 1 is positive.

Proof From Axiom 7 it follows that $1^2 = 1$. Since $1 \neq 0$, the desired conclusion follows from Theorem 1-9▲

Theorem 1-13 If a is any integer, then $a < a + 1$.

Proof The truth of this is an immediate consequence of Theorem 1-12 and Axiom 11▲

Exercises

1-5 In each of the following state and prove the relation existing between the integers a and c: (a) $a < b$ and $b \leq c$; (b) $a \leq b$ and $b < c$; (c) $a \leq b$ and $b \leq c$.

1-6 Prove that if a, b, c, and d are any integers, then the relations $a < b$ and $c < d$ imply that $a + c < b + d$.

1-7 Prove that if a, b, and c are any integers, then the relations $a<0$ and $b<c$ imply $ac<ab$.

1-8 Prove Theorem 1-9.

1-9 Prove Theorem 1-10.

1-10 Prove that if a, b, and c are integers with the properties $a=bc$, $b\geq 1$, and $c\geq 0$, then $a\geq c$.

1-11 Prove that if a and b are positive integers such that $a^2=b^2$, then $a=b$.

1-12 Prove that if a and b are integers such that $a^2+b^2=0$, then $a=b=0$.

1-13 Let J denote a system consisting of two members 0 and 1 $(0\neq 1)$ with addition and multiplication defined by

$$0+0=0 \qquad 1+0=1$$

$$1+1=0 \qquad 0+1=1$$

$$0\cdot 0=0 \qquad 0\cdot 1=0$$

$$1\cdot 1=1 \qquad 1\cdot 0=0.$$

(a) Show that this system satisfies Axioms 1 through 9. (b) Show that it is not possible to define an order relation for J in such a way that both Axioms 11 and 12 are satisfied.

1-14 Show, by means of a counterexample, that in I addition is not always distributive with respect to multiplication.

1-11 The Least Member Axiom

We shall see later that the rational and real number systems satisfy all the axioms set forth so far for the system of integers. A property of the system of integers not to be shared by either the rational or real number system is postulated now; this is the so-called *least member axiom.*

Axiom 15 Let A denote any set of integers with the following properties:

(i) A possesses one or more members.

(ii) An integer b exists such that for each member n of A it is the case that $b\leq n$.

Then there exists in the set A an integer n_0 such that $n_0\leq n$ for each n in A.

A set which possesses one or more members will be called *nonempty.* A set of integers with the property that all its members exceed or equal some fixed integer is said to be *bounded below.* With such terminology Axiom 15 may then be paraphrased: *Every nonempty set of integers that is bounded below possesses a least member.*

The set of axioms is now complete and the system of integers defined. At this stage the reader may wonder if the system I that has been formally defined is consistent with his intuitive notion of integers; he may be concerned that further axioms are needed to insure that I possess all the properties that he generally ascribes to the integers. This concern should vanish as we develop the various properties that follow from the axioms.

Theorem 1-14 There exists no integer between 0 and 1.

Proof Denote by A the set of all integers that lie between 0 and 1, and assume that A is nonempty. Then by Axiom 15 the set A possesses a least member, which we denote by a. Thus $0 < a < 1$ and $a \leq n$ for each member n of A. From $a < 1$ and Axiom 13 we conclude that $a^2 < a$. Moreover, from Theorem 1-9 we have $0 < a^2$. Thus is exhibited an integer a^2 that, since it lies between 0 and 1, is a member of A, yet is less than the least member of A. This obvious absurdity can stem only from a false premise, i.e., that the set A is nonempty. We conclude that A is *empty*; that is, no integer lies between 0 and 1▲

Corollary For each integer n there exists no integer between n and $n + 1$.

The above suggests the use of the term *consecutive* to describe a pair of integers of the form n and $n + 1$. The integer $n + 1$ is called the *follower* of the integer n and n the *predecessor* of $n + 1$.

Theorem 1-15 (Principle of Archimedes) If a and b are positive integers, then an integer n exists such that $b \leq na$.

Proof Since $0 < a$, it follows by Theorem 1-14 that $1 \leq a$, so that $b \leq ab$. Hence, with $n = b$, the theorem follows▲

1-12 The Conventional Symbols

So far only the integers 0 and 1 have been denoted by special symbols. Now we introduce some others. For the sum $1 + 1$ we employ the symbol 2. Also, we write $2 + 1 = 3$, $3 + 1 = 4$, and so forth. In this fashion the set of integers is expressed in familiar terms. It is left to the reader to prove the various arithmetic number facts which concern addition and multiplication. For example, to prove that $2 + 2 = 4$ we have

$$2 + 2 = 2 + (1 + 1) = (2 + 1) + 1 = 3 + 1 = 4.$$

Also, that $3 \cdot 2 = 6$ follows from

$$3 \cdot 2 = 3(1 + 1) = 3 + 3 = 3 + 1 + 2 = 4 + 1 + 1 = 5 + 1 = 6.$$

We can also establish certain inequalities, for example, that 3 and 5 are different integers. We have

$$3 < 3 + 1 = 4 < 4 + 1 = 5,$$

whence $3 < 5$. This implies $3 \neq 5$.

It is not our purpose here to pursue the manifold properties and characteristics of the system of integers. The reader is acquainted with much already and may familiarize himself further by consulting sources devoted to such various topics as factorability of integers, distribution and interrelation of perfect squares and cubes, and the nature and distribution of prime numbers.

1-13 Logical Principles

In proving mathematical propositions or theorems it is necessary to accept certain logical principles or rules.[5] For the most part these are familiar to the reader through frequent usage. One of these principles, the so-called *law of the excluded middle*, is to the effect that any meaningful statement is either true or false. For example, the statement "if $a \neq 0$, then $a^2 > 0$" has been proved true while the assertion "an integer a exists such that $0 < a < 1$" has been shown to be false. Acceptance of the principle of the excluded middle forces the conclusion that a proposition is true if it has been shown to be not false and that it is false if demonstrated to be not true.

The principle of *noncontradiction* is also assumed; that is, a statement cannot be both true and false. If an assertion has been shown to be true, then it may not also be false; if false, then it is not true.

Generally, propositions or theorems we wish to prove (or disprove) will be in the form of *if . . . then* statements. For example, "if $a > 0$, then $-a < 0$" or "if a and b are integers, then $(-a)b = -(ab)$." Such propositions may be represented symbolically in the form "if p, then q" or "p implies q," where p and q stand for certain assertions. In the first example, p is the assertion "$a > 0$" while q is "$-a < 0$." Conventionally p is styled the *hypothesis* of the proposition and q the *conclusion*.

Regarding the proposition "if p, then q" three questions may be asked: Is p true? Is q true? Is the proposition itself true? Usually it is with the last that we are concerned. For example, consider the assertion "if Axioms 1 through 15 are valid, then $0 < 1$." Ignoring the question of truth or falsity of the individual statements "Axioms 1 through 15 are valid" and "$0 < 1$," we may concern ourselves solely with proving that *if* we assume the axioms to be valid, *then* it necessarily follows that $0 < 1$.

As far as the axioms are concerned, we shall say that they are true in the sense that we simply accept them as such. We assert them to be the basic truths upon which all that follows may be justified. In this sense then, all consequences of the axioms, for example, $0 < 1$, will be spoken of as being true.

We designate the negation of a statement p by the symbol $\sim p$ (read "not p"). For example, if p denotes "$a > 0$," then $\sim p$ is the statement "$a \not> 0$." Because

[5] Here we include but a bare outline of logical principles employed in the course of the text. For a full treatment of the subject the reader is referred to any of the standard treatises on symbolic logic.

of the law of the excluded middle, a double negation is equivalent to the original assertion. Thus p and $\sim\sim p$ (read "not not p") are one and the same statement.

If p implies q, then q is described as a *necessary condition* for p; that is, p cannot be true without q being true also. On the other hand, if q implies p, then q may be regarded as a *sufficient condition* for p. Although many conditions or facts may be found that are necessary for some statement and many that are sufficient, it is often a difficult task to determine a single assertion that is both *necessary and sufficient*. It is clearly the case that if q is necessary and sufficient for p, then p is likewise necessary and sufficient for q. Other expressions for the necessary and sufficient relation are sometimes employed. Thus we may say "p implies q and conversely" or "q if and only if p." The latter is occasionally abbreviated "q iff p."

To prove a theorem of the form "p implies q," an *indirect proof* known as *proof by contradiction* or *reductio ad absurdum* is frequently effective. Suppose we know certain statements $p, r, s, \ldots$ to be true (for example, Axioms 1 through 15 and their consequences) and we wish to prove the statement "p implies q." We begin by assuming that q is false or, what is the same, that $\sim q$ is true. If it is then possible to show that this statement in turn implies a denial of one of the statements $p, r, s, \ldots$ or even of $\sim q$ itself, then it follows that the assumption was in error, and hence q, not being false, must be true. This type of reasoning was employed in the proof of Theorem 1-11, where it was shown that the integers 0 and 1 are not identical. There we assumed that 0 and 1 are equal and showed that as a consequence every integer is identical with 0. Since this is in contradiction to Axiom 14, the assumption is unwarranted and therefore $1 \neq 0$. In Theorem 1-14, also, an indirect proof was employed to demonstrate that no integer lies between 0 and 1. There it was assumed that at least one integer does lie between 0 and 1. From this followed a denial of the least member axiom and hence the truth of the theorem.

Another technique frequently used to prove the theorem "p implies q" is that of proving the theorem "$\sim q$ implies $\sim p$." This is valid since the statements are equivalent. To see this, suppose that we know that $\sim q$ implies $\sim p$. Then assume that p is true and q is false. From "$\sim q$ implies $\sim p$" and the falsity of q follows the truth of $\sim p$ and hence the falsity of p, which is a contradiction. From this we see that if p is true, then q must also be true. Similarly, from the truth of "p implies q" follows the truth of "$\sim q$ implies $\sim p$."

The equivalence of "p implies q" and "$\sim q$ implies $\sim p$" was employed in the proof in Section 1-4 that if a and b are different integers, then $-a$ and $-b$ are different also.

In the above, the statement "if p, then q" was looked upon as a theorem or implication to be proved or disproved. The reader is cautioned that this is not the only manner in which the statement is employed in mathematical literature; it is also used in the statement of definitions. For example, to define the term "positive" we might say "if $a > 0$, then a is *positive*." This should not be construed as a theorem to be proved or disproved. We simply signify that "$a > 0$" and "a is positive" have the same meaning. A much more satisfactory method of writing the definition would be "a is positive if and only if $a > 0$." This last convention is beginning to appear in the literature, but it is certainly not standard.

Exercises

1-15 Let B denote a set of integers which is (i) nonempty and (ii) bounded above (that is, an integer m exists such that $n \leq m$ for each n in B). Prove that B contains a greatest member.

1-16 Prove that if a and b are integers such that $a-1<b<a+1$, then $b=a$.

1-17 Prove that if a, b, and c are integers such that $a+bc=0$ and $0 \leq a < b$, then $a=c=0$.

1-18 Prove the following: (a) $3+4=7$; (b) $(-2)(-2)=4$; (c) $3 \cdot 3=9$.

1-19 (a) Find the fallacy in the following proof. *Theorem.* 1 is the largest integer. *Proof.* Let a denote the largest integer and assume $a \neq 1$. It follows that $a>1$ and hence $a^2>a$. Thus a^2 is an integer larger than a, which is a contradiction. We conclude that the assumption $a \neq 1$ is in error. Hence $a=1$. (b) Prove that there is no largest integer.

Let a and c denote any integers. Then we say that *a is divisible by c* or that *c divides a* or that *c is a factor of a*, and write $c \mid a$, if an integer q exists such that $a=cq$.

1-20 Prove that if a and c are positive integers such that $c \mid a$, then $c \leq a$.

1-21 Prove that if a and b are any positive integers, then there exists a largest integer d such that $d \mid a$ and $d \mid b$. (The integer d is called the *greatest common divisor* (*or factor*) of a and b.)

1-22 Prove that if $c \mid a$ and $c \mid b$, then $c \mid (a+b)$.

1-23 Prove that if a is any integer and d any positive integer, then integers q and r exist such that $a=dq+r$ and $0 \leq r < d$. (This is called the *division algorithm.*) Prove also that q and r are unique.

An integer p is said to be *even* if an integer s exists such that $p=2s$; it is *odd* if an integer t exists such that $p=2t+1$.

1-24 Prove that (a) an integer cannot be both even and odd, and (b) each integer is either even or odd.

1-25 Prove that (a) if p is even, then pq is even; (b) if p and q are both odd, then pq is odd; and (c) p is even (odd) if and only if p^2 is even (odd).

An integer p is said to be *prime* if $p>1$ and if p has no factors other than ± 1 and $\pm p$.

1-26 Prove that if a is not a prime integer and if $a>1$, then a prime p exists such that $p \mid a$ and $p^2 \leq a$.

1-27 Prove that if p is prime and divides ab where a and b are positive and $a<p$, then $p \leq b$.

1-28 Prove that if p is prime and p divides ab, then p divides a or p divides b.

1-29 Let a be a positive integer and p a prime integer that does not divide a. Prove that integers s and t exist such that $1=as+pt$.

1-14 Mathematical Induction

Theorem 1-16 Let A be a set of integers such that (i) an integer n_0 is in A and (ii) if n is a member of A, then so is $n + 1$. Then the set A contains, together with n_0, all integers greater than n_0.

Proof Denote by C the set of all integers that are greater than n_0 but that are not contained in A. Assume that C is nonempty. Then, by Axiom 15, C possesses a least member c and, by definition of C, $c > n_0$. Now $c - 1$ is less than the least member of C and hence is not a member of C. However, since $c - 1 \geq n_0$, it follows that $c - 1$ is in A. Part (ii) of the hypothesis then implies that $(c - 1) + 1 = c$ is in A. But this is in contradiction to the premise that c is a member of C. Thus our assumption that C is not empty has led to a contradiction. We conclude that C possesses no members and that consequently each integer n such that $n \geq n_0$ is a member of A▲

A method of proof which finds application from time to time is that referred to as *mathematical induction*. We consider a proposition (a formula, a statement of equality or of inequality, or the like) which is concerned with an integer n and whose truth (or falsity) for some or all values of n is to be investigated. Such a proposition is conveniently denoted by the symbol $P(n)$, enabling us to speak of individual cases such as $P(3)$ and $P(7)$ corresponding to integers 3 and 7, respectively. For example, consider the proposition

$$P(n):\quad n(n^2 + 5) \text{ is a multiple of } 6.$$

It is obvious that "$P(2)$: $2(4 + 5)$ is a multiple of 6" is a correct statement, as is "$P(5)$: $5(25 + 5)$ is a multiple of 6."

A rather obvious fact should be noted, namely, that the truth of a proposition for several values of n in no way implies truth for all n. For example, consider the statement

$$P(n):\quad 5 + 4n - n^2 \geq 0.$$

Clearly it is true for $n = 1, 2, 3, 4$, and 5 but false for all larger values of n.

An even more striking example is the proposition

$$P(n):\quad n^2 - n + 41 \text{ is a prime number.}$$

One is impressed by the preponderance of evidence in favor of the truth of this statement for all positive values of n. Indeed, it may be verified that $n^2 - n + 41$ is a prime number for all values of n in the range $n = 1$ through $n = 40$. For $n = 41$, however, the proposition is not true.

Our special concern will be with certain propositions that are true for all positive integers n or at least for all integers greater than or equal to some one integer n_0. The validity of such statements can often be established by using the following theorem, which states the *principle of mathematical induction*:

Theorem 1-17 Suppose a proposition $P(n)$ to be stated for each integer n such that $n \geq n_0$ where n_0 is a particular integer. If (i) the statement $P(n_0)$ is true and if (ii) the assumed truth of $P(k)$ implies the truth of $P(k+1)$ for each $k \geq n_0$, then the proposition is true for all $n \geq n_0$.

Proof Denote by A the set of all integers n for which $P(n)$ is true. Then A is such a set of integers as that defined in the hypothesis of Theorem 1-16. As a consequence, A contains all integers $n \geq n_0$, or, what is the same, $P(n)$ is true for all $n \geq n_0$ ▲

Example 1-1 For $n \geq 1$ we prove the proposition

$$P(n): \quad n(n+1) \text{ is an even integer (i.e., it is a multiple of 2).}$$

It is immediate that $P(1)$ is true since $1 \cdot 2 = 2$ is even. Assume then the truth[6] of $P(k)$; that is, that $k(k+1) = 2m$ for some integer m. As a consequence, we have $(k+1)(k+2) = (k+1)k + (k+1)2 = 2m + 2(k+1)$ or finally $(k+1)(k+2) = 2(m+k+1)$, which is the statement $P(k+1)$. It follows that $P(n)$ is true for all $n \geq 1$.

Mathematical induction can be employed to prove the familiar rules regarding exponents. We define powers of an integer a by

$$a^0 = 1 \tag{1}$$

and

$$a^{n+1} = a^n a \quad \text{for} \quad n \geq 0. \tag{2}$$

(A definition such as this is called a *recursive* definition.)

Theorem 1-18 If a and b are integers and if m and n are non-negative integers, then (a) $a^m a^n = a^{m+n}$, (b) $(a^m)^n = a^{mn}$, and (c) $(ab)^n = a^n b^n$.

Proof Proofs of parts (b) and (c) are left as exercises. We prove (a) by employing mathematical induction on n, that is, by showing that for each non-negative integer m the formula $a^m a^n = a^{m+n}$ is true for all $n \geq 0$.

For $n = 0$ we have, using definitions (1) and (2), $a^m a^0 = a^m 1 = a^m = a^{m+0}$. Thus (a) is true for $n = 0$. Now assume $a^m a^k = a^{m+k}$. Then, using (2) and the inductive hypothesis, we have $a^m a^{k+1} = a^m a^k a = a^{m+k} a = a^{m+k+1}$, as desired. This proves (a)▲

Example 1-2 (The Bernoulli Inequality) For $n \geq 2$ we prove the inequality

$$P(n): \quad (1+x)^n > 1 + nx,$$

where x is any positive integer.

Obviously $P(2)$ is true; that is,

$$(1+x)^2 = 1 + 2x + x^2 > 1 + 2x$$

[6] The assumption that $P(k)$ is true we call the *induction hypothesis*, and the proof that $P(k)$ implies $P(k+1)$ we call the *induction step*.

since $x^2 > 0$. Then, from the assumed truth of

$$P(k): \quad (1+x)^k > 1+kx$$

it follows for $k \geq 2$ that

$$(1+x)^{k+1} > (1+kx)(1+x) = 1+(k+1)x+kx^2$$

or

$$(1+x)^{k+1} > 1+(k+1)x,$$

which is $P(k+1)$. Hence $P(n)$ is true for all $n \geq 2$. (Note that $P(1)$ is not true since $1+x \not> 1+x$.)

The following theorem states the *second principle of mathematical induction.*

Theorem 1-19 Suppose a proposition $P(n)$ to be stated for all integers n such that $n \geq n_0$, where n_0 is a particular integer. If (i) the statement $P(n_0)$ is true and (ii) the assumed truth of $P(j)$ for all j satisfying $n_0 \leq j \leq k$ implies the truth of $P(k+1)$ for each $k \geq n_0$, then the proposition is true for all $n \geq n_0$.

Proof Denote by C the set of all integers that are greater than n_0 and for which the corresponding proposition is false. Assume that C is nonempty. Then C possesses a least member c and certainly $c > n_0$. Hence $P(j)$ is true for all j satisfying $n_0 \leq j \leq c-1$. But part (ii) of the hypothesis then implies $P(c)$ is true, contrary to c being a member of C. Thus we conclude that C is empty and the theorem is proved▲

Example 1-3 We prove the proposition $P(n)$: The sum of n integers does not depend on the manner in which the addition is performed, that is, upon the arrangement or grouping of the integers involved.

The statement $P(1)$ is trivially true. Incidentally, we see also that $P(2)$ is true since it is nothing more than Axiom 1.

Assume $P(j)$ to be true for all j satisfying $1 \leq j \leq k$. Then, in particular, the sum of k integers $a_1, a_2, \ldots, a_k$ does not depend on the order or grouping of the summands and therefore without ambiguity may be denoted by $a_1 + a_2 + \cdots + a_k$.

Now suppose that $k+1$ integers $a_1, a_2, \ldots, a_k, a_{k+1}$ are added in some manner to produce a sum s. Since exactly two integers are combined in any addition, it will require a total of k additions to produce s. We denote the last of these by $s_1 + s_2$, where s_1 is the sum of from 1 to k of the integers $a_1, a_2, \ldots, a_k, a_{k+1}$ and s_2 is the sum of the remaining. Without loss of generality,[7] we may suppose that a_{k+1} is among those integers which combine to yield s_2.

We consider two cases. First, suppose s_2 is the single integer a_{k+1}. Then $s = s_1 + a_{k+1}$, where s_1 is the sum of the integers $a_1, \ldots, a_k$. By the induction hypothesis, $s_1 = a_1 + \cdots + a_k$; hence $s = (a_1 + \cdots + a_k) + a_{k+1}$.

[7] The phrase "without loss of generality" should always be viewed with suspicion because it frequently indicates that the proof is incomplete and it is up to the reader to supply the missing details. For example, here the reader must prove that the final conclusion is valid in the event a_{k+1} is a term in the sum s_1. Can this be done without appealing to Axiom 1?

Next, suppose s_2 is a sum of from 2 to k integers. Then, by the induction hypothesis, the terms constituting s_2 may be combined in any manner without altering their sum. Hence we may write $s_2 = s_2' + a_{k+1}$, where s_2' is the sum of all integers of s_2 except for a_{k+1}. Thus $s = s_1 + s_2 = s_1 + (s_2' + a_{k+1}) = (s_1 + s_2') + a_{k+1}$. But $s_1 + s_2'$ is the sum of the k integers $a_1, \ldots, a_k$ and hence is equal to $a_1 + \cdots + a_k$. Thus in this case, too, $s = (a_1 + \cdots + a_k) + a_{k+1}$.

By using the induction hypothesis we have shown that any sum s of the integers $a_1, \ldots, a_{k+1}$ is equal to $(a_1 + \cdots + a_k) + a_{k+1}$. This establishes $P(k+1)$ and completes the proof that $P(n)$ holds for all $n \geq 1$.

The proposition proved in Example 1-3 is called the *general commutative and associative law for addition*. In a parallel manner we can prove the *general commutative and associative law for multiplication*: The product of n integers does not depend on the order or grouping of the factors. Thus it is established that the expressions $a_p + a_{p+1} + \cdots + a_q$ and $a_p a_{p+1} \cdots a_q$ are not ambiguous. For brevity we shall on occasion denote them by $\sum_{k=p}^{q} a_k$ and $\prod_{k=p}^{q} a_k$, respectively.

Example 1-4 By Axiom 9 we have $a(b+c) = ab + ac$ for any integers a, b, and c. By induction this may be extended as follows.

Let $P(n)$ be the statement of the *general distributive law*; that is,

$$P(n): \quad a(b_1 + b_2 + \cdots + b_n) = ab_1 + ab_2 + \cdots + ab_n.$$

Observe that $P(1)$ is trivially true while $P(2)$ is true because of Axiom 9. Now assume the proposition to be true for $n = k$; that is,

$$P(k): \quad a(b_1 + \cdots + b_k) = ab_1 + \cdots + ab_k.$$

Observe that by Example 1-3 and Axiom 9

$$\begin{aligned} a(b_1 + \cdots + b_k + b_{k+1}) &= a[(b_1 + \cdots + b_k) + b_{k+1}] \\ &= a(b_1 + \cdots + b_k) + ab_{k+1}. \end{aligned}$$

The induction assumption then implies

$$a(b_1 + \cdots + b_k + b_{k+1}) = ab_1 + \cdots + ab_k + ab_{k+1},$$

which is the statement of the given proposition for $n = k+1$. Thus the proposition $P(n)$ is true for all $n \geq 1$, and the general distributive law is established.

The following is typical of many propositions that can be proved by using mathematical induction.

Example 1-5 Consider the proposition

$$P(n): \quad 6(1^2 + 2^2 + \cdots + n^2) = n(n+1)(2n+1).$$

First, observe that $P(1)$ is true; that is,

$$P(1): \quad 6(1^2) = 1(2)(3).$$

Now assume the truth of $P(k)$; that is,

$$P(k):\quad 6(1^2+2^2+\cdots+k^2)=k(k+1)(2k+1),$$

and add $6(k+1)^2$ to both sides of the equation. We obtain

$$\begin{aligned}6[1^2+2^2+\cdots+k^2+(k+1)^2] &= k(k+1)(2k+1)+6(k+1)^2\\ &=(k+1)[(k+1)+1][2(k+1)+1],\end{aligned}$$

which is $P(k+1)$. Thus the assumed truth of $P(k)$ for any $k \geq 1$ implies the truth of $P(k+1)$. Hence $P(n)$ is true for all $n \geq 1$.

It is possible in the case of a proposition $P(n)$ for $P(k)$ to imply $P(k+1)$ for each k and yet for $P(n)$ to be false. For example, consider the proposition

$$P(n):\quad 2(1+2+\cdots+n)=(n-1)(n+2).$$

Assuming the truth of

$$P(k):\quad 2(1+2+\cdots+k)=(k-1)(k+2),$$

we obtain

$$\begin{aligned}2[1+2+\cdots+k+(k+1)] &= (k-1)(k+2)+2(k+1)\\ &=[(k+1)-1][(k+1)+2],\end{aligned}$$

which is $P(k+1)$. It does not follow, however, that $P(n)$ is true, since it has not been established that $P(n)$ is true for some particular value of n. Nor is it possible to find an n for which the proposition is true. We may, in fact, assert the negation of $P(n)$ as a new proposition and show it to be true. Thus, letting $Q(n)$ be the statement

$$2(1+2+\cdots+n)\neq(n-1)(n+2),$$

it is clear that $Q(1)$ is true. Also, the same simple algebraic manipulation that shows how $P(k)$ implies $P(k+1)$ may be employed to demonstrate that $Q(k)$ implies $Q(k+1)$. Thus $Q(n)$ is true for $n \geq 1$ and $P(n)$ false.

Exercises

1-30 Let B be a set of integers such that (i) an integer n_0 is in B and (ii) if n is a member of B, then so is $n-1$. Prove that B contains all integers less than n_0.

1-31 Prove parts (b) and (c) of Theorem 1-18.

1-32 Prove that if m is any integer and if a is an integer greater than 1, then an integer n exists such that $a^n > m$.

1-33 Prove that if n is any non-negative integer, then $n < 3^n$.

In Exercises 1-34 through 1-52 use mathematical induction.

1-34 Prove $2(1^3+2^3+\cdots+n^3)=[n(n+1)]^2$ for $n \geq 1$.

1-35 Prove $(1-x)(1+x+x^2+\cdots+x^n)=1-x^{n+1}$ for any integer x and all $n \geq 0$.

1-36 Prove $(1-x)[(1+x)(1+x^2)\cdots(1+x^{2^n})]=1-x^{2^{n+1}}$ for any integer x and all $n \geq 0$.

1-37 Prove $(1+x)^n \geq 1+nx$ for all integers x such that $x > -1$ and for all $n \geq 1$. (Compare with illustrative Example 1-2.)

1-38 Prove that the sum of the interior angles of a polygon of $n+2$ sides is $180n$ degrees.

1-39 Prove that 8 is a factor of $5^{2n}+7$ for all $n \geq 1$.

1-40 Prove that for each positive integer n the integer 5^n "ends in 5"; that is, for each n there exists an integer k_n such that $5^n = 10k_n + 5$.

1-41 Prove 5 is a factor of $7(16^n)+3$ for all $n \geq 0$.

1-42 Prove $n(n^2+5)$ is a multiple of 6 for all $n \geq 1$.

1-43 Prove 6 is a factor of $n(n^2-1)$ for all $n \geq 0$.

1-44 Prove $3[1 \cdot 2 + 2 \cdot 3 + \cdots + n(n+1)] = n(n+1)(n+2)$ for all $n \geq 1$.

1-45 Prove $3[1^2+3^2+\cdots+(2n+1)^2] = (n+1)(2n+1)(2n+3)$ for all $n \geq 1$.

1-46 Prove $1^3+3^3+\cdots+(2n+1)^3 = (n+1)^2(2n^2+4n+1)$ for all $n \geq 1$.

1-47 Prove that if m is any non-negative integer, then $1^m+2^m+\cdots+n^m \leq n^{m+1}$ for all $n \geq 1$.

1-48 Let $x_1, x_2, \ldots, x_n$ be integers such that $x_i < x_{i+1}$ for $i = 1, 2, \ldots, n-1$. Prove that $x_1 < x_n$.

1-49 Prove that if p is a prime integer that divides $a_1 a_2 \cdots a_n$, then p divides a_i for some i.

1-50 Prove that if x and y are any integers, then $x^n - y^n$ is divisible by $x-y$ if n is any positive integer and by $x+y$ if n is any even positive integer.

1-51 The formula $P(n)$: $2+2^2+\cdots+2^n = 2^{n+1}$ is incorrect. Show, however, that $P(k)$ implies $P(k+1)$. Then prove that $P(n)$ is incorrect for all $n \geq 1$.

1-52 Prove that if a and b are positive integers such that $a<b$, then $a^n < b^n$ for $n \geq 1$. Show that, as a consequence, if a and b are positive integers such that $a^n = b^n$ for some integer n, then $a=b$.

1-53 Point out the fallacy in the following inductive "proof." Consider the proposition

$P(n)$: If a and b are any non-negative integers such that $a+b \leq n$, then $a=b$.

First, observe that $P(0)$ is obviously true, for if $a+b \leq 0$, then necessarily $a=b=0$. Next, assume $P(k)$ to be true. Let a and b be integers such that $a+b \leq k+1$, and define c and d by $c=a-1$ and $d=b-1$. Then $c+d = a+b-2 \leq k+1-2 \leq k$. The truth of $P(k)$ implies $c=d$, whence $c+1=d+1$ or $a=b$. Thus from $a+b \leq k+1$ has followed $a=b$; that is, $P(k+1)$ is true. We conclude that $P(n)$ is true for all $n \geq 0$.

1-54 Prove the general commutative and associative law of multiplication of integers.

1-55 Prove that if n is any integer greater than 1, then n can be written as a product of prime factors. Further, show that such a factorization is unique, except for the order of the factors.

1-56 We define the symbol $n!$ (*factorial* n) by $0!=1$ and $n! = n(n-1)!$ for $n \geq 1$. Prove the following: (a) If $k! = a(r-1)!\,(k+1-r)!$ and $k! = br!\,(k-r)!$, then $(k+1)! = (a+b)r!\,(k+1-r)!$. (b) If $0 \leq r \leq n$, then $r!\,(n-r)!$ divides $n!$ for $n \geq 0$.

2

Finite and Infinite Sets

And thick and fast they came at last,
And more, and more, and more—
LEWIS CARROLL (1832–1898), *The Walrus and the Carpenter*

2-1 The Set Concept

The concept of a *set*[1] of *entities* or *members*[2] is basic to all that follows. It is essential, therefore, that we study it in some detail. We have already considered the set I of all integers and in passing have referred to the set of positive integers, a set of axioms, a set of operations (addition and multiplication), the set of zeros in I (consisting of exactly one member), etc. We can, of course, speak of sets whose members are not integers nor related to the system of integers. Thus we may define A to be the set of all redheaded human females over twenty-one, the set D of all known cases of a certain contagious disease occurring north of the thirty-seventh parallel in January 1729, or the set Σ of all blue-eyed centaurs. We shall consider a set to be *defined* if concerning any conceivable entity it is clear whether or not it is a member of the set.

A set is frequently defined by a statement describing properties common to all

[1] Also, *collection*, *family*, *aggregate*, and *class*.

[2] A common designation for what we here call a *member* of a set is the term *element*. However, since we shall employ the latter in another connection, we shall not use that term as a synonym for member.

its members. For such a statement we introduce a convenient and compact notation of the form

$$A = \{x;\quad \text{"properties of } x\text{"}\}.$$[3]

For example, the set A consisting of all even integers lying between -3 and 15 is represented by

$$A = \{x;\quad x \text{ an even integer, } -3 < x < 15\}.$$

Also, the set P of all prime numbers is described by

$$P = \{p;\quad p \text{ a prime number}\}.$$

A listing of members of a set also defines the set. Thus we write $B = \{2, +, \#\}$ to denote the set comprised of the members 2, +, and #, and $C = \{2,4,6, \ldots\}$ for the set of positive even integers.

Later we shall be interested in the structure of particular sets, and, for the most part, these will be sets whose members have a mathematical character. Just as we have defined and studied the set of integers, so shall we define and investigate sets of real numbers, families of continuous functions, collections of sets, etc. It is important, therefore, that we first attain a familiarity with the abstract concept of set itself.

We shall denote individual sets by capital letters such as A, B, C, Γ, and Ψ and entities which may be members of sets by lower-case symbols such as a, b, m, n_1, n_2, α, and β. Concerning a set A and an entity x we recognize only one thing as important, namely, that x either belongs or does not belong to A. Whether x is a number, a person, a disease, or an idea is of no consequence. Only its membership or nonmembership in A is of interest.

It is convenient to introduce a notation indicating membership in a set. Thus we employ the symbol $\in$ to represent such phrases as *is a member of* or *belongs to*. If x is a member of a set A, we write $x \in A$; if not, then $x \notin A$.

2-2 Equality and the Subset Relation

Two sets will be called *equal* if they are one and the same set. Thus $A = B$ if and only if A and B are comprised of the same members. It is possible, of course, for equal sets to be differently described. For example, the set A of solutions to the quadratic equation $x^2 - 6x + 8 = 0$ and the set B of even integers between 0 and 6 are equal. If two sets are not equal, we write $A \neq B$.

Trivially, if $A = B$ and $B = C$, then $A = C$.

[3] The symbol "=" employed here should not be confused with the equal sign for integers. Here it may simply be regarded as a symbol for the word "is" in the definition "*A is* the set whose members have such and such properties."

If every member of a set A is also a member of a set B, then we say that A is a *subset* of B or that A is *contained in* B, and we denote this relation by $A \subset B$. In symbols, if $x \in A$ implies $x \in B$, then $A \subset B$. For convenience, we shall also say that B *contains* A and write $B \supset A$. It is, of course, an immediate consequence of the definition that any set is a subset of itself; that is, $A \subset A$.

If A and B are such that $A \subset B$ and $B \subset A$, it is immediate that $A = B$. We shall have many occasions to prove the equality of two sets. One method is clear; we have only to prove that each of the sets is a subset of the other.

As examples of the subset relation we cite the following. The set of all positive integers is a subset of the set of all integers; in symbols, $I_+ \subset I$. The set of axioms that concern the order relation in Chapter 1 is a subset of the collection of all axioms in that chapter. The set of serious students in any university is a subset of all those enrolled. The sets $A = \{\pi,*\}$ and $B = \{3,*,\pi,?,+\}$ are subsets of $C = \{\pi,3,?,+,*\}$.

It is clear that the subset relation enjoys a *transitive* property; that is, from $A \subset B$ and $B \subset C$ follows $A \subset C$. For if $x \in A$, then the relation $A \subset B$ implies $x \in B$. In turn, $B \subset C$ implies $x \in C$. Therefore, from $x \in A$ follows $x \in C$, so that $A \subset C$.

If $A \subset B$ and if $A \neq B$, we shall refer to A as a *proper subset* of B or say that A is *properly contained in* B or that B *properly contains* A. Expressed differently, A is a proper subset of B if it is part but not all of B, that is, if every member of A is a member of B, yet B has at least one member which does not also belong to A. It is clear that if A is properly contained in B, then $A \subset B$, but not conversely. As an example, the set of positive integers is a proper subset of the set of all integers. Also, the set of all citizens of the United States properly contains the set of all its legal voters.

There is no trichotomy for the proper subset relation and equality of sets, as was seen to be the case for the order relation and equality of integers. For example, the set of positive integers and the set of inhabitants of the United States are not equal, nor is either a proper subset of the other. Also, while the set A of integers less than 7 and the set I_+ have members in common, neither the relationship of equality nor proper subset exists between them.

No special symbol has been introduced to denote that A is a proper subset of B. Instead, on those occasions when it is important to distinguish proper subset from subset we shall write simply $A \subset B$, $A \neq B$.

It is sometimes the case that a set is defined in such a way that it has no members at all. For example, the set of all integers x satisfying the equation $3x^2 + 2x + 1 = 0$ is such a set. It is desirable to define formally the notion of an *empty set* as a set possessing no members. Denoting it by the symbol $\varnothing$ it clearly is the case that $\varnothing \subset A$ no matter what the set A is. This is true *vacuously* or by default, as it were. That is, no member of $\varnothing$ exists which is not a member of A. Hence all members of $\varnothing$ are in A.

That there exists but one empty set follows trivially from the relation $\varnothing \subset A$. For if $\varnothing_1$ and $\varnothing_2$ are two empty sets (say $\varnothing_1$ is an empty set of integers and $\varnothing_2$ an empty set of people), then $\varnothing_1 \subset \varnothing_2$ and $\varnothing_2 \subset \varnothing_1$. Hence $\varnothing_1 = \varnothing_2$.

2-3 Union and Intersection of Sets

We define the *union* of two sets A and B to be the set each of whose members is a member of A or of B. Denoting the union of A and B by the symbol $A \cup B$ (read "A union B"), it follows that if $x \in A \cup B$, then $x \in A$ or $x \in B$ (or both), and conversely. Clearly $A \subset A \cup B$ and $B \subset A \cup B$.

A convenient graphical illustration of the union of two sets A and B is possible if we think of the members of A and B as points in a plane. Suppose, as in Figure 2-1, that A and B are points in a plane that lie in and on two closed

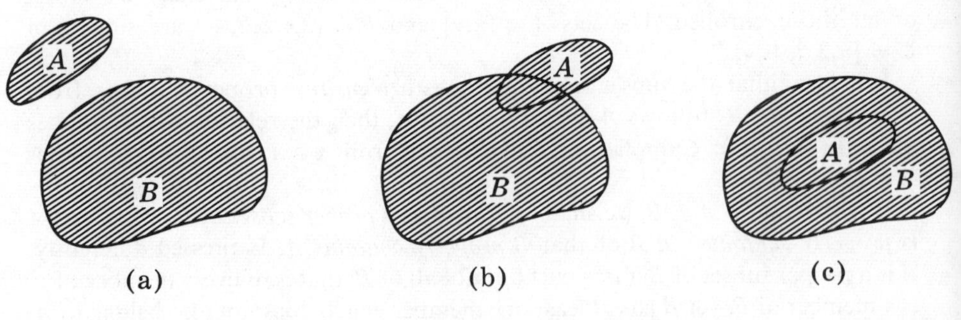

Figure 2-1

curves. Then the union of A and B presents three possibilities. In Figure 2-1(a) the two sets are *disjoint*; that is, they have no members in common. The union consists of all the members of A and all the members of B. In (b) the sets A and B have some members in common. Therefore some members of $A \cup B$ belong to both A and B. Finally, in (c) all members of A belong to B. Hence $A \cup B$ is identical with B.

Sketches like those of Figure 2-1 suggest certain truths which may be proved by recourse to definitions. For example, Figure 2-1(c) suggests the following general theorem.

Theorem 2-1 If $A \subset B$, then $A \cup B = B$, and conversely. In particular, $\varnothing \cup B = B$.

Proof If $x \in A \cup B$, then $x \in A$ or $x \in B$. Since $A \subset B$, it follows that $x \in A$ implies $x \in B$. Hence $x \in A \cup B$ implies $x \in B$. Thus $A \cup B \subset B$. However, from the definition of union, $B \subset A \cup B$. It follows that $A \cup B = B$. As for the converse, let $x \in A$. Then $x \in A \cup B$, or, since $A \cup B = B$, $x \in B$. Hence $A \subset B$▲

We define the *intersection* of two sets A and B to be the set each of whose members is a member of *both* A and B. We denote this set by the symbol $A \cap B$ (read "A intersect B"). Thus $x \in A \cap B$ if and only if $x \in A$ *and* $x \in B$. Again a sketch is instructive. In Figure 2-2(a) the sets A and B are disjoint. Hence, in this

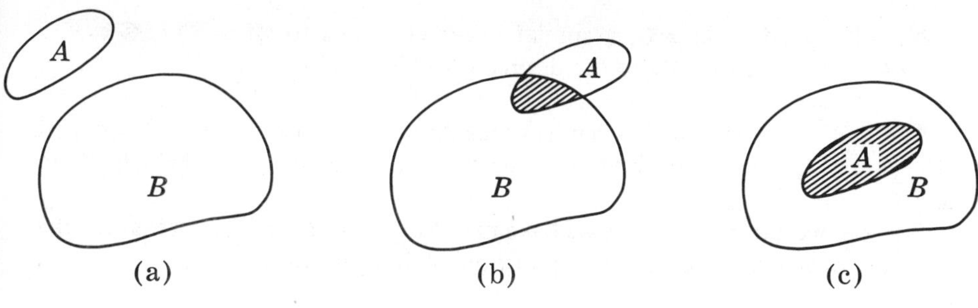

Figure 2-2

instance, $A \cap B = \varnothing$. In (b) sets A and B have some members in common. The shaded portion represents $A \cap B$. In (c) all of set A is contained in B, and $A \cap B$ is identical with A. The last statement suggests the following general truth, which the reader is encouraged to verify.

Theorem 2-2 If $A \subset B$, then $A \cap B = A$, and conversely. In particular, $\varnothing \cap B = \varnothing$.

2-4 The Algebra of Sets

It is convenient to look upon the notions of union and intersection as operations denoted by the symbols $\cup$ and $\cap$, respectively, in a system in which sets of entities are members. Then the various properties or facts which may be deduced concerning the operations form the basis for the so-called *algebra of sets*.[4] Obviously, among these properties, since they are inherent in the definitions of union and intersection, are the *commutative laws*

$$A \cup B = B \cup A \quad \text{and} \quad A \cap B = B \cap A$$

for sets A and B.

Each of the operations $\cup$ and $\cap$ also possesses the property of *associativity*. We state this in the form of a theorem.

Theorem 2-3 If A, B, and C are any sets, then

(a) $A \cup (B \cup C) = (A \cup B) \cup C$,

(b) $A \cap (B \cap C) = (A \cap B) \cap C$.

Proof We prove (a) and leave (b) to the reader.

Let $x \in A \cup (B \cup C)$. Then $x \in A$ or $x \in B \cup C$. If the latter is the case, then $x \in B$ or $x \in C$. Hence x is a member of at least one of the sets A, B, and C. In any event it follows that $x \in (A \cup B) \cup C$, whence $A \cup (B \cup C) \subset (A \cup B) \cup C$.

[4] Also called *Boolean algebra* after George Boole (1815–1864).

Now if $x \in (A \cup B) \cup C$, a similar argument leads to $(A \cup B) \cup C \subset A \cup (B \cup C)$ and consequently to the desired conclusion ▲

The truth of Theorem 2-3 renders meaningful the symbols $A \cup B \cup C$ and $A \cap B \cap C$ for the union and the intersection, respectively, of the sets A, B, and C.

For the operations $\cup$ and $\cap$ we have two *distributive laws.* (Recall that for the operations of addition and multiplication of integers only one such law was postulated.)

Theorem 2-4 If A, B, and C are any sets, then

(a) $A \cap (B \cup C) = (A \cap B) \cup (A \cap C)$,

(b) $A \cup (B \cap C) = (A \cup B) \cap (A \cup C)$.

Proof Again we include a proof for the first part only.

Let $x \in A \cap (B \cup C)$. Then $x \in A$ and $x \in B \cup C$. Hence $x \in B$ or $x \in C$. Suppose the former. Then from $x \in A$ and $x \in B$ it follows that $x \in A \cap B$ and therefore $x \in (A \cap B) \cup (A \cap C)$. The same result follows from $x \in C$. It has thus been shown that $A \cap (B \cup C) \subset (A \cap B) \cup (A \cap C)$.

Now let $x \in (A \cap B) \cup (A \cap C)$. Then $x \in A \cap B$ or $x \in A \cap C$. Suppose the former. Then $x \in A$ and $x \in B$. From $x \in B$ follows $x \in B \cup C$. Thus $x \in A \cap (B \cup C)$. The same is a consequence of the supposition $x \in A \cap C$. Therefore $(A \cap B) \cup (A \cap C) \subset A \cap (B \cup C)$, and the proof of (a) is complete ▲

The reader should provide an example to show that the sets $A \cup (B \cap C)$ and $(A \cup B) \cap C$ need not be equal. It follows that the symbols $A \cup B \cap C$ and $A \cap B \cup C$ are ambiguous and hence meaningless.

The empty set has the character of an *indentity with respect to the union operation.* Thus

$$A \cup \varnothing = A$$

for any set A.

In any particular discussion the sets encountered will be subsets of some set which we call the *universe* or *space* and denote by S. For example, if we are discussing collections of integers, then the space S may be the set I itself. If we discuss sets all of whose members are people, then the set S could be the set of all people. Of course, in a given discussion we have a choice as to what we shall regard as the space. If we discuss the set of positive integers, we might consider either I or I_+ as the set S.[5]

In this chapter and whenever we discuss abstract sets we shall regard them as subsets of a space S. The set S will then serve as an *identity with respect to the*

[5] We shall not venture to elaborate further on the subject of a space. The concept of all possible members of which sets may be composed, including such entities as sets themselves, sets of sets, etc., leads to bizarre complications, even to paradoxes.

intersection operation. For if A is any of the sets we are considering, then $A \subset S$ and

$$A \cap S = A.$$

Two further facts important in the structure of the algebra of sets follow immediately from definitions. These are the so-called *idempotent laws:*

$$A \cup A = A \quad \text{and} \quad A \cap A = A.$$

The set consisting of those members of a set B that are not members of a set A is called the *complement of A with respect to B* and is denoted by $B - A$. In particular, the complement of a set A with respect to S is simply called the *complement of A* and is denoted by $\mathscr{C}A$. Immediate consequences of the definitions are the properties

$$A \cup \mathscr{C}A = S, \quad A \cap \mathscr{C}A = \varnothing, \quad \text{and} \quad \mathscr{C}(\mathscr{C}A) = A$$

for any set A.

Two important properties relating the concept of complement to those of union and intersection are given in the following theorem.

Theorem 2-5 If A and B are any sets, then

(a) $\mathscr{C}(A \cup B) = \mathscr{C}A \cap \mathscr{C}B$,

(b) $\mathscr{C}(A \cap B) = \mathscr{C}A \cup \mathscr{C}B$.

Proof We leave the proof of (b) to the reader. To prove (a) let $x \in \mathscr{C}(A \cup B)$. Then $x \notin A \cup B$ so that $x \notin A$ and $x \notin B$. Consequently $x \in \mathscr{C}A$ and $x \in \mathscr{C}B$. Therefore $x \in \mathscr{C}A \cap \mathscr{C}B$, and it has been shown that $\mathscr{C}(A \cup B) \subset \mathscr{C}A \cap \mathscr{C}B$.

Now suppose $x \in \mathscr{C}A \cap \mathscr{C}B$. Then $x \in \mathscr{C}A$ and $x \in \mathscr{C}B$. Hence $x \notin A$ and $x \notin B$ and, as a consequence, $x \notin A \cup B$. It follows that $x \in \mathscr{C}(A \cup B)$ and therefore $\mathscr{C}A \cap \mathscr{C}B \subset \mathscr{C}(A \cup B)$. We conclude that $\mathscr{C}(A \cup B) = \mathscr{C}A \cap \mathscr{C}B$▲

We leave the proof of the following statement to the reader.

Theorem 2-6 If A and B are any sets, then the relation $A \subset B$ implies $A \cap \mathscr{C}B = \varnothing$, and conversely.

Having established certain fundamental identities concerning unions, intersections, and complements of sets, we can now prove other identities without direct recourse to the definition of equality of sets.

Example 2-1 We may prove the identity $A \cup \mathscr{C}(A \cup \mathscr{C}B) = A \cup B$ by observing that

$$\begin{aligned} A \cup \mathscr{C}(A \cup \mathscr{C}B) &= A \cup (\mathscr{C}A \cap B) \\ &= (A \cup \mathscr{C}A) \cap (A \cup B) \\ &= S \cap (A \cup B) \\ &= A \cup B. \end{aligned}$$

Exercises

2-1 Prove each of the following: (a) If A is a proper subset of B and B a proper subset of C, then A is a proper subset of C. (b) If B properly contains A and if $B \subset C$, then C properly contains A. (c) If $A \subset B$ and if B is properly contained in C, then A is properly contained in C. (d) A is not a proper subset of A.

2-2 Prove Theorem 2-2.

2-3 Prove part (b) of Theorem 2-3.

2-4 Prove part (b) of Theorem 2-4.

2-5 Prove part (b) of Theorem 2-5.

2-6 Prove Theorem 2-6.

In each of the following, the sets A, B, and C are to be regarded as subsets of a space S.

2-7 Prove that if $A \subset B$, then (a) $A \cap C \subset B \cap C$, and (b) $A \cup C \subset B \cup C$.

2-8 Prove that if $A \subset B$, then $A \cup C = (A \cup C) \cap (B \cup C)$.

2-9 Prove that $A \subset B$ if and only if $B \cap (A \cup C) = (B \cap C) \cup A$.

Prove the identities of Exercises 2-10 through 2-17 by using the algebra of sets, as in Example 2-1.

2-10 $A = A \cup (A \cap B)$.

2-11 $A = (A \cap B) \cup (A \cap \mathscr{C}B)$.

2-12 $A \cup B = (A \cap \mathscr{C}B) \cup B$.

2-13 $A \cup B = (A \cap B) \cup (A \cap \mathscr{C}B) \cup (\mathscr{C}A \cup B)$.

2-14 $(A \cup B) \cap (B \cup C) \cap (C \cup A) = (A \cap B) \cup (B \cap C) \cup (C \cap A)$.

2-15 $\mathscr{C}[(A \cap B) \cup (A \cap \mathscr{C}C)] = \mathscr{C}A \cup (\mathscr{C}B \cap C)$.

2-16 $(A \cup B) \cap \mathscr{C}(A \cup C) = \mathscr{C}A \cap B \cap \mathscr{C}C$.

2-17 $A = B$ if and only if $(A \cap \mathscr{C}B) \cup (\mathscr{C}A \cap B) = \varnothing$.

2-18 Express $\mathscr{C}[\mathscr{C}(A \cap B) \cap A] \cup (A \cap B)$ in simpler form.

2-19 Cite an example to show that $A \cup (B \cap C)$ need not equal $(A \cup B) \cap C$ and hence that $A \cup B \cap C$ is meaningless.

2-20 Cite an example to show that $(A \cup B) \cap (C \cup D)$ need not equal $A \cup (B \cap C) \cup D$ and hence that $A \cup B \cap C \cup D$ is meaningless.

2-21 Comment on the following argument: If A and B are any subsets of a space S, then $A \cup S = S$ and $B \cup S = S$. Hence $A \cup S = B \cup S$. By cancelling S, this gives $A = B$. Thus any two subsets of a space S are equal.

Prove each of the following:

2-22 $B - A = B \cap \mathscr{C}A$.

2-23 $B - B = \varnothing$.

2-24 A and B are disjoint if and only if $B - A = B$.

2-25 $(B - A) \cup A = B$ if and only if $A \subset B$.

2-26 $A \cap B = B - (B - A)$.

2-27 $A - (B \cup C) = (A - B) \cap (A - C)$.

2-28 $A - (B \cap C) = (A - B) \cup (A - C)$.

2-5 The Concept of Function

A rule which establishes an association or correspondence between all the members of a set A and some or all of the members of a set B is called a *function defined on A with values in B* or simply a *function*. The terms *mapping* and *transformation* are also employed in this connection. We shall use them interchangeably.

If a function has the property that to each member of the set A there is associated exactly one member of B, then the function is said to be *single valued*. This is not intended to suggest, however, that a member of B may be the associate of only one member of A. Indeed, one member of B may be the associate of many, or even all, members of A. Throughout this book we shall be concerned with single-valued functions only; accordingly, *the term function will be synonymous with single-valued function*. Functions will be denoted by literal symbols such as f, g, F, and φ.

The set A on which a function f is defined is to be known as the *domain of definition of f*, briefly, the *domain* of f. A typical or representative member of the domain of a function is frequently denoted by x and is called an *independent variable* while an individual member of A is referred to as a *value* of the independent variable. It is common to describe the relation between a function f and its domain of definition by saying that *f is a function of x*.

That member of the set B which the function f associates with a member x of A is denoted by the symbol $f(x)$. We call it the *image* or *map* or *transform* of x or the *value of the function* or *function value* at x. The set R of function values corresponding to all the members of A is a subset of the set B and is referred to as the *range of the function*. A typical or representative member of the range set of a function is called a *dependent variable*. Function values are particular values of the dependent variable.

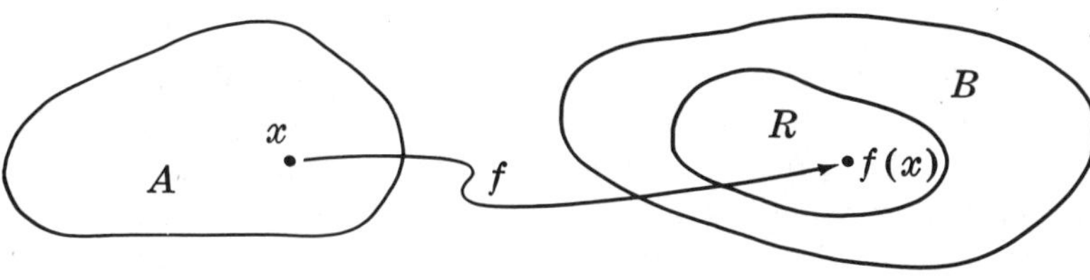

Figure 2-3

Figure 2-3 illustrates graphically the concept of a function for which A, B, and R are sets of points in a plane. Note that although the domain set A and the range R are unique attributes of a function, the set B referred to above may in reality be any set with the property $R \subset B$.

Example 2-2 Denote by f a function defined on the domain $A = \{2,1,7\}$ with values in the set $B = \{\%,\pi,\&,*\}$. Let the rule which defines f be given explicitly by the statements:

$$\% \text{ is the image of } 2,$$

$$\pi \text{ is the image of } 1,$$

$$\pi \text{ is the image of } 7.$$

It is clear that f is single valued since only one of the members $\%$, π, &, and $*$ of B is associated with each member of A. The range of f is the set $R = \{\pi, \%\}$. Consistent with the notation that associates the member $f(x)$ of R with x of A we may write $f(2) = \%$, $f(1) = \pi$, and $f(7) = \pi$.

Example 2-3 Consider the function which associates with each positive integer n the integer $2n$. If we denote the function by g, we may write $g(n) = 2n$. The domain of definition of g is the set I_+ and the range R the set of all positive even integers. The set B might reasonably be taken to be the set R itself or the set I_+ or, quite generally, the set I of all integers.

A function is completely defined only when its three attributes—domain, range, and rule of association—are given. Any change in one or more of the three gives rise to a different function. However, it is sometimes convenient to refer to functions defined on different domains by the same symbol as long as the rule of association is identical on the set common to the two domains. Suppose, for instance, that a function f maps a domain A onto a range R. Then if C denotes a subset of A, which by the rule of association of f is mapped onto a subset R_1 of R, we shall also denote the function defined on C with range R_1 by the symbol f. This function is called the *restriction* of the original function to the domain C.

Example 2-4 The function g, defined by $g(n) = 2n$ in Example 2-3, has for its domain A the set of positive integers. Consider the subset C consisting of odd integers only. Then the function which assigns to each odd integer n the image $2n$ is the restriction of g to C; we also denote this restriction by g and write $g(n) = 2n$.

Frequently we wish to extend the definition of a function to include a wider domain. Suppose f to be a given function defined on a domain A, and let E be such that $A \subset E$. Then any function g defined on E such that $g(x) = f(x)$ for all x in A is called an *extension* of f to the domain E. In the event there is no ambiguity we shall, in fact, often retain the symbol f for the extended function.

Example 2-5 If f denotes a function defined by

$$f(n) = 2n + 1 \quad \text{for } n \text{ even},$$

then a possible extension to the domain of all integers is that for which

$$f(n) = \begin{cases} 2n + 1 & \text{for } n \text{ even} \\ 3n - 5 & \text{for } n \text{ odd.} \end{cases}$$

Let f be a function defined on a domain A with values in a set B, and let g be a function defined on B with values in a set C. It is then possible to define a function on A with values in C by associating with a member a of A the member $g(f(a))$ of C. We denote such a function by the symbol $g \circ f$ and refer to it as a *function of a function*. Figure 2-4 depicts the notion graphically for sets of points A, B, and C.

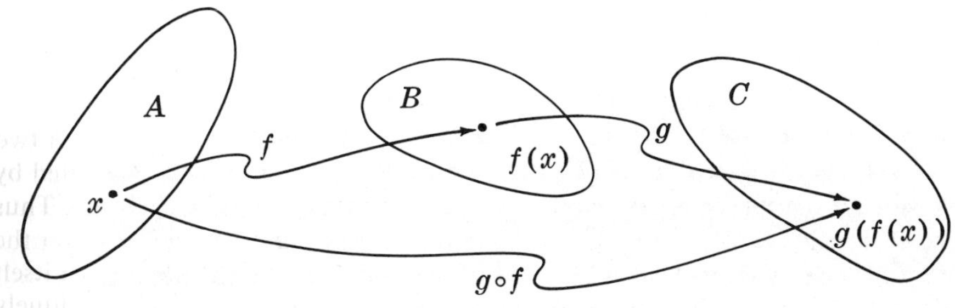

Figure 2-4

It is important to observe that the concepts of function and function value are quite distinct. A function is a rule of association whereas a function value is a member of the range set. The employment of different symbols f and $f(x)$ for function and function value is obviously desirable and is the prevalent convention in certain branches of mathematics, including real variable theory. It has the advantage of clearly distinguishing different concepts. However, in many areas—notably elementary and advanced calculus, applied mathematics, and complex variable theory—it is not unusual to find the symbol $f(x)$ denoting both function and function value. The symbol thereby acquires two meanings, the relevant one in any situation being singled out by context. This convention has the advantage of serving to minimize, and thus to simplify, notation. For example, if we understand the symbol $\sin x$ to represent a function as well as a function value, we avoid the need for a new symbol to stand for the rule of association commonly referred to as the "sine function." Actually, throughout the text we shall adhere to the notational distinction between f and $f(x)$ except when discussing special functions (such as $\sin x$ or x^2), in which case it is much more convenient to use one symbol for the two concepts.

2-6 One-to-One Correspondence

Let f be a function defined on a set A with values in a set B. Then f is called a *one-to-one correspondence between A and B* or a *one-to-one mapping of A onto B* if the range of f is B and if different members of A have as images different members of B. Expressed otherwise, a one-to-one correspondence is a pairing of the members of A and B in such a way that each member of each set is used *exactly once*.

Example 2-6 The two sets $A = \{1,2,5,7,9\}$ and $B = \{\#,\&,\%,*,\$\}$ may be placed in one-to-one correspondence in several ways. Two such are given by functions f and g explicitly defined by

$$\begin{aligned} f(1) &= \& \qquad & g(1) &= \# \\ f(2) &= * & g(2) &= \& \\ f(5) &= \# & g(5) &= * \\ f(7) &= \$ & g(7) &= \% \\ f(9) &= \% & g(9) &= \$. \end{aligned}$$

If it is possible to define or set up a one-to-one correspondence between two sets A and B, then A and B are called *equivalent* sets, and this fact is denoted by $A \sim B$. It is easy to see that any set A is equivalent to itself; that is, $A \sim A$. Thus equivalence of sets is a reflexive relation. To prove this we need only consider the special function which places the set A in one-to-one correspondence with itself by associating with each member x of A the member x itself. Such a uniquely defined function is called the *identity function* for the set A.

Let f denote a one-to-one correspondence defined on A with range B. It is then possible in the following way to define a function on B with A as its range. With any member y of B let there be associated that member x of A for which $f(x) = y$. Because of the one-to-one character of f there will be a unique x corresponding to each y. The newly defined function is also a one-to-one correspondence. We denote it by f^{-1} and refer to it as the *inverse* of the function f. It has the property that $f^{-1}(y) = x$ if and only if x and y are such that $f(x) = y$.

It is clear that if f is any function defined on a set A with the property that its inverse f^{-1} exists, then $f^{-1} \circ f$ is the identity function on A and $f \circ f^{-1}$ the identity function on B.

By using the inverse function we see at once that equivalence of sets is a symmetric relation; that is, $A \sim B$ implies $B \sim A$. For if f is a one-to-one mapping defined on A with range B, then f^{-1} is a one-to-one mapping defined on B with range A.

We may also show that equivalence of sets is a transitive relation; that is, $A \sim B$ and $B \sim C$ imply $A \sim C$. The truth of this follows from the notion of function of a function. For if f is the one-to-one correspondence defined on A and having range B and if g is that defined on B with range C, then the function $g \circ f$ is defined on A and has C as its range. Moreover, if x and y are different members of A, then $f(x)$ and $f(y)$ are different members of B since f is one-to-one. Then, since g is also one-to-one, it follows that $g(f(x))$ and $g(f(y))$ are different members of C. Consequently, different members of A have as images different members of C, and it follows that $g \circ f$ is a one-to-one mapping of A onto C.

2-7 Finite and Infinite Sets

For any positive integer n, consider the set $I_n = \{1,2, \ldots ,n\}$. Trivially, if $m = n$, the sets I_m and I_n are equivalent since they are identical. We shall assume without proof what may also seem intuitively clear to the reader; that is, if I_m and I_n are equivalent sets, then $m = n$.

If a set A can be placed in one-to-one correspondence with the set I_n for some n, we look upon such a correspondence as a *counting* or *enumeration* of A and refer to n as the *number of members* of A. The positive integer n may be regarded as the common measure of that class or collection of sets each of which is equivalent to I_n. It is frequently termed the *cardinal number* of the class. It is quite clear that the number of members of a set A is unique. For, if A has m members and also n members, then A is equivalent to I_m and to I_n. It follows that I_m and I_n are equivalent, and as a consequence $m = n$.

For convenience we shall define the number of members in the empty set to be zero.

We say that a set is *finite* if it is empty or if it is equivalent to the set I_n for some positive integer n.

If a set A is finite, then, for a particular one-to-one correspondence between A and I_n, that member of A which is associated with the integer 1 of I_n may be denoted conveniently by a_1, that corresponding to 2 by a_2, and in general that corresponding to k by a_k. Hence the set A may be denoted by $\{a_1,a_2, \ldots ,a_n\}$.

It is immediate from the foregoing that two finite sets are equivalent if and only if they have the same number of members. Moreover, if two sets are equivalent, then both are finite or neither is finite.

Theorems 2-7 through 2-10 assert fundamental properties of finite sets. The reader who feels that they express obvious truths should give special attention to the manner in which they follow from the definition of finite set as given above.

Theorem 2-7 Any subset of a finite set is finite.

Proof Because of the definition of a finite set it is sufficient to consider sets I_n of positive integers. Observe first that all subsets of $\varnothing$ and I_1 are finite. Then assume that each subset of I_k is finite, and let A denote any subset of I_{k+1}. If it is the case that A is also a subset of I_k, then A is finite by the induction hypothesis. On the other hand, if $A \not\subset I_k$, then $A = B \cup \{k + 1\}$, where $B \subset I_k$. Again by the induction hypothesis, B is finite, whereupon a one-to-one correspondence f exists between B and a set I_m. Extending the definition of f to include $f(k + 1) = m + 1$, it follows that A and I_{m+1} are equivalent; that is, A is finite▲

Theorem 2-8 If A and B are disjoint sets having m and n members, respectively, then $A \cup B$ has $m + n$ members.

Proof If either A or B or both are empty, the conclusion is immediate. Hence assume both A and B to be nonempty. Since A is equivalent to I_m and B to I_n, we may describe sets A and B by $A = \{a_1,a_2, \ldots ,a_m\}$ and $B = \{b_1,b_2, \ldots ,b_n\}$, respectively. From the hypothesis that $A \cap B$ is empty we may write $A \cup B = \{a_1,a_2, \ldots ,a_m,b_1,b_2, \ldots ,b_n\}$, which is clearly equivalent to I_{m+n}▲

Corollary The union of any two finite sets is finite.

Proof Let A and B be finite sets. Then $A \cup B$ is the union of the disjoint finite sets A and $B \cap \mathscr{C}A$▲

Theorem 2-9 A finite set cannot be placed in one-to-one correspondence with a proper subset of itself.

Proof Denote by B a proper subset of a finite set A. Then by Theorem 2-7 both B and $A \cap \mathscr{C}B$ are finite. Suppose that B has m members and $A \cap \mathscr{C}B$ has n members. From $A \cap \mathscr{C}B \neq \varnothing$ it follows that $n > 0$. Then by Theorem 2-8 the set $A = B \cup (A \cap \mathscr{C}B)$ has $m + n$ members. Now if A and B were equivalent sets, it would follow that $m + n = m$, which is not possible▲

If a set is not finite, we call it *infinite*. Because of Theorem 2-9 we conclude that a set is infinite if a one-to-one correspondence can be established between the set and a proper subset of it. Shortly we shall show that this, indeed, characterizes infinite sets; that is, infinite sets are those and only those which are equivalent to proper subsets of themselves.

Example 2-7 The set I_+ of positive integers is infinite since the function g defined by $g(n) = 2n$ on the domain I_+ clearly defines a one-to-one correspondence between the set I_+ and the set of positive even integers, a proper subset of I_+.

Just as sets which are equivalent to finite sets are themselves finite, so is it the case that sets equivalent to infinite sets are likewise infinite. In particular, having demonstrated that a set A is infinite by establishing a one-to-one correspondence with a proper subset B, it follows that the set B, too, is infinite.

Theorem 2-10 If $A \subset B$ and if A is infinite, then B is infinite.

Proof This follows directly from Theorem 2-7; if B is finite, then A is necessarily finite, which is a contradiction▲

2-8 Unions and Intersections of Any Collection of Sets

The concepts of union and intersection, defined earlier for a pair of sets A and B and easily extended to any finite family of sets, we now further generalize to apply to any collection of sets. To this end we let Ω designate any set, finite or infinite, and refer to it as an *index set*. Then we consider a family whose members are sets individually denoted by A_ν, where ν represents a member of or *varies over* the set Ω. Thus the collection of all sets A_ν is equivalent to Ω.

By the *union* of the sets A_ν we then mean the set, denoted by $\bigcup_{\nu \in \Omega} A_\nu$, consisting of all those entities each of which is a member of A_ν for at least one ν of Ω. The *intersection* of the sets A_ν is denoted by $\bigcap_{\nu \in \Omega} A_\nu$ and consists of those entities which are members of A_ν for all ν in Ω. It is clear that the definitions given earlier in Section 2-3 correspond to the case in which Ω is a set consisting of exactly two members. Also, it is evident that

$$\bigcap_{\nu\in\Omega} A_\nu \subset A_\nu \subset \bigcup_{\nu\in\Omega} A_\nu$$

for each ν in Ω.

If Ω is the set of all positive integers, the special notations $\bigcup_{\nu=1}^{\infty} A_\nu$ and $\bigcap_{\nu=1}^{\infty} A_\nu$ are customarily employed for union and intersection, respectively.[6] If Ω is the finite set $\{1,2,\ldots,n\}$, we use $\bigcup_{\nu=1}^{n} A_\nu$ and $\bigcap_{\nu=1}^{n} A_\nu$. The symbols $\bigcup_\nu A_\nu$ and $\bigcap_\nu A_\nu$ are frequently employed in the event that Ω is the set of positive integers or is the set I_n for some n.

Certain relations valid for unions and intersections of two sets admit of generalization in the light of the extended definitions. Thus, for example, the distributive laws

(1) $$B \cap \left(\bigcup_{\nu\in\Omega} A_\nu\right) = \bigcup_{\nu\in\Omega} (B \cap A_\nu)$$

and

(2) $$B \cup \left(\bigcap_{\nu\in\Omega} A_\nu\right) = \bigcap_{\nu\in\Omega} (B \cup A_\nu)$$

may be shown to be true. Similarly for the relations

(3) $$\mathscr{C}\left(\bigcap_{\nu\in\Omega} A_\nu\right) = \bigcup_{\nu\in\Omega} \mathscr{C}A_\nu$$

and

(4) $$\mathscr{C}\left(\bigcup_{\nu\in\Omega} A_\nu\right) = \bigcap_{\nu\in\Omega} \mathscr{C}A_\nu,$$

where the notion of complement is with respect to a space S of which A_ν is a subset for each ν of Ω. It is left to the reader to verify the above identities.

Exercises

2-29 By the *maximum* of a set B of integers is meant an integer b such that $b \in B$ and $x \leq b$ for each integer x in B. We write $b = \max B$. (a) Give an example of a set of integers that has no maximum. (b) Use the method of mathematical induction to show that every nonempty finite set of integers has a maximum.

2-30 By the *minimum* of a set A of integers is meant an integer a such that $a \in A$ and $a \leq x$ for each integer x in A. We write $a = \min A$. Prove that a is the minimum of a set of integers A if and only if $-a$ is the maximum of the set $B = \{x;\ \ -x \in A\}$.

[6] The symbol $\bigcup_{\nu=1}^{\infty} A_\nu$ is read "the union from one to infinity of A_ν." The other symbols in the paragraph are to be read in a similar manner.

2-31 Prove that the set of all prime integers is infinite. (*Hint:* If $p_1, p_2, \ldots, p_n$ are prime numbers, then none of them is a divisor of $p_1 p_2 \cdots p_n + 1$.)
2-32 Prove identities (1) and (2).
2-33 Prove identities (3) and (4).
2-34 What are the sets $\bigcup_{\nu=1}^{\infty} A_\nu$ and $\bigcap_{\nu=1}^{\infty} A_\nu$ if $A_\nu = \{x; \quad x \text{ an integer}, 0 < x \leq \nu\}$?
2-35 What are the sets $\bigcup_{\nu=1}^{\infty} B_\nu$ and $\bigcap_{\nu=1}^{\infty} B_\nu$ if $B_\nu = \{x; \quad x \text{ an integer}, \nu \leq x\}$?

2-9 Countable Sets

A set is said to be *countably infinite*[7] if it is equivalent to the set I_+ of positive integers. If a set A is countably infinite, then for a particular one-to-one correspondence between A and I_+ that member of A which is associated with the integer n may be denoted conveniently by a_n, as in the finite case. In effect, the members of a countably infinite set may be written in a progression or list in which each member occupies a position, say, first, second, or nth. We denote such a set A by the appropriate symbol $A = \{a_1, a_2, \ldots, a_n, \ldots\}$.

Trivially, the set I_+ is countably infinite, as is the set of positive even integers because of an equivalence already established. It is also easy to see that the set I of all integers is countably infinite. We have only to arrange the integers in an ordered progression such as $0, 1, -1, 2, -2, \ldots, n, -n, \ldots$ in which it is obvious that every integer occurs.

If a set is either finite or countably infinite, we shall describe it as *countable*.[8] If it has been determined of a set that it is not infinite, then for emphasis we may say it is *at most finite*. Further, if a set is found to be finite or countably infinite, it may be described as *at most countable*.

Theorems 2-11 through 2-15 concern important properties of countable sets.

Theorem 2-11 Any subset of a countable set is countable.

Proof Let B be a given countable set, and let A be such that $A \subset B$. If A is a finite set, the conclusion follows trivially. Thus we consider only the case in which A is an infinite set. By Theorem 2-10 this means that B is countably infinite.

Denote the set B by $\{b_1, b_2, \ldots, b_n, \ldots\}$, and define a one-to-one correspondence between the subset A and the set I_+ as follows. Let the integer 1 of I_+ correspond to that member of A with the smallest subscript, say, b_{i_1}. Then let 2 be associated with that member of A with the smallest subscript exceeding i_1; call this b_{i_2}. In general, let n correspond to that member of A with the smallest subscript that exceeds i_{n-1}, and denote this member by b_{i_n}. Since each member of A is associated with an integer by this mapping, it follows that A is countably infinite▲

[7] Also *enumerably infinite*.
[8] Also *enumerable* or *denumerable*.

Corollary A set is countable if and only if it is equivalent to a subset of I_+.

Theorem 2-12 The union of two countably infinite sets is a countably infinite set.

Proof Denote the given countably infinite sets by $A = \{a_1, a_2, \ldots, a_n, \ldots\}$ and $B = \{b_1, b_2, \ldots, b_n, \ldots\}$. The union $A \cup B$ of A and B is the set consisting of all members of A and all members of B. To demonstrate the countable character of $A \cup B$ we have only to show that a progression or listing of all its members is possible. This we do by writing the members in the arrangement

$$a_1, b_1, a_2, b_2, a_3, \ldots, a_n, b_n, \ldots,$$

deleting any symbol representing a member already listed (that is, if $a_i = b_j$, then delete b_j if $i \leq j$ and delete a_i if $i > j$). It is clear from the countably infinite character of A and B that every member of $A \cup B$ appears in the progression. That $A \cup B$ is countably infinite follows, of course, from the equivalence with I_+, which associates a_1 with 1, b_1 with 2, a_2 with 3, etc.▲

Corollary The union of any finite collection of countably infinite sets is countably infinite.

Theorem 2-13 The union of a countably infinite collection of countably infinite sets is countably infinite.

Proof Let $A_i = \{a_{i1}, a_{i2}, \ldots, a_{in}, \ldots\}$, where $i = 1, 2, 3, \ldots$, denote a countably infinite collection of countably infinite sets. The truth of the theorem then becomes evident when a progression is constructed in which all the members of all the sets A_i appear. Such a list is depicted by the following array together with a path indicating the ordering of the members in the progression:

$$\begin{array}{ccccccccc}
a_{11} & \rightarrow & a_{12} & & a_{13} & \rightarrow & a_{14} & & \cdots \\
 & \swarrow & & \nearrow & & \swarrow & & \nearrow & \\
a_{21} & & a_{22} & & a_{23} & & a_{24} & & \cdots \\
\downarrow & \nearrow & & \swarrow & & \nearrow & & & \\
a_{31} & & a_{32} & & a_{33} & & a_{34} & & \cdots \\
 & \swarrow & & \nearrow & & & & & \\
a_{41} & & a_{42} & & a_{43} & & a_{44} & & \cdots \\
\downarrow & \nearrow & & & & & & & \\
\cdots & \cdots & \cdots & \cdots & \cdots & \cdots & \cdots & \cdots & \cdots .
\end{array}$$

It is clear that every member of each set A_i lies somewhere on the path, and hence a one-to-one correspondence between the a_{in} and the integers of I_+ is implied. (Repetitions should be omitted, of course.)▲

Corollary The unior of a countable collection of countable sets is countable.

Proof If A_i should be a finite set for any set of values of i or if i itself should assume only a finite set of values, then the union of all the A_i would constitute a subset of the set of members in the above array. By Theorem 2-11 such a union is countable▲

Theorem 2-14 Every infinite set possesses a countably infinite subset.

Proof Let A be a given infinite set and a_1 some member of it. Let A_1 be the set of all members of A except for a_1. Choose some member of A_1 and call it a_2. Denote by A_2 the set of all members of A_1 except for a_2, and select from it a member a_3. Continue in this vein selecting from the set A_n a member to be called a_{n+1}, and denote by A_{n+1} the set of all members of A_n except for a_{n+1}. Either this process can be continued for all n, in which case a countable subset $\{a_1,a_2, \ldots, a_n, \ldots\}$ of A is constructed, or after a finite set of steps the procedure terminates because the set A_k, for some integer k, is empty. However, if the latter is the case, then the given set A is the finite set $\{a_1,a_2,\ldots,a_k\}$, which is contrary to hypothesis. Consequently the set $\{a_1,a_2,\ldots,a_n,\ldots\}$ exists and is a subset of A. (It should be noted that A was not assumed to be countably infinite. Were such the case, the truth of the theorem would, of course, be immediate.)▲

Theorem 2-15 A necessary and sufficient condition for a set to be infinite is that it be possible to place it in one-to-one correspondence with a proper subset of itself.

Proof The sufficiency of the stated condition is a consequence of Theorem 2-9. To prove necessity, assume the set A to be infinite. By Theorem 2-14, A contains a countably infinite subset B, which we may denote by $\{b_1, b_2, \ldots, b_n, \ldots\}$. Consider then the function f defined on A as follows:

$$\text{If } a \notin B, \text{ then } f(a) = a.$$
$$\text{If } a \in B \text{ (with } a = b_n), \text{ then } f(a) = b_{n+1}.$$

Observe that f establishes a one-to-one correspondence between the set A and that proper subset of A which consists of all members of A except b_1▲

2-10 Noncountable Sets

The reader may perhaps find it remarkable that the notions of infinite set and countably infinite set are not one and the same. It is inherently difficult to conceive of a set of entities that cannot be compiled in a list or regarded member by member with the assurance that each member will ultimately have attention fixed upon it. That such sets do actually exist is the substance of the following discussion.

Consider two different symbols + and ○ with no special meaning or property to be associated with either. From them construct arrangements or progressions such as

(5) $$+\circ++\circ\circ\circ+\circ++\circ\cdots$$

without asserting any special law of formation. The set of positions occupied by the symbols + and ○ should, in any case, be infinite. (This does not preclude the possibility that one of the two symbols may occur in only a finite set of positions.)

Consider now the set A of all possible arrangements of the type (5). We shall show that A is infinite but noncountable.

That A is infinite is easy to see. We have only to consider that subset B of A whose members are arrangements in each of which the symbol + occurs exactly once. These may be listed as follows:

$$\begin{array}{l} +\bigcirc\bigcirc\bigcirc\bigcirc\cdots \\ \bigcirc+\bigcirc\bigcirc\bigcirc\cdots \\ \bigcirc\bigcirc+\bigcirc\bigcirc\cdots \\ \bigcirc\bigcirc\bigcirc+\bigcirc\cdots \\ \cdots\cdots\cdots\cdots\quad . \end{array}$$

The symbol + occupies successively the first position, second position, third position, etc.; since such a listing obviously includes all members of B, it follows that B is countably infinite. By Theorem 2-10 it is then the case that A is infinite.

To prove that A is noncountable we assume the contrary and show that this implies a contradiction. Thus, assume that A is a countable set, and let an equivalence with I_+ be demonstrated by a listing of the members according to some rule. The following array is illustrative of such a listing:

$$\begin{array}{ll} 1 & ++\bigcirc\bigcirc+\bigcirc\bigcirc\bigcirc+\cdots \\ 2 & \bigcirc\bigcirc+\bigcirc+\bigcirc\bigcirc++\cdots \\ 3 & \bigcirc++\bigcirc+++\bigcirc\bigcirc\cdots \\ 4 & ++\bigcirc++\bigcirc\bigcirc+\bigcirc\cdots \\ 5 & \bigcirc\bigcirc\bigcirc+\bigcirc\bigcirc+\bigcirc+\cdots \\ \cdots & \cdots\cdots\cdots\cdots\cdots \\ n & \bigcirc+\bigcirc\bigcirc\bigcirc++\bigcirc\bigcirc\cdots \\ \cdots & \cdots\cdots\cdots\cdots\cdots \end{array}$$

Our assumption is that all members of A are to be found somewhere in the list. However, it is easily established that *at least one member of A is not present*; for consider the main diagonal of the array and define an arrangement (member of A) as follows. Let the symbol in the nth position be + or ○ according as the symbol in the nth position of the diagonal is ○ or +. Then it is clear that the newly constructed member does not appear anywhere in the list inasmuch as it differs from the first member in at least the first position, from the second member in at least the second position, and from the nth (for each n) in at least the nth position. That a member of A does not appear in a list containing all members of A is an obvious contradiction. Hence the assumption that A is countable is denied. It follows that A is a noncountable set.[9]

[9] The proof presented here of the existence of noncountable sets is due to Georg Cantor (1845–1918). It is familiarly known as *Cantor's diagonal proof.*

The noncountability of the set A above may be employed to show that if C is any countably infinite set, then the set of all subsets (finite and infinite) of C is noncountable. The reader should try to prove this as an exercise. It is clear that this fact may be employed as a device for the construction of noncountable sets. For example, the family whose members are all the possible subsets of the set of positive integers is noncountable. In particular, it is true that the set of infinite subsets of a countably infinite set is noncountable, while the family of all finite subsets is countable.

If a set is infinite but not countably infinite, then a one-to-one mapping cannot be set up between it and I_+. Expressed differently, it is not possible in such a case to construct a list of members which will include all. Thus not every set can be denoted correctly by such a symbol as $\{a_1, a_2, \ldots, a_n, \ldots\}$.

Exercises

2-36 Let f denote a function with domain A and range B. Prove that if A is countable, then so is B.

2-37 Let $A_1, A_2, \ldots, A_n, \ldots$ be given sets, and let $A = \bigcup_{n=1}^{\infty} A_n$. Define sets $B_1, B_2, \ldots, B_n, \ldots$ by $B_1 = A_1$ and $B_k = \mathscr{C}A_1 \cap \cdots \cap \mathscr{C}A_{k-1} \cap A_k$ for $1 < k$. Prove that (a) $B_n \cap B_m = \varnothing$ if $n \neq m$, and (b) $A = \bigcup_{n=1}^{\infty} B_n$. (This proves that the union of a countable family of sets may be expressed as a union of a countable family of disjoint sets.)

2-38 Prove that the set of all subsets of a countably infinite set is not countable.

2-39 Let F denote the set of all functions defined on the set $I_n = \{1, 2, \ldots, n\}$ with range in I_+. Prove that F is countably infinite.

2-40 Let G denote the set of all functions defined on I and having range in I. Prove that G and I are not equivalent. (*Hint*: Consider the subset G_1 of G consisting of all functions having range in the set whose members are 0 and 1. Compare G_1 with the set of all subsets of I.)

2-41 Let f denote a one-to-one mapping with domain I_{n+1} and range I_{m+1}. Define a mapping g on the set I_n by

$$g(x) = \begin{cases} f(x) & \text{if } f(x) \neq m+1 \\ f(n+1) & \text{if } f(x) = m+1. \end{cases}$$

(a) Prove that g is a one-to-one mapping of I_n onto I_m. (b) Prove by an induction on n that if I_n is equivalent to I_m, then $n = m$.

3

The Real Number System

"*This was the most unkindest cut of all.*"

WILLIAM SHAKESPEARE (1564–1616), *Julius Caesar*

3-1 Introduction

Having established the system of integers I in an axiomatic fashion, we proceed to the development of a more extensive number system. What we shall refer to as the *system of rational numbers* will consist of a set of members or *numbers* of which a proper subset will be equivalent to the set of integers and distinguished by possession of the same properties as the integers.

Definitions and properties relating to the rational number system will not require recourse to postulation of axioms. Rather, direct reference will be made to the properties and operations which are embodied in the system of integers. With one exception, the axioms listed in Chapter 1 for integers will be satisfied by the rational numbers. On the other hand, the rational numbers will enjoy a certain flexibility not possessed by the integers.

We shall discover that even the rational number system is inadequate for our needs. As a consequence, we shall define a still more extensive number system, the so-called *system of real numbers*. Unlike the systems of integers and rational numbers, this system will possess a certain *completeness property* which effectively places a limitation on further extension. The purpose of this text is the study of the real number system and of functions having domain and range in the set of real numbers.

3-2 Rational Numbers

If p is any integer and q any positive integer, then the ordered[1] arrangement (p,q) is called a *rational number*. We may denote it also by a simpler symbol r, that is, $r = (p,q)$,[2] and refer to p as the *numerator* and to q as the *denominator* of r. In accordance with the definition, the ordered pairs (4,3), (−2,5), and (0,3) are rational numbers. On the other hand, (2,−3) and (−1,0) are not rational numbers since the second integer in each pair is not positive. We denote the set or system of all rational numbers by the symbol R_*.

We say that two rational numbers (p_1,q_1) and (p_2,q_2) are *equal* and write $(p_1,q_1) = (p_2,q_2)$ if $p_1 = p_2$ and $q_1 = q_2$. That is, rational numbers are equal if and only if they are identical. As it happens, the relation of equality is too restrictive for all of our purposes. For this reason we proceed to define and study another relation between rational numbers.

3-3 Equivalence of Rational Numbers

Two rational numbers $r_1 = (p_1,q_1)$ and $r_2 = (p_2,q_2)$ are said to be *equivalent*, $r_1 \sim r_2$, if $p_1q_2 = p_2q_1$. (The latter equality is meaningful, of course, since p_1q_2 and p_2q_1 are integers.) It is immediate that equivalence of rational numbers is a *reflexive* relation; that is, $r \sim r$ for any rational number r. However, non-identical rational numbers may be equivalent, witness the numbers (8,12), (2,3), and (4,6). Quite generally, $(kp,kq) \sim (p,q)$ for any positive integer k.

Equivalence of rational numbers is also a *symmetric* relation; that is, $r_1 \sim r_2$ implies $r_2 \sim r_1$. This follows directly from the symmetric property of equality of integers.

A *transitive* property may also be demonstrated. That is, from $r_1 \sim r_2$ and $r_2 \sim r_3$ we can conclude $r_1 \sim r_3$. For if $r_1 = (p_1,q_1)$, $r_2 = (p_2,q_2)$, and $r_3 = (p_3,q_3)$, then the relation $r_1 \sim r_2$ means $p_1q_2 = p_2q_1$, while $r_2 \sim r_3$ signifies $p_2q_3 = p_3q_2$. It follows that $p_1p_2q_2q_3 = p_2p_3q_1q_2$. If $p_2 \neq 0$, this in turn implies that $p_1q_3 = p_3q_1$ or $r_1 \sim r_3$. On the other hand, if $p_2 = 0$, then $p_1q_2 = 0$ and $p_3q_2 = 0$, whence $p_1 = p_3 = 0$. The last imply $p_1q_3 = p_3q_1$ so that, again, $r_1 \sim r_3$.

The reader will recall that in Section 2-6 we discussed another reflexive, symmetric, and transitive relation and that it, too, was termed an equivalence. In general the term *equivalence relation* is reserved for any binary relation (that is, a relation involving two entities) that possesses these three properties.

By using the equivalence relation it is possible to divide R_* into disjoint sets called *equivalence classes*. Let r_1 be a member of R_*. Then the set $\{r;\quad r \sim r_1\}$

[1] The term *ordered* as employed here is not to be confused with the ordering established by the *less than* relation for integers. It signifies only that (p,q) and (q,p) do not mean the same thing.

[2] The *equality* here points simply to a common meaning for the symbols r and (p,q).

is called the *class of members of R_* which are equivalent to r_1* or, for short, the *equivalence class* determined by r_1. It is clear that if r and s are members of an equivalence class, then $r \sim s$. Also, it can be seen that equivalent rational numbers determine the same equivalence class. For assume that $r_1 \sim r_2$. If $r \sim r_1$, then $r \sim r_2$, from which $\{r;\ \ r \sim r_1\} \subset \{r;\ \ r \sim r_2\}$. Similarly, $\{r;\ \ r \sim r_2\} \subset \{r;\ \ r \sim r_1\}$ so that the equivalence classes are equal.

If two equivalence classes have a nonempty intersection, then the classes must be equal. For suppose that some rational number r_0 is a member of both $\{r;\ \ r \sim r_1\}$ and $\{r;\ \ r \sim r_2\}$. Then $r_0 \sim r_1$ and $r_0 \sim r_2$ so that $r_1 \sim r_2$, and it follows from the remarks in the last paragraph that $\{r;\ \ r \sim r_1\} = \{r;\ \ r \sim r_2\}$. Thus if two equivalence classes are not identical, they must be disjoint.

Finally, it is obvious that each member of R_* must belong to some (unique) equivalence class because it certainly belongs to the class which it itself determines.

In summarizing we can say that by using the equivalence relation we can partition R_* into disjoint equivalence classes in such a way that each member of R_* belongs to one and only one equivalence class, and two rational numbers belong to the same class if and only if they are equivalent.

Clearly this technique of partitioning a set is a general one and can be employed whenever an equivalence relation is defined on the set.

3-4 Lowest Terms Representatives

It is frequently convenient to single out one member of a class of equivalent rational numbers and refer to it as a *representative* of the class. Of the set of all rational numbers equivalent to a rational number r there is one, say (p,q), such that its denominator q is less than that of any other. This is assured by the least member axiom since all denominators are positive integers. The number (p,q) so defined is referred to as the *lowest terms* representative of the equivalence class of which it is a member. The uniqueness of such a representation is obvious since from the equivalence $(p_1,q) \sim (p,q)$ follows $p_1 = p$.

It is clear, too, that if (p,q) is a lowest terms representative, then p and q are *relatively prime*; that is, they have no common factors greater than 1. This is because if $p = kp_1$ and $q = kq_1$, where k is an integer exceeding 1, then $q_1 < q$ and $(p,q) = (kp_1,kq_1) \sim (p_1,q_1)$, denying the lowest terms character of (p,q).

It is left as an exercise for the reader to prove that if (p,q) is a lowest terms representative, then $(p_1,q_1) \sim (p,q)$ if and only if $p_1 = kp$ and $q_1 = kq$ for some positive integer k.

Example 3-1 Consider the equivalence class determined by (8,12). Clearly $(2,3) \sim (8,12)$ so that (2,3) belongs to this class. Also, if $(p,q) \sim (2,3)$, then $3p = 2q$. This implies that $3p$ is a positive even integer. Therefore $p = 2k$ for some positive integer k. But then $2q = 3(2k)$ so that $q = 3k$, and $(p,q) = (2k,3k)$. It follows that (2,3) is the lowest terms representative of the equivalence class and that the class is the set $\{(2k,3k);\ \ k \in I_+\}$.

3-5 Addition

By the *sum* of two rational numbers $r_1 = (p_1,q_1)$ and $r_2 = (p_2,q_2)$ we mean the rational number denoted by $r_1 + r_2$ or $(p_1,q_1) + (p_2,q_2)$ and defined by

$$r_1 + r_2 = (p_1q_2 + p_2q_1, q_1q_2).$$

That this truly defines a rational number is clear because the positive characters of q_1 and q_2 imply that q_1q_2 is positive also. It is obvious, too, that since sums and products of integers are unique, this is also the case for a sum of rational numbers.

Commutative and associative laws for addition of rational numbers follow from those for addition and multiplication of integers. To demonstrate commutativity we have

$$r_1 + r_2 = (p_1q_2 + p_2q_1, q_1q_2) = (p_2q_1 + p_1q_2, q_2q_1) = r_2 + r_1.$$

The reader may verify in a similar fashion that

$$(r_1 + r_2) + r_3 = r_1 + (r_2 + r_3)$$

for any rational numbers r_1, r_2, and r_3.

In connection with the operation of addition there arises the question of the possible existence of a number having the character of a zero, that is, a rational number x such that for each r of R_* it is the case that $r + x = r$. Actually it is a simple matter to verify that the number $x = (0,1)$ is unique in this respect and reasonably, therefore, should be termed the zero of R_*. It is curious, however, that the very uniqueness of (0,1) is a drawback. For if, given a rational number r, we propose to determine its negative, that is, a number $-r$ with the property $r + (-r) = (0,1)$, we find the situation impossible! The reader may easily establish that $(p_1,q_1) + (p_2,q_2) = (0,1)$ if and only if $q_1 = q_2 = 1$ and $p_2 = -p_1$. Thus, on the basis of such definitions only a rational number of the form $(p,1)$ can have a negative. The inadequacy of the unique zero (0,1) suggests the need for greater flexibility. This is achieved through the concept of equivalence.

With respect to the operation of addition, it happens that rational numbers of the same equivalence class behave in a similar manner. More precisely

$$r_1 \sim r_1' \text{ and } r_2 \sim r_2' \quad \text{imply} \quad r_1 + r_2 \sim r_1' + r_2'. \tag{1}$$

To see this, write $r_1 = (p_1,q_1)$, $r_1' = (p_1',q_1')$, $r_2 = (p_2,q_2)$, and $r_2' = (p_2',q_2')$, and suppose that $(p_1,q_1) \sim (p_1',q_1')$ and $(p_2,q_2) \sim (p_2',q_2')$ or, what is the same,

$$p_1q_1' = p_1'q_1 \quad \text{and} \quad p_2q_2' = p_2'q_2. \tag{2}$$

Form the sums

$$(p_1,q_1) + (p_2,q_2) = (p_1q_2 + p_2q_1, q_1q_2) \tag{3}$$

and

$$(p_1',q_1') + (p_2',q_2') = (p_1'q_2' + p_2'q_1', q_1'q_2'). \tag{4}$$

By (2) we may then prove the equivalence of the right members of (3) and (4). We have

$$\begin{aligned}(p_1q_2 + p_2q_1)(q_1'q_2') &= p_1q_2q_1'q_2' + p_2q_1q_1'q_2' \\ &= p_1'q_2q_1q_2' + p_2'q_1q_1'q_2 \\ &= (p_1'q_2' + p_2'q_1')(q_1q_2).\end{aligned}$$

From this we conclude the truth of (1).

Example 3-2 By definition, $(8,12)+(-4,2)=(-32,24)$ and $(2,3)+(-2,1)=(-4,3)$. Since $(8,12)\sim(2,3)$ and $(-4,2)\sim(-2,1)$, it follows from (1) that $(-32,24)\sim(-4,3)$. This fact may also be verified directly.

To circumvent the difficulty encountered earlier, we make the following definitions. Any rational number of the form $(0,q)$ is called a *zero*. This name is derived from the fact that

$$r + (0,q) \sim r$$

for any rational number r and any zero. Thus $(0,q)$ has the property which we associate with the word "zero," but it possesses this property relative to the equivalence relation rather than to the relation of equality.

It is immediate that the set of all zeros is an equivalence class. The lowest terms representative of this class is $(0,1)$.

A *negative* or *inverse with respect to addition* of a rational number r is defined to be any rational number s with the property that $r + s$ is a zero. If s is a negative of r and if $s \sim t$, then t is also a negative of r because $r + s \sim r + t$ by (1). Also, if s and t are both negatives of r, then $s \sim s + (r + t) \sim (s + r) + t \sim t$, or $s \sim t$. Thus the set of all negatives of r is an equivalence class, that is, unless it is merely an empty set.

That a number r actually possesses a negative follows at once. For if $r = (p,q)$, then it is easy to see that $(-p,q)$ is a negative of r. We shall single out this particular negative to be designated[3] by the symbol $-r$; that is, $-(p,q) = (-p,q) = -r$.

Exercises

3-1 Let (p,q) be a lowest terms respresentative of some equivalence class. Prove $(p_1,q_1)\sim(p,q)$ if and only if $p_1 = kp$ and $q_1 = kq$ for some positive integer k.

3-2 Prove the associative law of addition for the system R_*.

3-3 Prove that $r + x = r$ for every $r \in R_*$ if and only if $x = (0,1)$.

3-4 Prove that $(p_1,q_1) + (p_2,q_2) = (0,1)$ if and only if $p_1 = -p_2$ and $q_1 = q_2 = 1$.

[3] We do this so that the symbol $-r$ will have a unique meaning in the sequel.

3-6 Multiplication

By the *product* of two rational numbers $r_1 = (p_1,q_1)$ and $r_2 = (p_2,q_2)$ we mean the number denoted by r_1r_2 or $(p_1,q_1)(p_2,q_2)$ and defined by

$$r_1r_2 = (p_1p_2,q_1q_2).$$

That the product is rational follows from $q_1q_2 > 0$.

Multiplication, like addition, is unique in that any two rational numbers have exactly one product. The nature of the definition of product is such that the commutative and associative laws

$$r_1r_2 = r_2r_1$$

and

$$(r_1r_2)r_3 = r_1(r_2r_3),$$

respectively, are virtually immediate.

An investigation of the bond between addition and multiplication shows once again that in R_* it is more satisfactory to deal with the relation of equivalence than with that of equality. Thus the reader may verify that the distributive law *relative to equivalence*

$$r_1(r_2 + r_3) \sim r_1r_2 + r_1r_3$$

holds, but that it is not always valid if equivalence is replaced by equality.

Equivalent numbers exhibit similar behavior under multiplication. As the reader may prove,

$$r_1 \sim r_1' \text{ and } r_2 \sim r_2' \quad \text{imply} \quad r_1r_2 \sim r_1'r_2'.$$

Following the pattern of the last section, we use the equivalence relation as a guide in defining a unit. Thus, a rational number of the form (q,q) is called a *unit* because

$$r(q,q) \sim r$$

for any rational number r and any unit. The set of all units is an equivalence class of which $(1,1)$ is the lowest terms representative.

It is clear that a rational number cannot be both a unit and a zero. For if (p,q) is a zero, then $p = 0$; but if (p,q) is a unit, then $p = q > 0$.

So far we have discussed no properties of the rational system that are significantly different from those of the integers. In order to discover such, we investigate the equivalence

$$rx \sim (1,1) \tag{5}$$

and ask: If we are given a rational number r, can we find a rational number x which satisfies this equivalence?

Any rational number s such that rs is a unit will be called a *reciprocal* or *inverse* or *inverse with respect to multiplication* of r. If s is an inverse of r and $s \sim t$, then $rs \sim rt$ so that t is also an inverse of r. Also, if s and t are inverses of r, then $s \sim s(rt) \sim (sr)t \sim t$, so that $s \sim t$. Thus we see that for each r the set of inverses of r is an equivalence class unless, of course, it is an empty set. Thus our question can be rephrased: Given a rational number r, is the set of inverses of r nonempty?

First, observe that it is impossible to satisfy (5) in case r is a zero. For then, no matter what the number x is, rx is a zero and hence is not equivalent to a unit. For this reason we further restrict the proposal by requiring that r be nonzero.

Thus, suppose that r is nonzero and write $r = (p,q)$. Define the rational number r^{-1} to be the number (q,p) if $p > 0$ and $(-q,-p)$ if $p < 0$.[4] If $p > 0$ it follows that $rr^{-1} = (p,q)(q,p) = (pq,pq)$ so that rr^{-1} is a unit. A similar result holds in case $p < 0$. Hence we have proved that any nonzero rational number r has an inverse and, in particular, that the number denoted by r^{-1} is one.

The chief advantage of the rational system over the system of integers lies in the fact that a nonzero rational number possesses an inverse. Several important results follow at once. For example, a cancellation law for multiplication *relative to equivalence* holds in R_*. If r_1, r_2, and r_3 are such that $r_1 r_3 \sim r_2 r_3$ and if r_3 is nonzero, then $r_3{}^{-1}$ exists and $r_1 r_3 r_3{}^{-1} \sim r_2 r_3 r_3{}^{-1}$ or $r_1 \sim r_2$. That is,

(6) $$r_1 r_3 \sim r_2 r_3 \text{ and } r_3 \text{ nonzero} \quad \text{imply} \quad r_1 \sim r_2 .$$

The existence of a reciprocal also assures us that an x exists such that

(7) $$r_1 x \sim r_2$$

provided that r_1 is nonzero. That the number $r_1{}^{-1} r_2$ satisfies equivalence (7) follows by direct substitution. Moreover, any number that satisfies (7) must be equivalent to $r_1{}^{-1} r_2$ because if $r_1 x \sim r_2$ and $r_1 y \sim r_2$, then $r_1 x \sim r_1 y$ so that $x \sim y$ by (6).

The product $r_1{}^{-1} r_2$ is commonly referred to as the *quotient* of r_2 by r_1; and the operation of forming a quotient as *division*.

Exercises

3-5 Prove that $r_1(r_2 + r_3) \sim r_1 r_2 + r_1 r_3$ for any r_1, r_2, and r_3 in R_*.
3-6 Prove that $r_1(r_2 + r_3) = r_1 r_2 + r_1 r_3$ is not always valid in R_*.
3-7 Prove that $r_1 \sim r_1'$ and $r_2 \sim r_2'$ imply $r_1 r_2 \sim r_1' r_2'$.
3-8 Prove that $(r_1 r_2)^{-1} = r_1{}^{-1} r_2{}^{-1}$.

[4] The definition of r^{-1} is separated into two cases to ensure that the denominator of r^{-1} is positive.

3-7 The Order Relation

An order relation is defined for rational numbers as for integers. If $r_1 = (p_1,q_1)$ and $r_2 = (p_2,q_2)$, then we say that r_1 is *less than* r_2 and write $r_1 < r_2$ (or that r_2 is *greater than* r_1 and write $r_2 > r_1$) if $p_1q_2 < p_2q_1$. It is immediately clear that if $r_1 \sim r_1'$ and $r_2 \sim r_2'$ and if $r_1 < r_2$, then $r_1' < r_2'$.

From the trichotomy for integers follows a similar (but not identical) principle for rationals: Any two rational numbers r_1 and r_2 satisfy exactly one of the relations $r_1 < r_2$, $r_2 < r_1$, and $r_1 \sim r_2$.

Certain properties of the less than relation for integers also hold for that of the rational numbers. For example, the relation is transitive; that is,

$$r_1 < r_2 \text{ and } r_2 < r_3 \quad \text{imply} \quad r_1 < r_3.$$

To verify this, observe that from $r_1 < r_2$ and $r_2 < r_3$ follow $p_1q_2 < p_2q_1$ and $p_2q_3 < p_3q_2$. Therefore, $p_1q_2q_3 < p_2q_1q_3 < p_3q_1q_2$, and we conclude $p_1q_3 < p_3q_1$ or $r_1 < r_3$.

It is also the case that

$$r_1 < r_2 \quad \text{implies} \quad r_1 + r_3 < r_2 + r_3$$

for any r_3. For from $p_1q_2 < p_2q_1$ we have

$$p_1q_2q_3 + p_3q_1q_2 < p_2q_1q_3 + p_3q_1q_2$$

$$(p_1q_3 + p_3q_1)q_2 < (p_2q_3 + p_3q_2)q_1$$

or

$$(p_1q_3 + p_3q_1)(q_2q_3) < (p_2q_3 + p_3q_2)(q_1q_3).$$

A rational number r is called *positive* if $(0,1) < r$ and *negative* if $r < (0,1)$. Clearly, $r = (p,q)$ is positive or negative according as p is a positive or negative integer. It is left to the reader to show that if r_1, r_2, and r_3 are rational numbers, then

$$r_1 < r_2 \text{ and } r_3 \text{ positive} \quad \text{imply} \quad r_1r_3 < r_2r_3.$$

3-8 The Infinite Character of the Set of Rational Numbers

Let (p,q) denote the lowest terms representative of a class of equivalent rational numbers. Since the members of this class are of the form (kp,kq) where k is a positive integer, it is immediate that the class is countably infinite. Further, the set of all rational numbers is countably infinite. To see this, observe that for any one value of p the numbers (p,q) constitute a countably infinite set since q is an integer. Since p likewise ranges over a countably infinite set, the conclusion follows from Theorem 2-13.

The set R_* exhibits certain characteristics not possessed by the integers. Although it is the case for integers that between any two of the form n and $n + 1$ no other integer exists, for rational numbers it is easily shown that between any two nonequivalent rationals there exists another. Let $r_1 = (p_1,q_1)$ and $r_2 = (p_2,q_2)$ denote any two rationals with the property $r_1 < r_2$. Then, from $p_1q_2 < p_2q_1$ follows $2p_1q_2 < p_1q_2 + p_2q_1$, whence $p_1(2q_1q_2) < (p_1q_2 + p_2q_1)q_1$. Therefore, $r_1 < r_3$, where $r_3 = (p_1q_2 + p_2q_1, 2q_1q_2)$. In a similar fashion it follows that $r_3 < r_2$. Thus the existence of a number r_3 lying between r_1 and r_2 has been demonstrated. (It may be noted that r_3 as constructed is the so-called *average* of r_1 and r_2.) An immediate consequence, of course, is that between r_1 and r_3 lies still another rational number r_4 and between r_1 and r_4 still another. Since this argument can be continued indefinitely, by induction we may conclude that between any two nonequivalent rational numbers there lies an infinite set of nonequivalent rational numbers. We describe this property of the set R_* by saying that it is a *dense* set. In contrast, the set of integers may be termed *discrete*.

The positive integers were observed to allow *consecutive arrangement*; that is, any subset of I_+ may be written in a progression $a_1, a_2, \ldots, a_n, \ldots$ such that $a_n < a_{n+1}$ for each positive integer n. The same is not true, however, for dense subsets of the set of positive rational numbers. For, supposing it to be possible, let $r_1, r_2, \ldots, r_n, \ldots$ denote such a progression. Then, between r_n and r_{n+1} for each n lies an infinite set of rational numbers that, for a consecutive arrangement, should follow r_n and precede r_{n+1}. Since they appear elsewhere in the progression, this is a contradiction; hence the proposed arrangement is not possible. In effect, although the rational numbers may be counted, they cannot be counted in order of size.

Exercises

3-9 Prove that the order relation for rational numbers is (a) not reflexive, and (b) not symmetric.

3-10 Prove that $r_1 < r_2$ and r_3 positive imply $r_1r_3 < r_2r_3$.

3-11 Prove that if $(p_1,q_1) < (p_2,q_2)$, then $(p_1,q_1) < (p_1 + p_2, q_1 + q_2) < (p_2,q_2)$.

3-9 A Modified System of Rational Numbers

One fact we have discovered in working with R_* is that it is not feasible to design a unique zero or unit. More generally, we found that our discussion of addition, multiplication, and order led to the consideration of entire classes of numbers rather than to that of individual members. One way to simplify this situation is to consider a subset of R_* consisting of exactly one member from each equivalence class and to set aside the rest. This can be done in many ways, but perhaps the simplest is to consider the set R consisting of all rational numbers that are lowest terms representatives of their equivalence classes.

The set R so constructed has the property that if s is any rational number, then $s \sim r$ for some unique member r of R. This can be expressed by saying that each member of R_* is represented in R. For later use we note, too, that R contains every rational number of the form $(n,1)$.

Since R consists of exactly one member from each equivalence class, it follows that distinct members of R cannot be equivalent. Expressed differently, if r_1 and r_2 are members of R and $r_1 \sim r_2$, then $r_1 = r_2$.

Unfortunately a difficulty arises with the set R (and this is true of any set of similar construction), namely, that it is not a system of rational numbers in which the operations of addition and multiplication are meaningful! It is not always the case that the sum or the product of two rationals in lowest terms is again a rational number in lowest terms. For example, $(1,2) + (1,2) = (4,4)$ and $(3,8)(4,1) = (12,8)$. Indeed, it was this difficulty that prompted us from the beginning to develop R_* rather than to restrict our attention to R.

It is possible to overcome this deficiency in R by a rather simple device: *If, in any addition or multiplication, a number arises which is not in lowest terms, then let it be replaced by that number in lowest terms which is equivalent to it.* (The reader will recall the instruction in arithmetic to reduce all answers to lowest terms.) In this way the sum and the product of two members of R are again members of R.

It is easily verified that the set R of all lowest terms representatives, *subject to the simplifying device above*, is a system possessing the algebraic properties of R_* but with one important difference. Since equivalent members of R must be equal, *the equivalence symbol may always be replaced by the symbol of equality*. Thus it is the case that

$$r_1(r_2 + r_3) = r_1 r_2 + r_1 r_3$$

for all r_1, r_2, and r_3 in R even though this statement is not always true in R_*. In R the cancellation law for multiplication,

$$r_1 r_3 = r_2 r_3 \text{ and } r_3 \text{ nonzero} \quad \text{imply} \quad r_1 = r_2,$$

always holds as does solvability with respect to multiplication; that is,

$$r_1 x = r_2 \quad (r_1 \text{ nonzero})$$

always has a unique solution. Also, the trichotomy law applies; that is, exactly one of the relations

$$r_1 < r_2, \quad r_1 = r_2, \quad r_1 > r_2$$

holds.

We shall use the symbol R to denote the *system* as well as the *set* of lowest terms rationals. It is easy to see that R has the unique zero $(0,1)$ and the unique unit $(1,1)$. Any number r of R has a unique negative which is precisely that denoted earlier by $-r$. Also, a nonzero number r of R has a unique reciprocal which is the number we have denoted by r^{-1}.

Henceforth, by the *set or system of rational numbers we shall understand R rather than R_**. The elimination of a multiplicity of equivalent rationals will simplify references to existence, uniqueness, etc.

3-10 Comparing Two Systems

A number system is essentially the embodiment of the set of properties which define it, together with consequences of those properties. Accordingly, in a comparison of number systems one is interested in properties only and is indifferent to symbols, terminology, and genetic development. If two number systems are devised—by whatever means conceivable—with the end result that they have identical properties, then they are to be considered one and the same system. Or if two distinct systems have a certain set of properties in common, then consequences of these properties in one system are necessarily consequences in the other also. For example, if we once prove that associativity of addition, the existence of a zero, and the existence of a negative imply the cancellation law for addition (as we did in Chapter 1), then the law of cancellation must hold in any system which possesses these three properties.

As Section 3-8 proves, the rational number system and the system of integers are two different systems. However, we have already seen that they have much in common. Each possesses operations of addition and multiplication which are commutative and associative and to which the same distributive law applies; each possesses a unique zero and unit; and each is endowed with an order relation. In short, we have proved that the set R satisfies every property of the integers to be found in Axioms 1 through 14. As a consequence, it follows at once that R possesses every property that can be proved from these axioms. In particular, the cancellation law of addition, that is,

$$r_1 + r_3 = r_2 + r_3 \quad \text{implies} \quad r_1 = r_2,$$

follows as do the properties stated in Theorems 1-1 through 1-13. We shall make use of such properties without further comment.

Another general remark that can be made about a mathematical system concerns the introduction of new operations. If an operation of addition is defined in any mathematical system and if each number has a negative, then it is customary to write $a - b$ for the number $a + (-b)$ and to call this the *difference* of a and b. In this way the operation of *subtraction* is introduced in the system R or in any comparable system. Similar remarks, of course, apply to the *quotient* and to *division*.

3-11 The Rational Number System as an Extension of the System of Integers

The system of integers I and the system of rational numbers R exhibit certain similarities and differences which point to the manner in which the system R may be regarded as an extension or generalization of the system I. Regarding similarities, they amount to these. Each of the properties of integers postulated in Axioms 1 through 13 is also a property of the rational numbers (in the sense that the symbols a, b, c, etc. that appear in those axioms are regarded as rational numbers and the operations and relations have the meanings ascribed to them in the system R). If we consider that the first thirteen axioms characterize the so-called *algebraic* structure of the integer system, then we observe that the algebraic properties of the integers are properties of the rational numbers as well.

Axioms 14 and 15 postulate what we might term *set* properties of the integers. Of the two, Axiom 14 is certainly valid for the set of rational numbers also. Axiom 15, on the other hand, imposes a restriction on the system of integers which does not extend to the rationals. While it is true of rational numbers that some sets which are nonempty and bounded below do possess least members, others exist which are not so endowed. An example of these is the set of all rationals of the form $(1,n)$, where n ranges over all positive integers. Clearly, such rational numbers do exist; therefore the set is not empty. Also, every member of the set is positive; therefore the set is bounded below by $(0,1)$. However, no number in the set is less than all others since $(1, n+1) < (1, n)$ for each n.

Recall from Chapter 1 that the proof of the principle of Archimedes for integers made use of the least member axiom. Although such a principle is, in fact, also germane to the system of rational numbers, it must be proved without recourse to a least member principle. We leave it to the reader to establish the principle of Archimedes: If r_1 and r_2 are any positive rational numbers, then a rational number r of the form $(n,1)$ exists such that $r_2 \leq rr_1$.

Another important difference between the systems of integers and rational numbers is algebraic in character and concerns the operation of multiplication. For integers no inverse with respect to multiplication is postulated, and therefore solvability with respect to multiplication may not be concluded. Indeed, no integer x exists to satisfy such an equation as $2x = 3$. On the other hand, corresponding to any nonzero rational number r, there exists another, r^{-1}, such that $rr^{-1} = 1$. As a consequence, every equation of the form $r_1x = r_2$ $(r_1 \neq 0)$ is such that a rational number $r_1{}^{-1}r_2$ exists to satisfy it.

The fact that both the rational system and the system of integers have the properties described by Axioms 1 through 14 but that the rational system does not suffer limitations with regard to the least member principle and solvability suggests that the rational system R may be regarded as an *extension* of the system I. It is to be observed that such a description is based essentially on a comparison of *two distinct systems* with similar properties imposed or deduced. The development of the rationals has not consisted of starting with the integers and then incorporating more numbers simply to guarantee solvability with respect to multiplication. Analysis of the two systems, however, shows something very like this to be the case.

Consider that proper subset of the set R whose members are of the form $(n,1)$. Denote this set by J and note that between it and the set of integers I a one-to-one correspondence may be defined in which the integer n and the rational number

$(n,1)$ correspond. This equivalence has certain far-reaching consequences with respect to both the algebraic and set characteristics of I and J.

Let m and n denote any two integers. Then their images $(m,1)$ and $(n,1)$ are clearly seen to have the sum $(m,1) + (n,1) = (m + n, 1)$, which by the same correspondence is the image of $m + n$. Similarly, $(m,1)(n,1) = (mn,1)$ is the image of mn. Hence

(8) $\qquad m + n$ corresponds to $(m,1) + (n,1)$,

and

(9) $\qquad mn$ corresponds to $(m,1)(n,1)$.

Further, as the reader may verify,

(10) $\qquad m < n$ if and only if $(m,1) < (n,1)$.

We say that I and J are *isomorphic*, or that an *isomorphism* exists between them, to describe the fact that the correspondence associating n and $(n,1)$ is such that (8), (9), and (10) hold.

It is clear that because of the isomorphism the axioms of Chapter 1 are as meaningful for the system J as for I. In particular, consider the least member axiom. If some subset (nonempty and bounded below) of J should not possess a least member, then by the isomorphism this would also be the case for the equivalent set of integers, and this is not possible.

The systems I and J are distinguished only in the manner in which they have come into being and in the symbols n and $(n,1)$ used to denote their respective members. As a consequence, we look upon the distinction between I and J as artificial and regard the symbols n and $(n,1)$ as merely different symbols for the same number. This done, the integers become a proper subset of the set of rational numbers, and we say that I is *imbedded* in the system R.

At this point, too, we adopt the more familiar fraction notation p/q for the rational number previously denoted (p,q). In particular, if q is a power of 10, we employ the familiar decimal representation. Also, we shall assume the results of Exercise 3-15 and allow q to be either positive or negative.

Exercises

3-12 Prove the principle of Archimedes: If r_1 and r_2 are positive rational numbers, then a rational number $r = (n,1)$ exists such that $r_2 \leq rr_1$.

3-13 Let m and n denote integers. Prove that $m < n$ if and only if $(m,1) < (n,1)$.

3-14 Prove that the set I of integers and the set E of even integers are equivalent sets but not isomorphic systems.

3-15 Show how the definitions of addition, multiplication, and order for R_* must be revised if the number q in the pair (p,q) is merely required to be nonzero.

3-16 Denote the rational number (p,q) by the symbol p/q. Express the definitions of addition, multiplication, and order in this notation, and hence verify that they are the familiar rules of elementary algebra.

3-12 The Number Axis

Further insight into the structure of the rational number system R is gained through a geometrical representation of the members of R. Assuming familiarity with the concepts of line and line segment, we introduce the notion of a *number axis.* This we do by regarding an indefinite line as made up of a so-called *continuum of points* and setting up a one-to-one correspondence between a subset of this continuum and the set R. In this way, points on a line may be looked upon as further representations of rational numbers. It must be understood, of course, that a geometrical representation of a number system is not essential to its investigation or development. The introduction of geometric notions is simply in the nature of a convenient device designed to promote fuller understanding.

On a line L of indefinite length we select arbitrarily two distinct points P_0 and P_1 such that P_1 lies to the right of P_0 and define the segment $\overline{P_0 P_1}$ to be a *unit of length.* We then establish a correspondence between the rational numbers and the points on L in such a way that the point P_0 and the number 0 correspond, as do the point P_1 and the number 1. A segment $\overline{P_1 P_2}$ equal in length to $\overline{P_0 P_1}$ being laid off to the right of P_1, as in Figure 3-1, the point P_2 is then associated

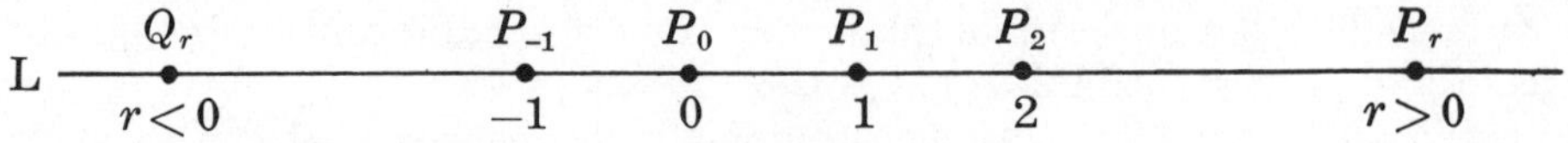

Figure 3-1

with the integer 2. Similarly, P_3, where $\overline{P_2 P_3}$ is a unit length, is the image of 3. Continuing in this fashion, the set of all positive integers is associated with the set of right endpoints of a succession of unit segments laid off to the right of P_0.

To depict nonintegers in the correspondence we divide each of the segments $\overline{P_0 P_1}, \overline{P_1 P_2}$, etc., into q equal subdivisions. Then the point P_r lying to the right of P_0 and such that the segment $\overline{P_0 P_r}$ contains exactly p of these subdivisions will be the correspondent of the number $r = p/q$. In this manner all positive rational numbers find images in points on the number axis. To represent negative numbers we simply mark off segments to the left rather than to the right of P_0.

The equivalence so defined is one in which each member of the set R corresponds to exactly one point of the number axis. The aggregate of all such points is referred to as the *set of rational points.* It is important to note that it does not follow that every point of the line L is a rational point. Indeed, we shall see that this is not the case.

It will be convenient to use the terms *rational number* and *rational point* or simply *number* and *point* interchangeably. For example, we shall not hesitate to say of two points that one is greater than another, meaning, of course, that this is true of their numerical counterparts. Actually, it is clear from the nature of the mapping that direction on the number axis corresponds to the order relation for numbers. Thus, if $r_1 < r_2$, then P_{r_1} lies to the left of P_{r_2}.

3-13 The Need for Further Extension

Having seen the rational number system to be an extension of the system of integers, we now look into the desirability and possibility of further extension. It is, in fact, not difficult to show that such extension is desirable; that is, the rational number system exhibits certain deficiencies which only the creation of new numbers can remedy.

We have seen it to be true that the equation $r_1 x = r_2$ possesses a solution for any pair of rational numbers r_1 and r_2, provided only that r_1 be nonzero. Let us propose, however, a slightly different question. If r is any given rational number, then does another rational number x necessarily exist with the property $x^2 = r$? If $r < 0$, it is immediate, of course, that no such number exists since $x^2 \geq 0$ for all rational x. On the other hand, it is easy to think of positive values of r for which x does exist. For example, $r = 4$ and $r = 9/25$ lead to $x = \pm 2$ and $x = \pm 3/5$, respectively. It is the case, however, that positive rational numbers do exist which cannot be written as squares of other rational numbers. In particular, the positive integer 2 is such a number. We show this to be the case by the following indirect proof.

Suppose a rational number p/q of R to exist with the property $(p/q)^2 = 2$. That is, we assume $(p/q)^2 = 2$, where the integers p and q are relatively prime. It follows that $p^2 = 2q^2$, whence p^2 is an even number. Therefore, p is even (see Exercise 1-25). The existence of an integer p' such that $p = 2p'$ is thus established, and the integers p' and q are related by $4p'^2 = 2q^2$ or $q^2 = 2p'^2$. From this it follows that q^2 is even and, hence, so is q. As a consequence, p and q are seen to possess the common factor 2, thereby denying the assumed lowest terms character of p/q. We conclude from this contradiction that no rational number exists whose square is 2.

In Figure 3-2 the number axis L is depicted together with a segment $\overline{P_1 T}$ of length equal to $\overline{P_0 P_1}$ constructed perpendicular to L. With center at P_0 an arc of radius $\overline{P_0 T}$ is described intercepting the number axis at Q. If Q is a rational point—that is, if $\overline{P_0 Q}$, and hence the segment $\overline{P_0 T}$, has a length which is measured by a rational number p/q—then that number must satisfy the pythagorean relation $(p/q)^2 = 1^2 + 1^2 = 2$. However, we have seen that no rational number exists with this property and therefore Q is not a rational point. We conclude that the linear geometric continuum L possesses points in addition to those which, in the correspondence defined at the outset, are associated with rational numbers.

The number 2 is not unique in the possession of the property deduced above. The reader may prove for himself that such numbers as 3, 5, and 6/11 also are not squares of rational numbers. Further, one may show that equations of the form $x^3 = r$, $x^4 = r$, and, in general, $x^n = r$ (n a positive integer) are such that rational solutions do not always exist.

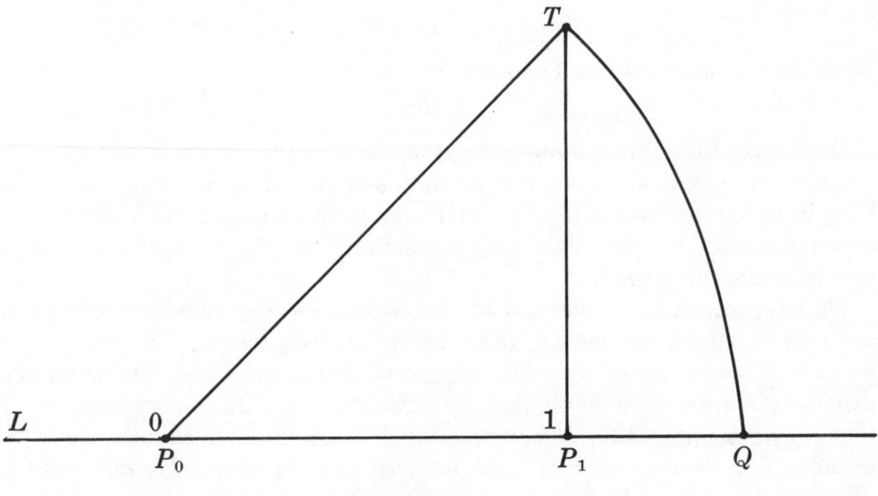

Figure 3-2

That the linear continuum possesses points which are not rational points or, what is the same, that equations such as $x^n = r$ exist which do not admit of rational solutions, suggests an inadequacy or incompleteness in the system R of rational numbers. The question then is how to overcome this deficiency. In principle, a new and more extensive number system is desired, one that has members defined in terms of rational numbers and that has the system of rational numbers imbedded in it.

It seems quite natural to propose an extension in a manner similar to that in which the rationals themselves grew out of the integers. Consider, for example, a system obtained from the set of all ordered pairs (r_1, r_2), where r_1 and r_2 are rational numbers, with addition, multiplication, equality, and order defined as in the case of rationals. It is then easy to show that the new system is isomorphic to the system of rational numbers, and accordingly no extension is achieved. The reader should prove this for himself.

The extension of the rational to the more extensive *real number system* may be presented in either of two ways: the first, due to Georg Cantor, concerns convergent sequences of rational numbers; and the second, the creation of Richard Dedekind, involves contemplation of a division of the set of rational numbers into classes. We shall give here the latter definition and refer the reader to other sources for the Cantor approach.

Exercises

3-17 Prove that no rational number r exists such that: (a) $r^2 = 3$; (b) $r^2 = 5$; (c) $r^2 = 2k$ for any odd integer k.

3-18 An integer a is called a *perfect square* if there exists an integer b such that $b^2 = a$. Prove that if m/n is a rational number in lowest terms with the property that m and n are not both perfect squares, then no rational number r exists such that $r^2 = m/n$.

3-19 Prove that a solution of the equation $x^n + a_{n-1}x^{n-1} + \cdots + a_0 = 0$, where a_k $(k = 0, \ldots, n-1)$ is an integer, is an integer or is irrational.

3-20 Let P_* denote the set of all ordered pairs (r_1,r_2), where r_1 and r_2 are rational numbers with r_2 nonzero. Let equivalence, addition, multiplication, and less than be defined as for R_*. Denote by P a subset of P_* consisting of one member from each class of equivalent members of P_* and define addition, multiplication, and order for P following the pattern in which R was obtained from R_*. Prove that P and R are isomorphic. (This shows that it is not possible to extend the system R by repeating the construction wherein the rational numbers were obtained from the integers.)

3-14 Dedekind Cuts

A *Dedekind cut* or, briefly, a *cut* is an ordered pair (R_1,R_2) of sets of rational numbers possessing the following properties:

1. $R_1 \cup R_2 = R$.
2. Neither R_1 nor R_2 is empty.
3. Every member of R_1 is less than every member of R_2.
4. R_2 does not contain a least member.

From property *1* we see that each rational number is either in R_1 or in R_2. However, no rational number can be in both R_1 and R_2 because of *3*. Thus, a cut may be described as a *division* or *separation* of the set of rational numbers R into two subsets R_1 and R_2 satisfying conditions *1* through *4*. The set R_1 we call the *left* or *lower class* and R_2 the *right* or *upper class* of the cut (R_1,R_2).

The set of all cuts may be separated into two categories or types according to whether

(i) R_1 has a greatest member, or
(ii) R_1 has no greatest member.

The existence of a cut of Type (i) is easily demonstrated. Let r be a particular rational number, and denote by R_1 the set consisting of all rational numbers less than or equal to r. Let R_2 be the set of all rational numbers which exceed r. It is then easily seen by examination of each of the four cut requirements that the ordered pair (R_1,R_2) is, in fact, a cut. *1.* By the trichotomy of the order relation each member of R finds itself in R_1 or R_2. Hence $R_1 \cup R_2 = R$. *2.* Since no rational number is either greater than or less than all other rational numbers, it follows that neither R_1 nor R_2 is empty. *3.* If r_1 is any member of R_1 and r_2 any member of R_2, then $r_1 \leq r$ and $r < r_2$ from which it follows that $r_1 < r_2$. *4.* If r_2 is in R_2, then $r < r_2$. By the dense character of the rationals there then exists a rational number r_2' such that $r < r_2' < r_2$. It follows that r_2' is in R_2 and that R_2 has no least member.

As we have just seen, corresponding to each rational number is a cut of Type (i) having that rational number as the greatest member of its lower class. Conversely, to each cut of Type (i) corresponds that rational number which is greatest in its lower class. It follows that a one-to-one correspondence exists between the set of all rational numbers on the one hand and the set of all cuts of Type (i) on the other. Because of this correspondence, cuts of Type (i) will be called *rational cuts*. If r is the greatest member of a lower class R_1, then the cut (R_1,R_2) is said to be *based on r*.

Next we construct an example of a cut of Type (ii). Let Z_1 be the set consisting of all negative rational numbers together with 0 and the set of all positive numbers whose squares are less than 2. Let Z_2 denote the set of all other rational numbers. First we show that (Z_1,Z_2) is truly a cut. *1.* For one thing, putting in the upper class Z_2 all rational numbers not in Z_1 guarantees that each rational number is in one class or the other. *2.* Neither class is empty since Z_1 contains all the negative numbers and Z_2 contains, for example, the number 3 since its square is not less than 2. *3.* Further, Z_2 contains only positive numbers since all nonpositive numbers are in Z_1. Trivially, the latter numbers are less than all numbers in Z_2. As far as the positive numbers in Z_1 are concerned, let z_1 be one of them. Then ${z_1}^2 < 2$. If z_2 is in Z_2, then $2 < {z_2}^2$. (Of course ${z_2}^2 = 2$ is impossible since no rational number exists with this property.) Hence ${z_1}^2 < {z_2}^2$, and it follows that $z_1 < z_2$. Therefore, every positive number in Z_1 is less than every number in Z_2. *4.* Finally, it may be shown that the upper class Z_2 does not possess a least member. For suppose x to denote any member of Z_2, and consider the rational number y defined by

$$y = \frac{2(1+x)}{2+x}.$$

Some simple algebra leads immediately to the two expressions

$$y - x = \frac{2 - x^2}{2 + x} \quad \text{and} \quad 2 - y^2 = \frac{2(2 - x^2)}{(2 + x)^2}. \tag{11}$$

Since x is in Z_2, we have $2 - x^2 < 0$ whence the right members of both expressions are negative. Therefore $y < x$ and $2 < y^2$. It follows that, corresponding to any number x in Z_2, there exists a number y which is less than x and which is also in Z_2. Hence Z_2 does not possess a least member. Thus it is established that (Z_1,Z_2) is a cut.

That it is a cut of Type (ii) follows also from the expressions (11). For if x denotes a non-negative member of Z_1, then y is such that $x < y$ and $y^2 < 2$, indicating that y is greater than x but is also in Z_1. Hence Z_1 has no greatest member.

A cut of Type (ii) is called an *irrational cut*.

Exercises

3-21 Let (R_1,R_2) denote a cut. Prove that if $r \in R_1$ and $s < r$, then $s \in R_1$. Also, if $t \in R_2$ and $t < u$, then $u \in R_2$.

3-22 Prove that the subsets R_1 and R_2 of R define a cut (R_1,R_2) if and only if the following conditions are satisfied:

1′. $R_1 \cup R_2 = R$ and $R_1 \cap R_2 = \varnothing$.

2. Neither R_1 nor R_2 is empty.

3′. If $t \in R_2$ and $t < s$, then $s \in R_2$.

4. R_2 does not contain a least member.

3-23 Let Y_1 be the set of all rationals which are less than or equal to 0 together with all positive rationals r such that $r^2 < 3$. Let Y_2 be the set of all other rationals. (a) Prove that (Y_1, Y_2) is a cut. (b) Prove that (Y_1, Y_2) is an irrational cut.

3-24 Let ε be any positive rational number, and let (R_1,R_2) denote a cut. Prove that there exist numbers r_1 and r_2 in R_1 and R_2, respectively, such that $r_2 - r_1 < \varepsilon$.

3-15 Equality and Order for Cuts

Two cuts (A_1,A_2) and (B_1,B_2) are said to be *equal,* $(A_1,A_2) = (B_1,B_2)$, if they are identical, that is, if $A_1 = B_1$ and $A_2 = B_2$. Otherwise, they are *different* or *unequal.* To establish the equality of two cuts it is necessary only to show that their lower classes are identical or that their upper classes are the same since from the definition of cut $A_1 = B_1$ implies $A_2 = B_2$, and conversely.

If A_1 is a proper subset of B_1 (or, what is the same, if B_2 is a proper subset of A_2), then the cut (A_1,A_2) is said to be *less than* the cut (B_1,B_2), and we write $(A_1,A_2) < (B_1,B_2)$. (Also, we write $(B_1,B_2) > (A_1,A_2)$ and say that (B_1,B_2) is *greater than* (A_1,A_2).) Because of the cut definition it follows that one and only one of the following set relations is valid: A_1 is a proper subset of B_1, B_1 is a proper subset of A_1, and A_1 is equal to B_1. From this follows a trichotomy for the order relation for cuts: For the two cuts (A_1,A_2) and (B_1,B_2) exactly one of the relations $(A_1,A_2) < (B_1,B_2)$, $(B_1,B_2) < (A_1,A_2)$, and $(A_1,A_2) = (B_1,B_2)$ is true.

The order relation for cuts is transitive. From $(A_1,A_2) < (B_1,B_2)$ and $(B_1,B_2) < (C_1,C_2)$ follows that A_1 is a proper subset of B_1 and that B_1 is a proper subset of C_1. Therefore A_1 is a proper subset of C_1, whence $(A_1,A_2) < (C_1,C_2)$.

We define the *zero cut* (O_1,O_2) to be the rational cut based on the rational number 0. A cut (A_1,A_2) is then said to be *positive* if $(O_1,O_2) < (A_1,A_2)$ and *negative* if $(A_1,A_2) < (O_1,O_2)$.

3-16 Addition of Cuts

By the *sum* of two cuts (A_1,A_2) and (B_1,B_2) we mean the cut (C_1,C_2) such that C_2 contains all rational numbers which can be expressed in the form $a_2 + b_2$, where a_2 and b_2 are any members of A_2 and B_2, respectively, and such that C_1 contains all other rational numbers. We write $(C_1,C_2) = (A_1,A_2) + (B_1,B_2)$.

It must be verified, of course, that the ordered pair (C_1,C_2) as defined is, in fact, a cut. It is convenient in this instance to employ the alternate cut definition of Exercise 3-22. *1′.* That each rational number is in either C_1 or C_2 but not both is inherent in the definition. *2.* C_2 is certainly not empty since A_2 and B_2

are not. To see that C_1 is not empty, let a_1 and b_1 be members of A_1 and B_1, respectively, and consider the number $c = a_1 + b_1$. If c is in C_2, then rational numbers a_2 and b_2 exist in A_2 and B_2, respectively, such that $c = a_2 + b_2$. However, from $a_1 < a_2$ and $b_1 < b_2$ follows $a_1 + b_1 < a_2 + b_2$ or $c < c$, which is impossible. Consequently, C_1 is not empty. *3'*. Let c_2 be a member of C_2. Then a_2 and b_2 exist in A_2 and B_2, respectively, such that $c_2 = a_2 + b_2$. Let x be such that $c_2 < x$. Then we may write $x = a_2 + (x - c_2) + b_2 = a_2' + b_2$. Since $a_2' = a_2 + (x - c_2) > a_2$, it follows that a_2' is in A_2, and hence x is in C_2. *4*. Finally, we have to show that C_2 does not possess a least member. If a_2 and b_2 are any members of A_2 and B_2, respectively, then members a_2' and b_2' of A_2 and B_2, respectively, are known to exist such that $a_2' < a_2$ and $b_2' < b_2$. Hence $a_2' + b_2' < a_2 + b_2$. Thus, for any c_2 in C_2 another number c_2' also exists in C_2 such that $c_2' < c_2$.

In the verification of *2* above, the sum of a member of A_1 and a member of B_1 was shown to be a member of C_1. We must be careful not to conclude that the converse is also true; that is, that each member of C_1 is the sum of two such numbers. Although this would indeed simplify matters, Exercise 3-30 shows that it is an invalid inference.

Addition of cuts is a commutative operation. If (A_1,A_2) and (B_1,B_2) are any two cuts, let $(C_1,C_2) = (A_1,A_2) + (B_1,B_2)$ and $(D_1,D_2) = (B_1,B_2) + (A_1,A_2)$. The set C_2 consists of all rational numbers of the form $a_2 + b_2$ with a_2 in A_2 and b_2 in B_2, while D_2 is the set of all numbers of the form $b_2 + a_2$. Since $a_2 + b_2 = b_2 + a_2$, it follows that C_2 and D_2 are identical sets. From this follows the equality of (C_1,C_2) and (D_1,D_2); that is,

$$(A_1,A_2) + (B_1,B_2) = (B_1,B_2) + (A_1,A_2).$$

The associative law

$$[(A_1,A_2) + (B_1,B_2)] + (C_1,C_2) = (A_1,A_2) + [(B_1,B_2) + (C_1,C_2)]$$

is proved with like facility.

It is left as an exercise to prove that if (A_1,A_2), (B_1,B_2), and (C_1,C_2) are any cuts, then

$$(A_1,A_2) < (B_1,B_2) \quad \text{implies} \quad (A_1,A_2) + (C_1,C_2) < (B_1,B_2) + (C_1,C_2).$$

Next we show that the zero cut (O_1,O_2) has the property

$$(A_1,A_2) + (O_1,O_2) = (A_1,A_2) \tag{12}$$

for any cut (A_1,A_2). For, if c_2 is a member of the upper class of $(A_1,A_2) + (O_1,O_2)$, then numbers a_2 and o_2 of A_2 and O_2, respectively, exist such that $c_2 = a_2 + o_2$. However, since $0 < o_2$, it follows that $a_2 < c_2$, and consequently c_2 is a member of A_2. On the other hand, if c_2' is a member of A_2, then a_2' also exists in A_2 such that $a_2' < c_2'$. It follows from $c_2' = a_2' + (c_2' - a_2')$ that c_2' may be expressed as the sum of a member of A_2, i.e., a_2', and a member of O_2, i.e., $c_2' - a_2'$. Hence c_2' is in the upper class of $(A_1,A_2) + (O_1,O_2)$. Therefore A_2 is identical with this upper class and (12) follows.

Let (A_1,A_2) be any cut and define a new cut (B_1,B_2) such that the lower class B_1 consists of all rational numbers b_1 of the form $b_1 = -a_2$, where a_2 is some member of A_2, together with the negative of the greatest member of A_1 if such exists. Also, B_2 should contain all numbers b_2 of the form $b_2 = -a_1$, where a_1 is a member of A_1 except if a_1 is the greatest member of A_1. We call (B_1,B_2) the *negative* of (A_1,A_2) and denote it by the symbol $-(A_1,A_2)$. It is left to the reader to verify that the negative satisfies the four cut requirements and that

$$(A_1,A_2) + [-(A_1,A_2)] = (O_1,O_2)$$

for any cut (A_1,A_2).

Examination of the several properties of cuts in the above paragraphs reveals that all of the additive properties postulated in the axioms for the integers are also satisfied by the system of cuts. As a consequence, the system of cuts possesses every property that can be proved from these axioms. In particular, it possesses a cancellation law of addition:

$$(A_1,A_2) + (C_1,C_2) = (B_1,B_2) + (C_1,C_2) \quad \text{implies} \quad (A_1,A_2) = (B_1,B_2).$$

Also, every equation of the form

$$(A_1,A_2) + (X_1,X_2) = (B_1,B_2)$$

has a unique solution $(X_1,X_2) = (B_1,B_2) - (A_1,A_2)$. Further, $-[-(A_1,A_2)] = (A_1,A_2)$, and $(A_1,A_2) < (O_1,O_2)$ if and only if $(O_1,O_2) < -(A_1,A_2)$.

Exercises

3-25 Let (A_1,A_2) denote a cut and r a rational number. Assume that for any positive rational number ε there is a rational number a_2 in A_2 such that $a_2 < r + \varepsilon$. Prove that this implies (A_1,A_2) is less than or equal to the rational cut based on r.

3-26 Let (Z_1,Z_2) be the cut denoted by that symbol in Section 3-14, and let (II_1,II_2) be the rational cut based on 2. Prove that $(Z_1,Z_2) < (II_1,II_2)$.

3-27 Prove that $(A_1,A_2) < (B_1,B_2)$ implies $(A_1,A_2) + (C_1,C_2) < (B_1,B_2) + (C_1,C_2)$.

3-28 Verify that the negative of a cut satisfies the four cut requirements.

3-29 Prove that $(A_1,A_2) + [-(A_1,A_2)] = (O_1,O_2)$.

3-30 Let $(A_1,A_2) + (B_1,B_2) = (C_1,C_2)$. Prove that there may be a member c_1 of C_1 that cannot be expressed as a sum $a_1 + b_1$ with $a_1 \in A_1$ and $b_1 \in B_1$. Do this by considering the sum $(Z_1,Z_2) + [-(Z_1,Z_2)] = (O_1,O_2)$, where (Z_1,Z_2) is the cut defined in Section 3-14.

3-17 Multiplication of Cuts

Let (A_1,A_2) and (B_1,B_2) be any two non-negative cuts. Then by the *product* of (A_1,A_2) and (B_1,B_2) we mean the cut (C_1,C_2) such that C_2 contains all numbers c_2 that may be expressed as $c_2 = a_2 b_2$ where a_2 and b_2 are members of A_2 and

B_2, respectively, and C_1 contains all other rational numbers. We write $(C_1,C_2) = (A_1,A_2)(B_1,B_2)$.

To prove that (C_1,C_2) is really a cut we again use Exercise 3-22. *1'*. The statement of the contents of C_1 and C_2 is such that every rational number is employed once and only once. *2*. Since a_2 and b_2 are necessarily positive, C_1 contains all negative numbers as well as zero and as such is not empty. Also, since A_2 and B_2 are not empty, neither is C_2. *3'*. If c_2 is in C_2, then $c_2 = a_2 b_2$, where a_2 and b_2 are members of A_2 and B_2, respectively. Let x be a rational number such that $c_2 < x$. Clearly, y exists such that $x = b_2 y$, and from $a_2 b_2 < b_2 y$ follows $a_2 < y$. Hence y is in A_2, and it follows that x is in C_2. *4*. If a_2 and b_2 are any members of A_2 and B_2, respectively, then a_2' and b_2' also exist in A_2 and B_2 such that $a_2' < a_2$ and $b_2' < b_2$. Hence $a_2'b_2' < a_2 b_2$, and it follows that C_2 has no least member.

The extension of the definition of product to negative cuts is immediate. If (A_1,A_2) is negative and (B_1,B_2) non-negative, then their product is defined to be $-\{[-(A_1,A_2)](B_1,B_2)\}$; if both are negative, then $[-(A_1,A_2)][-(B_1,B_2)]$.

To demonstrate the commutativity of multiplication of cuts, let $(C_1,C_2) = (A_1,A_2)(B_1,B_2)$ and $(D_1,D_2) = (B_1,B_2)(A_1,A_2)$, where (A_1,A_2) and (B_1,B_2) are (without loss of generality) any two non-negative cuts. The set C_2 consists of all numbers of the form $a_2 b_2$, where a_2 is in A_2 and b_2 in B_2. Also, D_2 is the set of all numbers of the form $b_2 a_2$. Since $a_2 b_2 = b_2 a_2$, it follows that C_2 and D_2 are identical. Therefore

$$(A_1,A_2)(B_1,B_2) = (B_1,B_2)(A_1,A_2).$$

Proved in a similar manner are the associative law

$$[(A_1,A_2)(B_1,B_2)](C_1,C_2) = (A_1,A_2)[(B_1,B_2)(C_1,C_2)]$$

and the distributive law

$$(A_1,A_2)[(B_1,B_2) + (C_1,C_2)] = (A_1,A_2)(B_1,B_2) + (A_1,A_2)(C_1,C_2)$$

for any cuts (A_1,A_2), (B_1,B_2), and (C_1,C_2).

It is left as an exercise to prove that if (C_1,C_2) is positive, then

$$(A_1,A_2) < (B_1,B_2) \quad \text{if and only if} \quad (A_1,A_2)(C_1,C_2) < (B_1,B_2)(C_1,C_2).$$

By the *unit cut* (I_1,I_2) we mean the rational cut based on the rational number 1. It is easily seen that if (A_1,A_2) is any cut, then

$$(A_1,A_2)(I_1,I_2) = (A_1,A_2).$$

A simple consequence of the difference of the rational numbers 0 and 1 is $(O_1,O_2) \neq (I_1,I_2)$.

If (A_1,A_2) is a positive cut, then its *reciprocal* or *inverse with respect to multiplication* $(A_1,A_2)^{-1}$ is defined to be the cut (B_1,B_2) such that B_1 contains all negative numbers and zero as well as the reciprocals of all numbers in A_2 and the reciprocal of the greatest member of A_1, if such exists, while B_2 is the set of reciprocals of all positive numbers in A_1, except for that of the greatest member of

A_1, if there be one. It is not difficult to verify that the reciprocal is, indeed, a cut.

If (A_1,A_2) is a negative cut, we define its reciprocal to be $-[-(A_1,A_2)]^{-1}$. For (O_1,O_2) no reciprocal is defined. The reader may prove the basic property of reciprocals, namely, that if (A_1,A_2) is any nonzero cut, then

$$(A_1,A_2)(A_1,A_2)^{-1} = (I_1,I_2).$$

Having established that each nonzero member of our system possesses a reciprocal, it follows at once that a cancellation law applies; thus,

$$(A_1,A_2)(C_1,C_2) = (B_1,B_2)(C_1,C_2) \quad \text{and} \quad (C_1,C_2) \neq (O_1,O_2)$$

imply

$$(A_1,A_2) = (B_1,B_2).$$

Also, we are assured that the equation

$$(A_1,A_2)(X_1,X_2) = (B_1,B_2) \quad [(A_1,A_2) \neq (O_1,O_2)]$$

has a unique solution $(B_1,B_2)(A_1,A_2)^{-1}$. Verification of these facts follows the pattern employed for the rational system.

In addition to the above it can be verified that the principle of Archimedes holds in the system of cuts; namely, if (A_1,A_2) and (B_1,B_2) are positive cuts, then a cut (N_1,N_2), based on an integer n, exists such that $(B_1,B_2) \leq (N_1,N_2)(A_1,A_2)$.

We have now established for the system of cuts all the properties postulated for the integers in Axioms 1 through 14. Hence such consequences as those expressed in Theorems 1-1 through 1-13 are also valid for cuts.

Exercises

3-31 Let (C_1,C_2) be positive. Prove that $(A_1,A_2) < (B_1,B_2)$ if and only if

$$(A_1,A_2)(C_1,C_2) < (B_1,B_2)(C_1,C_2).$$

3-32 Prove that $(A_1,A_2)(I_1,I_2) = (A_1,A_2)$.

3-33 Verify that the reciprocal of a cut satisfies the four cut requirements.

3-34 Let (A_1,A_2) be nonzero. Prove that $(A_1,A_2)(A_1,A_2)^{-1} = (I_1,I_2)$.

3-35 Prove the principle of Archimedes: If (A_1,A_2) and (B_1,B_2) are positive cuts, then a rational cut (N_1,N_2) based on the integer n exists such that $(B_1,B_2) \leq (N_1,N_2)(A_1,A_2)$.

3-18 Rational and Irrational Cuts

Some simple facts concern the rational and irrational characters of cuts which are combinations of other cuts. For one thing, both the sum and the product of two rational cuts are rational. To see this, let (A_1,A_2) and (B_1,B_2) be rational cuts based on rational numbers a and b, respectively. It readily follows that

$a+b$ and ab are the greatest members of the respective lower classes of (A_1,A_2) $+(B_1,B_2)$ and $(A_1,A_2)(B_1,B_2)$. Hence these are rational cuts.

If (A_1,A_2) is rational, it follows from definitions that the cut $-(A_1,A_2)$ is rational and so is $(A_1,A_2)^{-1}$ if defined. If (A_1,A_2) is irrational, then $-(A_1,A_2)$ and $(A_1,A_2)^{-1}$ are irrational also.

The sum of a rational cut (A_1,A_2) and an irrational cut (B_1,B_2) is irrational. To see this, note that the identity

$$[(A_1,A_2)+(B_1,B_2)]+[-(A_1,A_2)]=(B_1,B_2)$$

implies that if $(A_1,A_2)+(B_1,B_2)$ is rational, then so is (B_1,B_2). From the contradiction follows the irrational character of $(A_1,A_2)+(B_1,B_2)$.

If (A_1,A_2) is the rational cut (O_1,O_2), then, of course, $(A_1,A_2)(B_1,B_2)$ is rational regardless of the nature of (B_1,B_2). However, if (A_1,A_2) is rational but nonzero and if (B_1,B_2) is irrational, then an argument such as that used for the sum implies that the product $(A_1,A_2)(B_1,B_2)$ is irrational.

The sum and product of irrational cuts may be rational or irrational. Some examples follow.

Example 3-3 If (A_1,A_2) is irrational, then the sum $(A_1,A_2)+[-(A_1,A_2)]$ is the rational cut (O_1,O_2). Also, $(A_1,A_2)(A_1,A_2)^{-1}=(I_1,I_2)$ is rational.

Example 3-4 Consider the product $(C_1,C_2)=(Z_1,Z_2)(Z_1,Z_2)$ or, briefly, $(C_1,C_2)=(Z_1,Z_2)^2$, where (Z_1,Z_2) is the irrational cut defined in Section 3-14. We prove that (C_1,C_2) is the rational cut based on 2.

The upper class C_2 consists of members $c_2=z_2z_2'$, where z_2 and z_2' are members of Z_2. Since $2<z_2^2$ and $2<z_2'^2$, it follows that $4<c_2^2$; or, since c_2 is non-negative, $2<c_2$. Thus all members of the upper class C_2 exceed 2, from which it follows that (C_1,C_2) is greater than or equal to the rational cut based on 2.

Corresponding to each positive integer n, there exist rational numbers z_1 and z_2 in Z_1 and Z_2, respectively, such that $z_2<z_1+1/n$ (see Exercise 3-24). From the definition of (Z_1,Z_2) it is also the case that $z_1^2<2<z_2^2$, and this, in particular, implies $z_1<2$. Making use of these relations we obtain

$$z_2^2<\left(z_1+\frac{1}{n}\right)^2=z_1^2+\frac{2z_1}{n}+\frac{1}{n^2}<2+\frac{5}{n}.$$

Since z_2^2 is in C_2 and since n may be any positive integer, it follows that C_2 contains rational numbers arbitrarily close to 2. We conclude that (C_1,C_2) is not greater than the rational cut based on 2. Therefore it is precisely that cut. (See Exercise 3-25.)

Example 3-5 The cut $(D_1,D_2)=(Z_1,Z_2)+(Z_1,Z_2)$ is irrational.

By the distributive law we have

$$(D_1,D_2)=(Z_1,Z_2)[(I_1,I_2)+(I_1,I_2)]=(Z_1,Z_2)(II_1,II_2),$$

where (II_1,II_2) denotes the rational cut based on 2. Thus (D_1,D_2) equals the product of a nonzero rational cut and an irrational cut and is, therefore, irrational.

Example 3-6 Let (Z_1,Z_2) again be as above, and define another cut (Y_1,Y_2) such that Y_1 consists of all negative rationals and 0 as well as all positive rationals whose squares do not exceed 3, while Y_2 consists of all positive rational numbers whose squares are greater than 3. Exercise 3-23 shows that (Y_1,Y_2) is an irrational cut. We now prove that the product of the irrational cuts (Z_1,Z_2) and (Y_1,Y_2) is irrational.

Writing $(Z_1,Z_2)(Y_1,Y_2)=(X_1,X_2)$ it follows that if x_2 is in X_2, then $x_2=y_2z_2$, where $3<y_2{}^2$ and $2<z_2{}^2$. Hence $6<x_2{}^2$, and each member of X_2 is seen to be such that its square exceeds 6.

Corresponding to each positive integer n, there exist z_1 in Z_1, z_2 in Z_2, y_1 in Y_1, and y_2 in Y_2, such that $z_2<z_1+1/n$ and $y_2<y_1+1/n$. We have seen $z_2{}^2<2+5/n$, and by employing similar inequalities we may show that $y_2{}^2<3+7/n$. Hence

$$(y_2z_2)^2<\left(3+\frac{7}{n}\right)\left(2+\frac{5}{n}\right)<6+\frac{64}{n}.$$

Since y_2z_2 is in X_2, it follows that X_2 has members with squares arbitrarily close to 6. Hence X_2 contains all positive rationals whose squares exceed 6, and X_1 all others. Since no rational number exists whose square is 6, we conclude that $(Z_1,Z_2)(Y_1,Y_2)$ is irrational.

Exercises

3-36 Prove that $(Y_1,Y_2)^2$ is rational, where (Y_1,Y_2) has the meaning of Example 3-6.

3-37 Prove that $(Z_1,Z_2)+(Y_1,Y_2)$ is irrational, where the cuts have the meaning of Example 3-6.

3-19 Density of Cuts

If $(A_1,A_2)<(B_1,B_2)$, then B_1 contains an infinite set of rational numbers not also present in A_1. This can be seen as follows. Since A_1 is a proper subset of B_1, some rational number b exists which is in B_1 but is not in A_1. Thus b is in A_2; and, since A_2 has no least member, there is some rational number a in A_2 such that $a<b$. The latter inequality implies that a is in B_1 and, since it is in A_2, it cannot be in A_1. Hence both a and b are in B_1 and not in A_1. Finally, it follows that each of the infinite set of rational numbers between these two also has this property.

A direct consequence of the above is the following.

Theorem 3-1 If (A_1,A_2) and (C_1,C_2) are any two cuts with the property $(A_1,A_2)<(C_1,C_2)$, then there exists a rational cut (B_1,B_2) such that $(A_1,A_2)<(B_1,B_2)<(C_1,C_2)$.

Proof From the fact that A_1 is a proper subset of C_1 follows the existence of two numbers b and c, $b<c$, that are in C_1 but not in A_1. Denote by (B_1,B_2) the rational cut based on b. Then, since b is in B_1 but not in A_1, A_1 is a proper subset of B_1, whence $(A_1,A_2)<(B_1,B_2)$. Further, the number c, which is in C_1, is not in B_1 since $b<c$. Hence $(B_1,B_2)<(C_1,C_2)$, and the rational cut (B_1,B_2) is seen to lie between (A_1,A_2) and (C_1,C_2)▲

Theorem 3-2 If (A_1,A_2) and (C_1,C_2) are any two cuts with the property $(A_1,A_2) < (C_1,C_2)$, then there exists an irrational cut (B_1,B_2) lying between them.

Proof On the strength of the preceding theorem, suppose (D_1,D_2) and (E_1,E_2) to be rational cuts such that $(A_1,A_2) < (D_1,D_2) < (E_1,E_2) < (C_1,C_2)$. Let (S_1,S_2) denote some positive irrational cut (for example, the cut (Z_1,Z_2) defined earlier). Then, by the principle of Archimedes, a positive rational cut (N_1,N_2) exists such that

$$(S_1,S_2) \leq [(E_1,E_2) - (D_1,D_2)](N_1,N_2).$$

Following rules already established, it is then easily verified that

$$(B_1,B_2) = (S_1,S_2)(N_1,N_2)^{-1} + (D_1,D_2)$$

is an irrational cut with the property $(D_1,D_2) < (B_1,B_2) \leq (E_1,E_2)$▲

As in the case of rational numbers we employ the term *dense* to describe the property of the system of cuts deduced above; that is, that between any two distinct cuts there lies another. It is clear, of course, that between any two cuts there lies an infinite set of rational cuts and an infinite set of irrational cuts.

3-20 The Infinite Character of the Set of All Cuts

Theorem 3-3 The set of all Dedekind cuts is noncountable.

Proof Denote by A the set of all sequences or progressions which can be formed of the integers 0 and 1; for example, 011010010 . . . . With a slightly different symbolism this set was introduced in Section 2-10. It was shown there that A is noncountable.

To each member a of A let correspond a set B_a of rational numbers defined by placing decimal points before partial sequences of the members of A. For example, with the member $a = 011010010\ldots$ of A we associate the set of rational numbers.

$$B_a = \{.0, .01, .011, .01101, .01101001, \ldots\}.$$

Consider now a particular set B_a, and let C_a denote the set of all rational numbers each of which is greater than all the rational numbers in B_a. Certainly

(13) $$r \in C_a \text{ and } r < s \quad \text{imply} \quad s \in C_a.$$

Suppose that a_1 and a_2 are different members of A. We wish to show that $C_{a_1} \neq C_{a_2}$. If a_1 and a_2 differ, then their difference is first apparent in the nth place for some n. Let r_0 be the rational number determined by the partial sum of n terms taken from a_1 or a_2, whichever has a 1 in the nth place. To be definite, suppose that r_0 is constructed from a_2. Then $r_0 \in B_{a_2}$ so that $r_0 \notin C_{a_2}$. On the other hand, each member of B_{a_1} is less than r_0; hence $r_0 \in C_{a_1}$. This proves that different members of A give rise to different sets C_a. It follows from this that the collection of all such sets C_a is noncountably infinite since A has this property.

It may happen that a set C_a constructed in this manner has a least member. For example, if $a = 11111\ldots$, then C_a has the rational number 1/9 as its least member. However, since the set of rational numbers is countable, at most a countable collection of sets C_a can have this property; these we discard. The remaining collection $\mathscr{F}$—that is, those sets C_a which do not have a least member—is still noncountably infinite.

Finally, we observe that for each $C_a \in \mathscr{F}$ the pair of sets $(\mathscr{C}C_a, C_a)$ is a Dedekind cut. This follows at once from the definition of C_a and from (13). Different members of $\mathscr{F}$ obviously give rise to different cuts. Hence the collection of all cuts $(\mathscr{C}C_a, C_a)$ is noncountable. Consequently the set of all Dedekind cuts has this property ▲

3-21 The Real Number System

The foregoing sections and exercises establish a basis for comparison of the systems of rational numbers and cuts. Each consists of a set of members for which operations termed addition and multiplication are defined, as well as an order relation and concepts of equality, zero, and unit. It is clearly the case that those properties of rational numbers which involve the operations of addition and multiplication and the order relation are properties of cuts as well. Moreover, with respect to set properties, it has been noted that in each system the zero and the unit are different members and, too, that both the set of rationals and the set of cuts possess subsets denying the least member principle.

Differences between the systems are apparent also. For example, if r is a positive rational number and n a positive integer, it is not always the case that a rational number x exists such that $x^n = r$; whereas for cuts, if a positive cut (R_1,R_2) is given, then a cut (X_1,X_2) exists such that $(X_1,X_2)^n = (R_1,R_2)$. (This is proved in Chapter 5.) The several exercises left to the reader point to yet other properties of cuts not possessed by rational numbers.

A comparison of the infinite characters of the sets of rational numbers and cuts also emphasizes a marked difference between the two. While the rational numbers are countably infinite, the set of cuts is noncountably infinite.

Let r denote any rational number and (R_1,R_2) the rational cut based on r. Consider, then, an equivalence between the set R of rational numbers, on the one hand, and the set of all rational cuts, on the other, in which r and (R_1,R_2) correspond. With reference to the systems of rational numbers and cuts an isomorphism is immediately established. For, if r and s are any two rational numbers and (R_1,R_2) and (S_1,S_2) the rational cuts associated with them, then

$$r + s \quad \text{corresponds to} \quad (R_1,R_2) + (S_1,S_2),$$

$$rs \quad \text{corresponds to} \quad (R_1,R_2)(S_1,S_2),$$

and

$$r < s \quad \text{if and only if} \quad (R_1,R_2) < (S_1,S_2).$$

Thus it follows that all properties of rational numbers are properties of rational cuts, and conversely.

Just as a certain subset of the set of rational numbers was identified with the set I of all integers, so, now, the set of rational cuts may be identified with the set R of rational numbers. In the system of Dedekind cuts we have thus achieved the extension of the system of rational numbers proposed earlier. The system of cuts, possessing properties not shared by the system R of rational numbers, is such that a proper subset is equivalent to and, indeed, isomorphic with the system R. It is entirely proper that cuts should be called *numbers.* Henceforth a Dedekind cut shall also be referred to as a *real number* and the system of cuts the *system of real numbers.* We denote it by the symbol $\mathscr{R}$.

3-22 Dedekind Sections

Before going on to a more detailed study of the system of real numbers, we explore the possibility of further extension by applying the device of a cut to the real numbers themselves.

By a *Dedekind section* or, simply, a *section* we mean an ordered pair $[\mathscr{R}_1,\mathscr{R}_2]$ of sets of real numbers with the following properties:

1. $\mathscr{R}_1 \cup \mathscr{R}_2 = \mathscr{R}$.
2. Neither $\mathscr{R}_1$ nor $\mathscr{R}_2$ is empty.
3. Every member of $\mathscr{R}_1$ is less than every member of $\mathscr{R}_2$.
4. $\mathscr{R}_2$ does not possess a least member.

We call $\mathscr{R}_1$ the *lower class* of the section and $\mathscr{R}_2$ the *upper class* and recognize two possible types of sections, namely, those for which

(i) $\mathscr{R}_1$ has a greatest member,
(ii) $\mathscr{R}_1$ has no greatest member.

As an example of a section of Type (i), let (R_1,R_2) denote some cut, and place in $\mathscr{R}_1$ all cuts less than (R_1,R_2) together with (R_1,R_2) itself; in $\mathscr{R}_2$ put all cuts greater than (R_1,R_2). It is easily verified that this is, indeed, a Dedekind section of Type (i).

Next we proceed to show that sections of Type (ii) do not, in fact, exist! That is, in any section the lower class necessarily has a greatest member. The implication is obvious. If only sections of Type (i) exist, then the set of all such may be placed in one-to-one correspondence with the set of all real numbers in such a way that each section corresponds to the real number which is greatest in its lower class. Accordingly, with appropriate definitions of sum, product, equality, less than, zero, and unit, the system of Dedekind sections is isomorphic with the system of real numbers. The dearth of sections of Type (ii), in effect, implies that by the device of a Dedekind section no new numbers come into being.

Theorem 3-4 (The Completeness Theorem) In any Dedekind section the lower class possesses a greatest member.

Proof Let $[\mathscr{R}_1,\mathscr{R}_2]$ denote a Dedekind section, and let R_1 be the union of the lower classes of all cuts in $\mathscr{R}_1$. That is, R_1 is the set of all rational numbers each of which belongs to the lower class of at least one cut in $\mathscr{R}_1$. Let R_2 be the set of all other rational numbers.

For purposes of this proof we shall let (S_1,S_2) and (T_1,T_2) be rational cuts based on rational numbers s and t, respectively.

From the definition of R_1 it is clear that $s \in R_1$ if $(S_1,S_2) \in \mathscr{R}_1$. The converse of this statement is also true. For suppose that $s \in R_1$. Then there is some cut (A_1,A_2) in $\mathscr{R}_1$ such that $s \in A_1$. Thus $(S_1,S_2) \leq (A_1,A_2)$, and from the properties of a Dedekind section it follows that (S_1,S_2) also belongs to $\mathscr{R}_1$.

It is now easy to verify that (R_1,R_2) satisfies the four cut requirements. Property *1* obviously holds and *2* is a consequence of the last paragraph and the fact that $\mathscr{R}_2$ obviously contains at least one rational cut. To prove *3*, let $s \in R_1$ and $t \in R_2$. Then, by the above property, $(S_1,S_2) \in \mathscr{R}_1$ and $(T_1,T_2) \in \mathscr{R}_2$. Hence $(S_1,S_2) < (T_1,T_2)$ from which $s < t$. Finally, to prove *4*, assume that $t \in R_2$ so that the cut (T_1,T_2) is in $\mathscr{R}_2$. Now $\mathscr{R}_2$ does not have a least member, and, in particular, there must be a rational cut (S_1,S_2) in $\mathscr{R}_2$ with $(S_1,S_2) < (T_1,T_2)$. From this $s \in R_2$ and $s < t$.

Since (R_1,R_2) is a cut, it must fall in one of the two classes of $[\mathscr{R}_1,\mathscr{R}_2]$. Consider first the possibility that (R_1,R_2) is a member of the upper class $\mathscr{R}_2$. Since $\mathscr{R}_2$ does not possess a least member, it follows that a rational cut (S_1,S_2) exists in $\mathscr{R}_2$ such that $(S_1,S_2) < (R_1,R_2)$. This inequality implies that $s \in R_1$, but this is impossible since $(S_1,S_2) \in \mathscr{R}_2$. We conclude that (R_1,R_2) is a member of $\mathscr{R}_1$.

Now let (A_1,A_2) be any cut in $\mathscr{R}_1$. Then, from the definition of R_1, it follows that $A_1 \subset R_1$. Hence $(A_1,A_2) \leq (R_1,R_2)$, so that (R_1,R_2) is necessarily the greatest member of $\mathscr{R}_1$, and $[\mathscr{R}_1,\mathscr{R}_2]$ is a section of Type (i), as was to be proved▲

3-23 Notation and Summary

At this point it becomes convenient to simplify notation by recalling symbolism familiar to the reader from earlier study of mathematics. We shall denote cuts (real numbers) by such symbols as a, b, x, y, and n. For cuts with special properties the appropriate notation is usually obvious. In particular, we shall write 0 for the zero cut (O_1,O_2), 1 for the unit cut (I_1,I_2), and, in general, denote by the symbol r the rational cut based on the rational number r. For irrational cuts we employ standard notation. For example, for the cut z, earlier denoted by (Z_1,Z_2) and having the property $z^2 = 2$, we employ the symbol $\sqrt{2}$. Symbols such as $1/x$, x^{-2}, and $\sqrt[n]{a}$ will have obvious significance.

The following is a summary of properties of the real number system $\mathscr{R}$ with its operations of addition and multiplication and relation of order. We employ the simplified notation where a, b, and c represent real numbers.

1. $a + b = b + a$.
2. $(a + b) + c = a + (b + c)$.

3. There is a real number 0 such that $a + 0 = a$.
4. Corresponding to each real number a there is a real number $-a$ such that $a + (-a) = 0$.
5. $ab = ba$.
6. $(ab)c = a(bc)$.
7. There is a real number 1 such that $a1 = a$.
8. If $a \neq 0$, there is a real number a^{-1} such that $aa^{-1} = 1$.
9. $a(b + c) = ab + ac$.
10. $0 \neq 1$.
11. If $a < b$ and $b < c$, then $a < c$.
12. If $a < b$, then $a + c < b + c$.
13. If $a < b$ and $0 < c$, then $ac < bc$.
14. a and b are related by exactly one of the following: $a < b$, $a = b$, $a > b$.
15. Let $\mathscr{R}_1$ and $\mathscr{R}_2$ be sets of real numbers such that
    1. $\mathscr{R}_1 \cup \mathscr{R}_2 = \mathscr{R}$.
    2. Neither $\mathscr{R}_1$ nor $\mathscr{R}_2$ is empty.
    3. Every member of $\mathscr{R}_1$ is less than every member of $\mathscr{R}_2$.
    4. $\mathscr{R}_2$ does not possess a least member.

    Then $\mathscr{R}_1$ has a greatest member.

3-24 Absolute Value

We denote the *absolute value* of a real number x by the symbol $|x|$ and define it by

$$|x| = \begin{cases} x & \text{if } x \geq 0 \\ -x & \text{if } x < 0. \end{cases}$$

It is clear that $0 \leq |x|$, $x \leq |x|$, and $-x \leq |x|$ for any value of x. Also, $|x|^2 = x^2$ from which it follows that $|x|$ and $-|x|$ are the positive and negative square roots, respectively, of the number x^2. The following theorems point to important properties of the absolute value of a real number. We shall make frequent use of these properties throughout the text.

Theorem 3-5 If x and y are any real numbers, then

(a) $|x + y| \leq |x| + |y|$, and

(b) $|xy| = |x|\,|y|$.

Proof Since, for any x and y, $x \leq |x|$ and $y \leq |y|$, it follows that $x + y \leq |x| + |y|$. Also, from $-x \leq |x|$ and $-y \leq |y|$ follows $-(x + y) \leq |x| + |y|$. Therefore, from the definition of absolute value $|x + y| \leq |x| + |y|$.

As for (b), we have $|xy|^2 = (xy)^2 = x^2y^2 = |x|^2\,|y|^2 = (|x|\,|y|)^2$. From the non-negative character of absolute value follows $|xy| = |x|\,|y|$ ▲

In particular, with $y = -1$ it follows from (b) of Theorem 3-5 that $|-x| = |x|$.

Theorem 3-6 Let x, y, z, a, and b be any real numbers. Then

(a) $|x| - |y| \leq \big| |x| - |y| \big| \leq |x - y|$,

(b) $|x - y| \leq |x - z| + |y - z|$,

(c) $|x - b| < a$ if and only if $b - a < x < b + a$.

Proof Writing $x = (x - y) + y$ we have $|x| = |(x - y) + y| \leq |x - y| + |y|$. Hence $|x| - |y| \leq |x - y|$. Also, from $y = x - (x-y)$ follows $|y| = |x - (x-y)| \leq |x| + |-(x - y)| = |x| + |x - y|$ or $-(|x| - |y|) \leq |x - y|$. The two results then imply $\big| |x| - |y| \big| \leq |x - y|$. That $|x| - |y| \leq \big| |x| - |y| \big|$ is immediate.

Writing $x - y = (x - z) - (y - z)$ it follows that $|x - y| = |(x - z) - (y - z)| \leq |x - z| + |-(y - z)| = |x - z| + |y - z|$, or $|x - y| \leq |x - z| + |y - z|$.

From $x - b \leq |x - b| < a$ follows $x < b + a$, whereas $-(x - b) \leq |x - b| < a$ leads to $x > b - a$. Hence $|x - b| < a$ implies $b - a < x < b + a$. Conversely, $x < b + a$ implies $x - b < a$, while from $x > b - a$ follows $-(x - b) < a$. Hence $|x - b| < a$ is a consequence of $b - a < x < b + a$▲

The non-negative number $|a - b|$ is sometimes referred to as the *distance* between the points a and b or as the *length* of the *line segment* joining a and b.

Exercises

3-38 Assume that $y \neq 0$ and prove that (a) $\left|\dfrac{1}{y}\right| = \dfrac{1}{|y|}$, and (b) $\left|\dfrac{x}{y}\right| = \dfrac{|x|}{|y|}$.

3-39 Describe in simpler terms the sets of numbers x for which (a) $|x|\,|x + 2| < 4$, and (b) $|x - 1| + |x + 1| < 4$.

3-40 Prove that $|\sum_{k=1}^{n} a_k| \leq \sum_{k=1}^{n} |a_k|$.

3-41 Prove that $|\prod_{k=1}^{n} a_k| = \prod_{k=1}^{n} |a_k|$.

3-25 The Extended Real Number System

We have seen how the rational number system is an extension of the system of integers and has the integers imbedded in it; we have seen also that the rational number system is imbedded in the more extensive system of real numbers. On the other hand, an attempt at further extension by the device of a cut has been seen to fail. The question arises: Is it possible to achieve an extension by some other means? That is, does a system of numbers exist which has all the properties of the real numbers and which contains a proper subset isomorphic to the latter? The answer is no. Rather, it may be shown that if $\mathscr{R}'$ denotes a system in which, as in $\mathscr{R}$, addition, multiplication, and order are defined and are governed by properties 1 through 15 of Section 3-23, then the systems $\mathscr{R}$ and $\mathscr{R}'$ are isomorphic. We outline a proof of this fact and further demonstrate that the only isomorphism between $\mathscr{R}$ and a subset of $\mathscr{R}'$ necessarily involves all members of $\mathscr{R}'$. The reader should supply missing details.

First, in any isomorphism the zeros 0 and 0′ and the units 1 and 1′ of $\mathscr{R}$ and $\mathscr{R}'$, respectively, must correspond. Further, in order that sums correspond properly $1 + 1 = 2, 2 + 1 = 3, \ldots$ must correspond with $1' + 1' = 2', 2' + 1' = 3', \ldots$. In general, the positive integer n of $\mathscr{R}$ must correspond with the positive integer n' of $\mathscr{R}'$. As for the correspondence between negative integers, this follows from the definitions of $-n$ and $-n'$. So far, then, the integers I in $\mathscr{R}$ and the integers I' in $\mathscr{R}'$ constitute isomorphic systems.

Turning to noninteger rational numbers, it is clear from the relations $(n)(1/n) = 1$ and $(n')(1'/n') = 1'$ (where $n \neq 0$ and $n' \neq 0'$) that the numbers $1/n$ and $1'/n'$ must correspond. Because a rational number p/q may be written as a product $(p)(1/q)$, it then follows that p/q and p'/q' are associated in the correspondence. Thus is established an isomorphism between the rational system R in $\mathscr{R}$ and what may be styled the system R' of rational numbers in $\mathscr{R}'$.

Next we extend the domain of the one-to-one mapping to the set of all real numbers $\mathscr{R}$. Suppose a to be an irrational number in $\mathscr{R}$. Corresponding to it is a certain subset $\mathscr{R}_a'$ of $\mathscr{R}'$ consisting of those rational numbers in $\mathscr{R}'$ which are images of rationals in $\mathscr{R}$ which exceed a. In symbols, with r and r' indicating corresponding rationals, $\mathscr{R}_a' = \{r'; \quad r > a\}$. Now denote by $\mathscr{R}_2'$ the set of all members of $\mathscr{R}'$ each of which exceeds at least one member of $\mathscr{R}_a'$. Then, if $\mathscr{R}_1'$ represents the set of all members of $\mathscr{R}'$ which are not in $\mathscr{R}_2'$, it is not difficult to verify that the ordered pair $[\mathscr{R}_1',\mathscr{R}_2']$ is a Dedekind section; that is, $\mathscr{R}_1'$ and $\mathscr{R}_2'$ satisfy the conditions of property 15 in Section 3-23. As a consequence, $\mathscr{R}_1'$ possesses a greatest member. Denoting the latter by a', the demands of the isomorphism are such that a and a' must correspond.

It is thus to be concluded that each member of $\mathscr{R}$ has an image in $\mathscr{R}'$ and that as a consequence of the mapping a unique isomorphism is established between $\mathscr{R}$ and a subset of $\mathscr{R}'$. It remains to show that each member of $\mathscr{R}'$ has been employed in the mapping. Actually, this follows from an argument based on the completeness theorem, as in the preceding paragraph.

Suppose a' to be any member of $\mathscr{R}'$. Then the set $\{r; \quad r' > a'\}$ defines the upper class $\mathscr{R}_2$ of a Dedekind section $[\mathscr{R}_1,\mathscr{R}_2]$. Let a be the greatest member of $\mathscr{R}_1$ and assume that a' is not the image of a. Without prejudice to the argument, suppose the image b' of a to exceed a'. Then properties 1 through 15 of Section 3-23 provide that a rational number r' of $\mathscr{R}'$ exists such that $a' < r' < b'$. It follows that the rational number r in $\mathscr{R}$ of which r' is the image necessarily falls in $\mathscr{R}_2$. Hence $a < r$. However, since $r' < b'$ and since b' is the image of a we conclude, too, that $r < a$. The obvious contradiction implies that a' is the image of a. Thus each member of $\mathscr{R}'$ is the image of a member of $\mathscr{R}$, and it is established that the two systems are isomorphic.

The reader will recall that the number axis of Section 3-12 was described as a continuum of points. Essentially we now define that continuum to be a line whose points are in one-to-one correspondence with the real numbers.

If anything in the nature of a more extensive system is to be achieved, it will have to come about through relaxation of some of the properties of reals. This, of course, is exactly what is done in defining the *complex number system*. The latter, although it contains the real numbers as a proper subset and shares many properties with that system, fails to possess the order properties of the reals.

It is possible in still another way to obtain a system which is more extensive than the reals. Let $\mathscr{R}^*$ consist of the set $\mathscr{R}$ together with two additional members denoted by ∞ and $-\infty$ (read *plus infinity* and *minus infinity*, respectively). The definitions listed below serve to extend the operations of addition and multiplication and the order relation of $\mathscr{R}$ to the set $\mathscr{R}^*$. The symbol a denotes any real number, that is, any member of $\mathscr{R}$.

1. $-\infty < a < \infty$.
2. $a + \infty = \infty + a = a - (-\infty) = \infty$.
3. $a + (-\infty) = -\infty + a = a - \infty = -\infty$.
4. $a \cdot \infty = \infty \cdot a = \begin{cases} \infty & \text{if } a > 0 \\ -\infty & \text{if } a < 0. \end{cases}$
5. $a \cdot (-\infty) = -\infty \cdot a = \begin{cases} -\infty & \text{if } a > 0 \\ \infty & \text{if } a < 0. \end{cases}$
6. $\infty + \infty = \infty - (-\infty) = \infty$.
7. $-\infty + (-\infty) = -\infty - \infty = -\infty$.
8. $\infty \cdot \infty = -\infty \cdot (-\infty) = \infty$.
9. $\infty \cdot (-\infty) = -\infty \cdot \infty = -\infty$.

Observe that although the order relation has been extended to all of $\mathscr{R}^*$, the same is not true of multiplication and addition. In particular, we do not define such symbols as $0 \cdot \infty$ and $\infty - \infty$. Therefore, while we shall refer to $\mathscr{R}^*$ as the *extended real number system*, it is not an extension of $\mathscr{R}$ in the sense discussed earlier in this section.

The notion of absolute value can be extended to $\mathscr{R}^*$ by defining $|\infty| = |-\infty| = \infty$. By using this convention, Theorems 3-5 and 3-6 are valid in $\mathscr{R}^*$ with the understanding, of course, that all sums and products employed must have meaning.

We can look upon the adjoining of ∞ and $-\infty$ to the system $\mathscr{R}$ as a relaxation of the second cut requirement. For, if it be allowed that either class of a Dedekind cut be empty, then cuts $(R,\varnothing)$ and $(\varnothing,R)$ may be identified with the appended symbols. The reader should investigate the effect this alteration has on the definitions of addition, multiplication, and order.

We caution the reader that throughout this text the term *real number* will signify a member of $\mathscr{R}$ and not the symbol ∞ or $-\infty$. These last symbols will be referred to as *infinite numbers* while, for emphasis, the members of $\mathscr{R}$ may be called *finite numbers*.

Exercises

3-42 Prove that if a and b are positive real numbers, then $a < b$ if and only if $a^2 < b^2$.

3-43 Prove that if a and b are positive real numbers and n any positive integer, then $a < b$ if and only if $a^n < b^n$.

3-44 Prove that if a and b are positive real numbers such that $a^n = b^n$ for some positive integer n, then $a = b$.

3-45 Prove that if x_1 denotes a solution of $x^2 = a$ ($a \geq 0$), then no solutions exist other than x_1 and $-x_1$.

3-46 Prove that $\sqrt{3}\sqrt{12} = 6$.

3-47 Prove that the equation $x^2 = d$ has a real solution if and only if $d \geq 0$.

3-48 Determine a necessary and sufficient condition for $ax^2 + bx + c = 0$, where a, b, and c are real, to have a real solution.

3-49 Determine a necessary and sufficient condition for $ax^2 + bx + c = 0$, where a, b, and c are real, to have two distinct solutions.

3-50 (a) Prove that if a and b are any real numbers, then $a^2 + b^2 = 0$ if and only if $a = b = 0$. (b) Generalize this to any finite sum $\sum_{k=1}^{n} a_k{}^2$.

3-51 Prove that, corresponding to any real number a, there exists a unique integer m such that $m \leq a < m + 1$.

3-52 (a) Prove that if a and b are real numbers such that $a \neq 0$ and $ab = 1$, then $b = a^{-1}$. (b) Prove that $1^{-1} = 1$.

3-53 Prove that if a is any nonzero real number, then (a) $a^{-1} \neq 0$, and (b) $(a^{-1})^{-1} = a$.

3-54 (a) Prove that if a and b are real numbers such that $ab \neq 0$, then $(ab)^{-1} = a^{-1}b^{-1}$. (b) Generalize this to any finite product $\prod_{k=1}^{n} a_k$.

3-55 Prove that if a is a real number, then $a > 0$ if and only if $a^{-1} > 0$.

3-56 Prove that if a and b are real numbers such that $0 < a < b$, then $b^{-1} < a^{-1}$.

3-57 For any nonzero real number a we define rules of exponents by $a^0 = 1$, $a^n = a^{n-1}a$ ($n = 1, 2, \ldots$), and $a^n = a^{n+1}/a$ ($n = -1, -2, \ldots$). Prove that if a and b are real and nonzero and if m and n are any integers, then (a) $a^m a^n = a^{m+n}$, (b) $(ab)^n = a^n b^n$, and (c) $(a^m)^n = a^{mn}$. (Also we define $0^n = 0$ for $n = 1, 2, \ldots$ and observe that (a), (b), and (c) hold trivially if $m > 0$ and $n > 0$.)

3-58 Prove that if a and b are nonzero real numbers, then $(ab^{-1})^n = a^n b^{-n}$ for any integer n.

3-59 Assume that $a > 0$ and that m and n are integers such that $m < n$. Prove that (a) $a^m < a^n$ if $a > 1$, (b) $a^m = a^n$ if $a = 1$, and (c) $a^m > a^n$ if $a < 1$.

3-60 Let ε denote any positive real number. Prove that an integer n exists such that $0 < 1/n < \varepsilon$.

3-61 Let ε denote any positive real number. Prove that an integer n exists such that $1/3^n < \varepsilon$.

3-62 Prove that $\sqrt{2} + \sqrt{3}$ is irrational.

3-63 Determine for which positive integers n the number $\sqrt{n-1} + \sqrt{n+1}$ is rational and for which irrational.

3-64 Prove that $\sqrt{n} + \sqrt{n+1}$ is irrational for each positive integer n.

3-65 Is $\sqrt[3]{\sqrt{5}+2} - \sqrt[3]{\sqrt{5}-2}$ rational or irrational?

3-66 Is $\sqrt{2} + \sqrt{3} + \sqrt{5}$ rational or irrational?

3-67 Is $\dfrac{\sqrt{3-\sqrt{5}}}{\sqrt{2}+\sqrt{7-3\sqrt{5}}}$ rational or irrational?

3-68 By induction prove that every nonempty finite set of real numbers has a maximum and a minimum member.

3-69 By induction prove that if a and b are any real numbers and n any positive integer, then $(a+b)^n = \sum_{r=0}^{n} \dfrac{n!}{r!(n-r)!} a^{n-r} b^r$.

3-70 Prove by induction that if n is any positive integer and x any real number such that $x > -1$, then $(1 + x)^n \geq 1 + nx$.

3-71 Prove by induction that if n is any positive integer and x any real number, then $(1 + x)^{2n} \geq 1 + 2nx$.

3-72 Prove by induction that if n is any positive integer and x any real number such that $x < 1$, then $(1 - x)^n \geq 1 - nx$.

3-73 By induction prove that $\sum_{k=1}^{n} (a_k - a_{k-1}) = a_n - a_o$.

3-74 Prove that $[(a + b)/2]^2 \geq ab$.

3-75 Prove the *Cauchy-Schwartz inequality* $(\sum_{k=1}^{n} a_k b_k)^2 \leq (\sum_{k=1}^{n} a_k^2)(\sum_{k=1}^{n} b_k^2)$, where a_k and b_k $(k = 1,2,\ldots,n)$ are any real numbers. Show also that the equality sign holds if and only if a real number λ exists such that $\lambda a_k + b_k = 0$ for $k = 1,2,\ldots,n$. (*Hint:* Show that the non-negative sum $\sum_{k=1}^{n} (a_k x + b_k)^2$ has the value 0 for at most one real value of x.)

3-76 Show that the *Minkowski inequality* $[\sum_{k=1}^{n} (a_k \pm b_k)^2]^{1/2} \leq [\sum_{k=1}^{n} a_k^2]^{1/2} + [\sum_{k=1}^{n} b_k^2]^{1/2}$, where a_k and b_k $(k = 1,2,\ldots,n)$ are real, is a consequence of the Schwartz inequality.

3-77 Prove that the sets $\{x;\ 0 < x < 1\}$ and $\{y;\ 0 < y < 2\}$ are equivalent.

3-78 Prove that the set $\{x;\ 0 < x < 1\}$ is equivalent to the set $\{y;\ 0 < y\}$.

3-79 Prove that the set $\{x;\ 0 < x < 1\}$ is equivalent to the set $\mathscr{R}$ of all real numbers.

3-80 Let C denote the set of all functions with domain and range in $\mathscr{R}$. Prove that C and $\mathscr{R}$ are not equivalent. (*Hint:* Assume an equivalence to exist, with the member a of $\mathscr{R}$ and a function f_a of C corresponding. Then define a function h by $h(x) = 1 + f_x(x)$, and show that a contradiction is inevitable.)

3-81 A real number is said to be an *algebraic number* if it is a zero of a polynomial with integer coefficients. (For example, $\sqrt{2}$, 3, and $\sqrt[3]{7}$ are algebraic numbers.) Prove that the set of all algebraic numbers is countably infinite. (Assume it known that a polynomial of degree n has at most n zeros.) If a real number is not algebraic, we call it a *transcendental number*.

3-82 Supply the missing details of the proof in Section 3-25 that the system of real numbers is unique.

4
Topological Concepts

... some gathered at six, some at seven, some at eight, ...

JOHN DONNE (1571–1631), *Eternity*

4-1 Intervals

To this point we have been concerned with those properties of real numbers which are primarily *algebraic* in character. In the present chapter we shall fix attention on *geometrical* or *topological* properties of sets of real numbers. It is these topological properties that are basic to the definitions of such important notions as limit and continuity of a function.

By the term *bounded interval*[1] is meant the set of all real numbers x lying between and possibly including one or both of two different real numbers a and b. If a and b are excluded, that is, if $a < x < b$, the bounded interval is said to be *open*, and we denote it also by the symbol (a,b). If a and b are included, we say the bounded interval is *closed* and employ the notation $[a,b]$.

It is possible for a bounded interval to be neither open nor closed. This is the case if one of the numbers a and b is included in the set and one excluded. If $a \leq x < b$, we denote the interval by $[a,b)$; if $a < x \leq b$, we write $(a,b]$. For all bounded intervals, open, closed, or neither, the numbers a and b will be referred to as *endpoints*. Note that from the stipulation $a \neq b$ it follows that a bounded interval cannot consist of one point only, nor can it be empty.

[1] The term *finite interval* is sometimes employed in this connection. It should not be confused, however, with *finite set*.

For any real number a, sets such as $\{x;\ \ x < a\}$, $\{x;\ \ x \leq a\}$, $\{x;\ \ x > a\}$, and $\{x;\ \ x \geq a\}$ are called *unbounded intervals*[2]. Those for which the single endpoint a is excluded are called *open*, and the two which include a are *closed*. It is convenient, also, to look on the set of all real numbers as an *unbounded interval.*[2]

In all that follows the term *interval* will be used to signify either a bounded or an unbounded interval. If reference to one is intended to the exclusion of the other, this will be clearly stated.

4-2 Neighborhoods

An open bounded interval $(c - \varepsilon, c + \varepsilon)$, where ε is a positive real number, is called a *neighborhood* or, in particular, an *ε-neighborhood* of the real point c. We employ the symbol $\mathcal{N}(c,\varepsilon)$ to denote such a neighborhood and refer to ε as the *radius* of $\mathcal{N}(c,\varepsilon)$. The point c itself is the *midpoint* or *center* of $\mathcal{N}(c,\varepsilon)$. Clearly, $x \in \mathcal{N}(c,\varepsilon)$ iff $|x - c| < \varepsilon$. We note the following elementary, yet important, properties of neighborhoods.

Theorem 4-1 If b is any point of the neighborhood $\mathcal{N}(c,\varepsilon)$, then a neighborhood of b exists which is entirely contained in $\mathcal{N}(c,\varepsilon)$.

Proof Let $\varepsilon_1 = \varepsilon - |c - b|$. Then $\varepsilon_1 > 0$ since $|c - b| < \varepsilon$. Let x denote any point of $\mathcal{N}(b,\varepsilon_1)$. Then $|x - b| < \varepsilon_1$ or $|x - b| + |c - b| < \varepsilon$ so that $|x - c| < \varepsilon$. This implies that x is a point of the neighborhood $\mathcal{N}(c,\varepsilon)$. Thus $\mathcal{N}(b,\varepsilon_1) \subset \mathcal{N}(c,\varepsilon)$▲

Theorem 4-2 If c_1 and c_2 are any two distinct points, then neighborhoods of c_1 and c_2 exist which are disjoint.

Proof We have only to observe that if $\varepsilon = \frac{1}{3}|c_1 - c_2|$, then the neighborhoods $\mathcal{N}(c_1,\varepsilon)$ and $\mathcal{N}(c_2,\varepsilon)$ are such that no real number exists which is a member of both▲

Theorem 4-3 If a denotes any point of the intersection A of neighborhoods $\mathcal{N}(c_1,\varepsilon_1)$ and $\mathcal{N}(c_2,\varepsilon_2)$, then a neighborhood of a exists which is entirely contained in A.

Proof Let ε denote the lesser of the two positive numbers $\varepsilon_1 - |c_1 - a|$ and $\varepsilon_2 - |c_2 - a|$, and let x denote any point of $\mathcal{N}(a,\varepsilon)$. Then $|x - a| < \varepsilon \leq \varepsilon_1 - |c_1 - a|$, from which $|x - c_1| < \varepsilon_1$; therefore $x \in \mathcal{N}(c_1,\varepsilon_1)$. In a similar fashion, $x \in \mathcal{N}(c_2,\varepsilon_2)$. Hence $x \in A$, and it follows that $\mathcal{N}(a,\varepsilon) \subset A$▲

The set of all real numbers x such that $x \in \mathcal{N}(c,\varepsilon)$ but $x \neq c$ is called a *deleted neighborhood* of c. Precisely, it is the union of the two open bounded intervals

[2] The term *infinite interval* is also employed.

$(c - \varepsilon, c)$ and $(c, c + \varepsilon)$. We denote such by $\mathcal{N}^*(c,\varepsilon)$. It is left to the reader to verify that each of the theorems above concerning neighborhoods is true for deleted neighborhoods as well.

The concept of neighborhood is basic to the material that follows. The reader should observe that the definitions in the next few sections depend on the notion of neighborhood.

4-3 Closed Sets of Points

A real number c is said to be a *cluster point*[3] of a set A if each neighborhood of c contains an infinite subset of A. Observe that a cluster point of a set A may, but need not, be a member of A.

Example 4-1

(a) Let $A = \{x;\ 0 < x \leq 1\}$. The cluster points of A constitute the set $\{x;\ 0 \leq x \leq 1\}$.

(b) Let B be the set of all rational numbers. Then every real number is a cluster point of B.

(c) Let C be any finite set of points. Then no real number is a cluster point of C.

(d) Let D be the set of all integers. Then no real number is a cluster point of D.

(e) Let E be the set of rational numbers of the form $1/n$, where n is a positive integer. Then the number 0 is the only cluster point of E.

The following theorem suggests a possible alternate definition of cluster point.

Theorem 4-4 A point c is a cluster point of a set A if and only if each deleted neighborhood of c contains at least one member of A.

Proof Suppose that c is not a cluster point of A. Then a deleted neighborhood $\mathcal{N}^*(c,\varepsilon_1)$ of c exists which contains at most a finite subset B of A. If B is empty, then $\mathcal{N}^*(c,\varepsilon_1)$ is devoid of members of A. If B is nonempty, denote its members by $b_1, b_2, \ldots, b_n$, and consider the finite set of positive numbers $|c - b_1|, |c - b_2|, \ldots, |c - b_n|$. Representing the least of these by ε_2, it follows that the deleted neighborhood $\mathcal{N}^*(c,\varepsilon_2)$ contains no members of A. Consequently, c is a cluster point of A if each deleted neighborhood of c contains at least one member of A.

For the converse, if c is a cluster point of A, then each deleted neighborhood of c contains an infinite subset of A and hence, trivially, at least one member ▲

If for a point c and each positive value of ε the interval $(c - \varepsilon, c)$ contains an infinite set of points of a set A, then the point c is called a *cluster point of A from the left*. Similarly, if for each positive ε the interval $(c, c + \varepsilon)$ contains an

[3] Also *limit point* and *point of accumulation*.

infinite set of points of A, then c is a *cluster point of A from the right*. It follows immediately from these definitions and from that of cluster point that if c is a cluster point of A, then it is either a cluster point from the left or a cluster point from the right or both. Conversely, if it is a cluster point from the left or from the right, then it is also a cluster point.

Example 4-2

(a) The point b is a cluster point of the interval (a,b) from the left but not from the right, and a is a cluster point from the right but not from the left. Every point of the interval itself is a cluster point from both left and right.

(b) The point 0 is a cluster point from the right of the set of rational numbers of the form $1/n$, where n is any positive integer.

The set comprised of all the cluster points of a set A is called the *derived set* of A and is denoted by A'. The union of A and its derived set is termed the *closure* of A and is represented by $\bar{A}$. Thus, $\bar{A} = A \cup A'$. It is immediate that $A \subset \bar{A}$. The next two theorems state additional simple facts concerning the concept of closure.

Theorem 4-5 If $A \subset B$, then $\bar{A} \subset \bar{B}$.

Proof Let x denote a point of $\bar{A}$. If x belongs to A, then it is a point of B and hence of $\bar{B}$. If $x \notin A$, then it is a cluster point of A. Since every point of A is also a member of B, it then follows that x is a cluster point of B. Hence, again, $x \in \bar{B}$. Therefore $\bar{A} \subset \bar{B}$ ▲

Theorem 4-6 If A is any set, then $(\bar{A})' \subset \bar{A}$.

Proof Assume that x does not belong to $\bar{A}$. Then some neighborhood $\mathcal{N}(x,\varepsilon)$ exists which contains no points of A. Further, this neighborhood contains no cluster points of A because if $y \in \mathcal{N}(x,\varepsilon)$, then some neighborhood $\mathcal{N}(y,\varepsilon_1)$ is contained in $\mathcal{N}(x,\varepsilon)$ and hence possesses no points of A. This implies that y is not a cluster point of A. Since $\mathcal{N}(x,\varepsilon)$ contains no points of $\bar{A}$, x does not belong to $(\bar{A})'$ ▲

Corollary If A is any set, then $\bar{\bar{A}} = \bar{A}$.

Proof From Theorem 4-5 it is immediate that $\bar{A} \subset \bar{\bar{A}}$, and from Theorem 4-6, $\bar{\bar{A}} = \bar{A} \cup (\bar{A})' \subset \bar{A}$. Hence $\bar{\bar{A}} = \bar{A}$ ▲

A set is said to be *closed* if it contains all of its cluster points. We have just proved that the closure of a set A is a closed set. It is quite clear, too, that a set is closed if and only if it is identical with its closure since $A \cup A' = A$ if and only if $A' \subset A$.

Example 4-3

(a) The closed-bounded interval $[a,b]$ is a closed set.

(b) The closed unbounded intervals $\{x;\quad x \geq a\}$ and $\{x;\quad x \leq a\}$ are closed sets.

(c) The set of all real numbers is closed.

(d) Any finite set (including the empty set) is devoid of cluster points. Hence such a set is closed since, vacuously, all of its cluster points belong to it.

A member of a set A which is not a cluster point of A is called an *isolated* point of A. A set is termed a *discrete* set if all of its points are isolated points.

Example 4-4

(a) Any finite set is discrete.

(b) The set of all integers is a discrete set.

(c) The set of rational numbers of the form $1/n$, where n is a positive integer, is discrete.

It is clear from the definitions that each point of a set A is either an isolated point of A or a cluster point of A. If a set A possesses no isolated points or, what is the same, if $A \subset A'$, then it is said to be *dense in itself*. Further, a set is styled *perfect* if it is both dense in itself and closed. Since A is closed if $A' \subset A$, it follows that A is perfect if and only if $A = A'$.

Example 4-5

(a) The empty set is perfect.

(b) The closed bounded interval $[a,b]$ is a perfect set.

(c) The closed unbounded intervals $\{x; \quad x \geq a\}$ and $\{x; \quad x \leq a\}$ are perfect sets.

(d) The set of all real numbers is perfect.

A less trivial example of a perfect set will be considered later.

4-4 Open Sets of Points

A set of points G is said to be *open* if its complement $\mathscr{C}G$ (with respect to the set of all real numbers) is a closed set. The concepts of open and closed sets play a most important role in topology.

Example 4-6

(a) The open bounded interval (a,b) is an open set; any neighborhood or deleted neighborhood is open.

(b) The open unbounded intervals $\{x; \quad x > a\}$ and $\{x; \quad x < a\}$ are open sets.

(c) The set of all real numbers is open.

(d) The empty set is open.

Observe that the empty set and the set of all real numbers are both closed and open. (Cf. Exercise 4-33.) It is also the case that sets exist which are neither closed nor open. The interval $(a,b]$ is an example of such a set.

The following theorem concerns an important and very useful characterization of open sets.

Theorem 4-7 A necessary and sufficient condition that a set G be open is that each member of G have a neighborhood which is contained in G.

Proof Assume G to be an open set, and let c denote any member of G. It follows that $\mathscr{C}G$ is closed and that $c \notin \mathscr{C}G$. As a consequence, c is not a cluster point of $\mathscr{C}G$, and this in turn implies the existence of a neighborhood $\mathscr{N}(c,\varepsilon)$ which contains no points of $\mathscr{C}G$. Hence $\mathscr{N}(c,\varepsilon) \subset G$.

To prove the sufficiency of the condition, assume that each point of a set G possesses a neighborhood which is entirely contained in G. It follows that no point of G is a cluster point of $\mathscr{C}G$. Hence every cluster point of $\mathscr{C}G$ is a member of $\mathscr{C}G$. Therefore $\mathscr{C}G$ is closed, and as a consequence the set G is open▲

Exercises

4-1 Prove that the set of all bounded intervals is not countable.

4-2 Prove that the set of all intervals with rational endpoints is countable.

4-3 Prove that Theorems 4-1, 4-2, and 4-3 are true for deleted neighborhoods as well as for neighborhoods.

4-4 Cite an example to show that A' need not be a subset of $(A')'$.

4-5 Prove that if $A \subset B$, then $A' \subset B'$.

4-6 Prove that $(A \cup B)' = A' \cup B'$.

4-7 Prove that $(\bar{A})' = A'$.

4-8 Construct a set A with the property that A' consists of the number 0 together with the set of all rational numbers of the form $1/n$, where n is an integer such that $n \geq 1$.

4-9 Prove that no set A exists for which A' is the set R of all rational numbers.

4-10 Show by an example that a discrete set may possess cluster points.

4-11 Let F be a closed set and assume $A \subset F$. Prove that $\bar{A} \subset F$. This proves that F is the least (relative to set inclusion) closed set containing A.

4-12 Prove that $\overline{A \cup B} = \bar{A} \cup \bar{B}$.

4-13 Prove that $\overline{A \cap B} \subset \bar{A} \cap \bar{B}$ but that sets A and B exist such that $\overline{A \cap B} \neq \bar{A} \cap \bar{B}$.

4-14 Prove that if A is any set, then A' is closed.

4-15 Prove that if A is any set and G any open set, then (a) $\bar{A} \cap G \subset \overline{A \cap G}$, and (b) $A \cap G = \varnothing$ implies $\bar{A} \cap G = \varnothing$.

4-16 Prove that if G is open, then $G \subset G'$.

4-17 Let F be a closed set and assume $a \notin F$. Prove that disjoint open sets G_1 and G_2 exist such that $F \subset G_1$ and $a \in G_2$.

4-18 Can an open set be countable? Noncountable? What about a closed set? Cite examples to support affirmative answers. If the answer is no, give the reason.

4-19 Determine the set of cluster points in each of the following:

(a) $\left\{\frac{1}{m}+\frac{1}{n};\quad m,n = 1,2,\ldots\right\}$; (b) $\left\{(-1)^n+\frac{1}{m};\quad m,n = 1,2,\ldots\right\}$;

(c) $\{2^{-m}+3^{-n};\quad m,n = 1,2,\ldots\}$; (d) $\left\{\frac{(-1)^n}{1+\frac{1}{n}};\quad n = 1,2,\ldots\right\}$.

4-5 Unions and Intersections of Closed and Open Sets

Unions and intersections of families of closed or open sets sometimes retain the closed or open character of their components, sometimes not. The theorems and examples in this section show exactly when each characteristic is retained.

Theorem 4-8 The intersection of a finite collection of open sets is open.

Proof Let $\{G_1,G_2,\ldots,G_n\}$ denote a finite family of open sets, and consider the set $G = \bigcap_{\nu=1}^{n} G_\nu$. If c is any member of G, then c is necessarily a member of each of the G_ν. It follows then, from Theorem 4-7, that neighborhoods $\mathcal{N}(c,\varepsilon_\nu)$ exist such that $\mathcal{N}(c,\varepsilon_\nu) \subset G_\nu$ for $\nu = 1, 2, \ldots, n$. Denoting the minimum of the set of numbers $\varepsilon_1, \varepsilon_2, \ldots, \varepsilon_n$ by ε, observe that the neighborhood $\mathcal{N}(c,\varepsilon)$ is a subset of each of the neighborhoods $\mathcal{N}(c,\varepsilon_\nu)$ and hence is contained in each of the sets G_ν. As a consequence, $\mathcal{N}(c,\varepsilon) \subset G$ so that by Theorem 4-7 the set G is open ▲

Theorem 4-9 The union of a finite collection of closed sets is closed.

Proof Let $\{F_1,F_2,\ldots,F_n\}$ denote a family of n closed sets, and consider the set $F = \bigcup_{\nu=1}^{n} F_\nu$. Since $\mathscr{C}F = \bigcap_{\nu=1}^{n} \mathscr{C}F_\nu$ and since $\mathscr{C}F_\nu$ for each ν is an open set, it follows from Theorem 4-8 that $\mathscr{C}F$ is an open set. As a consequence, the set F is closed ▲

Theorem 4-10 The union of any collection of open sets is open.

Proof Let G_ν be an open set for each member ν of some index set Ω, and consider the set $G = \bigcup_{\nu\in\Omega} G_\nu$. If c is any member of G, then $c \in G_\alpha$ for some member α of Ω. Therefore, since G_α is open, a neighborhood $\mathcal{N}(c,\varepsilon)$ exists such that $\mathcal{N}(c,\varepsilon) \subset G_\alpha$. From $G_\alpha \subset G$ it then follows that $\mathcal{N}(c,\varepsilon) \subset G$. Therefore G is open ▲

Theorem 4-11 The intersection of any collection of closed sets is closed.

Proof The truth of this theorem follows from Theorem 4-10 by considering complements. However, we present here an alternate proof.

Let F_ν denote a closed set for each ν of some index set Ω, and consider the set $F = \bigcap_{\nu\in\Omega} F_\nu$. From $F \subset F_\nu$ it follows that $\bar{F} \subset \bar{F}_\nu = F_\nu$ for each ν in Ω. This implies that $\bar{F} \subset F$, whence F is closed ▲

It is not the case that the union of any family of closed sets is closed, nor is it true that the intersection of any family of open sets is open. Some simple counterexamples serve to make this clear.

Example 4-7 Let $\{F_1,F_2,\ldots,F_n,\ldots\}$ denote a countably infinite family of closed sets with members defined by $F_n = \left[\frac{1}{2^n}, \frac{3}{2^{n+1}}\right]$ for each positive integer n. Consider the set $F = \bigcup_{n=1}^{\infty} F_n$. Since numbers of the form $1/2^n$ may be found in any deleted neighborhood of the number 0, it follows that 0 is a cluster point of F. However, 0 is not a member of any of the sets F_n and consequently does not belong to F. It follows that F is not closed.

Example 4-8 Let $\{G_1, G_2, \ldots, G_n, \ldots\}$ denote a countably infinite family of open sets with members defined by $G_n = \left(-\frac{1}{n}, \frac{1}{n}\right)$ for each positive integer n. It is evident that only the real number 0 is contained in all of the sets G_n, and as a consequence the set $\bigcap_{n=1}^{\infty} G_n$ consists of exactly one member. Such a set is not open because of Theorem 4-7. This, then, is an example of a family of open sets whose intersection is not open.

4-6 Interior, Exterior, and Boundary Points

A point c is called an *interior point* of a set A if a neighborhood of c exists which is contained in A. A point c is an *exterior point* of A if c possesses a neighborhood entirely contained in the complement of A. By a *boundary point* of a set A we mean a point c such that each neighborhood of c contains at least one point of A and at least one point of $\mathscr{C}A$. The set of all interior points of a set A is said to constitute the *interior* of A, the set of exterior points the *exterior* of A, and the set of boundary points the *boundary* of the set A.

It is clear that for any set A the interior, exterior, and boundary are disjoint; that is, no real number is a member of more than one of the three. It is equally plain that each real number is a member of one of the sets.

Theorem 4-7 implies that the interior and exterior of any set A are both open sets. The boundary of A, since it is the complement of the union of two open sets, is closed. It is easily verified, also, that a set is open only if all its points are interior points, that only points of A can be interior points of A, and that only points of $\mathscr{C}A$ can be exterior points of A. Boundary points may be members of A or of $\mathscr{C}A$. Some examples of exterior, interior, and boundary points follow.

Example 4-9

(a) The interior of the set $(0,1]$ is the set $(0,1)$; the boundary consists of the two points 0 and 1; and the exterior is made up of all other real numbers.

(b) The set R of all rational numbers is such that its boundary consists of all real numbers. Both interior and exterior are empty.

(c) Let C be the set of all rational numbers of the form $1/n$, where n is a positive integer. The boundary of C consists of the set C together with the number 0. All other points constitute the exterior of C. The interior is empty.

4-7 Bounded Sets of Points

A set A is said to be *bounded above* if a number c exists such that $x \leq c$ for all points x of A, and to be *bounded below* if a number b exists such that $b \leq x$

for all members x of A. The number c is called an *upper bound* of A and b a *lower bound.* A set is said to be *bounded* if it is bounded above and bounded below. It is immediately evident that a set A is bounded if and only if there exists a number d with the property that $|x| \leq d$ for all members x of A.

Upper and lower bounds, if they exist, are not unique. For if c is an upper bound of a set A, then all numbers which exceed c are likewise upper bounds. Also, if b is a lower bound, then all numbers less than b are also.

In a discussion of upper and lower bounds we are aware of a kind of symmetry of properties or *duality*, as we shall call it. Certain statements concerning upper bounds compare with similar or dual statements about lower bounds and conversely. An example is the pair of statements concerning uniqueness immediately above. Recognizing this, we save time by omitting virtual repetition for lower bounds of a statement concerning upper bounds. Henceforth, when convenient, we shall simply make one statement pertaining either to upper or to lower bounds and refer to the other as being true *dually*.

We have already employed such terminology as the *greatest* or *maximum* member of a set as well as *least* or *minimum* member. We have seen that all finite sets possess such members. Also, in connection with the theory of cuts, the existence or nonexistence of greatest or least members in upper and lower classes, respectively, was made to be of considerable importance. It is now apparent that the maximum member of a set is a point which is at once a member and an upper bound of the set. Dually, the minimum is a member which is also a lower bound. We shall denote the maximum and minimum members of a set A, if such exist, by max A and min A, respectively.

It is evident that a set may have at most one maximum and one minimum point. Also, it may contain one or the other or neither. A closed bounded interval has both a maximum and a minimum point; the bounded interval $(0,1]$ has a maximum point but no minimum point. The same is true for the lower class of a rational cut. On the other hand, the upper class of any cut is devoid of both maximum and minimum points. A further example is the set of rational numbers of the form $1/n$, where n is any positive integer. It has the number 1 as a maximum point but possesses no minimum.

It is clear from the above examples that boundedness of a set is not sufficient to the existence of maximum and minimum members. Somewhat more useful concepts are the following. Given a set A, denote by U the set of all upper bounds of A, and by L the set of all lower bounds. If U possesses a minimum member, it is called the *supremum*[4] of the set A; if L has a maximum member, it is termed the *infimum*[5] of A. The former we denote conveniently by sup A and the latter by inf A.

It is plain that if a set has a maximum point, then it is the supremum of the set. Also, if it has a minimum point, then it is the infimum. In this sense the concepts of supremum and infimum may be regarded as generalizations of the notions of maximum and minimum points.

[4] Also, *least upper bound* or lub A.

[5] Also, *greatest lower bound* or glb A.

It follows from earlier remarks that a set has at most one supremum and one infimum. Such points need not exist at all, of course, as in the case of the set of all integers or of all real numbers. We have, however, the following important theorem.

Theorem 4-12 If a set is nonempty and is bounded below, then an infimum exists. (Dually, if a set is nonempty and bounded above, then a supremum exists.)

Proof Let A be a nonempty set which is bounded below. Denote by C_1 the set of all real numbers which are lower bounds of A, and let C_2 consist of all other real numbers. We may then show that $[C_1,C_2]$ is a Dedekind section. *1.* The nature of the definition of C_2 is assurance that each real number is in C_1 or C_2. *2.* Since A is bounded below, C_1 is not empty. Since A is not empty, not all real numbers are in C_1. Hence C_2 is not empty. *3.* Let c_1 be any member of C_1, and let c be any number such that $c < c_1$. It follows that c is less than a lower bound of A and hence is also a lower bound of A. Therefore c is a member of C_1. This implies that all members of C_2 exceed all members of C_1. *4.* Suppose C_2 to have a least member. Denote it by c. Then c is not a lower bound of A, and, as a consequence, a number x of A exists such that $x < c$. Also, between x and c exists another real number y; $x < y < c$. Since y is less than c, it is necessarily in C_1. Also, since y exceeds x, a member of A, it is not a lower bound and so is not in C_1. The contradiction implies that C_2 does not possess a least member.

It has thus been shown that $[C_1,C_2]$ is indeed a Dedekind section. By the completeness theorem it then follows that C_1 possesses a greatest member. Since C_1 is the set of all lower bounds of A, we have thereby proved that A possesses an infimum.

Proof of the dual theorem by the same procedure is awkward because of the lack of symmetry in the definition of Dedekind section. It follows, however, from the above if we consider the set B defined as the set $\{x; \quad -x \in A\}$ and employ the fact that the negative of the infimum of B is necessarily the supremum of A. The reader should supply details▲

The following theorem establishes a significant property of the supremum and the infimum of a set: members of a set exist which are arbitrarily close to the supremum and to the infimum.

Theorem 4-13 Let b denote the infimum of a nonempty set A, and let ε be any positive number. Then a member a of A exists such that $b \leq a < b + \varepsilon$. (The dual statement is also true.)

Proof Since b is the infimum of A and $b + \varepsilon > b$, it follows that $b + \varepsilon$ is not a lower bound of A. Hence some member a of A exists such that $a < b + \varepsilon$. But $b \leq a$ because b is a lower bound, so that $b \leq a < b + \varepsilon$ as required▲

Corollary The infimum of a nonempty set A is either a member of A or a cluster point of A. (The dual statement is also true.)

Proof Let $b = \inf A$. If b is a member of A, it is the minimum point of A. If b is not a member of A, then, corresponding to each positive ε, the deleted neighborhood $\mathscr{N}^*(b,\varepsilon)$ contains a point a of A. Hence b is a cluster point of A. (Actually it is a cluster point from the right.)▲

It is to be observed, of course, that the infimum or the supremum of a set may be a member of that set and a cluster point as well.

Theorem 4-14 A nonempty closed set A which is bounded below possesses a minimum point. (The dual statement is also true.)

Proof By Theorem 4-12 the infimum of the set A exists; call it b. By the corollary to Theorem 4-13, b is either a point of A or a cluster point of A. In the latter case it also follows that b is in A since A is closed. Hence the infimum of A belongs to A and is, of course, the minimum member of A▲

Exercises

4-20 For each positive integer n, let F_n, be a closed set such that $F_n \subset \{x;\ \ n-1 \leq |x| < n\}$. Prove that $F = \bigcup_{n=1}^{\infty} F_n$ is closed.

4-21 Prove that a set is open if and only if all of its points are interior points.

4-22 Prove that the interior and exterior of any set are open.

4-23 Prove that a cluster point of a set is either an interior point or a boundary point.

4-24 For any set A, prove that (a) the interior is the set $\mathscr{C}(\overline{\mathscr{C}A})$ and (b) the boundary is the set $\overline{A} \cap \overline{\mathscr{C}A}$.

4-25 Must a boundary point of a set be a cluster point? Explain.

4-26 Can a set have only interior points? Only exterior points? Only boundary points? Support affirmative answers with examples.

4-27 Let A be a nonempty set which is bounded above, and let $B = \{x;\ \ -x \in A\}$. Prove that $\inf B = -\sup A$.

4-28 Prove that the supremum of a set is a boundary point of the set.

4-29 Cite an example to show that Theorem 4-13 is no longer true if $b \leq a < b + \varepsilon$ is revised to read $b < a < b + \varepsilon$.

4-30 Prove that if $b = \sup B$, then $b = \sup \overline{B}$.

4-31 Let A and B be nonempty sets of real numbers, and assume that no member of A exceeds any member of B. Prove that $\sup A \leq \inf B$.

4-32 Prove that if A is bounded and nonempty, then a smallest closed interval I exists which contains A. (In other words, I has the property that if J is any closed interval containing A, then $I \subset J$.) Give an example of a set A such that no smallest open interval exists containing it.

4-33 (a) Prove that no two nonempty open sets G_1 and G_2 exist such that $G_1 \cap G_2 = \varnothing$ and $G_1 \cup G_2 = \mathscr{R}$. (b) Prove that $\varnothing$ and $\mathscr{R}$ are the only two sets that are both open and closed.

4-34 Prove that every closed set is the intersection of a countable family of open sets.

4-35 Determine the infimum and supremum of each of the following sets:

(a) $\left\{\frac{1}{m}+\frac{1}{n};\quad m,n=1,2,\ldots\right\}$; (b) $\left\{\left(\frac{1}{m}+\frac{1}{n}\right)^{m+n};\quad m,n=1,2,\ldots\right\}$;

(c) $\{2^{-m}+3^{-n};\quad m,n=1,2,\ldots\}$; (d) $\{r;\quad r\in R, r^2\leq 5\}$.

4-8 The Nature of Open Sets

The concepts of supremum and infimum may be employed to demonstrate the essential composition of open sets. First, however, we state Theorem 4-15, which concerns disjoint intervals.

Theorem 4-15 Any family of disjoint intervals is countable.

Proof Let Δ denote a family of disjoint intervals. (Members of Δ may be open, closed, both or neither, bounded or unbounded, and need not all be of the same type.) From each of the intervals of Δ select a rational point (it is obvious that such exists), and consider the correspondence between the set Δ, on the one hand, and the set of rational points so chosen, on the other. To each interval corresponds the single rational number selected in it. Also, from the disjoint character of the intervals, it follows that none of the chosen rational numbers is a member of more than one of the intervals. Hence the correspondence is one-to-one. The truth of the theorem then follows from the countability of any set of rational numbers▲

The intimate relation between open sets and open intervals has already been revealed in Theorem 4-7, where it was shown that each member of an open set is accompanied by an open interval also contained in the set. The following theorem shows further how an open set may be resolved into components which are open intervals.

Theorem 4-16 Any nonempty open set is the union of a unique countable collection of disjoint open intervals.

Proof Let G be a nonempty open set and p any point of G. Consider the closed sets $L_p=\mathscr{C}G\cap\{x;\quad x\leq p\}$ and $U_p=\mathscr{C}G\cap\{x;\quad x\geq p\}$. The point p, although it belongs to neither, is an upper bound of L_p and a lower bound of U_p. One or both of the sets L_p and U_p may be empty. If, however, L_p is not empty, then $\max L_p$ exists; denote it by a. Similarly, if U_p is nonempty, then $b=\min U_p$ exists.

Consider now the open interval I_p defined to be (a,b) if $L_p\neq\varnothing$ and $U_p\neq\varnothing$; $\{x;\quad x>a\}$ if $L_p\neq\varnothing$ and $U_p=\varnothing$; $\{x;\quad x<b\}$ if $L_p=\varnothing$ and $U_p\neq\varnothing$; and $\mathscr{R}$ if L_p and U_p are both empty.

In all cases the set I_p is a subset of G, but its endpoints, if it has such, do not belong to G. Further, if p_1 and p_2 are distinct points of G, then the corresponding intervals I_{p_1} and I_{p_2} are either identical or disjoint. As a consequence, the collection of all possible disjoint intervals constructed as above is such that each point of G is contained in exactly one of them, and in turn each interval is a subset of G. Thus G is the union of a collection of disjoint open intervals, which, by Theorem 4-15, is necessarily countable.

Uniqueness of the collection follows easily. For suppose $G = \bigcup_\nu I_\nu$ and $G = \bigcup_\kappa J_\kappa$, where the I_ν and J_κ are open intervals. Since $\bigcup_\nu I_\nu = \bigcup_\kappa J_\kappa$, it follows that each interval of one collection must intersect some interval of the other. Imagine that an interval I_p of the one collection and J_q of the other are such $I_p \cap J_q \neq \varnothing$. It follows, since the endpoints of I_p are not in G, that $J_q \subset I_p$. Similarly, $I_p \subset J_q$; whence $I_p = J_q$. Thus the collections are identical▲

With the revelation in Theorem 4-16 of the relatively simple composition of an open set, it is natural to conjecture that closed sets may be of similar construction. Since finite sets are closed, it is clear that closed sets are not unions of closed intervals. However, one wonders if a closed set might be composed of a countable collection of disjoint closed intervals together with isolated points. Perhaps the reader will be surprised to learn that this is not the case. An example of a closed set not so composed is the Cantor set to be discussed later in this chapter.

The fact that a closed set need not be as simple as we might at first expect sheds further light on the character of open sets. Although every open set is the union of a countable collection of open intervals, these intervals need not be such that they can be counted from left to right. For whenever the open intervals can be so counted, the closed complementary set does consist of a countable collection of closed intervals or points.

Exercises

4-36 Cite an example to show that the following statement is false: Any family of disjoint sets is countable.

4-37 State and prove a generalization of Theorem 4-15 which involves relaxing the condition that the sets concerned be intervals.

4-38 Exhibit an open set that is the union of a countable collection of disjoint open intervals but is such that these intervals cannot be counted from left to right.

4-9 Sequences of Closed Sets

If a collection or family of sets of real numbers is countably infinite, an ordering of the members based on a particular one-to-one correspondence with the set of positive integers is called a *sequence of sets*. Denoting that member of the family which corresponds to the integer n by A_n, we may represent the sequence by $A_1, A_2, \ldots, A_n, \ldots$ or, briefly, by the symbol $\{A_n\}$.

If a sequence of sets of points has the property that $A_{n+1} \subset A_n$ for each n, then the sequence is referred to as *decreasing* or *nested*. Concerning a nested sequence of sets, each of which is closed, we prove the following useful property.

Theorem 4-17 (Nested Sequences) If $\{F_n\}$ is a nested sequence of nonempty, closed, and bounded sets, then the set $F = \bigcap_{n=1}^{\infty} F_n$ is nonempty.

Proof By Theorem 4-14 each of the sets F_n possesses a maximum point. Let $x_n = \max F_n$ for each n. From the hypothesis $F_{n+1} \subset F_n$ it follows that $x_{n+1} \leq x_n$, and for each positive integer q it is the case that $x_n \in F_q$ for all $n \geq q$.

Denote the set of numbers x_n by T. Because $T \subset F_1$ and F_1 is bounded, it follows that T is bounded and, in particular, is bounded below. Denoting the infimum of T by c, we shall show that $c \in F_n$ for all n.

Let q denote an arbitrary positive integer, and consider the subset of T defined by $T_q = \{x_n;\ \ n \geq q\}$. Since c is the infimum of T, it is also the infimum of T_q. By the corollary to Theorem 4-13, it follows that $c \in \overline{T}_q$. Also, since $T_q \subset F_q$, we have $\overline{T}_q \subset \overline{F}_q = F_q$. As a consequence, $c \in F_q$▲

All the conditions imposed on the sets F_n in the statement of Theorem 4-17 are essential to the conclusion of the theorem. The following examples show that this is the case.

Example 4-10

(a) Let $F_n = \varnothing$ for all positive integers n. Then F_n is closed and bounded, and $\{F_n\}$ is nested. However, the set $F = \bigcap_{n=1}^{\infty} F_n$ is empty since the F_n are not nonempty.

(b) Let $F_n = \{x;\ \ x \geq n\}$. Here the sequence $\{F_n\}$ is nested, and F_n is nonempty and closed for each n. However, each F_n is unbounded. It is easy to see that $F = \bigcap_{n=1}^{\infty} F_n$ is empty.

(c) Let $F_n = (0,1/n]$. The sequence $\{F_n\}$ is obviously nested, each F_n is nonempty and bounded but not closed. The set $F = \bigcap_{n=1}^{\infty} F_n$ is seen to be empty.

(d) Let $F_n = [2n,2n+1]$. Here each F_n is nonempty, closed, and bounded, but the sequence $\{F_n\}$ is not nested. The set $F = \bigcap_{n=1}^{\infty} F_n$ is empty.

4-10 Coverings of Sets of Real Numbers

Let Γ denote a family of sets of real numbers. Then Γ is called a *covering* of a set A of real numbers or is said to *cover A* if each member of A is a member of at least one of the sets which belong to Γ. Expressed differently, the family Γ covers A if and only if A is contained in the union of the members of Γ. We shall have particular interest in coverings of which each member is an open set. We refer to such as *open coverings*. Some examples of coverings follow.

Example 4-11

(a) Let $F_n = [-n,n]$, and let Γ be the collection of sets F_n corresponding to all positive integer values of n. Then Γ covers the set of all real numbers.

(b) Let $G_n = (1/n,1)$, and let Γ be the family of sets G_n corresponding to all positive integers n. Then Γ is an open covering of the interval (0,1). The family Γ is also an open covering of the set $(\frac{1}{2},\frac{2}{3}]$.

(c) Let Γ be the family of all neighborhoods $\mathcal{N}(r,\varepsilon)$, where r and ε are rational numbers. Then Γ is an open covering of the set of all real numbers.

A set A is said to be *compact* if each open covering of A possesses a finite subfamily which also covers A. Compact sets play an important role in certain aspects of real variable theory which occur in later chapters. The notion of compactness is not an easy one to grasp. It is difficult to demonstrate that a set is compact directly from the definition since one must consider "all possible open coverings." An exception is any finite set, which is obviously compact. Noncompactness, on the other hand, is often easier to prove.

Example 4-12 Consider the open interval $A = (0,1)$ and the open covering Γ of that interval defined as the family of sets G_n, where $G_n = (1/n,1)$ for each positive integer n.

Although it is clear that Γ covers A, it is not possible for a finite subfamily of Γ to cover A. For if Γ' denotes any finite subfamily of Γ, then each member of Γ' possesses an infimum of the form $1/n$, and the set of such bounds is finite. Consequently one value of $1/n$, say $1/N$, is less than all the others. It is then obvious that the subset $(0,1/N]$ of A is not covered by Γ'. Having demonstrated the existence of *one* covering Γ that does not admit of a finite subcovering, it follows A is not compact.

To facilitate demonstration of compactness of a set, it is important to obtain a characterization of compact sets which is more easily applied than the definition itself. A step in this direction is provided by the following theorem.

Theorem 4-18 (Lindelöf) If Γ is any open covering of a set, then there exists a countable subfamily of Γ which also covers the set.

Proof Let a denote any member of the given set A. Then a set G_a of Γ exists such that $a \in G_a$. Further, since G_a is open, a neighborhood $\mathcal{N}(a,\varepsilon)$ exists such that $\mathcal{N}(a,\varepsilon) \subset G_a$. Now let r_1 and r_2 designate two rational points with the property that $a - \varepsilon < r_1 < a < r_2 < a + \varepsilon$. It is then the case that the interval $I_a = (r_1,r_2)$ is such that $a \in I_a$ and $I_a \subset G_a$. Hence in this manner we may associate with each member a of the set A an open interval I_a with rational endpoints. Since the set of all possible intervals with rational endpoints is countable, it follows that the set $B = \{I_a;\ \ a \in A\}$ is also countable.

Each interval I_a is contained in at least one of the open sets of Γ. Denote one such by G_a'. In this way a subfamily Γ' of Γ is constructed with the property that with each interval I_a of the countable set B is associated exactly one member G_a' of Γ'. Consequently Γ' is countable and, moreover, covers A since for each member a of A we have $a \in I_a$ and $I_a \subset G_a'$ ▲

The following most important theorem constitutes the first part of the desired characterization of compact sets.

Theorem 4-19 (Heine-Borel) A closed and bounded set is compact.

Proof Let A be a closed and bounded set and Γ an open covering of A. Because of the Lindelöf theorem, we may assume, without loss of generality, that Γ is countable and thereby denote its members by $G_1, G_2, \ldots, G_n, \ldots$. Define the sets

$$K_n = \bigcup_{i=1}^{n} G_i \quad \text{and} \quad L_n = A \cap \mathscr{C}K_n$$

for $n = 1, 2, \ldots$, and observe that, because of the theorems on unions and intersections of closed and open sets, the sets K_n are open and the sets L_n closed for all values of n. Further, it is the case that $K_n \subset K_{n+1}$, and from this follows $L_{n+1} \subset L_n$ for all n.

Assume now that none of the sets L_n is empty. Since $L_n \subset A$ and A is bounded, it follows then that the sequence $\{L_n\}$ satisfies the hypotheses of the nested sequence theorem. Therefore a point c exists which is a member of every set L_n. Because of the definition of the L_n, the point c is necessarily a point of A as well as of each of the sets $\mathscr{C}K_n$. The latter implies that c is not a member of K_n and hence not a member of G_n for any value of n. Thus c is a point of A not covered by Γ. This contradiction to the hypothesis of the theorem denies the assumption that all sets L_n are nonempty. Therefore, for some positive integer q the set $L_q = A \cap \mathscr{C}K_q$ is empty. Hence $A \subset K_q = \bigcup_{i=1}^{q} G_i$. We conclude that A is covered by the finite subfamily $\{G_1, G_2, \ldots, G_q\}$ of Γ▲

The converse to the Heine-Borel theorem completes the characterization of compact sets.

Theorem 4-20 A compact set is closed and bounded.

Proof Let Γ denote the family of open sets of points G_n defined by $G_n = (-n,n)$ for each positive integer n. It is obvious that Γ is an open covering of the set of all real numbers and hence of any set of real numbers. Suppose now that A is some compact set. Then, since any open covering of A possesses a finite subfamily which also covers A, it follows, in particular, that this is true of Γ. Consequently, a finite collection of intervals G_n covers the set A; and, if n_0 denotes the maximum subscript for this finite family, then clearly the open interval $G_{n_0} = (-n_0, n_0)$ covers A. This implies that a compact set is bounded.

To show that A is necessarily closed, let c be a real number such that $c \notin A$, and consider the family of closed sets $F_n = [c - 1/n, c + 1/n]$ for each positive integer n. The sets $H_n = \mathscr{C}F_n$, $n = 1, 2, \ldots$, then constitute a family of open sets, which we denote by Δ. It is obvious that the set $\bigcap_{n=1}^{\infty} F_n$ consists of the single point c; and, since c is not in A, it follows that $A \subset \mathscr{C} \bigcap_{n=1}^{\infty} F_n$, whence $A \subset \bigcup_{n=1}^{\infty} H_n$. Thus the set A is covered by the family Δ.

The compactness of A implies that a finite subfamily of Δ exists which also covers A. Therefore a positive integer n_1 exists such that each point of A is contained in at least one of the open sets $H_1, H_2, \ldots, H_{n_1}$. Necessarily, then, no point of A is contained in $F_{n_1} = [c - 1/n_1, c + 1/n_1]$, and from this it follows that the point c is not a cluster point of the set A. Thus it is proved that any point which is not a point of the set A is also not a cluster point of A. All cluster points of A are, therefore, points of A itself. Hence A is closed ▲

4-11 The Bolzano-Weierstrass Theorem

It is intuitively reasonable that the points of a bounded and infinite set will necessarily be crowded together to a degree that at least one cluster point must exist. The formal statement of this fact is presented in the following theorem. We shall have frequent occasion to refer to it.

Theorem 4-21 (Bolzano-Weierstrass) If a set of points is both bounded and infinite, then a real number exists which is a cluster point of the set.

Proof Let A denote a bounded set of points, and assume that no real number exists which is a cluster point of A. Two consequences of this assumption are immediate. First, the set A is closed (since $A' = \varnothing$); second, each point of A is an isolated point. From the latter it follows that if a is any point of A, then a neighborhood $\mathcal{N}(a,\varepsilon)$ exists which contains no points of A other than a itself. Let Γ be the family of all such neighborhoods of all points of A. Then Γ is an open covering of the closed and bounded set A whence, by the Heine-Borel theorem, a finite subfamily of Γ exists which covers A. Since each of the neighborhoods which make up this subfamily contains exactly one point of the set A, it follows that A is a finite set. Thus, a bounded set with no cluster points must be a finite set▲

The conditions of bounded and infinite imposed on the set A in the Bolzano-Weierstrass theorem are obviously not superfluous since a finite set can have no cluster points while the set of all integers, for example, is infinite, unbounded, and possesses no cluster points. It is clear, of course, that unbounded sets may have cluster points; witness the set of all real numbers.

4-12 Compact Sets and Cluster Points

A second characterization of compact sets concerns the existence of cluster points.

Theorem 4-22 A set is compact if and only if every infinite subset has a cluster point which is a member of the set.

Proof Let A be a compact set and hence a closed and bounded set. Then any infinite subset of A is bounded and, because of the Bolzano-Weierstrass theorem, possesses a cluster point a. Since a is also a cluster point of A and since A is closed, it follows that a is a point of A.

For the converse, assume that every infinite subset of a set A has a cluster point which is a point of A. Then to show that A is compact, we prove that A is both bounded and closed.

Suppose that A is not bounded. In particular, and without any loss of generality, suppose that A is not bounded above. Define an infinite subset B of points x_n of A by $x_1 > 1$ and $x_{n+1} > x_n + 1$ for each positive integer n. Since any two members of this set are such that the absolute value of their difference exceeds 1, it follows that an infinite set of them cannot occur in an arbitrarily small neighborhood of any real number. Hence B is without cluster points, and, since this contradicts the hypothesis on A, we conclude that A is bounded.

Suppose now that a number c is a cluster point of the set A. Define a set C of points y_n of A such that $|y_n - c| < 1/n$ for each positive integer n. Then C is an infinite subset of A, and c is a cluster point of C. Now denote by b any real number different from c, and consider the set of all integers n such that $1/n < \frac{1}{2}|b - c|$. For such n we have $|y_n - c| < \frac{1}{2}|b - c|$, whence

$$|b - y_n| \geq |b - c| - |y_n - c| > \tfrac{1}{2}|b - c|.$$

Therefore only a finite set of points of C are present in the neighborhood $\mathcal{N}(b,\frac{1}{2}|b - c|)$, and from this it follows that b is not a cluster point of C. Consequently c is the only cluster point of C, and from the hypothesis it follows that c is a member of the set A. Therefore all cluster points of A belong to A, so that A is closed ▲

Exercises

4-39 Cite an example of a sequence of sets $\{F_n\}$ satisfying the hypotheses of Theorem 4-17 and having the property that the point c as described in the proof of that theorem is (a) in T, (b) not in T.

4-40 Show that the set $A = \{x; \ \ x > 0\}$ is covered by the family of neighborhoods $\mathcal{N}(r,r/2)$, where r is any positive rational number, but that no finite subset of this family exists which also covers A.

4-41 Cite an example of a collection Γ of closed sets with the property that it covers the interval $[0,1]$, yet no finite subset of Γ exists which covers $[0,1]$.

4-42 Prove that a discrete set is countable.

4-43 Prove that if A' is countable, then A is countable.

4-44 Prove that if F is a closed set and K a compact set, then $F \cap K$ is compact.

4-45 Prove that if A and B are compact sets, then so are $A \cap B$ and $A \cup B$.

4-46 Prove the following variation of the Heine-Borel theorem: Let A be a closed and bounded set and Γ a family of sets with the property that each point of A is an interior point of at least one set of Γ. Then a finite subset Γ' of Γ exists which covers A and is such that each point of A is an interior point of at least one set of Γ'.

4-47 (a) Prove that if F_1 and F_2 are nonempty disjoint compact sets, then a positive number k exists such that $|x_1 - x_2| \geq k$ for all $x_1 \in F_1$ and all $x_2 \in F_2$. (b) Cite an example to show that the last statement is not true if the nonempty sets F_1 and F_2 are closed but not compact.

4-48 (a) Prove that if F_1 and F_2 are nonempty disjoint compact sets, then disjoint open sets G_1 and G_2 exist such that $F_1 \subset G_1$ and $F_2 \subset G_2$. (b) Investigate the validity of (a) if the sets F_1 and F_2 are closed but not compact.

4-49 Investigate the validity of the following statement: If G_1 and G_2 are nonempty disjoint open sets, then disjoint closed sets F_1 and F_2 exist such that $G_1 \subset F_1$ and $G_2 \subset F_2$.

4-50 With respect to the set R of rational numbers, define a rational number r to be a *cluster point* of a subset A of R if, corresponding to each $\varepsilon > 0$, there exists a member a of A such that $0 < |a - r| < \varepsilon$. Cite an example to show that the Bolzano-Weierstrass theorem is not true in R.

4-51 Let $\mathscr{F}$ be a family of compact sets, and assume $\mathscr{F}$ has the *finite intersection* property: Any nonempty finite subfamily $\{F_1, F_2, \ldots, F_n\}$ is such that $\bigcap_{\nu=1}^{n} F_\nu$ is not empty. Prove that a real number exists which is common to all members of $\mathscr{F}$.

4-13 Cantor's Perfect Set

As promised, we present a nontrivial example of a perfect set together with a theorem reporting the infinite character of such a set. This example, which is due to Cantor, has many remarkable properties.

Let F_0 denote the closed bounded interval $[0,1]$. Remove from this set of points the open interval $(\frac{1}{3},\frac{2}{3})$, and denote the residue by F_1. Thus

$$F_1 = [0,\tfrac{1}{3}] \cup [\tfrac{2}{3},1].$$

Now remove the middle thirds $(\frac{1}{9},\frac{2}{9})$ and $(\frac{7}{9},\frac{8}{9})$ of the components of F_1. The remainder F_2, is the set

$$F_2 = [0,\tfrac{1}{9}] \cup [\tfrac{2}{9},\tfrac{1}{3}] \cup [\tfrac{2}{3},\tfrac{7}{9}] \cup [\tfrac{8}{9},1].$$

Repeat, removing middle thirds of each of the four intervals which make up F_2. What is left is the set

$$F_3 = [0,\tfrac{1}{27}] \cup [\tfrac{2}{27},\tfrac{1}{9}] \cup [\tfrac{2}{9},\tfrac{7}{27}] \cup [\tfrac{8}{27},\tfrac{1}{3}] \cup [\tfrac{2}{3},\tfrac{19}{27}] \cup [\tfrac{20}{27},\tfrac{7}{9}] \cup [\tfrac{8}{9},\tfrac{25}{27}] \cup [\tfrac{26}{27},1].$$

Continue in this fashion, generating a sequence of sets $\{F_n\}$, where F_{n+1} is obtained from F_n by removing the (open) middle thirds of the disjoint closed intervals of which F_n is composed. The accompanying diagram depicts F_0, F_1, F_2, F_3, and F_4. We define the *Cantor set* K to be the set of points common to all the sets F_n; thus,

$$K = \bigcap_{n=1}^{\infty} F_n.$$

Each of the sets F_n is nonempty, closed, and bounded. Moreover, the construction of the successive sets is such that $F_{n+1} \subset F_n$ for all n. Hence the sequence $\{F_n\}$ satisfies the conditions of the nested sequence theorem. As a

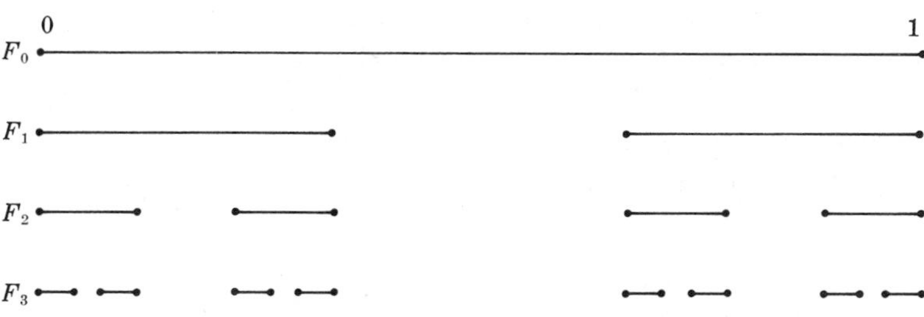

Figure 4-1

consequence, the set K is not empty. The nonempty character of K follows also from the observation that the endpoints of the disjoint closed intervals at any stage of the construction of the sequence $\{F_n\}$ are not deleted at any subsequent stage. Hence such points are members of the set K.

We prove now that the Cantor set K is a perfect set; that is, $K = K'$. First, that K is closed is immediate since it is defined as the intersection of a family of closed sets. Second, to show that every point of K is a cluster point of the set, let x denote a point of K. Then x is a point of each of the sets F_n and, for each n, is contained in one of the 2^n disjoint closed and bounded intervals, each of length $1/3^n$, which go to make up the set F_n.

Now, if ε denotes a positive number, then a positive integer n_0 exists such that $1/3^{n_0} < \varepsilon$. It follows that that closed interval of F_{n_0} which contains x is entirely contained in the neighborhood $\mathscr{N}(x,\varepsilon)$. In particular, the two endpoints of that interval are in $\mathscr{N}(x,\varepsilon)$. At least one of the two endpoints is not the point x; denote it by y. Since y is a member of K, it follows that each deleted neighborhood of a point x of K contains another point of K. Hence each point of K is a cluster point of K. From the fact that K is closed and that every point is a cluster point, it follows that K is a perfect set.

Some reflection on the construction of the set K will reveal the following interesting fact: Between any two points of the set K there exists an open interval which belongs to the complement of K. Thus the set K is nowhere dense (see Exercise 4-62).

Concerning perfect sets in general and the Cantor set in particular, it is obvious that if any such set is nonempty, then it is infinite. This follows from the observation that a finite set has no cluster points. In the following theorem we go a step further to describe the infinite character of perfect sets.

Theorem 4-23 A nonempty perfect set is noncountably infinite.

Proof Let P denote a nonempty perfect set, and suppose it to be countable; that is, $P = \{x_1,x_2,\ldots,x_n,\ldots\}$. Denoting the point x_1 also by the symbol x_{n_1} (that is, $n_1 = 1$), let $\mathscr{N}_1$ denote some neighborhood of x_{n_1}. Since x_{n_1} is a cluster point of P, other points of P are contained in $\mathscr{N}_1$. Let n_2 be the least integer greater than n_1 for which x_{n_2} is contained in $\mathcal{N}_1$. (Then any point x_j of P for which $n_1 < j < n_2$ is not a point of $\mathcal{N}_1$.) Denote by $\mathcal{N}_2$ a neighborhood of x_{n_2} with the special properties that $\mathcal{N}_2 \subset \mathcal{N}_1$ and $x_{n_1} \notin \overline{\mathcal{N}_2}$. Since the point x_{n_2} is also a cluster point of P, the neighborhood $\mathcal{N}_2$ contains points of P other than x_{n_2}. Let n_3 be the least integer greater than n_2 such that $x_{n_3} \in \mathcal{N}_2$. Let $\mathcal{N}_3$ be a neighborhood of x_{n_3} such that $\mathcal{N}_3 \subset \mathcal{N}_2$ and $x_{n_2} \notin \overline{\mathcal{N}_3}$. Continue in this fashion, thereby generating a progression $x_{n_1}, x_{n_2}, \ldots, x_{n_i}, \ldots$ of points which constitute a subset Q of the set P, together with a nested sequence of open intervals $\mathcal{N}_1 \supset \mathcal{N}_2 \supset \cdots \supset \mathcal{N}_i \supset \cdots$ with the properties

$$x_{n_{i+1}} \in \mathcal{N}_i \quad \text{and} \quad x_{n_i} \notin \overline{\mathcal{N}_{i+1}}$$

for each positive integer i. Also, if $n_i < p < n_{i+1}$, then $x_p \notin \mathcal{N}_i$. The set Q is infinite, and, since $\mathcal{N}_i \subset \mathcal{N}_1$ for all i, it follows that Q is bounded. As a consequence, Q possesses a cluster point x. It is immediate that x is also a cluster point of P, and hence $x \in P$. Therefore x appears somewhere in the progression $x_1, x_2, \ldots, x_n, \ldots$; that is, $x = x_m$ for some positive integer m.

Necessarily $x_m \in \mathcal{N}_i$ for all i. Therefore $x_m \notin Q$ since each member of Q is contained in at most a finite set of neighborhoods $\mathcal{N}_i$. As a consequence, points x_{n_j} and $x_{n_{j+1}}$ of Q exist such that $n_j < m < n_{j+1}$, and x_m and $x_{n_{j+1}}$ are both contained in $\mathcal{N}_j$; but this denies the role of the number n_{j+1} as the least integer p greater than n_j for which $x_p \in \mathcal{N}_j$. We conclude that the perfect set P is a noncountable infinite set▲

4-14 The Topology of $\mathscr{R}^*$

The topology of the extended real number system $\mathscr{R}^*$ is similar in many respects to that of $\mathscr{R}$. The definition of neighborhood of ∞ cannot be given in precisely the same manner as for finite numbers since the set $\{x;\ \infty - \varepsilon < x < \infty + \varepsilon\}$ is empty. For any k we define a neighborhood of ∞ to be the set $\{x;\ k < x \leq \infty\}$. Similarly, the set $\{x;\ -\infty \leq x < k\}$ is a neighborhood of $-\infty$. Further, the sets $\{x;\ k < x < \infty\}$ and $\{x;\ -\infty < x < k\}$ are deleted neighborhoods of ∞ and $-\infty$, respectively.

Since the concepts of cluster point, closed and open sets, etc., all depend on the notion of neighborhood, they are immediately generalized by means of the extended definition of neighborhood. For example, ∞ is a *cluster point* of a set A if each deleted neighborhood of ∞ contains an infinite subset of A. With such definitions Theorems 4-1 through 4-11 remain valid.

Some of the following examples seem to dispute statements made earlier. It must be kept in mind, of course, that they concern $\mathscr{R}^*$, not $\mathscr{R}$.

Example 4-13

(a) Both ∞ and $-\infty$ are cluster points of I, the set of all integers; the set I is neither open nor closed.

(b) $\mathscr{R}^*$ is open and closed; $\mathscr{R}$ is open but not closed.

(c) Both $\{x;\ \ 1 < x \leq \infty\}$ and $\{x;\ \ 1 < x < \infty\}$ are open. Each will be called an *open interval*.

A significant difference between $\mathscr{R}$ and $\mathscr{R}^*$ lies in the concept of boundedness. *In the set $\mathscr{R}^*$ all sets are bounded.* Each member x of any set A is such that $-\infty \leq x$ and $x \leq \infty$; that is, every set is bounded below by $-\infty$ and above by ∞. Because of this, the topology of $\mathscr{R}^*$ may be regarded as simpler than that of $\mathscr{R}$. This is one of the principal advantages of $\mathscr{R}^*$ over $\mathscr{R}$. Actually, all theorems stated earlier which concern bounded sets in $\mathscr{R}$ are true for all sets in $\mathscr{R}^*$. In particular, every nonempty set in $\mathscr{R}^*$ possesses an infimum and a supremum (cf. Theorem 4-12). Other theorems remain valid with obvious modifications made necessary by the extended definition of neighborhood. For example, if $-\infty$ is the infimum of a set A, then for any number k there exists a member a of A such that $-\infty \leq a < k$ (cf. Theorem 4-13). Similar modifications are sometimes necessary in proofs. The reader may verify the validity of such theorems as Heine-Borel, Bolzano-Weierstrass, and the theorem on nested sequences, where the hypotheses of boundedness are redundant.

Exercises

A point c is a *condensation point* of a set A if each neighborhood of c contains a noncountable subset of A. In Exercises 4-52 through 4-59, C denotes the set of condensation points of A.

4-52 What is the set of condensation points of (a) the interval $[a,b]$? (b) the set R of rational numbers? (c) the set of irrational numbers?

4-53 Cite an example to show that C need not be a subset of A.

4-54 Prove that C is a closed set.

4-55 Prove that A is countable if and only if C is empty.

4-56 Prove that C is a perfect set.

4-57 Prove that $A - C$ is countable.

4-58 Prove that $A \cap C$ is noncountable if and only if A is noncountable.

4-59 Prove that every closed set is the union of a countable set and a perfect set.

We shall say that a set A is *dense in a set* B if $B \subset A'$. In particular, if A is dense in $\mathscr{R}$, we say simply it is *dense*.

4-60 Reconcile the above usage of the term dense with that of Section 3-8, where the rational numbers are described as a dense set.

4-61 Cite an example of sets A and B such that A is dense in B and $A \cap B = \varnothing$.

A set A will be called *nowhere dense in* B if, for each point b of B and each neighborhood $\mathscr{N}(b,\varepsilon)$, there is a neighborhood $\mathscr{N}(x,\varepsilon')$ such that $\mathscr{N}(x,\varepsilon') \subset \mathscr{N}(b,\varepsilon)$ and $\mathscr{N}(x,\varepsilon') \cap A = \varnothing$. In particular, A is said to be *nowhere dense* if it is nowhere dense in $\mathscr{R}$.

4-62 Prove that the Cantor set is nowhere dense in the interval [0,1].

4-63 Prove that for A to be nowhere dense in B it is necessary and sufficient that $B \cap \mathscr{C}\overline{A}$ be dense in B.

4-64 Prove that a set is nowhere dense if and only if its closure has no interior points.

4-65 Prove that a set which is closed and nowhere dense has for its complement a dense set consisting of a countable collection of disjoint open intervals.

We shall say that a set is of the *first category* if it is the union of a countable family of nowhere dense sets. A set not of the first category is of the *second category*.

4-66 Prove that the set R of rational numbers is of the first category.

4-67 Prove that (a) any countable set is of the first category, and (b) not every set of the first category is countable.

4-68 Prove that the union of a countable family of first category sets is also of the first category.

4-69 Prove that if A is of the first category, then $\mathscr{C}A$ is dense.

4-70 Prove that the set $\mathscr{R}$ of real numbers and the interval $[0,1]$ are each of the second category.

4-71 Prove that if A is of the first category, then $\mathscr{C}A$ is of the second category.

4-72 Prove that the set of irrational numbers is of the second category.

4-73 Cite an example to show that if A is of the second category, the complement of A need not be of the first category.

4-74 Prove that the set of rational numbers in the interval $(0,1)$ cannot be expressed as the intersection of a countable collection of open sets.

For any collection $\mathscr{E}$ of subsets of $\mathscr{R}$, define

$$\mathscr{E}_\sigma = \left\{\bigcup_{\nu=1}^{\infty} E_\nu;\quad E_\nu \in \mathscr{E},\ \nu = 1,2,\ldots\right\},$$

$$\mathscr{E}_\delta = \left\{\bigcap_{\nu=1}^{\infty} E_\nu;\quad E_\nu \in \mathscr{E},\ \nu = 1,2,\ldots\right\}.$$

In Exercises 4-75 through 4-78, $\mathscr{G}$ and $\mathscr{F}$ denote, respectively, the collections of open and closed subsets of $\mathscr{R}$.

4-75 Prove that $\mathscr{E} \subset \mathscr{E}_\sigma$ and $\mathscr{E} \subset \mathscr{E}_\delta$.

4-76 Prove that (a) $\mathscr{E}_\sigma = \mathscr{E}_{\sigma\sigma}$, where $\mathscr{E}_{\sigma\sigma} = (\mathscr{E}_\sigma)_\sigma$, and (b) $\mathscr{E}_\delta = \mathscr{E}_{\delta\delta}$.

4-77 Prove that (a) $\mathscr{G}_\sigma = \mathscr{G}$, and (b) $\mathscr{F}_\delta = \mathscr{F}$.

4-78 Prove that (a) $\mathscr{F} \subset \mathscr{G}_\delta$, and (b) $\mathscr{G} \subset \mathscr{F}_\sigma$.

5
Sequences of Real Numbers

... naturalists observe a flea
Has smaller fleas that on him prey
And these have smaller still to bite 'em
And so proceed ad infinitum.

JONATHAN SWIFT (1667–1745), *On Poetry*

5-1 Definition of Sequence

The term *sequence* has been employed in preceding chapters to designate the concept of consecutive presentation or ordering of members of sets. It was seen to be a notion on which was based the distinction between countability and noncountability. The importance, too, of the concept of a sequence of nested sets will be recalled. In this chapter we give attention to a detailed analysis of characteristics and properties of sequences of real numbers.

Let n_0 denote some integer, and let f be a function that is defined for all integers n such that $n \geq n_0$ and whose range is a subset of the set of all real numbers. Such a function is called a *sequence of real numbers*, a *sequence of constants*, or simply a *sequence*. Denoting function values by $a_n = f(n)$, it is convenient also to represent a sequence by the symbols $\{a_n\}$ (or $\{f(n)\}$) and $a_{n_0}, a_{n_0+1}, \ldots, a_n, \ldots$. The number a_{n_0} is called the *first element* of the sequence $\{a_n\}$, a_{n_0+1} the *second element*, and a_n the *general* or *nth element*. The value of n_0 does not play an important role; as a consequence, for convenience we usually arrange that n_0 have the value 1. In any case, if nothing is said to the contrary, it will be assumed that the leading element does correspond to $n = 1$.

Example 5-1

(a) $\left\{\frac{1}{n}\right\}$ or $1, \frac{1}{2}, \frac{1}{3}, \ldots, \frac{1}{n}, \ldots$.

(b) $\left\{\frac{1}{2^n}\right\}$ or $1, \frac{1}{2}, \frac{1}{4}, \ldots, \frac{1}{2^n}, \ldots$.

(c) $\left\{\frac{5n-4}{2-3n}\right\}$ or $-1, -\frac{3}{2}, -\frac{11}{7}, \ldots, \frac{5n-4}{2-3n}, \ldots$.

(d) $\left\{\frac{n}{a^n}\right\}$ or $\frac{1}{a}, \frac{2}{a^2}, \frac{3}{a^3}, \ldots, \frac{n}{a^n}, \ldots \quad (a > 1)$.

(e) $\{3^n\}$ or $3, 9, 27, \ldots, 3^n, \ldots$.

(f) $\{(-1)^n\}$ or $1, -1, 1, -1, \ldots, (-1)^n, \ldots$.

(g) $\{a\}$ or $a, a, a, \ldots, a, \ldots$.

In sequence (b) the leading element corresponds to $n = 0$; in (f) any even integer (say 0) suffices; and in (g), n_0 may have any value. The remaining sequences have first elements corresponding to $n = 1$.

The *range of a sequence* $\{f(n)\}$, that is, the range of the function f, may be a finite or an infinite set. In the sequences (a) through (e) of Example 5-1 the range is infinite; in (f) it consists of the numbers 1 and -1; and in (g) it is the single real number a. The range of a sequence should not be confused with the elements of the sequence. The range is a set; a particular number is either a member of the set or not—it is meaningless to question if the number is "repeated" in the set. An element of a sequence, on the other hand, is the number appearing in a particular position in the sequence. It is quite possible for a member of the range of the sequence to appear more than once as an element.

5-2 Convergence of a Sequence

A sequence $\{a_n\}$ is said to possess a *limit* a or to *converge* or to be *convergent* to a if, corresponding to each positive number ε, there exists an integer N such that

$$|a_n - a| < \varepsilon \tag{1}$$

for all $n > N$.[1] If a sequence is not convergent, we say that it *diverges* or is *divergent*.

A sequence which converges to the limit 0 is specially designated a *null*

[1] The substitution of $|a_n - a| \leq \varepsilon$ for $|a_n - a| < \varepsilon$ or of $n \geq N$ for $n > N$ would not alter the sense of the definition. It would *not* be proper, however, to employ the relation $\varepsilon \geq 0$ instead of $\varepsilon > 0$, for then convergence would be confined to sequences of the type of Example 5-1(g).

sequence. We shall show presently that the sequences (a), (b), and (d) in Example 5-1 are null sequences, that (c) converges to $-\frac{5}{3}$, that (g) converges to a, and that (e) and (f) are divergent.

If a sequence $\{a_n\}$ converges to a limit a, we shall frequently employ the symbolism

$$\lim \{a_n\} = a$$

to express this fact. The symbols $\lim_{n\to\infty} \{a_n\}$ and $\lim_{\infty} \{a_n\}$ are also standard notations.

The concept of convergence of a sequence to a limit may be paraphrased in the following way. A sequence has a limit a if an element $a_{n'}$ of the sequence exists such that it and all following elements (that is, those having subscripts which exceed n') differ from the number a by as little as one may propose, zero excepted, simply by choosing n' sufficiently large.

A consequence of the definition of convergence is the following. If $\{a_n\}$ and $\{b_n\}$ are two sequences with the property that integers n_0 and m exist such that $a_{n+m} = b_n$ for all $n \geq n_0$, then both sequences converge to the same limit or both diverge, regardless of differences between a_{n+m} and b_n for $n < n_0$. Proof is left to the reader.

The role of the positive number ε in the definition of convergence warrants attention. For one thing, it should be noted that a restriction of values of ε to any open bounded interval of which zero is the left endpoint does not entail a loss of generality in the definition. For example, if the conditions of the definition of convergence are satisfied for all values of ε in the range $0 < \varepsilon < K$, where K is any prescribed positive number, then they are automatically satisfied also for larger values of ε. For, if $|a_n - a| < \varepsilon < K$ and if ε' denotes any number such that $\varepsilon' \geq K$, then $|a_n - a| < \varepsilon'$.

Further, consider the effect of replacing inequality (1) in the definition of convergence by the inequality

$$(2) \qquad |a_n - a| < k\varepsilon,$$

where k is some positive constant (that is, k does not depend on ε). Actually this does not alter the import of the definition. For, suppose ε_1 to be any positive number. We may then define $\varepsilon_2 = \varepsilon_1/k$ and look upon ε_2 as another value of ε. Since (2) is to be valid for each positive value of ε, there is some N such that $|a_n - a| < k\varepsilon_2$ or $|a_n - a| < \varepsilon_1$ for $n > N$. Hence, if (2) is true for each value of ε, then so is inequality (1).

A simple but important role of the concept of *each positive number* ε is apparent in the following theorem.

Theorem 5-1 If a constant c has the property that $|c| < \varepsilon$ for each positive number ε, then $c = 0$. (Expressed otherwise, the only non-negative real number which is less than all positive numbers is 0.)

Proof Suppose $c \neq 0$ and choose $\varepsilon = \frac{1}{2}|c|$. It follows that $|c| < \frac{1}{2}|c|$, which is absurd. Hence $c = 0$▲

Corollary If $|c| < k\varepsilon$ for some positive number k and for each positive number ε, then $c = 0$.

It is important in the foregoing theorem and corollary that c be a constant. It would not do, for example, to argue that $\varepsilon = 0$ because $|\varepsilon/2| < \varepsilon$ for each positive value of ε. Nor should the meaning of the inequality $|c| < \varepsilon$ be confused with that of $|a_n - a| < \varepsilon$ appearing in the definition of convergence. In the latter $a_n - a$ is not a constant; rather it depends upon n.

We employ the above to show that a sequence may not have more than one limit.

Theorem 5-2 If a sequence converges to a limit, then that limit is unique.

Proof Suppose a sequence $\{a_n\}$ to converge to two limits a and b. Then, corresponding to each $\varepsilon > 0$, there exist integers N_1 and N_2 such that $|a_n - a| < \varepsilon$ for all $n > N_1$ and $|a_n - b| < \varepsilon$ for all $n > N_2$. Clearly, then, if n_0 designates an integer which exceeds both N_1 and N_2, we have $|a_{n_0} - a| < \varepsilon$ and $|a_{n_0} - b| < \varepsilon$. It follows for the constant $a - b$ that $|a - b| \leq |a_{n_0} - a| + |a_{n_0} - b| < 2\varepsilon$. Since this is true for each positive value of ε, we conclude from the corollary to Theorem 5-1 that $a = b$. Hence $\{a_n\}$ converges to at most one limit▲

5-3 Examples of Convergent and Divergent Sequences

The next theorem shows a useful comparison test for convergence of a sequence.

Theorem 5-3 If $\{a_n\}$ is a null sequence and if $\{b_n\}$ is a sequence such that, for some integer N, $|b_n| \leq |a_n|$ for all $n > N$, then $\{b_n\}$ is a null sequence also.

Proof Corresponding to each positive value of ε there exists an integer N_1 such that $|a_n| < \varepsilon$, and hence $|b_n| < \varepsilon$ for all n exceeding both N and N_1. Therefore $\{b_n\}$ is a null sequence▲

To prove that a sequence $\{a_n\}$ does not possess a given number a as a limit we have only to demonstrate the existence of a value of ε to which no integer N corresponds. This may be accomplished by exhibiting a value of ε for which $|a_n - a| \geq \varepsilon$ for an infinite set of values of n. If, moreover, this can be done regardless of the value of a, then it follows that $\{a_n\}$ is divergent.

The following examples demonstrate in some simple cases the technique of proof of convergence and divergence. With regard to those examples illustrating proof of convergence, observe that the value of the limit is always declared at the outset. In a later section we shall see how to establish convergence of certain sequences without presupposing knowledge of the actual limit value.

Example 5-2 We show that $\{1/n\}$ is a null sequence.

Let ε designate any prescribed positive number, and consider the possibility of finding an integer N such that

(3) $$\left|\frac{1}{n} - 0\right| < \varepsilon$$

for all $n > N$. It is clear that whatever the value of ε, the number $1/\varepsilon$ is exceeded by an infinite set of integers. Denote any one of them by N. (For instance, N may be chosen to be the least integer greater than $1/\varepsilon$.) It is then obviously the case that for all values of n which exceed N, the inequality $n > 1/\varepsilon$ is true. Since it in turn implies $1/n < \varepsilon$, which is precisely inequality (3), it follows that (3) is true for all $n > N$. Consequently, the sequence $\{1/n\}$ converges to 0.

Example 5-3 The sequence $\{1/2^n\}$ is also a null sequence.

To show this we first observe that from the Bernoulli inequality $(1 + x)^n > 1 + nx$ it follows, by setting $x = 1$, that $2^n > 1 + n > n$. Hence $1/2^n < 1/n$. Since $\{1/n\}$ has been shown in Example 5-2 to be a null sequence, it follows immediately by Theorem 5-3 that $\{1/2^n\}$ is a null sequence also.

Example 5-4 The sequence $\left\{\frac{5n-4}{2-3n}\right\}$ converges to $-\frac{5}{3}$.

To see that this is the case, consider the inequality

(4) $$\left|\frac{5n-4}{2-3n} - \left(-\frac{5}{3}\right)\right| < \varepsilon.$$

Because n is positive this is easily shown to be exactly equivalent to (i.e., implies and is implied by) the inequality

(5) $$n > \frac{1}{3}\left(\frac{2}{3\varepsilon} + 2\right).$$

For a given $\varepsilon > 0$, choose N to be any integer greater than the number $\frac{1}{3}\left(\frac{2}{3\varepsilon} + 2\right)$. Then for all $n > N$ inequality (5) is obviously true. Therefore (4) is also true for such n. It follows that the given sequence converges to $-\frac{5}{3}$.

Example 5-5 For $a > 1$, $\{n/a^n\}$ is a null sequence.

We observe that from $a > 1$ we have $a = 1 + b$, where $b > 0$. By the binomial theorem

$$a^n = (1 + b)^n = 1 + nb + \frac{n(n-1)}{2}b^2 + \cdots + b^n;$$

and, since each term on the right is positive, it follows for all $n \geq 2$ that $a^n > \frac{n(n-1)}{2}b^2$ or, what is the same, $\frac{n}{a^n} < \frac{2}{(n-1)b^2}$. Now, corresponding to a given $\varepsilon > 0$, let N be an integer exceeding $1 + 2/b^2\varepsilon$. Then, for all $n > N$, it follows that $n > 1 + 2/b^2\varepsilon$ or $2/(n-1)b^2 < \varepsilon$. Consequently, $n/a^n < \varepsilon$ for all $n > N$, proving that $\{n/a^n\}$ converges to 0.

Example 5-6 The sequence $\{3^n\}$ diverges.

If a is any given number and ε any positive number, then an integer N exists such that $3^n \geq a + \varepsilon$ for all $n > N$ (see Exercise 1-33). Hence $|3^n - a| < \varepsilon$ is valid for at most a finite set of values of n. Thus, the given sequence does not converge to a, whatever the value of a.

Example 5-7 The sequence $\{(-1)^n\}$ diverges. To see this, suppose the sequence to converge to a limit a. Then a number N exists such that for all even values of n which exceed N we have $|1 - a| < \frac{1}{4}$; and for all odd n greater than N, $|-1 - a| < \frac{1}{4}$. These imply, respectively, that $\frac{3}{4} < a < \frac{5}{4}$ and $-\frac{5}{4} < a < -\frac{3}{4}$. Since no number a exists with both these properties, we have a contradiction and conclude that $\{(-1)^n\}$ does not converge.

Example 5-8 The sequence $\{a\}$ converges to a.

This is an immediate consequence of the definition of convergence since, for any positive ε and for all values of n, we have $|a_n - a| = |a - a| = 0 < \varepsilon$.

Exercises

5-1 Let $\{a_n\}$ and $\{b_n\}$ be two sequences with the property that integers n_o and m exist such that $a_{n+m} = b_n$ for all $n \geq n_o$. Prove that both sequences converge to the same limit or both diverge.

5-2 Prove that if $a < b + \varepsilon$ for each positive number ε, then $a \leq b$. Cite an example to show that the equality may hold.

5-3 Use the definition of convergence to prove that $\left\{\dfrac{2 - 7n}{1 - 5n}\right\}$ has the limit $\dfrac{7}{5}$.

5-4 Use the definition of convergence to prove that $\left\{\dfrac{3n}{1 + 2n^2}\right\}$ is a null sequence.

5-5 Prove that $\{\sqrt{n}\}$ diverges.

5-6 Prove that $\{a^n\}$ converges to 0 if $|a| < 1$, converges to 1 if $a = 1$, and diverges if $a = -1$ or if $|a| > 1$.

5-7 Prove that if a sequence converges and if some real number appears infinitely often as an element of the sequence, then that number is the limit of the sequence.

5-8 Let $\{a_n\}$ and $\{b_n\}$ converge to a and b, respectively, and assume that $a_n \leq b_n$ for all n. Prove that $a \leq b$.

5-9 Prove that if $\{a_n{}^2\}$ is a null sequence, then so is $\{a_n\}$.

5-10 Prove that $\{na^n\}$ is a null sequence if $|a| < 1$ and diverges if $|a| \geq 1$.

5-11 Prove that $\{n^2/2^n\}$ is a null sequence.

5-12 Show that $\{\sqrt{n^2 + 1} - n\}$ is a null sequence.

5-13 Show that $\{\sqrt{n + 1} - \sqrt{n}\}$ is a null sequence.

5-14 A sequence $\{b_n\}$ is called a *rearrangement* of $\{a_n\}$ if a sequence $\{m_n\}$ exists in which each positive integer appears exactly once and if $\{m_n\}$ is such that $b_{m_n} = a_n$. Prove that $\{a_n\}$ has limit a if and only if any rearrangement has limit a.

5-4 Bounded Sequences

A sequence $\{a_n\}$ is said to be *bounded* if its range is bounded, that is, if a number M exists such that $|a_n| \leq M$ for all values of n. Similarly, it is *bounded above* or *bounded below* if its range is bounded above or bounded below.

Theorem 5-4 If a sequence is convergent, it is bounded.

Proof Let $\{a_n\}$ be convergent with limit a. Then an integer N exists such that $|a_n - a| < 1$, and hence $|a_n| < |a| + 1$ for all $n > N$. Denote by M the maximum of the set of numbers $|a_1|, |a_2|, \ldots, |a_N|, |a| + 1$. Then $|a_n| \leq M$ for all values of n. Thus $\{a_n\}$ is bounded▲

It is easily noted that a bounded sequence is not necessarily convergent. Witness the bounded but divergent sequence $\{(-1)^n\}$.

A variation of the notion of boundedness is treated in the following theorem.

Theorem 5-5 If a sequence $\{a_n\}$ converges to a nonzero limit a, then a number $k > 0$ and an integer N exist such that $0 < k < |a_n|$ for all $n > N$.

Proof Let $k = \frac{1}{2}|a|$. Then $k > 0$ since $a \neq 0$. Because of the convergence of $\{a_n\}$ to a, a number N exists such that $|a_n - a| < k$ for all $n > N$. Hence $|a| - |a_n| < k$ or, since $|a| = 2k$, $k < |a_n|$ for all $n > N$▲

Theorem 5-5 may be paraphrased as follows. If a sequence $\{a_n\}$ converges to a nonzero limit a, then a neighborhood of 0 exists which contains a_n for at most a finite set of values of n. We say that the sequence is *bounded away from zero*. This principle may be generalized: If $\{a_n\}$ converges to a and if b is any number different from a, then $\{a_n\}$ is bounded away from b. The reader should supply the simple proof.

5-5 Operations with Sequences

Let $\{a_n\}$ and $\{b_n\}$ be any two sequences. From their elements we may construct yet other sequences such as $\{a_n + b_n\}$, $\{a_n - b_n\}$, $\{a_n b_n\}$, and $\{a_n/b_n\}$, to be known as their *sum*, *difference*, *product*, and *quotient*, respectively. In the case of the quotient it is necessary, of course, to stipulate that 0 not appear as an element in the sequence $\{b_n\}$. We may consider, too, such sequences as $\{ka_n\}$, where k is any constant, as well as the sequence $\{|a_n|\}$ whose elements are the absolute values of the elements of $\{a_n\}$. The question of convergence or divergence of these sequences arises. The following theorems provide a partial answer.

Theorem 5-6 If $\{a_n\}$ and $\{b_n\}$ converge to a and b, respectively, then $\{a_n + b_n\}$ converges to $a + b$.

Proof Let $\varepsilon > 0$ be given. As in the proof of Theorem 5-2, there exists an integer N such that $|a_n - a| < \varepsilon$ and $|b_n - b| < \varepsilon$ for all $n > N$. Hence

$$|(a_n + b_n) - (a + b)| \leq |a_n - a| + |b_n - b| < 2\varepsilon$$

for all $n > N$, and it follows that $\{a_n + b_n\}$ converges to $a + b$▲

The converse of this theorem is not true, as may be seen from the following counterexample. The sequences $\{(-1)^n\}$ and $\{(-1)^{n+1}\}$ both diverge. Their sum, however, is the convergent sequence $\{0\}$.

Theorem 5-7 If $\{a_n\}$ and $\{b_n\}$ converge to a and b, respectively, then the sequence $\{a_n b_n\}$ converges to ab.

Proof We may write

$$\begin{aligned} |a_n b_n - ab| &= |a_n b_n - a_n b + a_n b - ab| \\ &= |a_n(b_n - b) + b(a_n - a)| \\ &\leq |a_n|\,|b_n - b| + |b|\,|a_n - a|\,. \end{aligned}$$

Now, let $\varepsilon > 0$ be given. Then an integer N exists such that $|a_n - a| < \varepsilon$ and $|b_n - b| < \varepsilon$ for all $n > N$. From the convergence of $\{a_n\}$ it also follows that M exists such that $|a_n| \leq M$ for all n. Consequently

$$|a_n b_n - ab| < M\varepsilon + |b|\,\varepsilon = (M + |b|)\,\varepsilon$$

for all $n > N$. Since $M + |b|$ is a constant, we conclude that $\{a_n b_n\}$ has the limit ab ▲

Again the converse is false, because although $\{(-1)^n\}$ diverges, the sequence $\{(-1)^n(-1)^n\} = \{1\}$ converges to the limit 1.[2]

Corollary If $\{b_n\}$ converges to b, then for any constant k, the sequence $\{kb_n\}$ converges to kb.

Proof This follows from the theorem inasmuch as $\{kb_n\}$ may be regarded as the product of the convergent sequences $\{k\}$ and $\{b_n\}$ ▲

If the notion of limit of a sequence be construed in an operational sense in which "lim" is viewed as an *operator* acting upon a sequence $\{a_n\}$, then it is clear from Theorem 5-6 and the corollary to Theorem 5-7 that lim is a *linear* operator; thus

$$\lim \{k_1 a_n + k_2 b_n\} = k_1 \lim \{a_n\} + k_2 \lim \{b_n\},$$

for any constants k_1 and k_2 and any convergent sequences $\{a_n\}$ and $\{b_n\}$.

A simple consequence of linearity is that $\lim \{a_n - b_n\} = a - b$ if $\{a_n\}$ and $\{b_n\}$ converge to a and b, respectively. Further, it becomes clear that convergence of $\{a_n\}$ to a and of $\{a_n - a\}$ to 0 subsist together. From linearity follows $\lim \{a_n - a\} = \lim \{a_n\} - a = 0$ if $\lim \{a_n\} = a$, while $\lim \{a_n\} = \lim \{(a_n - a) + a\} = a$ if $\{a_n - a\}$ is a null sequence.

A further consequence of Theorem 5-7 should be noted, namely, that if $\{a_n\}$

[2] Two sequences $\{a_n\}$ and $\{b_n\}$ are understood to be *equal*, that is, $\{a_n\} = \{b_n\}$, if $a_n = b_n$ for all values of n.

converges to a and if m is any positive integer, then $\{a_n{}^m\}$ converges to a^m. The reader should supply a simple proof by induction.

Theorem 5-8 If $\{a_n\}$ and $\{b_n\}$ converge to a and b, respectively, and if $b_n \neq 0$ for all n, and if $b \neq 0$, then $\{a_n/b_n\}$ converges to a/b.

Proof Since a_n/b_n may be written as a product $(a_n)(1/b_n)$, it suffices, by virtue of Theorem 5-7, to prove that $\{1/b_n\}$ converges to $1/b$.

Since $b \neq 0$, the sequence $\{b_n\}$ is bounded away from zero. That is, a number $k > 0$ and an integer N_1 exist such that $k < |b_n|$ for all $n > N_1$.

Let $\varepsilon > 0$ be given. From the convergence of $\{b_n\}$ to b then follows the existence of an integer N_2 such that $|b_n - b| < \varepsilon$ for all $n > N_2$. Denoting by N the maximum of the numbers N_1 and N_2, it is then true that $k < |b_n|$ and $|b_n - b| < \varepsilon$ for all $n > N$. Therefore

$$\left|\frac{1}{b_n} - \frac{1}{b}\right| = \frac{|b - b_n|}{|b_n|\,|b|} < \frac{\varepsilon}{k\,|b|}$$

for all $n > N$. Since $1/k\,|b|$ is a positive constant, it follows that $\{1/b_n\}$ converges to $1/b$▲

The reader will have no difficulty in providing a counterexample to the converse of Theorem 5-8. It is also interesting to note that the condition $b \neq 0$ is not necessary to the convergence of $\{a_n/b_n\}$ since this may converge if $\{a_n\}$ as well as $\{b_n\}$ is a null sequence. For example, let $\{a_n\} = \{b_n\} = \{1/n\}$. Then $\{a_n/b_n\}$ is the convergent sequence $\{1\}$.

Theorem 5-9 If $\{a_n\}$ converges to a, then $\{|a_n|\}$ converges to $|a|$.

Proof With $\varepsilon > 0$ given, an integer N exists such that $|a_n - a| < \varepsilon$ for all $n > N$. From $|\,|a_n| - |a|\,| \leq |a_n - a|$ then follows the convergence of $\{|a_n|\}$ to $|a|$▲

Theorem 5-10 If $\{|a_n|\}$ is a null sequence, then $\{a_n\}$ is also.

Proof Immediate from $|a_n| = |\,|a_n|\,|$▲

Observe that if $\{|a_n|\}$ converges to a nonzero limit, then $\{a_n\}$ need not converge. For example, while $\{(-1)^n\}$diverges, the sequence$\{|(-1)^n|\} = \{1\}$ converges.

Exercises

5-15 Prove that if $\{a_n\}$ converges to a and if $b \neq a$, then $\{a_n\}$ is bounded away from b.

5-16 Prove that if the difference of two sequences is a null sequence, then they both converge to the same limit or both diverge.

5-17 Cite an example to show that $\{a_n b_n\}$ may be a null sequence even though neither $\{a_n\}$ nor $\{b_n\}$ is a null sequence.

5-18 Prove by induction that if $\{a_n\}$ converges to a and if m is any positive integer. then $\{a_n{}^m\}$ converges to a^m.

5-19 Provide an example to show that the converse of Theorem 5-8 is false.

5-20 Is it possible for sequences $\{a_n + b_n\}$ and $\{a_n\}$ to converge and yet for $\{b_n\}$ to diverge? Support your answer by an argument or an example.

5-21 Can $\{a_n b_n\}$ and $\{a_n\}$ converge if $\{b_n\}$ diverges? Support your answer by an argument or an example.

5-22 Prove that the hypothesis in Theorem 5-8 is not necessary to the conclusion by considering the sequences $\{1/n^2\}$ and $\{1/(n+1)\}$.

5-23 Find the limits of the following sequences with the aid of Theorems 5-6, 5-7, and 5-8: (a) $\left\{\dfrac{2n+3n^2}{1+n^2}\right\}$; (b) $\left\{\dfrac{n+3}{2^n}\right\}$; (c) $\left\{\dfrac{n^2}{4^n}\right\}$.

5-24 Let $\{a_n\}$ be a null sequence and $\{b_n\}$ a bounded sequence. Prove that $\{a_n b_n\}$ is a null sequence.

5-25 Determine the limit of the sequence $\left\{\dfrac{1}{n^2} + \dfrac{2}{n^2} + \cdots + \dfrac{n}{n^2}\right\}$.

5-26 Determine the limit of the sequence $\left\{\dfrac{\left(\dfrac{1}{n}\right)^2 + \left(\dfrac{2}{n}\right)^2 + \cdots + \left(\dfrac{n}{n}\right)^2}{n}\right\}$.

5-27 Investigate possible convergence of $\left\{\dfrac{1}{\sqrt{n^2+1}} + \cdots + \dfrac{1}{\sqrt{n^2+n}}\right\}$.

5-28 Let a_1, b, and c be any positive real numbers, and define $a_{n+1} = \sqrt{2ba_n + c}$ for all $n > 1$. Prove that $\{a_n\}$ converges to $b + \sqrt{b^2 + c}$.

5-29 Determine a positive root of the equation $x^3 + 2x - 1 = 0$ by considering the sequence $\{x_n\}$ defined by $x_1 = 1$ and $x_{n+1} = \dfrac{1}{2 + x_n^{\,2}}$ for $n > 1$.

5-6 Further Examples

Example 5-9 In Chapter 3, in connection with the construction of the system of real numbers, the existence of a number a with the property $a^2 = 2$ was demonstrated. Omitted, however, was a general proof that if b is any positive number and m any nonzero integer, then a unique positive number a exists such that $a^m = b$. At this point we interject a proof of this fact, employing some of the properties of sequences of real numbers established in the foregoing sections. Since $a^{-m} = (1/a)^m$, it is sufficient to restrict attention to positive values of m.

Let b be any given positive real number and m a given positive integer, and consider the set of real numbers $B = \{x;\ \ x \geq 0,\ x^m \leq b\}$. Since $0^m < b$, it is clear that B is not empty. Also, if $c \in B$ and $0 \leq a < c$, then $a^m < c^m$ so that $a \in B$. The set B is bounded above, for if $b \leq 1$, then 1 is an upper bound, and if $b > 1$, then b itself is an upper bound. Hence the supremum of B exists; we denote it by a. We shall show that $a^m = b$.

Consider the sequence $\{a_n\}$ defined by $a_n = a + 1/n$ for $n = 1, 2, \ldots$. It is clear that $a_n > a$ for all n. Since $\{1/n\}$ is a null sequence, it follows that $\{a_n\}$ has the limit a and hence that the sequence $\{a_n{}^m\}$ converges to a^m.

Suppose, now, that $a^m < b$ so that $b - a^m$ is positive. Then, because $\{a_n{}^m\}$ converges to a^m, an integer n_1 exists such that $|a_{n_1}{}^m - a^m| < b - a^m$. This implies that $a_{n_1}{}^m - a^m < b - a^m$, whence $a_{n_1}{}^m < b$. Therefore $a_{n_1} \in B$, and, since a is an upper bound of the set B, it is then the case that $a_{n_1} \leq a$. This, however, contradicts the observation above that $a_n > a$ for all n. As a consequence, the supposition $a^m < b$ is denied, and we conclude that $a^m \geq b$.

Consider now the sequence $\{c_n\}$, where $c_n = a - 1/n$. We see that $c_n < a$ and that $\{c_n\}$ converges to a. Hence $\{c_n{}^m\}$ has the limit a^m. Assume that $a^m > b$. Then an integer n_2 exists such that $|a^m - c_{n_2}{}^m| < a^m - b$. It follows that $b < c_{n_2}{}^m$. Thus $c_{n_2} \notin B$ and, since a is the least of the upper bounds of B, we have $a \leq c_{n_2}$. This is a contradiction (to $c_n < a$ for all n) denying the assumption $a^m > b$. We conclude that $a^m \not> b$ and, as a consequence, $a^m = b$. We write, of course, $a = \sqrt[m]{b}$. The unique character of the positive number a follows from Exercise 3-44.

Example 5-10 We show that the sequence $\{\sqrt[n]{a}\}$, where $a > 0$, converges to the limit 1.

If $a > 1$, then $\sqrt[n]{a} > 1$; so we may write $\sqrt[n]{a} = 1 + b_n$, where $b_n > 0$ for each n. It follows that $a = (1 + b_n)^n > 1 + nb_n$, whence $b_n < (a - 1)/n$. Since $\{(a - 1)/n\}$ is a null sequence, the sequence $\{b_n\}$ is likewise, and hence $\{\sqrt[n]{a}\}$ converges to 1.

If $a < 1$, let $c = 1/a$. Then $\sqrt[n]{a} = 1/\sqrt[n]{c}$ and $\{\sqrt[n]{c}\}$ converges to 1 since $c > 1$. By Theorem 5-8 we conclude that $\{\sqrt[n]{a}\}$ converges to 1.

The case $a = 1$ is trivial.

Example 5-11 The sequence $\{\sqrt[n]{n}\}$ converges to the limit 1.

For $n \geq 2$ we have $\sqrt[n]{n} = 1 + b_n$, where $b_n > 0$. Hence

$$n = (1 + b_n)^n = 1 + nb_n + \frac{n(n-1)b_n{}^2}{2} + \cdots + b_n{}^n;$$

and, since each term on the right is positive, we see that $n > \dfrac{n(n-1)b_n{}^2}{2}$ or, what is the same, $b_n{}^2 < 2/(n - 1)$. Since $\{2/(n-1)\}$ is a null sequence, so is $\{b_n{}^2\}$, and, consequently (by Exercise 5–9), $\{b_n\}$ converges to 0. Therefore $\{\sqrt[n]{n}\}$ has the limit 1.

5-7 Monotone Sequences

A sequence $\{a_n\}$ is said to be *increasing* if $a_n \leq a_{n+1}$ for all values of n and *decreasing* if $a_n \geq a_{n+1}$. If, for all n, $a_n < a_{n+1}$, it is *strictly increasing*, and if $a_n > a_{n+1}$, *strictly decreasing*. A sequence falling into any of these four categories is said to be *monotone*. Because of the following theorem, monotone sequences assume a certain importance.

Theorem 5-11 Every bounded monotone sequence is convergent.

Proof Let $\{a_n\}$ denote a bounded increasing sequence. Since it is bounded, its range has a supremum a, and, as a consequence, corresponding to a given positive ε, an element a_N of the sequence exists such that $a - \varepsilon < a_N \leq a$. From the monotone character of $\{a_n\}$ it then follows that $a_N \leq a_n \leq a$ for all $n > N$. Hence $|a_n - a| < \varepsilon$ for all $n > N$, demonstrating the convergence of $\{a_n\}$ to the limit a. Similar reasoning proves the theorem for monotone decreasing sequences▲

Since, as we have already shown, every convergent sequence is bounded, we may conclude that a monotone sequence is convergent if and only if it is bounded. Many of the examples considered in foregoing sections are monotone sequences. Thus $\{1/n\}$ is a monotone strictly decreasing null sequence while $\{\sqrt[n]{a}\}$ $(a < 1)$ is monotone increasing with limit 1. The sequence $\{a\}$, which converges to a, falls into the two categories of increasing and decreasing sequences. A further example involving monotone sequences follows.

Example 5-12 Let $\{s_n\}$ be an increasing sequence and $\{t_n\}$ a decreasing sequence such that $s_n \leq t_n$ for all values of n. We may show that both sequences converge and that their respective limits s and t satisfy the relation $s \leq t$.

From the chain of relations $s_1 \leq s_n \leq t_n \leq t_1$, it follows that both $\{s_n\}$ and $\{t_n\}$ are bounded; hence both are convergent. Assume for their respective limits s and t that $t < s$. Then a number u exists such that $t < u < s$. Because $\{s_n\}$ converges to s and $\{t_n\}$ to t, an integer n_1 exists such that $u < s_{n_1} < s$ and $t < t_{n_1} < u$. Hence $t_{n_1} < s_{n_1}$, which is a contradiction. Therefore $s \leq t$.

Example 5-13 Illustrative of the principle established in Example 5-12 are the sequences $\{s_n\}$ and $\{t_n\}$, defined for $n > 1$ by

$$s_n = \left(1 + \frac{1}{n}\right)^n \quad \text{and} \quad t_n = \left(1 + \frac{1}{n}\right)^{n+1}.$$

It is immediate that $s_n < t_n$ for all n. We next show the monotone character of each of the given sequences. Consider the ratio

$$\frac{s_n}{s_{n-1}} = \left(\frac{1 + \dfrac{1}{n}}{1 + \dfrac{1}{n-1}}\right)^n \left(1 + \frac{1}{n-1}\right) = \left(1 - \frac{1}{n^2}\right)^n \frac{n}{n-1}.$$

By the Bernoulli inequality we obtain

$$\left(1 - \frac{1}{n^2}\right)^n > 1 - \frac{n}{n^2} = \frac{n-1}{n}.$$

It follows that $s_{n-1} < s_n$ so that $\{s_n\}$ is an increasing sequence.

For the sequence $\{t_n\}$ we consider the ratio

$$\frac{t_{n-1}}{t_n} = \left(\frac{1 + \dfrac{1}{n-1}}{1 + \dfrac{1}{n}}\right)^n \left(1 + \frac{1}{n}\right)^{-1} = \left(1 + \frac{1}{n^2 - 1}\right)^n \frac{n}{n+1}.$$

By Bernoulli's inequality we have

$$\left(1 + \frac{1}{n^2 - 1}\right)^n > 1 + \frac{n}{n^2 - 1} > 1 + \frac{1}{n} = \frac{n+1}{n}.$$

This implies that $\{t_n\}$ is a decreasing sequence.

Since the conditions of Example 5-12 are satisfied, it follows that $\{s_n\}$ and $\{t_n\}$ both converge. Moreover,

$$0 \le t_n - s_n = \left(1 + \frac{1}{n}\right)^n \frac{1}{n} \le \frac{4}{n}.$$

Consequently $\{t_n - s_n\}$ is a null sequence so that $\lim \{t_n\} = \lim \{s_n\}$. We shall not demonstrate here that, in fact, the common limit is the familiar number $e = 2.7182818\ldots$.

5-8 Subsequences

If some of the elements of a sequence are deleted, what remains constitutes a subsequence. Formally we define this concept as follows. Let $\{a_n\}$ be a given sequence and $\{i_n\}$ a strictly increasing sequence of positive integers. For each positive integer n, define b_n by $b_n = a_{i_n}$. Then the sequence $\{b_n\}$ is called a *subsequence* of the sequence $\{a_n\}$.

Example 5-14

(a) The sequence $\{1/n\}$ possesses subsequences $\{1/2n\}$, $\{1/(5n+1)\}$, $\{1/n^2\}$, and $\{1/n!\}$.

(b) The sequence $\{(-1)^n\}$ has $\{1\}$ and $\{-1\}$ as subsequences.

(c) The sequence of prime numbers 2, 3, 5, 7, 11, 13, ... is a subsequence of the sequence $\{n\}$ of all positive integers.

(d) The sequence $\{b_n\} = \{a_{n^2}\}$ is a subsequence of $\{a_n\}$.

(e) Any sequence is a subsequence of itself.

Theorem 5-12 If $\{a_n\}$ converges to a limit a and if $\{a_{i_n}\}$ is any subsequence of $\{a_n\}$, then $\{a_{i_n}\}$ likewise converges to a.

Proof Corresponding to a given positive ε, an integer N exists such that $|a_n - a| < \varepsilon$ for all $n > N$. Since the elements a_{i_n} of $\{a_{i_n}\}$ are all elements of $\{a_n\}$, it follows that $|a_{i_n} - a| < \varepsilon$ for all integers i_n which exceed N. Thus $\{a_{i_n}\}$ converges to a▲

The converse of this theorem is easily seen to be untrue. We may again call on $\{(-1)^n\}$; although divergent, it clearly possesses the convergent subsequence $\{1\}$

Theorem 5-13 Every bounded sequence possesses a convergent subsequence.

Proof Let $\{a_n\}$ be a bounded sequence, and consider two possibilities. First, suppose that for an infinite set of values of n it is the case that $a_n = a$ (that is, the number a appears infinitely often as an element of the sequence). Then $\{a\}$ is a convergent subsequence of $\{a_n\}$. Second, suppose that no one real number occurs as an element of $\{a_n\}$ infinitely often. It necessarily follows that the range of $\{a_n\}$ is an infinite set. Because it is bounded, a point a exists which is a cluster point. Consequently, in each deleted neighborhood of the form $\mathcal{N}^*(a,1/m)$, where m is a positive integer, there exists an infinite set of elements of $\{a_n\}$. Corresponding to each value of m, choose an element of the sequence in $\mathcal{N}^*(a,1/m)$ and denote it by a_{n_m}. It is clear that this may be done in such a way that $n_m < n_{m+1}$. The subsequence $\{a_{n_m}\}$ is then seen to converge to a▲

Example 5-15

(a) The sequence $\{a_n\}$ is bounded and divergent, where $a_n = 1/n$ if n is even and $a_n = 1$ if n is odd; that is, the sequence $1, \frac{1}{2}, 1, \frac{1}{4}, \ldots$. The subsequence $\{a_{2n}\} = \{1/2n\}$ converges to 0, while $\{a_{2n+1}\}$ has limit 1.

(b) A sequence need not be bounded to possess a convergent subsequence. Thus the sequence $1, \frac{1}{2}, 2, \frac{1}{3}, 3, \frac{1}{4}, \ldots$ is not bounded, but the subsequence $\{1/n\}$ converges to 0.

(c) The sequence $\{n\}$ is unbounded and possesses only unbounded, and hence divergent, subsequences.

(d) Let the set of all rational numbers be written in a sequence $\{r_n\}$. Then, because every real number is a cluster point of the set of rational numbers, it follows that given any real number a, a subsequence of $\{r_n\}$ exists which converges to a.

Exercises

5-30 Prove that if $\{a_n\}$ is monotone increasing with limit a, then $a_n \leq a$ for all n.

5-31 Show that the sequence $\left\{\dfrac{1}{n+1} + \dfrac{1}{n+2} + \cdots + \dfrac{1}{2n}\right\}$ is convergent.

5-32 Prove that the sequence $\{a_n\}$ defined by $a_1 = 4$ and $a_{n+1} = 3 - 2/a_n$ converges to 2.

5-33 Prove that if a sequence is bounded and divergent, then two subsequences exist converging to different limits.

5-9 Cauchy Convergence

In order to prove the convergence of a sequence directly from the definition, it is evident that we must guess at or have some previous indication of the value of the limit. If such knowledge is not forthcoming, the definition of convergence, of itself, is awkward as a mechanism of proof. To obviate this situation we introduce a criterion which is equivalent to the definition of convergence but which does not make reference to the limit value.

A sequence $\{a_n\}$ is said to be *Cauchy convergent* if, corresponding to each positive number ε, there exists an integer N such that $|a_{n+k} - a_n| < \varepsilon$ for all $n > N$ and all $k > 0$ (or, equivalently, $|a_p - a_q| < \varepsilon$ for all $p > N$ and all $q > N$).

To the end of demonstrating the equivalence of the concepts of *convergence* and *Cauchy convergence*, we first prove the following property.

Theorem 5-14 Every Cauchy convergent sequence is bounded.

Proof Let $\{a_n\}$ be Cauchy convergent. Then an integer N exists such that

$$|a_{n+k} - a_n| < 1 \tag{6}$$

for all $n > N$ and all $k > 0$. Denote by M_1 the maximum of the finite set of numbers $|a_1|, |a_2|, \ldots, |a_{N+1}|$. Then $|a_n| \leq M_1$ for all n such that $1 \leq n \leq N + 1$. Since (6) is valid for all $n > N$, it is true, in particular, for $n = N + 1$. Thus $|a_{N+1+k} - a_{N+1}| < 1$ for all $k > 0$, implying that $|a_{N+1+k}| < |a_{N+1}| + 1 \leq M_1 + 1$. Thus $|a_n| \leq M_1 + 1$ for all $n > N + 1$. With $M = M_1 + 1$ it follows that $|a_n| \leq M$ for all values of n▲

The next two theorems show that convergence and Cauchy convergence are different expressions of one and the same property of sequences.

Theorem 5-15 If a sequence is convergent, it is Cauchy convergent.

Proof If $\{a_n\}$ is convergent to a limit a, then, corresponding to each $\varepsilon > 0$, there exists an integer N such that $|a_n - a| < \varepsilon$ for all $n > N$. If k is any positive integer and if $n > N$, then $n + k > N$, and it follows that $|a_{n+k} - a| < \varepsilon$. Consequently, for $n > N$ and $k > 0$ we have $|a_{n+k} - a_n| \leq |a_{n+k} - a| + |a_n - a| < 2\varepsilon$. Hence $\{a_n\}$ is Cauchy convergent▲

Theorem 5-16 If a sequence is Cauchy convergent, it is convergent.

Proof If $\{a_n\}$ is Cauchy convergent, it is bounded. As a consequence, a subsequence $\{a_{i_n}\}$ exists which converges to a limit a. Thus, corresponding to a given $\varepsilon > 0$, an integer N_1 exists such that

$$|a_{i_n} - a| < \varepsilon \tag{7}$$

for all $i_n > N_1$. Directly from the Cauchy convergence of $\{a_n\}$ it also follows that N_2 exists such that

$$|a_{n+k} - a_n| < \varepsilon \tag{8}$$

for all $n > N_2$ and all $k > 0$. Now, let N denote a subscript of an element of $\{a_{i_n}\}$ with the property $N > N_1$ and $N > N_2$. Then, by (7) and (8),

$$|a_N - a| < \varepsilon \quad \text{and} \quad |a_{N+k} - a_N| < \varepsilon.$$

Combining these yields $|a_{N+k} - a| < 2\varepsilon$ for all $k > 0$ or, what is the same, $|a_n - a| < 2\varepsilon$ for all $n > N$. We conclude that $\{a_n\}$ converges to a▲

We include now some examples demonstrating how convergence or divergence of a sequence may be established by showing that the sequence is or is not Cauchy convergent.

Example 5-16 Consider the sequence $\{a_n\}$ defined by

$$a_0 = 0, \quad a_1 = 1, \quad a_n = \frac{a_{n-1} + a_{n-2}}{2} \text{ for } n \geq 2.$$

The first several terms of the sequence are

$$0, 1, \frac{1}{2}, \frac{3}{4}, \frac{5}{8}, \frac{11}{16}, \frac{21}{32}, \frac{43}{64}, \ldots,$$

from which it is not immediately apparent what is the limit of the sequence or if, indeed, it is convergent at all.

The reader may show by a simple induction proof that for $n \geq 0$

$$a_{n+1} - a_n = \left(-\frac{1}{2}\right)^n, \tag{9}$$

from which it follows that

$$|a_{n+1} - a_n| = \frac{1}{2^n}.$$

It is clear from the definition of the sequence that a_{n+k}, for any n and for all $k > 1$, lies between the elements a_{n+1} and a_n. (This is also easily proved formally by induction.) As a consequence, $|a_{n+k} - a_n| \leq 1/2^n$ for all $n \geq 0$ and all $k > 0$. Because the sequence $\{1/2^n\}$ is a null sequence it then follows that $\{a_n\}$ is Cauchy convergent and hence convergent. It is left to the reader to discover the limit.

Example 5-17 Consider the sequence $\{s_n\}$ defined, for $n \geq 1$, by $s_n = 1 + 1/2 + 1/3 + \cdots + 1/n$. Observe that

$$s_{2n} - s_n = \frac{1}{n+1} + \frac{1}{n+2} + \cdots + \frac{1}{2n}$$

$$\geq \frac{1}{2n} + \frac{1}{2n} + \cdots + \frac{1}{2n} = n\left(\frac{1}{2n}\right).$$

Therefore $s_{2n} - s_n \geq \frac{1}{2}$, and it follows that $|s_{n+k} - s_n| \geq \frac{1}{2}$ for $k = n$ and all $n \geq 1$. Thus the given sequence is not Cauchy convergent, and therefore it diverges.

Example 5-18 A simple example shows the force of the condition "all $k > 0$" in the definition of Cauchy convergence. Consider the divergent sequence $\{a_n\} = \{\sqrt{n}\}$. We have

$$|a_{n+k} - a_n| = \sqrt{n+k} - \sqrt{n} = \frac{k}{\sqrt{n+k} + \sqrt{n}} < \frac{k}{2\sqrt{n}}.$$

Corresponding to given positive values of ε and k there exists an integer N such that $k/2\sqrt{n} < \varepsilon$ for all $n > N$ and hence $|a_{n+k} - a_n| < \varepsilon$ for all $n > N$. This falls short of the condition for Cauchy convergence in that a fixed value of k is assumed instead of "all $k > 0$." If k is not assigned a particular value, then the situation is this. No integer N exists such that $k/2\sqrt{n} < \varepsilon$ for all $n > N$ and all $k > 0$. Expressed otherwise, if k is "any positive integer," the quantity $k/2\sqrt{n}$ cannot be made arbitrarily small by the simple choice of a "sufficiently large value of n."

Exercises

5-34 Establish equation (9) by induction.

5-35 For Example 5-16, establish by induction that a_{n+k} lies between a_{n+1} and a_n for all n and all $k > 0$.

5-36 Find the limit of the sequence defined in Example 5-16.

5-37 Let a sequence $\{a_n\}$ be defined by $a_1 = 1$ and $a_n = a_{n-1} + 1/a_{n-1}$ for $n > 1$. Prove that $\{a_n\}$ diverges.

5-38 Show that the sequence $\{a_n\}$ where $a_0 = 0$, $a_1 = 1$, and $a_n = \frac{1}{3}(a_{n-1} + 2a_{n-2})$ for $n > 1$ is Cauchy convergent. What is its limit?

5-39 Prove divergence of the sequence $\{n^2/(2n+1)\}$ by showing it is not Cauchy convergent.

5-40 Prove that if $\{a_n\}$ converges, then $\{a_{n+1} - a_n\}$ is a null sequence. Cite an example to show that the converse is not true.

5-41 Prove by induction that $\frac{n}{2} < \sum_{k=1}^{2^n-1} \frac{1}{k}$. Use this fact to prove that the sequence $\left\{\sum_1^n \frac{1}{k}\right\}$ diverges.

5-10 Limiting Point

A number is called a *limiting point* of a sequence $\{a_n\}$ if it is the limit of a subsequence of $\{a_n\}$. Obviously, not all sequences possess limiting points; for example, the divergent and unbounded sequence $\{n\}$. If a sequence is bounded, then, by virtue of Theorem 5-13, it possesses at least one limiting point. It is possible, too, for a sequence to be unbounded and yet possess limiting points. Concerning the set of limiting points of a bounded sequence we have the following theorems.

Theorem 5-17 If $\{a_n\}$ is such that $m \le a_n \le M$ for all n and if b is a limiting point of $\{a_n\}$, then $m \le b \le M$.

Proof Let $\{a_{i_n}\}$ denote a subsequence of $\{a_n\}$ that converges to b. Since $m \le a_{i_n} \le M$, it follows from Exercise 5-8 that $m \le b \le M$▲

Theorem 5-18 The set of limiting points of any bounded sequence is closed.

Proof Let b denote a cluster point of the set B of limiting points of a sequence $\{a_n\}$. Then, corresponding to each positive integer m, an infinite set of points of B is contained in the neighborhood $\mathcal{N}(b,1/m)$. Corresponding to each m, choose one such point and denote it by b_m; thus

$$|b_m - b| < \frac{1}{m}. \tag{10}$$

(In effect the points b_m define a sequence $\{b_m\}$ converging to b as a limit.) The point b_1, since it is a point of B, is the limit of a subsequence of $\{a_n\}$. As a consequence, an infinite set of elements of $\{a_n\}$ exists in the neighborhood $\mathcal{N}(b_1,1)$. Let a_{n_1} denote any one of them. Similarly, b_2 is a limit of a subsequence of $\{a_n\}$, and hence in the neighborhood $\mathcal{N}(b_2,\frac{1}{2})$ lies an infinite set of elements of $\{a_n\}$. Let a_{n_2} be one such, chosen, in particular, with the property that $n_1 < n_2$. In general, corresponding to each m, an element a_{n_m} of $\{a_n\}$ exists such that $n_{m-1} < n_m$ and

$$|a_{n_m} - b_m| < \frac{1}{m}. \tag{11}$$

The numbers a_{n_m} $(m = 1,2,\ldots)$ define a subsequence $\{a_{n_m}\}$ of $\{a_n\}$ with the property that, by (10) and (11), $|a_{n_m} - b| < 2/m$. From this we conclude that $\{a_{n_m}\}$ converges to the limit b. Thus b is a limiting point of $\{a_n\}$ and therefore belongs to B. Since b is any cluster point of B, it follows that B is closed ▲

The set B of limiting points of a bounded sequence is, of course, bounded. Hence it possesses an infimum and a supremum. Moreover, since B is closed, it follows that these points belong to B, so that subsequences of $\{a_n\}$ exist which converge to the numbers $\inf B$ and $\sup B$. We call $\inf B$ the *lower limiting point*[3] of $\{a_n\}$, denoting it also by $\underline{\lim}\,\{a_n\}$. The point $\sup B$ is called the *upper limiting point*[4] of $\{a_n\}$ and is denoted by $\overline{\lim}\,\{a_n\}$.

Example 5-19 For the sequence $\{(-1)^n\}$ the only limiting points are the lower limiting point -1 and the upper limiting point 1.

Example 5-20 Let $\{r_n\}$ denote a sequence in which each of the rational numbers between 0 and 1 occurs as an element. Then each point of the closed interval $[0,1]$ is a limiting point of the sequence. The number 0 is the lower and 1 the upper limiting point.

Theorem 5-19 If ε is any positive number and $\{a_n\}$ any bounded sequence, then an integer N exists such that $a_n < \overline{\lim}\,\{a_n\} + \varepsilon$ for $n > N$. (Dually, a_n is exceeded by $\underline{\lim}\,\{a_n\} - \varepsilon$ for at most a finite set of values of n.)

Proof Suppose $\overline{\lim}\,\{a_n\} + \varepsilon \le a_n$ for an infinite set of values of n. Then a subsequence $\{a_{i_n}\}$ of $\{a_n\}$ exists such that $\overline{\lim}\,\{a_n\} + \varepsilon \le a_{i_n}$ for all i_n. Since $\{a_n\}$ is bounded, so is $\{a_{i_n}\}$; hence $\{a_{i_n}\}$ possesses a limiting point b which, because of Theorem 5-17, is such that $\overline{\lim}\,\{a_n\} + \varepsilon \le b$. But b is also a limiting point of $\{a_n\}$, and this is in contradiction to the definition of $\overline{\lim}\,\{a_n\}$ ▲

Theorem 5-20 A bounded sequence $\{a_n\}$ is convergent if and only if $\underline{\lim}\,\{a_n\} = \overline{\lim}\,\{a_n\}$. Moreover, the limit of $\{a_n\}$ is the common value of the two numbers.

Proof The convergence of $\{a_n\}$ implies the convergence of all subsequences of $\{a_n\}$ to the same limit. Since subsequences exist converging to $\underline{\lim}\,\{a_n\}$ and $\overline{\lim}\,\{a_n\}$, it follows that $\lim\,\{a_n\} = \underline{\lim}\,\{a_n\} = \overline{\lim}\,\{a_n\}$.

On the other hand, suppose that $\underline{\lim}\,\{a_n\}$ and $\overline{\lim}\,\{a_n\}$ have the common value a. Then by Theorem 5-19, corresponding to each positive ε, the elements of $\{a_n\}$ are such that $a - \varepsilon < a_n < a + \varepsilon$ for all but at most a finite set of values of n. Hence $\{a_n\}$ converges to a ▲

[3] Also, *lower limit* and *limit inferior*, written $\lim\inf\,\{a_n\}$.

[4] Also, *upper limit* and *limit superior*, written $\lim\sup\,\{a_n\}$.

Exercises

5-42 Find the upper and lower limiting points of the following sequences:

$$\text{(a)}\ \left\{(-1)^n + \frac{1}{n}\right\};\quad \text{(b)}\ \left\{\frac{1}{n} + \frac{(-1)^n}{n^2}\right\};\quad \text{(c)}\ \left\{2 + \frac{(-1)^n}{n}\right\}.$$

5-43 Prove that if $a_n \le b_n$ for all n, then (a) $\overline{\lim}\,\{a_n\} \le \overline{\lim}\,\{b_n\}$ and (b) $\underline{\lim}\,\{a_n\} \le \underline{\lim}\,\{b_n\}$.

5-44 For bounded sequences $\{a_n\}$ and $\{b_n\}$, prove that $\underline{\lim}\,\{a_n\} + \underline{\lim}\,\{b_n\} \le \underline{\lim}\,\{a_n + b_n\} \le \underline{\lim}\,\{a_n\} + \overline{\lim}\,\{b_n\} \le \overline{\lim}\,\{a_n + b_n\} \le \overline{\lim}\,\{a_n\} + \overline{\lim}\,\{b_n\}$. Cite examples to show that the less than signs may actually apply.

5-45 Prove that if $\{b_n\}$ is a null sequence, then $\underline{\lim}\,\{a_n + b_n\} = \underline{\lim}\,\{a_n\}$ and $\overline{\lim}\,\{a_n + b_n\} = \overline{\lim}\,\{a_n\}$.

5-46 Prove that $\overline{\lim}\,\{a_n\} = -\underline{\lim}\,\{-a_n\}$.

5-47 (a) Cite examples to show that a limiting point of a sequence may, but need not, be a cluster point of the range of the sequence. (b) Prove that each cluster point of the range of a sequence is a limiting point of the sequence.

5-48 Show that a sequence may diverge even though the range of the sequence has exactly one cluster point.

5-49 Let $\{a_n\}$ be a bounded sequence no two of whose elements are equal. Prove that if the range of the sequence has a single cluster point a, then the sequence converges to a.

5-11 Convergence in $\mathscr{R}^*$

A generalization of the concept of convergence is possible if appeal is made to the extended system of real numbers $\mathscr{R}^*$. A sequence $\{a_n\}$, with elements in $\mathscr{R}$, is said to *converge to* ∞ if, corresponding to each real number k, there exists an integer N such that $a_n > k$ for all $n > N$. We write $\lim \{a_n\} = \infty$. Similarly, if $a_n < k$ for all $n > N$, then $\{a_n\}$ *converges to* $-\infty$ and we write $\lim \{a_n\} = -\infty$. For example, the sequence $\{n\}$, which is divergent in $\mathscr{R}$, converges to ∞.

The notions of convergence to finite and infinite limits are seen to be one and the same if definitions are stated in terms of neighborhoods. Thus, a sequence $\{a_n\}$ converges to a limit a of $\mathscr{R}^*$ if, corresponding to each neighborhood of a, there exists an integer N such that a_n is in that neighborhood for all $n > N$.

Of course, even with this extended concept of convergence, it is not the case that every sequence converges. For example, $\{(-n)^n\}$ and $\{(-1)^n\}$ do not converge to any number in $\mathscr{R}^*$. The distinction between convergence and divergence depends on the notion of limiting point. Following the pattern of the finite case, we say that ∞ or $-\infty$ is a *limiting point* of a sequence $\{a_n\}$ if it is the limit of a subsequence. If a sequence is not bounded above by a real number, it is clear that ∞ is a limiting point and is, indeed, the upper limiting point; if it is not bounded below by a real number, then $-\infty$ is the lower limiting point. Thus, every sequence possesses an upper limiting point and a lower limiting point in $\mathscr{R}^*$. These are equal if and only if the sequence converges. If a sequence does not converge, it must possess at least two different limiting points, either or both of which may be ∞ or $-\infty$. In such a case a sequence is said to *oscillate*.

One could further extend the discussion by considering sequences in which ∞ or $-\infty$ appear as elements. However, we shall not have occasion to do this.

In the sections to come, we shall not make use of the notion of convergence in $\mathscr{R}^*$ very frequently. Whenever a sequence is said to converge to a limit a, it is understood that a is finite unless otherwise stated.

Exercises

Given a sequence $\{a_n\}$, we write $b_n = \dfrac{a_1 + a_2 + \cdots + a_n}{n}$ for $n \geq 1$ and refer to $\{b_n\}$ as the sequence of *arithmetic means* of $\{a_n\}$.

5-50 Show that the sequence of arithmetic means of the divergent sequence $\{1-(-1)^n\}$ converges to 1.

5-51 Prove that if $\{a_n\}$ is a null sequence, then so is $\left\{\dfrac{a_1 + a_2 + \cdots + a_n}{n}\right\}$.

5-52 Prove that if $\{a_n\}$ converges to a, then so does $\left\{\dfrac{a_1 + a_2 + \cdots + a_n}{n}\right\}$.

5-53 Prove that $\left\{\dfrac{1}{n} + \dfrac{1}{2n} + \cdots + \dfrac{1}{n^2}\right\}$ is a null sequence.

5-54 Determine the limit of the sequence $\left\{\dfrac{1}{n} + \dfrac{\sqrt{2}}{n} + \cdots + \dfrac{\sqrt[n]{n}}{n}\right\}$.

5-55 Let $\{a_n\}$ and $\{b_n\}$ be sequences converging to a and b, respectively. Define $c_n = \dfrac{a_1 b_n + a_2 b_{n-1} + \cdots + a_n b_1}{n}$ for $n \geq 1$. Prove that $\{c_n\}$ converges to ab.

Let $\{a_n\}$ be a sequence such that $a_n > 0$ for $n \geq 1$, and define $b_n = \sqrt[n]{a_1 a_2 \cdots a_n}$. Then $\{b_n\}$ is called the sequence of *geometric means* of $\{a_n\}$.

5-56 Cite an example of a divergent sequence for which the sequence of geometric means converges.

5-57 Prove that if $\{a_n\}$ is a null sequence, then so is $\{\sqrt[n]{a_1 a_2 \cdots a_n}\}$.

5-58 Prove that if $\{a_n\}$ converges to 1, then so does $\{\sqrt[n]{a_1 a_2 \cdots a_n}\}$.

5-59 Prove that if $\{a_n\}$ has a positive limit a, then $\{\sqrt[n]{a_1 a_2 \cdots a_n}\}$ converges to a.

5-60 Prove that $\{1/\sqrt[n]{n!}\}$ is a null sequence.

5-61 Prove that if $\{b_{n+1} - b_n\}$ converges to b, then $\{b_n/n\}$ converges to b also. Cite an example to show that the converse is false.

5-62 (a) Suppose $b_n > 0$ for $n \geq 1$ and that $\{b_{n+1}/b_n\}$ converges to b. Prove that $\{\sqrt[n]{b_n}\}$ converges to b. (b) Use the result in (a) to prove that $\{\sqrt[n]{n}\}$ converges to 1. (c) Cite an example to show that the converse of (a) is false.

5-63 Let $\{a_n\}$ be a null sequence with $a_n \geq -1$ for $n \geq 1$, and let m be a positive integer. Prove that $\{\sqrt[m]{1 + a_n}\}$ converges to 1.

5-64 Prove that $\left\{\dfrac{n+1}{\sqrt[n]{n!}}\right\}$ has the limit e.

5-65 Let $\{a_n\}$ and $\{b_n\}$ be sequences with $b_n > 0$ for $n \geq 1$. Assume $\left\{\frac{a_n}{b_n}\right\}$ to be increasing and convergent to k. Define $c_n = \dfrac{a_1 + a_2 + \cdots + a_n}{b_1 + b_2 + \cdots + b_n}$ for $n \geq 1$. (a) Prove that $\{c_n\}$ is increasing and convergent to $c \leq k$. (b) Give an example in which $c < k$.

5-66 Prove that if b is a positive real number and m a negative integer, then a positive number a exists such that $a^m = b$.

5-67 Let p, q, and k be integers, q and k positive, and let b be any positive real number. Prove that (a) $(\sqrt[q]{b})^p = \sqrt[q]{b^p}$, (b) $(\sqrt[q]{b})^p = (\sqrt[kq]{b})^{kp}$, and (c) $\sqrt[q]{b^p} = \sqrt[kq]{b^{kp}}$. Because of these results we may define $b^r = \sqrt[q]{b^p}$ for any positive number b and any rational number $r = p/q$, where p and q are integers. (Parts (b) and (c) show that b^r does not depend on the representation employed for r.)

5-68 Let a and b be positive real numbers, and let r and s be rational numbers. Prove that (a) $a^r a^s = a^{r+s}$, (b) $(ab)^r = a^r b^r$, and (c) $(a^r)^s = a^{rs}$.

5-69 Let r and s be rational numbers with $r < s$, and let a be a positive real number. Prove that (a) $a^r < a^s$ if $a > 1$, and (b) $a^s < a^r$ if $a < 1$.

5-70 Let $\{r_n\}$ be a sequence of rational numbers which converges to the limit 0. Prove that if a is any positive real number, then $\{a^{r_n}\}$ converges to 1.

5-71 Let $\{r_n\}$ be a sequence of rational numbers which converges to a limit b (rational or irrational). Prove that if a is any positive real number, then $\{a^{r_n}\}$ converges. Further prove that if b is rational, then the limit of $\{a^{r_n}\}$ is a^b.

5-72 Let $\{r_n\}$ and $\{s_n\}$ be sequences of rational numbers converging to a common limit b (rational or irrational). Prove that if a is any positive real number, then the sequences $\{a^{r_n}\}$ and $\{a^{s_n}\}$ have the same limit. If b is irrational, we denote this limit by a^b and in this way define irrational powers of real numbers.

5-73 Let a, b, x, and y be real numbers, with a and b positive. Prove that (a) $a^x a^y = a^{x+y}$, (b) $(ab)^x = a^x b^x$, and (c) $(a^x)^y = a^{xy}$.

5-74 Let a, x, and y be real numbers with $x < y$. Prove that (a) $a^x < a^y$ if $a > 1$, and (b) $a^y < a^x$ if $0 < a < 1$.

The remaining exercises in this set concern the Cantor definition of a real number. All numbers mentioned are rational; and the notions of convergence, limit, and Cauchy convergence are understood to be with respect to the set of rational numbers. For example, a sequence $\{r_n\}$ of rational numbers converges if a rational number r exists such that, corresponding to each (rational number) $\varepsilon > 0$, there may be found an integer N with the property that $|r_n - r| < \varepsilon$ for all $n > N$.

5-75 Cite an example of a sequence $\{r_n\}$ of rational numbers which is Cauchy convergent but not convergent.

5-76 Determine which of Theorems 5-1 through 5-15 remain valid in the system of rational numbers. Support statements with arguments or counterexamples.

5-77 Two Cauchy convergent sequences $\{r_n\}$ and $\{s_n\}$ of rational numbers are said to be *equivalent*, $\{r_n\} \sim \{s_n\}$, if the sequence $\{r_n - s_n\}$ is a null sequence. Prove that this relation is (a) reflexive, (b) symmetric, and (c) transitive.

Let $\mathscr{R}_s$ be a set consisting of exactly one member from each set of equivalent Cauchy convergent sequences of rational numbers. Thus, each

Cauchy convergent sequence of rational numbers is equivalent to a unique member of the set $\mathscr{R}_s$, and two members of $\mathscr{R}_s$ are equivalent if and only if they are equal.

Corresponding to two members $\{r_n\}$ and $\{s_n\}$ of $\mathscr{R}_s$, we define the *sum* $\{r_n\}+\{s_n\}$ to be the (unique) member of $\mathscr{R}_s$ which is equivalent to $\{r_n+s_n\}$, the *product* $\{r_n\}\{s_n\}$ to be the (unique) member of $\mathscr{R}_s$ which is equivalent to $\{r_n s_n\}$, and the *less than* relation $\{r_n\}<\{s_n\}$ to signify that $\{s_n-r_n\}$ is not a null sequence and that an integer N exists such that $s_n-r_n>0$ for $n>N$.

Cantor defined the *system of real numbers* as the set $\mathscr{R}_s$ with the relations of equality and order and the operations of addition and multiplication as defined above.

5-78 Prove that the Cantor and Dedekind definitions of real number are the same by showing that the system $\mathscr{R}_s$ is isomorphic with the system of cuts.

5-79 Without employing the isomorphism between the system $\mathscr{R}_s$ and the system of cuts, prove that the system $\mathscr{R}_s$ possesses properties 1 through 15 of Section 3-23.

6

Series of Constants

...and all these things shall be added... .

Gospel according to St. Matthew

6-1 Definitions and Notation

Let $\{a_n\}$ be a given sequence of real numbers, and suppose there to be formed of its elements another sequence $\{s_n\}$ such that $s_1 = a_1$, $s_2 = a_1 + a_2$, and, in general, $s_n = a_1 + a_2 + \cdots + a_n$. The sequence $\{s_n\}$ is called an *infinite sum* or *infinite series* or simply a *series* of the elements of $\{a_n\}$. Various notations, involving explicit reference to the elements of $\{a_n\}$, are employed for the infinite series $\{s_n\}$. Among these are $\{a_1 + a_2 + \cdots + a_n\}$, $a_1 + a_2 + \cdots + a_n + \cdots$, and $\sum a_n$; also $\sum_{n=1}^{\infty} a_n$ and $\sum_1^{\infty} a_n$ or, if the range of the subscripts is the set of all n such that $n \geq n_0$, $\sum_{n=n_0}^{\infty} a_n$ and $\sum_{n_0}^{\infty} a_n$.

The elements a_n of the sequence $\{a_n\}$ are referred to as the *terms* of the series $\sum a_n$. Further, the elements s_n of the series are called its *partial sums.* The terms and the partial sums of an infinite series are obviously related by $a_1 = s_1$ and $a_n = s_n - s_{n-1}$ for $n \geq 2$.

Example 6-1 For the series $\sum a_n = 1 + 1/4 + \cdots + 1/n^2 + \cdots$ the terms are 1, $1/4, \ldots, 1/n^2, \ldots$ and the partial sums, 1, 5/4, 49/36, ...

Example 6-2 If the sequence of partial sums of a series $\sum_1^{\infty} a_n$ is $\{s_n\} = \{n/(2n+1)\}$, then $\sum\limits_1^{\infty} a_n = \sum\limits_1^{\infty} \left(\dfrac{n}{2n+1} - \dfrac{n-1}{2n-1}\right) = \sum\limits_1^{\infty} \dfrac{1}{4n^2-1}$.

6-2 Convergence and Divergence

Since the series $\sum a_n$ is, by definition, the sequence of its partial sums, much of the theory of infinite series concerns simple restatement, with explicit reference to the terms of the series, of the many definitions and properties already established for sequences in general. In particular, the concept of convergence of a series $\sum a_n$ is expressible directly in terms of the a_n. Thus, $\sum a_n$ converges to a limit S if, corresponding to each positive number ε, there exists an integer N such that $|s_n - S| < \varepsilon$ or, what is the same, such that

$$|a_1 + a_2 + \cdots + a_n - S| < \varepsilon$$

for all $n > N$. Divergence, of course, signifies the existence of at least one value of ε to which no N corresponds.

The limit S, if it exists, we call the *sum* of the series $\sum a_n$ and write, for convenience, $S = \sum a_n$, thereby employing the symbol $\sum a_n$ in a dual capacity in the event of convergence.

For a given infinite series, two basic questions present themselves: Does the series converge? If the series converges, what is its sum? Of the two, the first is the more readily answered. Certain tests exist, involving simple calculations or comparisons with other series, which enable one to ascertain convergence or divergence in many cases. The second question is often difficult to answer. It is not always a simple step from the established existence of a sum S to a determination of its actual value. We shall concern ourselves principally with the question of existence of sums and leave the actual calculation of values to particular examples.

Example 6-3 An infinite series of the form

$$\sum_1^\infty ar^{n-1} = a + ar + \cdots + ar^{n-1} + \cdots$$

(or $\sum_0^\infty ar^n$), where $a \neq 0$, is called a *geometric* series. It has certain important applications. The constant r is the *common ratio* of each pair of consecutive terms, and a is called the *initial* term.

From the nth partial sum

$$(1) \qquad s_n = a + ar + \cdots + ar^{n-1}$$

it follows that if $r = 1$, then $s_n = na$. Therefore the series diverges when $r = 1$ since $\{s_n\} = \{na\}$ is clearly a divergent sequence.

If $r \neq 1$, then we have $rs_n = ar + ar^2 + \cdots + ar^n$, which upon subtraction from (1) and rearranging yields $s_n = a(1 - r^n)/(1 - r)$. If $|r| < 1$, the sequence $\{r^n\}$ is a null sequence. It follows then that the sequence $\{s_n\}$ converges to $a/(1 - r)$. On the other hand, if $|r| \geq 1$ $(r \neq 1)$, then $\{r^n\}$ diverges and so does $\{s_n\}$. We conclude that the geometric series diverges if $|r| \geq 1$ and that

$$\sum_1^\infty ar^{n-1} = \frac{a}{1-r}$$

if $|r| < 1$. Thus, for example, the series $\sum_1^\infty (\frac{4}{3})^{n-1}$ diverges since $r = \frac{4}{3} > 1$. Also, the series $\sum_0^\infty (-\frac{1}{2})^n$ is geometric, with $r = -\frac{1}{2}$. It converges, therefore, to the sum $S = 1/(1 + \frac{1}{2}) = \frac{2}{3}$.

Example 6-4 The series

$$\sum_1^\infty \frac{1}{n} = 1 + \frac{1}{2} + \frac{1}{3} + \cdots + \frac{1}{n} + \cdots$$

is called the *harmonic* series. Here

$$s_n = 1 + \frac{1}{2} + \frac{1}{3} + \cdots + \frac{1}{n}.$$

It has already been shown (Example 5-17) that the sequence $\{s_n\}$ diverges. Hence $\sum 1/n$ is divergent.

Example 6-5 Consider the series $\sum 1/n(n+1)$, with

$$s_n = \frac{1}{1 \cdot 2} + \frac{1}{2 \cdot 3} + \cdots + \frac{1}{n(n+1)}.$$

Since for any number k we have $1/k(k+1) = 1/k - 1/(k+1)$, it follows that

$$s_n = \left(1 - \frac{1}{2}\right) + \left(\frac{1}{2} - \frac{1}{3}\right) + \cdots + \left(\frac{1}{n} - \frac{1}{n+1}\right) = 1 - \frac{1}{n+1} = \frac{n}{n+1}.$$

Hence $\{s_n\}$ converges to 1, or $\sum 1/n(n+1) = 1$. Because of the manner in which the sum s_n simplifies, we describe $\sum 1/n(n+1)$ as a *telescoping* series.

A series of the form $\sum (-1)^{n+1} a_n$, where $a_n > 0$ for all n, is called an *alternating* series. Convergence of a series of this type is frequently determined as a consequence of the following.

Theorem 6-1 The alternating series $\sum (-1)^{n+1} a_n$ is convergent if $\{a_n\}$ is a decreasing null sequence.

Proof From

$$s_{2n} = (a_1 - a_2) + (a_3 - a_4) + \cdots + (a_{2n-1} - a_{2n})$$

and from the monotone character of $\{a_n\}$ it follows that $\{s_{2n}\}$ is an increasing sequence. Rearranging the right side yields

$$s_{2n} = (a_1 - a_{2n}) - (a_2 - a_3) - \cdots - (a_{2n-2} - a_{2n-1}),$$

which shows $\{s_{2n}\}$ to be a bounded sequence; that is, $s_{2n} \leq a_1 - a_{2n} \leq a_1$ for all n. As a consequence $\{s_{2n}\}$ converges to a limit. In a similar fashion, it may be shown that $\{s_{2n+1}\}$ converges. Then, since $\{s_{2n+1} - s_{2n}\} = \{a_{2n+1}\}$ is a null sequence, it follows that $\{s_{2n+1}\}$ and $\{s_{2n}\}$ converge to the same limit; and as a consequence $\{s_n\}$ converges to that limit ▲

Example 6-6 The *alternating harmonic* series

$$\sum_{n=1}^{\infty} \frac{(-1)^{n+1}}{n} = 1 - \frac{1}{2} + \frac{1}{3} - \frac{1}{4} + \cdots + \frac{(-1)^{n+1}}{n} + \cdots$$

is a convergent alternating series since $\{1/n\}$ is a decreasing null sequence.

Example 6-7 Consider the series

$$\sum_{n=1}^{\infty} (-1)^{n+1} a_n = 1 - \frac{1}{1 \cdot 2} + \frac{1}{3} - \frac{1}{3 \cdot 4} + \frac{1}{5} - \frac{1}{5 \cdot 6} + \cdots,$$

where $a_n = 1/n$ if n is odd and $a_n = 1/(n-1)n$ if n is even.

This is an alternating series whose terms constitute a null, but not monotone, sequence. The partial sum s_{2n} is given by

$$s_{2n} = \sum_{k=1}^{n} \left[\frac{1}{2k-1} - \frac{1}{(2k-1)(2k)}\right] = \frac{1}{2} \sum_{k=1}^{n} \frac{1}{k}.$$

We have already seen that $\{s_{2n}\}$ diverges; hence $\sum a_n$ diverges. This illustrates that the hypothesis of monotony cannot be dispensed with in Theorem 6-1.

A series $\sum a_n$ with the property that $a_n \geq 0$ for all n converges if and only if the sequence of partial sums $\{s_n\}$ is bounded. This is an immediate consequence of the fact that $\{s_n\}$ is an increasing sequence, and we have seen that such converges if and only if it is bounded. Further, if M denotes an upper bound of $\{s_n\}$, then $s_n \leq \sum a_n \leq M$.[1] This, too, is a simple consequence of properties of bounded monotone sequences.

Example 6-8 $\sum_{n=0}^{\infty} \left[\frac{2}{5+(-1)^n}\right]^n$. We have

$$s_n = 1 + \frac{1}{2} + \frac{1}{3^2} + \frac{1}{2^3} + \cdots + \left[\frac{2}{5+(-1)^n}\right]^n$$
$$\leq 1 + \frac{1}{2} + \frac{1}{2^2} + \frac{1}{2^3} + \cdots + \frac{1}{2^n};$$

and since the latter is geometric, we have $s_n \leq 2$ for all n. We conclude that the given series converges and, moreover, that its sum does not exceed 2.

Let $\sum a_n$ be a given series with $\{s_n\}$ its sequence of partial sums. Then for some positive integer k consider the series $\sum b_n$ obtained by deleting the first k terms of $\sum a_n$. If $\{t_n\}$ denotes the sequence of partial sums of $\sum b_n$, then, clearly, $t_n = s_{n+k} - s_k$ for all n. It follows that if $\sum a_n$ converges to S, then $\sum b_n$ converges to $T = S - s_k$. On the other hand, if $\sum a_n$ diverges, then so does $\sum b_n$.

[1] Note that the symbol n plays two roles here. This is emphasized by comparison with the equivalent expression $s_n \leq \sum a_k \leq M$ in which the symbol k appears as the *summation index*.

Given a series $\sum_{i=1}^{\infty} a_i$, we refer to the series $\sum_{i=n+1}^{\infty} a_i$ as the *remainder after n terms.* Since such a remainder series is obtained from a given series by deleting the first n terms, it follows that $\sum_{i=1}^{\infty} a_i$ converges or diverges according as the remainder series converges or diverges. If the given series converges to S and the remainder series $\sum_{i=n+1}^{\infty} a_i$ to r_n for each n, then, clearly, $S = s_n + r_n$. Since $\{s_n\}$ has limit S, it follows that $\{r_n\}$ is a null sequence.

The concept of Cauchy convergence of the sequence of partial sums $\{s_n\}$ of an infinite series $\sum a_n$ is directly expressible in terms of the numbers a_n. A series $\sum a_n$ is *Cauchy convergent* if, corresponding to each positive number ε, there exists an integer N such that

$$\left|\sum_{i=n+1}^{n+k} a_i\right| = |a_{n+1} + a_{n+2} + \cdots + a_{n+k}| < \varepsilon \tag{2}$$

for all $n > N$ and all $k > 0$. This follows from the simple observation that $s_{n+k} - s_n = a_{n+1} + a_{n+2} + \cdots + a_{n+k}$.

That an infinite series is convergent if and only if it is Cauchy convergent is a simple restatement of Theorems 5-15 and 5-16.

Example 6-9 The series $\sum_1^{\infty} 1/(n+1)(n+2)$ may be shown to be Cauchy convergent. We have

$$\sum_{i=n+1}^{n+k} a_i = \sum_{i=n+1}^{n+k} \frac{1}{(i+1)(i+2)} = \sum_{i=n+1}^{n+k} \left(\frac{1}{i+1} - \frac{1}{i+2}\right)$$

$$= \left(\frac{1}{n+2} - \frac{1}{n+3}\right) + \left(\frac{1}{n+3} - \frac{1}{n+4}\right) + \cdots + \left(\frac{1}{n+k+1} - \frac{1}{n+k+2}\right)$$

$$= \frac{1}{n+2} - \frac{1}{n+k+2}.$$

Hence

$$\left|\sum_{i=n+1}^{n+k} a_i\right| < \frac{1}{n+2}$$

for all n and all $k > 0$. Cauchy convergence of the given series follows from the convergence of $\{1/(n+2)\}$.

By setting $k = 1$ in inequality (2) a necessary condition for convergence of a series $\sum a_n$ is obtained, namely, that the sequence $\{a_n\}$ be a null sequence. The condition is not sufficient, however. Witness the series $\sum 1/n$, which diverges even though $\{1/n\}$ is a null sequence.

If $\sum a_n$ and $\sum b_n$ are convergent series with sums S and T, respectively, then the series $\sum (a_n + b_n)$ converges to $S + T$; and, k being any constant, $\sum ka_n$ converges to kS. These facts follow immediately from the theorems of Chapter 5. For if s_n and t_n denote the nth partial sums of $\sum a_n$ and $\sum b_n$, respectively, then $s_n + t_n$ and ks_n are the respective nth partial sums of $\sum (a_n + b_n)$ and $\sum ka_n$. Then from $\lim \{s_n + t_n\} = \lim \{s_n\} + \lim \{t_n\}$ and $\lim \{ks_n\} = k \lim \{s_n\}$ follow $\sum (a_n + b_n) = \sum a_n + \sum b_n$ and $\sum ka_n = k \sum a_n$.

A corresponding statement concerning products of series is postponed until later. Multiplication of series presents much greater complication than addition or multiplication by a constant.

A very useful technique for proving convergence or divergence of a series of positive terms is provided by the next theorem.

Theorem 6-2 (The Comparison Test) If $\sum b_n$ is a convergent series and if $0 \leq a_n \leq b_n$ for all n, then $\sum a_n$ converges; if $\sum b_n$ diverges and if $0 \leq b_n \leq a_n$ for all n, then $\sum a_n$ diverges.

Proof To prove the first statement observe that the partial sums of the series $\sum a_n$ constitute an increasing sequence which is bounded above by $\sum b_n$. Hence $\sum a_n$ converges to a sum not exceeding $\sum b_n$. As for the second part, note that convergence of $\sum a_n$ would imply convergence of $\sum b_n$, by the first statement. From the contradiction follows divergence of $\sum a_n$▲

Example 6-10 From Example 6-5 it follows that the series $\sum 2/n(n+1)$ converges. Since $1/n^2 \leq 1/n(n+1)$ for all n, it is a consequence of the comparison test that $\sum 1/n^2$ converges as well.

Example 6-11 The series $\sum 1/n^p$, where p is any constant, is known as the *p-series*. We have seen that for $p = 1$, that is, $\sum 1/n$, it is divergent whereas for $p = 2$, that is, $\sum 1/n^2$, it converges. With the aid of the comparison test we may draw conclusions for certain other values of p.

Thus, suppose $p \leq 1$. Then $1/n^p \geq 1/n$ for all n. It follows that $\sum 1/n^p$ diverges. On the other hand, if $p \geq 2$, then $1/n^p \leq 1/n^2$ for all n, and we conclude that $\sum 1/n^p$ converges. Thus $\sum 1/n^p$ is divergent for $p \leq 1$ and convergent for $p \geq 2$. The situation for values of p between 1 and 2 will be decided in Example 6-32.

It is quite apparent that, as in the case of sequences, a finite set of terms may be ignored in making a comparison. Further, if all terms of a series are negative, a comparison test such as the one for series of positive terms is obviously valid. However, a comparison between series of mixed terms (an infinite set of terms of each sign) is, in general, insufficient to establish convergence or divergence. Example 6-12 serves to illustrate this point, which is often overlooked.

Example 6-12 Consider the series

$$\sum_{n=1}^{\infty} a_n = 1 - \frac{1}{1 \cdot 2} + \frac{1}{3} - \frac{1}{3 \cdot 4} + \frac{1}{5} - \frac{1}{5 \cdot 6} + \cdots,$$

where $a_n = 1/n$ if n is odd and $a_n = -1/(n-1)n$ if n is even, and

$$\sum_{n=1}^{\infty} b_n = 1 - \frac{1}{2} + \frac{1}{3} - \frac{1}{4} + \cdots + \frac{(-1)^{n+1}}{n} + \cdots.$$

It is immediate that $|a_n| \leq |b_n|$ for all n. However, as we have seen in Examples 6-6 and 6-7, the series $\sum a_n$ diverges whereas $\sum b_n$ converges.

Exercises

6-1 Prove that a series of non-negative terms converges if and only if its sequence of partial sums is bounded.

6-2 Let $\sum a_n$ be a convergent series of positive terms and let $b_n = \sum_{i=n}^{\infty} a_i$. Prove that the series $\sum (\sqrt{b_n} - \sqrt{b_{n+1}})$ is convergent.

6-3 If $\sum a_n$ and $\sum b_n$ $(b_n \neq 0)$ are convergent series with the property that $\{a_n/b_n\}$ is a null sequence, then the series $\sum b_n$ is said to *converge more slowly* than $\sum a_n$. Prove that if $\sum a_n$ is any convergent series, then a series $\sum b_n$ exists which converges more slowly than $\sum a_n$.

6-4 Let $\sum a_n$ denote a divergent series of positive terms. Prove that if s_n is the nth partial sum of $\sum a_n$, then $\sum (\sqrt{s_{n+1}} - \sqrt{s_n})$ also diverges.

6-5 If $\sum a_n$ $(a_n \neq 0)$ and $\sum b_n$ are divergent series with the property that $\{b_n/a_n\}$ is a null sequence, then the series $\sum b_n$ is said to *diverge more slowly* than $\sum a_n$. Prove that no series exists which diverges more slowly than all others.

6-6 Let r_n denote the remainder after n terms of the alternating series $\sum (-1)^{n+1} a_n$ where $\{a_n\}$ is a decreasing null sequence. Prove that $0 < (-1)^n r_n < a_{n+1}$.

6-7 Given a series $\sum a_n$, let $\sum b_n$ be a new series obtained from $\sum a_n$ by adding, deleting, or altering a finite set of terms. Prove that $\sum a_n$ and $\sum b_n$ both converge or both diverge.

6-8 (a) Show that if $\sum a_n$ and $\sum b_n$ are series of positive terms and if $\lim \{a_n/b_n\} = k > 0$, then the series both converge or both diverge. (b) Show that the above conclusion is not valid if $k = 0$. (c) Prove that $\sum \dfrac{n+3}{2n^3 - 7n}$ converges and $\sum \dfrac{2-7n}{n^2+4}$ diverges.

6-9 Prove or disprove: If $\sum a_n$ converges, then $\sum a_n^2$ converges.

6-10 Prove that $\sum 1/n^2 < 2$.

6-11 (a) Use induction to prove that if m and n are positive integers such that $m < n$, then $\dfrac{1}{m^2} + \dfrac{1}{(m+1)^2} + \cdots + \dfrac{1}{n^2} < \dfrac{m+2}{m^2} - \dfrac{1}{n}$. (b) Use the above to estimate the value of $\sum 1/n^2$ with an accuracy of at least $\frac{1}{4}$.

6-12 Prove that $\sum_{n=0}^{\infty} 1/n! < 3$ by comparing it with the series $\sum_{n=0}^{\infty} 1/2^n$.

6-13 Show that if $\sum a_n$ is convergent and if $\{a_n\}$ is decreasing, then $\{na_n\}$ is a null sequence.

6-14 Prove that $(\sum a_n b_n)^2 \leq \sum a_n^2 \sum b_n^2$ if $\sum a_n^2$ and $\sum b_n^2$ both converge. (Compare Exercise 3-75.)

6-3 Grouping of Terms of a Series

The designation *infinite sum* often employed for infinite series is somewhat misleading since it may invite unwarranted comparison with finite sums. While similarities exist, certain essential differences do also. It is of utmost importance that these differences be made clear in order to curb reckless handling of series. For finite sums the associative law for addition allows us to bracket or group the terms in any fashion whatever, thereby reducing a finite sum to a sum of sums.

Moreover, the commutative law permits rearrangement of the order of the summands without affecting the value of the sum. In this section we investigate the conditions under which the terms of an infinite series may be grouped without altering the state of convergence or divergence or the value of the sum in the event of convergence.

Example 6-13 We have already seen that the geometric series

$$\sum_{n=0}^{\infty}\left(-\frac{1}{2}\right)^n = 1 - \frac{1}{2} + \frac{1}{4} - \frac{1}{8} + \cdots + \left(-\frac{1}{2}\right)^n + \cdots$$

converges to $\frac{2}{3}$. By combining consecutive terms we obtain the series

$$\left(1-\frac{1}{2}\right)+\left(\frac{1}{4}-\frac{1}{8}\right)+\cdots+(-1)^{2n}\left(\frac{1}{2^{2n}}-\frac{1}{2^{2n+1}}\right)+\cdots$$

or

$$\frac{1}{2}+\frac{1}{8}+\frac{1}{32}+\cdots+\frac{1}{2^{2n+1}}+\cdots.$$

This is a geometric series with initial term $\frac{1}{2}$ and common ratio $\frac{1}{4}$. It converges to $\frac{1}{2}/(1-\frac{1}{4})=\frac{2}{3}$. We conclude that the grouping of terms of a given convergent series need not affect either the fact of convergence or the value of the sum.

Example 6-14 Consider the series

$$\sum_{n=0}^{\infty}(-1)^n = 1-1+1-1+\cdots+(-1)^n+\cdots,$$

which is clearly divergent. If we presume to group consecutive pairs of terms in two different ways, we obtain the series

$$(1-1)+(1-1)+\cdots=0$$

and

$$1-(1-1)-(1-1)-\cdots=1.$$

Thus are constructed two convergent series forcing the absurd conclusion $0=1$! It follows that grouping of terms in this instance is not a valid operation.

The key to the feasibility of grouping of terms lies in the concept of convergence. We put this in the form of a theorem.

Theorem 6-3 If, without changing their order, the terms of a convergent series $\sum a_n$ are grouped to form a series of finite sums, then the series $\sum b_n$ so formed is also convergent and has the same sum.

Proof Denote the partial sums of $\sum a_n$ and $\sum b_n$ by s_n and t_n, respectively. Then, corresponding to each integer n, there exists an integer k_n such that $t_n = s_{k_n}$. Trivially, $k_n < k_{n+1}$. Hence $\{t_n\}$ is a subsequence of $\{s_n\}$ and consequently converges to the same limit as $\{s_n\}$ ▲

The converse of the last theorem is not true. That is, if a convergent series $\sum b_n$ consists of terms which are themselves written as finite sums, then removal of the parentheses distinguishing these sums may yield a series which is divergent.

Example 6-15 Consider the series $\sum_1^\infty 1/(n+1)(n+2)$, which in Example 6-9 was shown to converge. If we write it in the form

$$\sum_{n=1}^{\infty} \frac{1}{(n+1)(n+2)} = \sum_{n=1}^{\infty} \left(\frac{n+1}{n+2} - \frac{n}{n+1}\right)$$

$$= \left(-\frac{1}{2} + \frac{2}{3}\right) + \left(-\frac{2}{3} + \frac{3}{4}\right) + \cdots + \left(-\frac{n}{n+1} + \frac{n+1}{n+2}\right) + \cdots$$

and remove parentheses, we have the new series

$$-\frac{1}{2} + \frac{2}{3} - \frac{2}{3} + \frac{3}{4} - \cdots - \frac{n}{n+1} + \frac{n+1}{n+2} - \cdots$$

which does not converge since the sequence of terms does not have limit zero.

If removal of parentheses from a given series $\sum b_n$ gives rise to a series $\sum a_n$ which is convergent, then, by virtue of Theorem 6-3, the sequence $\sum b_n$ is necessarily convergent.

Example 6-16 To show that $\sum_1^\infty 1/2n(2n-1)$ converges, we write

$$\sum_{n=1}^{\infty} \frac{1}{2n(2n-1)} = \sum_{n=1}^{\infty} \left(\frac{1}{2n-1} - \frac{1}{2n}\right)$$

$$= \left(1 - \frac{1}{2}\right) + \left(\frac{1}{3} - \frac{1}{4}\right) + \cdots + \left(\frac{1}{2n-1} - \frac{1}{2n}\right) + \cdots.$$

Removing parentheses yields the alternating harmonic series; since this converges, it follows that the given series does also.

A somewhat more general statement can be made apropos removal of parentheses.

Theorem 6-4 Let the terms of a series $\sum b_n$ be constructed by grouping terms of a given series $\sum a_n$ in such a way that no group contains more than a fixed number M of terms of $\sum a_n$. Then, if $\sum b_n$ converges and if $\{a_n\}$ is a null sequence, the series $\sum a_n$ converges to the same sum as $\sum b_n$.

Proof Let s_n and t_n denote the partial sums of $\sum a_n$ and $\sum b_n$, respectively. Then, corresponding to each n, there exists an integer k_n such that $t_n = s_{k_n}$. Moreover, $n \le k_n$, $k_n < k_{n+1}$, and, by the conditions of the theorem, $k_{n+1} - k_n \le M$ for all n. Observe, too, that $\{k_n\}$ increases without bound.

Let $\varepsilon > 0$ be given. Since $\{t_n\}$ converges to a limit T and since $\{a_n\}$ is a null sequence, an integer N exists such that $|t_n - T| < \varepsilon$ and $|a_n| < \varepsilon$ for all $n > N$. Denote by p an integer exceeding N. Then k_p exists such that $t_p = s_{k_p}$. Let m

be any integer exceeding k_p. Because $\{k_n\}$ is unbounded, successive elements k_q and k_{q+1} of $\{k_n\}$ exist such that $k_q \le m < k_{q+1}$. It follows that $k_p \le k_q$ from which $p \le q$. Therefore $N < q$ and $|t_q - T| < \varepsilon$. Now

$$\begin{aligned} s_m &= s_{k_q} + a_{k_q+1} + a_{k_q+2} + \cdots + a_m \\ &= t_q + a_{k_q+1} + a_{k_q+2} + \cdots + a_m. \end{aligned}$$

Also, from $N < p \le k_p \le k_q$ follows

$$|a_{k_q+1}| < \varepsilon, \quad |a_{k_q+2}| < \varepsilon, \quad \ldots, \quad |a_m| < \varepsilon.$$

Combining these we obtain

$$|s_m - t_q| \le |a_{k_q+1}| + |a_{k_q+2}| + \cdots + |a_m| < M\varepsilon.$$

Therefore

$$|s_m - T| \le |s_m - t_q| + |t_q - T| < (M + 1)\varepsilon.$$

As a consequence the sequence $\{s_n\}$, that is, the series $\sum a_n$, converges to T ▲

We may apply Theorem 6-4 to the formation of the sum of two convergent series. Suppose $\sum a_n$ and $\sum b_n$ converge. Then, as shown in Section 6-2, the series $\sum (a_n + b_n)$, that is, the series

$$(a_1 + b_1) + (a_2 + b_2) + \cdots + (a_n + b_n) + \cdots$$

converges also. By virtue of Theorem 6-4 it is immediate that the series

$$a_1 + b_1 + a_2 + b_2 + \cdots + a_n + b_n + \cdots$$

is convergent as well.

Example 6-17 Consider the series

$$1 + \frac{1}{2} + \frac{1}{4} + \frac{1}{6} + \frac{1}{9} + \frac{1}{12} + \cdots + \frac{1}{n^2} + \frac{1}{n(n+1)} + \cdots.$$

The series $\sum 1/n^2$ and $\sum 1/n(n+1)$ both converge. Hence so does their sum

$$\left(1 + \frac{1}{2}\right) + \left(\frac{1}{4} + \frac{1}{6}\right) + \left(\frac{1}{9} + \frac{1}{12}\right) + \cdots + \left(\frac{1}{n^2} + \frac{1}{n(n+1)}\right) + \cdots.$$

Since the parentheses may be removed, we conclude that the given series converges.

6-4 Absolute and Conditional Convergence

An infinite series $\sum a_n$ is said to be *absolutely convergent* if the series $\sum |a_n|$ is convergent. If $\sum a_n$ is convergent but not absolutely convergent, we say it is *conditionally convergent*.

Example 6-18 The geometric series $\sum (-\frac{1}{2})^n$ is absolutely convergent since $\sum (\frac{1}{2})^n$ converges. The former converges to $\frac{2}{3}$ and the latter to 2. This illustrates, too, that $\left|\sum a_n\right|$ need not equal $\sum |a_n|$.

Example 6-19 The alternating harmonic series $\sum (-1)^{n-1}/n$ is conditionally convergent since it converges while the harmonic series $\sum 1/n$ diverges.

Theorem 6-5 An absolutely convergent series is convergent.

Proof From the convergence of $\sum |a_n|$ it follows that, corresponding to a given $\varepsilon > 0$, an integer N exists such that

$$|a_{n+1}| + |a_{n+2}| + \cdots + |a_{n+k}| < \varepsilon$$

for all $n > N$ and all $k > 0$. Since $|a_{n+1} + a_{n+2} + \cdots + a_{n+k}| \leq |a_{n+1}| + |a_{n+2}| + \cdots + |a_{n+k}|$, it follows that $\sum a_n$, also, converges ▲

The usefulness of absolute convergence stems in part from the fact that it is often easier to establish absolute convergence than convergence.

Example 6-20 Consider the series $\sum c_n/2^n$, where $c_n = 1$ if n is a prime number and $c_n = -1$ otherwise. Proof of convergence directly from the definition offers some difficulties. However, since $\sum |c_n/2^n| = \sum 1/2^n$ is convergent, it follows that $\sum c_n/2^n$ is absolutely convergent, hence convergent.

It is clear that a convergent series $\sum a_n$ consisting only of non-negative terms or only of nonpositive terms is absolutely convergent; also, that $|\sum a_n| = \sum |a_n|$ for such series.

For a series $\sum a_n$, define numbers p_n and q_n by

$$p_n = \begin{cases} a_n & \text{if } a_n \geq 0 \\ 0 & \text{if } a_n < 0, \end{cases} \qquad q_n = \begin{cases} 0 & \text{if } a_n \geq 0 \\ -a_n & \text{if } a_n < 0. \end{cases}$$

Then the series $\sum p_n$ and $\sum q_n$, each consisting of non-negative terms only, are referred to as the *positive* and *negative parts*, respectively, of $\sum a_n$. The relations

$$p_n = \frac{1}{2}(|a_n| + a_n), \qquad q_n = \frac{1}{2}(|a_n| - a_n),$$

and

$$|a_n| = p_n + q_n, \qquad a_n = p_n - q_n,$$

are easily verified. In the case of the alternating harmonic series, for example, the positive and negative parts are the series

$$\sum p_n = 1 + 0 + \frac{1}{3} + 0 + \frac{1}{5} + \cdots$$

and

$$\sum q_n = 0 + \frac{1}{2} + 0 + \frac{1}{4} + 0 + \cdots.$$

Insight into the structure of an absolutely convergent series is gained by examining the positive and negative parts of the series.

Theorem 6-6 An infinite series is absolutely convergent if and only if its positive and negative parts converge.

Proof Suppose that $\sum |a_n|$ converges. Then, since $0 \le p_n \le |a_n|$ and $0 \le q_n \le |a_n|$, it follows by the comparison test that $\sum p_n$ and $\sum q_n$ are both convergent. On the other hand, if $\sum p_n$ and $\sum q_n$ both converge, then so does $\sum p_n + \sum q_n = \sum |a_n|$ ▲.

Incidentally, Theorem 6-6 provides an alternate proof of Theorem 6-5. This follows from the relation $\sum a_n = \sum p_n - \sum q_n$.

A conditionally convergent series must possess an infinite set of positive terms and an infinite set of negative terms. For suppose that the set of positive terms is finite. Then $\sum p_n$ is convergent, whence $\sum q_n = \sum p_n - \sum a_n$ is convergent also. But then $\sum a_n$ is absolutely convergent by Theorem 6-6.

We consider now the effect on the state of convergence or divergence of a series (or on its sum, if convergent) of permuting or rearranging the order of its terms. Under what conditions may the commutative law for finite sums be extended to infinite sums? The answer, briefly, lies in the inherent difference between absolute and conditional convergence. The next two theorems are in this direction.

Theorem 6-7 Let $\sum a_n$ be an absolutely convergent series with sum S, and let $\sum b_n$ be a series obtained by rearranging or permuting the terms of $\sum a_n$. Then $\sum b_n$ is also absolutely convergent with sum S.

Proof Denote by s_n and t_n the partial sums of $\sum a_n$ and $\sum b_n$, respectively, and let $\varepsilon > 0$ be given. Then, since $\sum |a_n|$ is convergent, there exists an integer N such that

$$(3) \qquad |a_{n+1}| + |a_{n+2}| + \cdots + |a_{n+k}| < \varepsilon$$

for all $n > N$ and all $k > 0$.

Let n_1 be such that among the first n_1 terms of $\sum b_n$ are to be found the first $N+1$ terms of $\sum a_n$. Then certainly $n_1 \ge N+1$, and for $n > n_1$ the difference $t_n - s_n$ is a sum of terms of $\sum a_n$ among which $a_1, a_2, \ldots, a_{N+1}$ are not to be found. Thus $t_n - s_n$ is a sum of terms of $\sum a_n$ lying beyond the $(N+1)$th term in that series. Consequently, for some value of k, it is the case that

$$|t_n - s_n| \le |a_{N+2}| + |a_{N+3}| + \cdots + |a_{N+k+1}|.$$

By (3) with $n = N+1$, we obtain $|t_n - s_n| < \varepsilon$ for all $n > n_1$. It follows that $\{t_n - s_n\}$ is a null sequence.

So far we have proved that any rearrangement of an absolutely convergent series is convergent with the same sum. To complete the proof of the theorem we must show further that the rearranged series is absolutely convergent. Actually, this follows at once from the above conclusion when applied to the series $\sum |a_n|$. Since the latter is absolutely convergent, it follows that $\sum |b_n|$ is convergent or, what is the same, $\sum b_n$ is absolutely convergent ▲

Before stating quite a different property for conditionally convergent series, we give a classical example to show how careful we must be in handling series.

Example 6-21 Beginning with the alternating harmonic series[2]

$$S = 1 - \frac{1}{2} + \frac{1}{3} - \frac{1}{4} + \cdots + \frac{(-1)^{n+1}}{n} + \cdots,$$

suppose it to be divided term by term by 2 to yield

$$\frac{S}{2} = \frac{1}{2} - \frac{1}{4} + \frac{1}{6} - \frac{1}{8} + \cdots + \frac{(-1)^{n+1}}{2n} + \cdots,$$

and this rewritten with some 0 terms inserted. Thus

$$(4) \qquad \frac{S}{2} = 0 + \frac{1}{2} - \frac{1}{4} + 0 + \frac{1}{6} - \frac{1}{8} + 0 + \frac{1}{10} - \frac{1}{12} + 0 + \frac{1}{14} - \frac{1}{16} + 0 + \cdots.$$

Now rearrange the terms of the given series as follows.

$$(5) \qquad S = 1 - \frac{1}{2} - \frac{1}{4} + \frac{1}{3} - \frac{1}{6} - \frac{1}{8} + \frac{1}{5} - \frac{1}{10} - \frac{1}{12} + \frac{1}{7} - \frac{1}{14} - \frac{1}{16} + \frac{1}{9} + \cdots.$$

(The pattern here is reasonably clear. Each term with odd denominator is followed by two terms with consecutive even denominators.) Adding the series (4) and (5) then yields

$$\frac{3S}{2} = 1 + 0 - \frac{1}{2} + \frac{1}{3} + 0 - \frac{1}{4} + \frac{1}{5} + 0 - \frac{1}{6} + \frac{1}{7} + 0 - \frac{1}{8} + \frac{1}{9} + \cdots$$

or, what is the same,

$$\frac{3S}{2} = 1 - \frac{1}{2} + \frac{1}{3} - \frac{1}{4} + \frac{1}{5} - \cdots = S.$$

From the patently absurd conclusion we deduce the performance of at least one invalid operation. Since division of series term by term by any constant is within the rules, as is the insertion or deletion of 0 terms at will, we conclude that the rearrangement of (5) is not valid. Actually, although the right member of (5) may be shown to be a convergent series, its sum is not S. The following theorem is in the nature of an explanation.

Theorem 6-8 Let $\sum a_n$ be a conditionally convergent series, and let $\sum b_n$ denote a series constructed by a rearrangement of the terms of $\sum a_n$. Then, depending upon its construction, $\sum b_n$ may converge to any prescribed sum T or it may diverge.

This rather surprising theorem, demonstrating as it does the remarkably fickle character of conditionally convergent series, requires a somewhat lengthy proof. Inasmuch as the theorem has no special application (except in the nature of a warning) in anything that follows in the text, we omit a proof here.

[2] At a later stage it may be shown that the sum of this series is log 2. However, this is not relevant here.

Exercises

6-15 Does the series $1 - \frac{1}{2 \cdot 3} - \frac{1}{4 \cdot 5} - \cdots - \frac{1}{(2n)(2n+1)} - \cdots$ converge or diverge? Explain.

6-16 Determine the sum of the series $\frac{7}{12} - \frac{25}{72} + \cdots + (-1)^{n+1}\left(\frac{3^n + 4^n}{2(6^n)}\right) + \cdots$.

6-17 Prove that if $\sum a_n$ is absolutely convergent, then $\left|\sum a_n\right| \le \sum |a_n|$.

6-18 Prove that if $\sum a_n$ is absolutely convergent, then (a) $\sum a_{2n}$ and $\sum a_{2n+1}$ converge, and (b) $\sum a_{2n} + \sum a_{2n+1} = \sum a_n$.

6-19 Determine the sum of $\frac{1}{2} + \frac{1}{2^3} - \frac{1}{2^2} + \frac{1}{2^5} + \frac{1}{2^7} - \frac{1}{2^4} + \frac{1}{2^9} + \frac{1}{2^{11}} - \frac{1}{2^6} + \cdots$.

6-20 By rearranging the convergent series $\sum (-1)^{n+1}/\sqrt{n}$, obtain the series

$$1 + \frac{1}{\sqrt{3}} - \frac{1}{\sqrt{2}} + \frac{1}{\sqrt{5}} + \frac{1}{\sqrt{7}} - \frac{1}{\sqrt{4}} + \cdots + \frac{1}{\sqrt{4n-3}} + \frac{1}{\sqrt{4n-1}} - \frac{1}{\sqrt{2n}} + \cdots.$$

Prove that this series diverges.

6-5 Tests for Convergence of a Series

The question of whether a given infinite series is convergent or divergent is often difficult to answer. No single formula or criterion other than the definition of convergence itself or its equivalent, Cauchy convergence, is sufficient to handle all series. We include here some simple tests that find frequent application, beginning with one that is quite general in scope and then specializing to others with which the reader should already have some familiarity.

Theorem 6-9 (Kummer's Test) (a) The series $\sum a_n$ $(a_n \ne 0)$ is absolutely convergent if there exist a sequence of positive numbers $\{b_n\}$ and a positive constant c such that

$$c \le b_n - b_{n+1}\left|\frac{a_{n+1}}{a_n}\right|$$

for all n. (b) The series $\sum a_n$ $(a_n > 0)$ is divergent if a sequence of positive numbers $\{b_n\}$ exists such that $\sum 1/b_n$ diverges and

$$b_n - b_{n+1}\left(\frac{a_{n+1}}{a_n}\right) \le 0$$

for all n.

Proof (a) If $\{b_n\}$ and c exist with the stated properties, then

$$c\,|a_1| \le b_1\,|a_1| - b_2\,|a_2|,$$

$$c\,|a_2| \le b_2\,|a_2| - b_3\,|a_3|,$$

$$\cdots\cdots\cdots\cdots\cdots\cdots\cdots\cdots$$

$$c\,|a_n| \le b_n\,|a_n| - b_{n+1}\,|a_{n+1}|.$$

Adding yields

$$c\sum_{i=1}^{n}|a_i| \le b_1|a_1| - b_{n+1}|a_{n+1}| \le b_1|a_1|.$$

Therefore

$$\sum_{i=1}^{n}|a_i| \le \frac{b_1|a_1|}{c}.$$

Thus $\{\sum_{i=1}^{n}|a_i|\}$ is increasing and bounded and it follows that $\sum_1^\infty |a_n|$ is convergent.

(b) From

$$b_1a_1 \le b_2a_2,$$
$$b_2a_2 \le b_3a_3,$$
$$\cdots\cdots\cdots\cdots\cdots$$
$$b_{n-1}a_{n-1} \le b_na_n,$$

we obtain

$$\frac{b_1a_1}{b_n} \le a_n$$

for all n. Since $\{1/b_n\}$ diverges, it follows by the comparison test that $\sum a_n$ diverges▲

With respect to (b) of Theorem 6-9 it may be noted that the given series must be a series of positive terms, while in (a) this is not essential. Also, it should be observed in (a) that no condition of convergence or divergence is imposed.

Example 6-22 The series $\sum \frac{[(2n)!]^3}{2^{6n}(n!)^6}$ converges. For by choosing $b_n = n$, we obtain

$$b_n - b_{n+1}\left|\frac{a_{n+1}}{a_n}\right| = \frac{1}{2} - \frac{6n+5}{8(n+1)^2}.$$

The sequence $\left\{\frac{6n+5}{8(n+1)^2}\right\}$ is easily seen to be a decreasing null sequence with the first element having the value $\frac{11}{32}$. It follows that

$$b_n - b_{n+1}\left|\frac{a_{n+1}}{a_n}\right| > \frac{1}{2} - \frac{11}{32} = \frac{5}{32} > 0$$

for all n. Hence the given series converges.

Example 6-23 We use Kummer's test to show that the series $\sum \frac{(2n)!}{2^{2n}(n!)^2}$ is divergent. Again choose $b_n = n$. (Observe that $\sum 1/b_n = \sum 1/n$ diverges.) Then

$$b_n - b_{n+1}\left(\frac{a_{n+1}}{a_n}\right) = -\frac{1}{2} \le 0$$

for all n. The series diverges.

The Kummer test requires some ingenuity in the choice of the sequence $\{b_n\}$. A test less general in scope, but somewhat more easy to apply, may be derived from it.

Theorem 6-10 (The Cauchy Ratio Test) Let $\sum a_n$ be a series of nonzero terms.

(a) If $\overline{\lim}\left\{\left|\dfrac{a_{n+1}}{a_n}\right|\right\} < 1$, the series is absolutely convergent.

(b) If $\underline{\lim}\left\{\left|\dfrac{a_{n+1}}{a_n}\right|\right\} > 1$, the series is divergent.

(c) If $\underline{\lim}\left\{\left|\dfrac{a_{n+1}}{a_n}\right|\right\} \leq 1 \leq \overline{\lim}\left\{\left|\dfrac{a_{n+1}}{a_n}\right|\right\}$, the test is inconclusive.

(In the event that $\{a_{n+1}/a_n\}$ is unbounded, the limiting points are to be considered in the extended real number system $\mathscr{R}^*$.)

Proof (a) A number k exists such that

$$\overline{\lim}\left\{\left|\frac{a_{n+1}}{a_n}\right|\right\} < k < 1.$$

It follows that N exists such that $|a_{n+1}/a_n| < k$ for all $n > N$, and hence $0 < 1 - k < 1 - |a_{n+1}/a_n|$. By Kummer's test, with $b_n = 1$, it follows that $\sum_{N+1}^{\infty} a_n$, and hence $\sum_1^{\infty} a_n$, converges absolutely.

(b) From $\underline{\lim}\{|a_{n+1}/a_n|\} > 1$ it follows that N exists such that $1 - |a_{n+1}/a_n| < 0$ for all $n > N$. With $b_n = 1$ it follows, by Kummer's test, that $\sum_{N+1}^{\infty} a_n$, and hence $\sum_1^{\infty} a_n$, diverges.

(c) To see that no decision is forthcoming in this case, we have only to observe that $\lim\{|a_{n+1}/a_n|\} = 1$ for both the divergent series $\sum 1/n$ and the convergent series $\sum (-1)^{n-1}/n$▲

Example 6-24 Consider the series

$$\sum_{n=1}^{\infty} a_n = 1 + \frac{1}{2} + \frac{1}{3} + \frac{1}{6} + \frac{1}{9} + \frac{1}{18} + \frac{1}{27} + \cdots,$$

where $a_{2n-1} = 1/3^{n-1}$ and $a_{2n} = a_{2n-1}/2$. Since $a_{2n}/a_{2n-1} = 1/2$ and $a_{2n+1}/a_{2n} = 2/3$, we have $\overline{\lim}\{|a_{n+1}/a_n|\} = 2/3$. Hence $\sum a_n$ converges.

With respect to divergence, the Cauchy ratio test says nothing very remarkable. The relation $|a_{n+1}/a_n| > 1$ simply implies $|a_n| < |a_{n+1}|$ and hence denies that $\{a_n\}$ is a null sequence, a condition already seen to be necessary for convergence. However, it must not be concluded that $|a_{n+1}/a_n| \geq 1$ for every divergent series. For example, the divergent series $\sum 1/n$ is such that $a_{n+1}/a_n = n/(n+1) < 1$ for all n.

Example 6-25 Consider the series

$$\sum_{n=1}^{\infty} a_n = \frac{1}{2} + 1 + \frac{1}{4} + \frac{1}{2} + \frac{1}{16} + \frac{1}{8} + \frac{1}{128} + \frac{1}{64} + \cdots,$$

where $a_{2n-1} = (1/2)^{(n^2-n+2)/2}$ and $a_{2n} = (1/2)^{n(n-1)/2}$. Here $a_{2n}/a_{2n-1} = 2$ and $a_{2n+1}/a_{2n} = 1/2^{n+1}$. Hence $\overline{\lim}\{|a_{n+1}/a_n|\} = 2$ and $\underline{\lim}\{|a_{n+1}/a_n|\} = 0$. The test, therefore, confirms neither convergence nor divergence.

A modification of the ratio test is that found in most calculus books. It is easy to use but is indecisive in many instances.

Theorem 6-11 (Modified Ratio Test) Let $\sum a_n$ be a series of nonzero terms and define $\rho = \lim \left\{\left|\dfrac{a_{n+1}}{a_n}\right|\right\}$ if such a limit exists, finite or infinite.

(a) The series converges absolutely if $\rho < 1$.
(b) The series diverges if $\rho > 1$.
(c) The test is inconclusive if $\rho = 1$.

Proof This is an immediate consequence of Theorem 6-10▲

Example 6-26 The series $\sum 2^n/n^2$ diverges since

$$\lim\{|a_{n+1}/a_n|\} = \lim\{2n^2/(n+1)^2\} = 2.$$

Example 6-27 The series $\sum \sqrt{n}/2^n$ converges since

$$\lim\{|a_{n+1}/a_n|\} = \lim\{\tfrac{1}{2}\sqrt{(n+1)/n}\} = \tfrac{1}{2}.$$

Another important test for convergence or divergence is one that makes use of the sequence $\{\sqrt[n]{|a_n|}\}$.

Theorem 6-12 (The Root Test) Let $\sum a_n$ be a given series, and define $\sigma = \overline{\lim}\{\sqrt[n]{|a_n|}\}$, finite or infinite.

(a) The series converges absolutely if $\sigma < 1$.
(b) The series diverges if $\sigma > 1$.
(c) The test is inconclusive if $\sigma = 1$.

Proof (a) From $\sigma < 1$ it follows that N exists such that $\sqrt[n]{|a_n|} < k < 1$, or, what is the same, $|a_n| < k^n < 1$ for all $n > N$. Since $\sum k^n$ converges, it follows by the comparison test that $\sum |a_n|$ converges also.

(b) Suppose $\sum a_n$ converges. Then N exists such that $|a_n| < 1$ for $n > N$. Hence $\sqrt[n]{|a_n|} < 1$ for $n > N$, so that $\sigma \leq 1$.

(c) Observe that $\sigma = 1$ for both $\sum 1/n$ and $\sum (-1)^{n-1}/n$▲

Example 6-28 The series $\sum 2^n/n^n$ converges because $\sigma = \overline{\lim}\{2/n\} = \lim\{2/n\} = 0$.

Example 6-29 Consider the series $\sum a_n = 1/2 + 1/4 + 1/4 + 1/16 + 1/8 + 1/64 + \cdots$, where $a_{2n-1} = 1/2^n$ and $a_{2n} = 1/4^n$. The sequence $\{\sqrt[n]{|a_n|}\}$ has two limiting points:

$$\lim \left\{ \sqrt[2n-1]{\frac{1}{2^n}} \right\} = \frac{1}{\sqrt{2}} \quad \text{and} \quad \lim \left\{ \sqrt[2n]{\frac{1}{4^n}} \right\} = \frac{1}{2}.$$

The series converges because $\overline{\lim} \{\sqrt[n]{|a_n|}\} = 1/\sqrt{2} < 1$.

Example 6-30 The series $\sum 2^n/n^4$ diverges since $\overline{\lim} \{\sqrt[n]{|a_n|}\} = \lim \{2/(n^{1/n})^4\} = 2$.

It is not surprising that the root test indicates divergence when $\sigma > 1$, since the latter simply implies that $\{a_n\}$ does not converge to zero.

Some indication of the comparative merits of the ratio and root tests is given by the following theorem.

Theorem 6-13 If $\sum a_n$ is any series, then

(a) $\overline{\lim} \{\sqrt[n]{|a_n|}\} \leq \overline{\lim} \left\{ \left| \frac{a_{n+1}}{a_n} \right| \right\}$, and

(b) $\underline{\lim} \left\{ \left| \frac{a_{n+1}}{a_n} \right| \right\} \leq \underline{\lim} \{\sqrt[n]{|a_n|}\}$

(where infinite limiting points are allowed).

Proof We prove (a) and leave the proof of (b) to the reader. Obviously the theorem is true if $\overline{\lim} \{|a_{n+1}/a_n|\}$ is infinite. If $\overline{\lim} \{|a_{n+1}/a_n|\}$ is finite, we conclude that, corresponding to each number c with the property $c > \overline{\lim} \{|a_{n+1}/a_n|\}$. there exists an integer N such that $|a_{n+1}/a_n| \leq c$ for all $n > N$. Hence, for any $k > 0$, we have

$$|a_{N+2}| \leq c\,|a_{N+1}|,$$
$$|a_{N+3}| \leq c\,|a_{N+2}|,$$
$$\cdots\cdots\cdots\cdots\cdots\cdots$$
$$|a_{N+k+1}| \leq c\,|a_{N+k}|.$$

Multiplying yields

$$|a_{N+k+1}| \leq c^k\,|a_{N+1}|$$

or, what is the same, $|a_n| \leq c^{n-N-1}\,|a_{N+1}|$ for $n > N$. Hence

$$\sqrt[n]{|a_n|} \leq c\,\sqrt[n]{c^{-N-1}\,|a_{N+1}|}\,.$$

Therefore $\overline{\lim} \{\sqrt[n]{|a_n|}\} \leq c\,\overline{\lim} \{\sqrt[n]{c^{-N-1}\,|a_{N+1}|}\}$. Since $c^{-N-1}\,|a_{N+1}|$ is a positive constant, it follows (from Example 5-10) that $\lim \{\sqrt[n]{c^{-N-1}\,|a_{N+1}|}\} = 1$. Inequality (a) follows from the arbitrary nature of c ▲

The import of the above is that, whenever the ratio test yields information about a series, the root test does likewise. That is, if $\overline{\lim}\{|a_{n+1}/a_n|\} < 1$, indicating convergence, then $\overline{\lim}\{\sqrt[n]{|a_n|}\} < 1$ also, pointing to the same conclusion. Similarly, if the ratio test indicates divergence, the root test does also. That the root test is actually more powerful than the ratio test is seen in the following example.

Example 6-31 For the series $\sum a_n = \frac{1}{4} + \frac{1}{4} + \frac{1}{16} + \frac{1}{16} + \cdots$, where $a_{2n-1} = a_{2n} = 1/2^{2n}$, we have $a_{2n}/a_{2n-1} = 1$ and $a_{2n+1}/a_{2n} = \frac{1}{4}$. Therefore $\overline{\lim}\{|a_{n+1}/a_n|\} = 1$. Consequently, as far as the ratio test is concerned, the question of convergence is still open. On the other hand, $\sqrt[2n]{|a_{2n}|} = \frac{1}{2}$ and $\sqrt[2n-1]{|a_{2n-1}|} < \sqrt[2n-1]{1/2^{2n-1}} = \frac{1}{2}$. Consequently, $\overline{\lim}\{\sqrt[n]{|a_n|}\} = \frac{1}{2}$. We conclude from the root test that the given series converges.

Another test, due to Cauchy, elicits from a series formed from a sparsely distributed subset of terms of $\sum a_n$ the state of convergence or divergence of $\sum a_n$. Appropriately, it is called *Cauchy's condensation test*. With this it is possible to settle the issue of convergence of the p-series.

Theorem 6-14 Let $\{a_n\}$ be a decreasing sequence of non-negative numbers. Then the series $\sum_{n=0}^{\infty} a_n$ and

$$\sum_{n=0}^{\infty} 2^n a_{2^n} = a_1 + 2a_2 + 4a_4 + 8a_8 + \cdots$$

both converge or both diverge.

Proof For the sum of a set of 2^n terms of the series $\sum a_n$ beginning with a term of the form a_{2^n}, we have

$$2^n a_{2^{n+1}} \leq a_{2^n} + a_{2^n+1} + \cdots + a_{2^{n+1}-1} \leq 2^n a_{2^n}. \tag{6}$$

Now, the partial sum $s_{2^{n+1}-1}$ of $\sum a_n$ may be written

$$s_{2^{n+1}-1} = a_1 + \sum_{2^1}^{2^2-1} a_i + \sum_{2^2}^{2^3-1} a_i + \cdots + \sum_{2^n}^{2^{n+1}-1} a_i,$$

so that by (6) we have

$$a_1 + \frac{1}{2}\sum_{2}^{n+1} 2^i a_{2^i} \leq s_{2^{n+1}-1} \leq a_1 + \sum_{1}^{n} 2^i a_{2^i}. \tag{7}$$

If $\sum a_n$ is convergent, then $\{s_{2^{n+1}-1}\}$ is bounded. From (7) follows boundedness and hence convergence of $\{a_1 + \frac{1}{2}\sum_2^n 2^i a_{2^i}\}$. As a consequence, $\sum_0^{\infty} 2^n a_{2^n}$ converges. Similarly, if $\sum_0^{\infty} 2^n a_{2^n}$ converges, then also from (7) follows boundedness and hence convergence of $\{s_{2^{n+1}-1}\}$. Therefore $\{s_n\}$ is bounded and so $\sum a_n$ converges▲

Example 6-32 The state of convergence or divergence of the p-series $\sum 1/n^p$ for different values of p has not been completely settled. (See Example 6-11.) With the aid of Theorem 6-14 we can reach a decision for all $p > 0$.

The series $\sum 1/n^p$ and $\sum 2^n(1/2^n)^p$ converge or diverge together. The latter may be written $\sum (1/2^{p-1})^n$, which is geometric; it converges for $p > 1$ and diverges for $p \leq 1$.

Exercises

6-21 Use Kummer's test to prove *Raabe's test*: A series $\sum a_n$ of positive terms converges if $\overline{\lim} \left\{n\left(\frac{a_{n+1}}{a_n} - 1\right)\right\} < -1$, and diverges if $\underline{\lim} \left\{n\left(\frac{a_{n+1}}{a_n} - 1\right)\right\} > -1$.

6-22 Prove *Gauss' test*: Let $\sum a_n$ be a series of positive terms, and assume that $\frac{a_n}{a_{n+1}} = 1 + \frac{k}{n} + \frac{f(n)}{n^2}$, where k is a constant and $f(n)$ is bounded for all n. The series $\sum a_n$ converges if $k > 1$ and diverges if $k \leq 1$.

6-23 Test the following series for convergence or divergence:

$$\text{(a)} \sum \frac{(1)(3) \cdots (2n-1)}{(2)(4) \cdots (2n)}; \qquad \text{(b)} \sum \frac{(1)(3) \cdots (2n-1)}{(2)(4) \cdots (2n)} \cdot \frac{1}{2n+1}.$$

6-24 Prove Theorem 6-10 directly without using Kummer's test.

6-25 (a) Prove that if an integer N exists such that $|a_{n+1}/a_n| \geq 1$ for all $n > N$, then the series $\sum a_n$ is divergent. (b) Cite an example of a series $\sum a_n$ for which $|a_{n+1}/a_n| < 1$ for all n and $\underline{\lim} \{|a_{n+1}/a_n|\} = 1$.

6-26 Cite an example to show that a series may converge and yet be such that $\overline{\lim} \{|a_{n+1}/a_n|\} = \infty$.

6-27 Let $a_{2n-1} = 1/(n+1)^n$ and $a_{2n} = 1/(n+2)^n$. Prove that $\sum a_n$ converges.

6-28 Prove part (b) of Theorem 6-13.

6-29 Use Theorem 6-13 to prove that $\lim \{\sqrt[n]{n}\} = 1$.

6-30 Use Cauchy's condensation test to prove that $\sum 1/n(\log n)^p$ converges if $p > 1$ and diverges if $p \leq 1$.

6-31 Prove that if $\{a_n b_n\}$ converges, then the series $a_0 b_0 + \sum_{n=1}^{\infty} (a_n - a_{n-1})b_n$ and $\sum_0^{\infty} a_n (b_n - b_{n+1})$ both converge or both diverge.

6-32 Prove that if $\sum a_n$ and $\sum b_n$ are such that the sequence $\{\sum_{k=1}^{n} a_k\}$ is bounded, $\{b_n\}$ is a null sequence, and $\sum(b_n - b_{n+1})$ is absolutely convergent, then the series $\sum a_n b_n$ converges.

6-33 Prove that if $\sum a_n$ and $\sum b_n$ are such that $\sum a_n$ is convergent and $\sum(b_n - b_{n+1})$ is absolutely convergent, then $\sum a_n b_n$ is convergent.

6-6 Multiplication of Series

Multiplication of two infinite series may be looked upon as a generalization of multiplication of polynomials. Thus, if $\sum a_n$ and $\sum b_n$ are two series with partial sums s_n and t_n, respectively, we may define the product of $\sum a_n$ and $\sum b_n$ to be the sequence

$$\{s_n t_n\} = \{(a_1 + a_2 + \cdots + a_n)(b_1 + b_2 + \cdots + b_n)\}.$$

Let the product series, with $s_n t_n$ as its nth partial sum, be denoted by $\sum p_n$. Then

$$\begin{aligned} p_n &= s_n t_n - s_{n-1} t_{n-1} \\ &= (a_1 + \cdots + a_{n-1} + a_n)(b_1 + \cdots + b_{n-1} + b_n) \\ &\quad - (a_1 + \cdots + a_{n-1})(b_1 + \cdots + b_{n-1}) \\ &= (a_1 + \cdots + a_{n-1}) b_n + a_n (b_1 + \cdots + b_n) \\ &= a_n t_n + b_n s_{n-1}. \end{aligned}$$

As an immediate consequence of the theorem on products of sequences, we may state the following.

Theorem 6-15 If $\sum a_n$ and $\sum b_n$ converge to S and T, respectively, and if s_n and t_n are the respective partial sums, then the product series $\sum p_n = \sum (a_n t_n + b_n s_{n-1})$ converges to ST.

A definition of the product of two series may be given in an essentially different yet equally reasonable manner, again in the nature of an extension of the distributive law of multiplication of polynomials. Thus, if $\sum a_n$ and $\sum b_n$ are given series, we may define their product to be an infinite sum of all possible products of the form $a_j b_k$, where j and k independently assume all positive integral values. Here, of course, the question of ordering or arranging terms arises and with it the possibility that absolute convergence of $\sum a_n$ and $\sum b_n$ may be important to the convergence of the product series. Indeed, we have the following theorem.

Theorem 6-16 If $\sum a_n$ and $\sum b_n$ are absolutely convergent to sums S and T, respectively, then, regardless of ordering, the product series $\sum a_j b_k$ is also absolutely convergent and has sum ST.

Proof Let $\sum q_n$ denote a product series with terms $a_j b_k$ in some given order. Then consider the partial sum

$$|q_1| + |q_2| + \cdots + |q_n| \tag{8}$$

of the series $\sum |q_n|$. Each of the numbers q_i is a product $a_j b_k$ for some j and k. Let m denote the largest value of j or k for terms occurring in (8). Then

$$|q_1| + |q_2| + \cdots + |q_n| \leq (|a_1| + \cdots + |a_m|)(|b_1| + \cdots + |b_m|).$$

Since $\sum |a_n|$ and $\sum |b_n|$ converge to, say, A and B, respectively, it follows that

$$|q_1| + |q_2| + \cdots + |q_n| \leq AB$$

for all values of n. The boundedness of the sequence of partial sums, together with its monotony, implies the convergence of $\sum |q_n|$.

Having shown $\sum q_n$ to be absolutely convergent, it follows that its terms may be rearranged and parentheses inserted without altering its sum. In particular, the product series $\sum p_n$ can be constructed from $\sum q_n$ by such operations. Since $\sum p_n$ was seen to converge to ST, it follows that $\sum q_n$ does likewise ▲

Still another definition of the product of two series is reasonable. Because of properties of a special type of series known as a *power series*, which we shall discuss in Chapter 12, it is desirable to define the product of $\sum a_n$ and $\sum b_n$ to be the series $\sum c_n$ such that

$$c_n = \sum_{i=1}^{n} a_i b_{n-i+1} = a_1 b_n + a_2 b_{n-1} + \cdots + a_n b_1.$$

Thus

$$\sum_{n=1}^{\infty} c_n = a_1 b_1 + (a_1 b_2 + a_2 b_1) + (a_1 b_3 + a_2 b_2 + a_3 b_1) + \cdots,$$

where in any one term the sum of the subscripts is constant throughout the summands of the term. The series $\sum c_n$ is known as the *Cauchy product* of $\sum a_n$ and $\sum b_n$. *From this point on, unless otherwise specified, whenever we speak of the product of two series we shall mean the Cauchy product and denote it by* $\sum a_n \sum b_n$.

Since the Cauchy product (with parentheses removed) is again but a particular arrangement of all products of the form $a_j b_k$, we may draw the following conclusion.

Theorem 6-17 If $\sum a_n$ and $\sum b_n$ converge absolutely, then the Cauchy product $\sum c_n$ is absolutely convergent and $\sum c_n = (\sum a_n)(\sum b_n)$.

The condition of absolute convergence placed upon $\sum a_n$ and $\sum b_n$ in Theorem 6-17 may be relaxed somewhat. As proved in Theorem 6-18, only one of the two series needs to be absolutely convergent in order to guarantee convergence (but not absolute convergence) of the product.

Theorem 6-18 If $\sum a_n$ is absolutely convergent with $\sum a_n = S$ and if $\sum b_n$ is convergent with $\sum b_n = T$, then the Cauchy product converges to ST.

Proof Introduce the symbols $s_n = a_1 + a_2 + \cdots + a_n$, $t_n = b_1 + b_2 + \cdots + b_n$, $u_n = c_1 + c_2 + \cdots + c_n$, $\tau_n = T - t_n$, and $\sigma_n = a_1\tau_n + a_2\tau_{n-1} + \cdots + a_n\tau_1$. The reader may easily verify that $u_n = Ts_n - \sigma_n$.

Inasmuch as $\lim \{s_n\} = S$, it follows that $\lim \{Ts_n\} = ST$. To show, then, that $\lim \{u_n\}$ exists and has the value ST, we have only to show that $\{\sigma_n\}$ is a null sequence. To this end, observe that since $\sum |a_n|$ is convergent, say, $\sum |a_n| = A$, it follows that

$$|a_1| + |a_2| + \cdots + |a_n| \leq A$$

for all values of n. Also, since $\sum b_n$ converges to T, the sequence $\{\tau_n\}$ is a null sequence and, of necessity, is bounded. Thus a number M exists such that $|\tau_n| \leq M$ for all n.

Now, let $\varepsilon > 0$ be given. Then, from the convergence of $\{\tau_n\}$ to 0, it follows that N_1 exists such that $|\tau_n| < \varepsilon$ for all $n > N_1$. Hence for such values of n we have

$$
\begin{aligned}
|\sigma_n| &\le (|a_1|\,|\tau_n| + \cdots + |a_{n-N_1}|\,|\tau_{N_1+1}|) + (|a_{n-N_1+1}|\,|\tau_{N_1}| + \cdots + |a_n|\,|\tau_1|) \\
&< (|a_1| + \cdots + |a_{n-N_1}|)\varepsilon + (|a_{n-N_1+1}| + \cdots + |a_n|)M \\
&< A\varepsilon + (|a_{n-N_1+1}| + \cdots + |a_n|)M.
\end{aligned}
$$

Since $\sum |a_n|$ converges, a number N_2 exists such that

$$|a_{n+1}| + \cdots + |a_{n+k}| < \varepsilon$$

for all $n > N_2$ and all $k > 0$. Let $N = N_1 + N_2$. Then $n > N$ implies $n > N_1$ and $n - N_1 > N_2$. Hence for $n > N$ we have

$$|a_{n-N_1+1}| + \cdots + |a_n| < \varepsilon.$$

It follows that

$$|\sigma_n| < A\varepsilon + M\varepsilon = (A + M)\varepsilon$$

for all $n > N$. Thus $\{\sigma_n\}$ has been shown to be a null sequence and the truth of the theorem follows ▲

Example 6-33 serves to demonstrate that we cannot conclude absolute convergence of the product of two series satisfying the hypothesis of Theorem 6-18.

Example 6-33 Consider the series $\sum (-1)^{n+1}/n^{3/2}$ and $\sum (-1)^{n+1}/n^{1/2}$. Since the first is absolutely convergent and the second conditionally convergent, it follows by Theorem 6-18 that the product series $\sum c_n$ is convergent, where

$$c_n = (-1)^{n+1} \sum_{k=1}^{n} \frac{1}{k^{3/2}(n+1-k)^{1/2}}.$$

It is easily seen, however, that it is not absolutely convergent. For from $k^{3/2}(n+1-k)^{1/2} \le n^2$ (since $k \le n$) we have

$$|c_n| = \sum_{k=1}^{n} \frac{1}{k^{3/2}(n+1-k)^{1/2}} \ge \frac{n}{n^2} = \frac{1}{n}.$$

From the divergence of $\sum 1/n$ follows that of $\sum |c_n|$.

Example 6-34 It is easy to see that if two series converge but do not converge absolutely, then their product need not converge. For suppose $\sum a_n = \sum b_n = \sum (-1)^{n+1}/\sqrt{n}$, which is conditionally convergent. We have

$$c_n = (-1)^{n+1} \sum_{k=1}^{n} \frac{1}{\sqrt{k(n-k+1)}},$$

and since $1/\sqrt{k(n-k+1)} \ge 1/n$ for all n and k, it follows that $|c_n| \ge 1$. Since the sequence $\{c_n\}$ is not a null sequence, $\sum c_n$ diverges. Thus the product of two conditionally convergent series may be divergent.

The converse of Theorem 6-18 is not true. A simple counterexample shows that the product of two series may converge even if neither of the factors does.

Example 6-35 Consider the series $\sum a_n$, where $a_1 = 1$ and $a_n = 2$ for $n > 1$, and the series $\sum b_n$, defined by $b_1 = 1$ and $b_n = (-1)^{n+1}2$ for $n > 1$. Each is clearly divergent. For the product series $\sum c_n$ we have $c_1 = 1$ and for $n > 1$

$$c_n = 2 - (-1)^{n+1}2 + 4\sum_{k=2}^{n}(-1)^{k+1} = 0.$$

Thus $\sum c_n$ is the convergent series $1 + 0 + 0 + 0 + \cdots$.

Exercises

6-34 (a) Find the Cauchy product $\sum_{n=0}^{\infty} r^n \sum_{n=0}^{\infty} r^n$. (b) What is the sum of the product series?

6-35 Let $\sum a_n$, $\sum b_n$, and $\sum c_n$ be series such that $c_n = a_1 b_n + a_2 b_{n-1} + \cdots + a_n b_1$. Prove that if the series converge to A, B, and C, respectively, then $AB = C$.

6-36 Prove that each real number a is the sum of a series of the form $\sum_{n=0}^{\infty} a_n/10^n$, where a_0 is an integer and where a_n is an integer such that $0 \leq a_n < 9$ for $n = 1, 2, \ldots$. (Such a series is called a *decimal representation* of the number a.)

6-37 Show that the series $1 + 0/10 + 0/10^2 + \cdots$ and $9/10 + 9/10^2 + \cdots$ both converge to 1. (It follows that the decimal representation of a number is not necessarily unique.)

6-38 Prove that a is a rational number if and only if its decimal representation $\sum_{n=1}^{\infty} a_n/10^n$ is *repeating*; that is, for some integers k and N it is the case that $a_{n+k} = a_n$ for all $n > N$.

6-39 In Example 5-13 it was shown that the sequence $\{(1 + 1/n)^n\}$ converges to a limit which was denoted by e. (a) Use the binomial expansion of $(1 + 1/n)^n$ to prove that $e \leq \sum_{n=0}^{\infty} 1/n!$. (b) Prove that $\sum_{n=0}^{\infty} 1/n! \leq e$ and conclude $e = \sum_{n=0}^{\infty} 1/n!$.

6-40 Prove that $0 < e - \sum_{t=0}^{n} 1/t! < 1/n(n!)$.

6-41 Prove that e is irrational.

6-42 Let $\{s_n\}$ be the sequence of partial sums of the series $\sum a_n$, and let $\sigma_n = (s_1 + \cdots + s_n)/n$. If $\lim \{\sigma_n\}$ exists, the series $\sum a_n$ is said to be *Cesáro summable*, and the limit is called the *Cesáro sum* of the series. (a) Prove that if $\sum a_n$ has a sum, then it is *Cesáro* summable and that the sum and Cesáro sum agree (compare Exercise 5-52). (b) Exhibit a divergent series that is Cesáro summable.

7
Limit of a Function

From whatever side we approach our principle, we reach the same conclusion, ...

JEAN JACQUES ROUSSEAU (1712–1778), *The Social Contract*

7-1 Functions of a Real Variable

In an earlier chapter we introduced the concept of a function that is defined on a set A and that has range in a set B. We now particularize and consider functions for which both A and B are sets of real numbers. Such *real-valued functions of a real variable* will be the principal object of our attention in the chapters to come.

The device of a sketch or graph of a function has certain instructive value. Accordingly, we consider two real number axes, styled the x and y coordinate axes, and look upon any given function as a mapping of points of the x-axis onto points of the y-axis. The plotting of a point (x,y) in the usual Cartesian plane is then indicative of the association of the numbers x and y by the given function. The set of points (x,y) so plotted constitute the *graph* of the function, and the relation $y = f(x)$ is referred to as the *equation* of the graph.

Following are examples of real functions together with domains of definition, ranges, and, in some cases, graphs.

Example 7-1 Let f be a function defined on the domain $(-1, 3]$ by $f(x) = x^2$. The range of f is the bounded interval $[0, 9]$. (See Figure 7-1.)

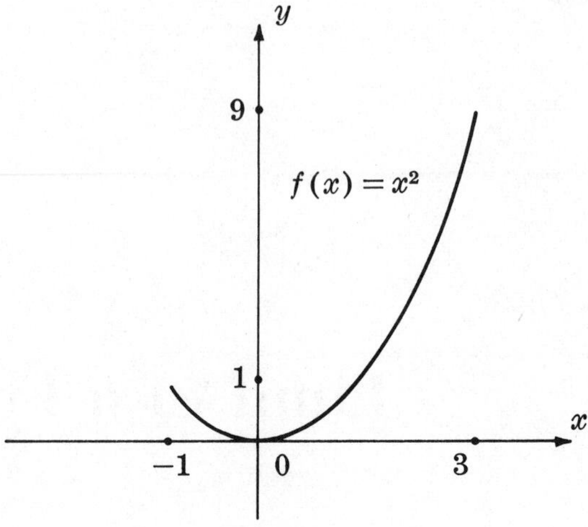

Figure 7-1

Example 7-2 Let h be a function defined on the interval $[0, \pi/2]$ by $h(x) = \sin x$.[1] Here the range is the interval $[0, 1]$. (See Figure 7-2.)

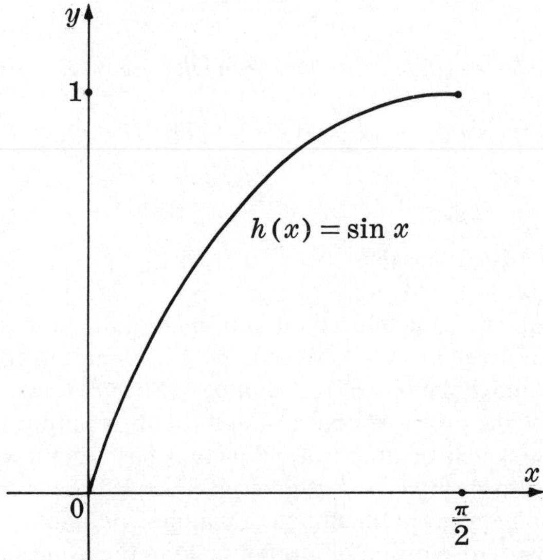

Figure 7-2

Example 7-3 The function G defined by $G(x) = (x/2) - 1$ if $-2 < x < 0$, $G(0) = 1$, and $G(x) = 3 - (x/3)$ if $1 \leq x < 3$ has for its range the set of real numbers $\{y;\ -2 < y < -1,\ y = 1,\ 2 < y \leq 8/3\}$. (See Figure 7-3.)

[1] We shall assume knowledge of the definitions and some properties of certain so-called elementary functions such as $\sin x$, $\cos x$, e^x, and $\log x$. Formal discussion is postponed until a later chapter.

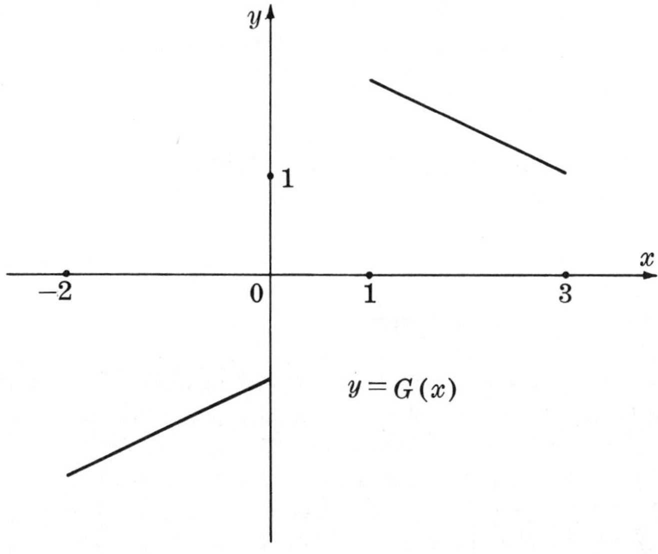

Figure 7-3

Example 7-4 Let Ψ denote the function defined on $[a, b]$ by

$$\Psi(x) = \begin{cases} 0 & \text{if } x \text{ is rational} \\ 1 & \text{if } x \text{ is irrational.} \end{cases}$$

Its range is the finite set consisting of the two numbers 0 and 1. It is apparent that Figure 7-4 is, at best, an unsatisfactory sketch of the graph of Ψ. We shall have frequent occasion to look into properties of this function and shall reserve the symbol Ψ for it exclusively.

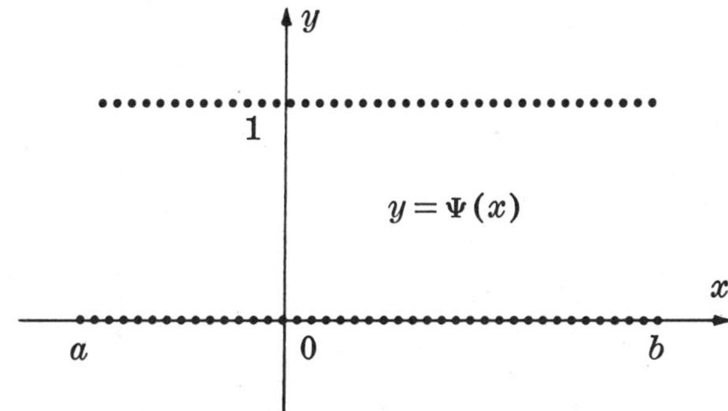

Figure 7-4

The above examples do not exhibit the fullest generality of the function concept with respect to possible domains of definition. All concern functions defined on domains consisting of a single interval or set of intervals. For the most part, discussions in the ensuing chapters will center about functions of this type. However, we must be aware of the variety of complications that may arise.

Example 7-5 Let p be defined on the set R of rational numbers by $p(x) = x/2$. The range of p, also, is the set of all rational numbers.

Example 7-6 Let q be defined on the set of rational numbers of the form $1/n$, where n is a positive integer, by $q(x) = x^2$. The range of q is, of course, the set of all rational numbers y such that $\sqrt{y}$ is of the form $1/n$.

In all that follows, unless some statement is made to the contrary, the term function is always to be construed as a real-valued function of a real variable.

Suppose f and g to denote functions defined on a common domain D. Then, by the *sum* of f and g is meant the function $f + g$ defined on D by $(f + g)(x) = f(x) + g(x)$. The function $-g$ that has the value $-g(x)$ at x is called the *negative* of g, and the *difference* of f and g is the function defined by $f - g = f + (-g)$. By the *product* of f and g we mean the function fg defined by $(fg)(x) = f(x)g(x)$. The function having the value $1/g(x)$ at x is the *reciprocal* of g and is denoted by $1/g$. It has meaning, of course, only at points of D for which $g(x) \neq 0$. The *quotient* $f/g = (f)(1/g)$ is such that $(f/g)(x) = f(x)/g(x)$ for all x for which $g(x) \neq 0$. We introduce, too, the notation $|f|$ to denote the *absolute value* function defined on D by $|f|(x) = |f(x)|$.

We shall have occasion to use certain special functions. We define the *minimum* (or *meet*) of f and g by $(f \wedge g)(x) = \min\{f(x), g(x)\}$ and the *maximum* (or *join*) by $(f \vee g)(x) = \max\{f(x), g(x)\}$. Further, we define the *positive* and *negative parts* of a function f by

$$f^+(x) = \begin{cases} f(x) & \text{if } f(x) \geq 0, \\ 0 & \text{if } f(x) < 0; \end{cases} \qquad f^-(x) = \begin{cases} 0 & \text{if } f(x) \geq 0, \\ -f(x) & \text{if } f(x) < 0. \end{cases}$$

It is easily seen that $f = f^+ - f^-$ and that $f^+ = f \vee 0$ and $f^- = (-f) \vee 0$, where 0 is the zero function on the domain of f.

Example 7-7 If f and g are defined on $\mathscr{R}$ by $f(x) = 1 + x$ and $g(x) = 2x^2$, then $(f + g)(x) = 1 + x + 2x^2$, $(f - g)(x) = 1 + x - 2x^2$, $(fg)(x) = 2(1 + x)x^2$, and $(f/g)(x) = (1 + x)/2x^2$ (for $x \neq 0$). Also, $|f|(x) = |1 + x|$. Further, we have

$$(f \wedge g)(x) = \begin{cases} 2x^2 & \text{if } -\frac{1}{2} < x < 1, \\ 1 + x & \text{otherwise;} \end{cases} \qquad (f \vee g)(x) = \begin{cases} 1 + x & \text{if } -\frac{1}{2} < x < 1, \\ 2x^2 & \text{otherwise;} \end{cases}$$

$$f^+(x) = \begin{cases} 0 & \text{if } x < -1, \\ 1 + x & \text{if } -1 \leq x; \end{cases} \qquad f^-(x) = \begin{cases} -1 - x & \text{if } x < -1, \\ 0 & \text{if } -1 \leq x; \end{cases}$$

$$g^+(x) = g(x); \qquad g^-(x) = 0.$$

If f and g are such that $f(x) \leq g(x)$ for all x in D, then we say that f is *less than or equal to* g and write $f \leq g$. The order relation $\leq$ for functions is obviously transitive. If f and g are not identical functions, that is, $f \neq g$, and if $f(x) \leq g(x)$ for all x in D, then we say that f is *less than* g and write $f < g$. Thus $f < g$ means $f(x) \leq g(x)$, with at least one point of D having the property that $f(x) < g(x)$.

It is clearly important to the above definitions that the functions involved have a common domain of definition. In the future we shall always assume this to be the case without explicit comment.

If a function is defined on a domain D such that its range consists of a single number k, then we call it a *constant* function on D and denote it by k. Note that in this context the symbol k assumes a dual role: in the statement $f = k$, it represents a function; in $f(x) = k$, a number. If $k = 0$ or $k = 1$, we refer to it as the *zero* function or the *unit* function, respectively, on D. It is clear that if f is any function defined on D, then $f - f = 0$ and $(f)(1/f) = 1$ (the last, provided f does not vanish on D).

Exercises

7-1 Let f and g be defined on a common domain D by $f(x) = x^3 + 3x$ and $g(x) = x + 2^x$. What are the functions: (a) $f - 2g$; (b) $g/4f$; (c) fg; (d) $|f+g|$; (e) f^+, f^-, g^+, g^-; (f) $f \wedge g, f \vee g$?

7-2 Let f and g be functions defined on a common domain D. Write $f <_1 g$ if and only if $f(x) < g(x)$ for all x in D. (a) Describe the difference between the statements $f < g$ and $f <_1 g$. (b) Discuss the statements (i) $f \leq g$ and $f \neq g$ imply $f < g$, and (ii) $f \leq g$ and $f \neq g$ imply $f <_1 g$.

7-2 Limit of a Function at a Point

Suppose f to denote a function defined on a domain D, and let c be a cluster point of D. Then f is said to have a *limit L at c* if, corresponding to each positive number ε, there exists a positive number δ with the property that $|f(x) - L| < \varepsilon$ for all x in the set $D \cap \mathscr{N}^*(c,\delta)$. For the number L, should it exist, it is also well to employ a notation exhibiting symbolically the function f and the point c. Thus we write $L = \lim_c f$. If the function is denoted by $f(x)$, we write $L = \lim_c f(x)$ or $L = \lim_{x \to c} f(x)$.

The concept of limit of a function at a point will dominate a considerable portion of our study of functions of a real variable. It is important, therefore, that we analyze it in great detail. Before presenting some examples, let us first dispose of any notion that a function may possess more than one limit at any one point.

Theorem 7-1 If a function has a limit at a point, then that limit is unique.

Proof Suppose a function f, defined on a domain D, to possess limits L_1 and L_2 at a point c. Corresponding to each positive number ε, there exist two positive numbers δ_1 and δ_2 with the properties $|f(x) - L_1| < \varepsilon$ for $x \in D \cap \mathcal{N}^*(c,\delta_1)$, and $|f(x) - L_2| < \varepsilon$ for $x \in D \cap \mathcal{N}^*(c,\delta_2)$. Then both inequalities are true for all points x in D such that $0 < |x - c| < \delta$, where $\delta = \min\{\delta_1,\delta_2\}$.

Because the point c is a cluster point of D, it follows that the deleted neighborhood $\mathcal{N}^*(c,\delta)$ contains at least one point x_1 of D. Therefore

$$|L_1 - L_2| \le |f(x_1) - L_1| + |f(x_1) - L_2| < 2\varepsilon,$$

and since $L_1 - L_2$ is a constant, it follows that $L_1 = L_2$ ▲

It should be observed that Theorem 7-1 would be false except for the restriction that c be a cluster point of the domain D of the function f. For, were c not a cluster point of D, then a deleted neighborhood $\mathcal{N}^*(c,\delta)$ would exist containing no point of D. Hence, for each $\varepsilon > 0$ and for any number L, the inequality $|f(x) - L| < \varepsilon$ would be true for all x in $D \cap \mathcal{N}^*(c,\delta)$ in the sense that no x would exist in $D \cap \mathcal{N}^*(c,\delta)$ to deny it. It is the desirability of uniqueness that prompts the inclusion of the restriction on c in the definition of limit.

The definition of limit of f at c has been so worded as not to require that c itself be a point of the domain D. Thus, a function may have a limit at a point at which the function is not defined. Further, if f is defined at c, the number $f(c)$ has no relevance in the definition of the number $\lim_c f$. In particular, the numbers $\lim_c f$ and $f(c)$ need not be equal.

The definition of limit of a function may be stated in terms of neighborhoods. A function f, defined on D, has a limit L at a cluster point c of D if, corresponding to any neighborhood $\mathcal{N}(L,\varepsilon)$, there exists a deleted neighborhood $\mathcal{N}^*(c,\delta)$ such that images of all points of $D \cap \mathcal{N}^*(c,\delta)$ lie in $\mathcal{N}(L,\varepsilon)$.

A consequence of the definition of the limit L of a function f at c is that L is in the closure of the range of f. This follows from the fact that, for a given $\varepsilon > 0$, a number x exists in the domain of f such that $x \ne c$ and $|f(x) - L| < \varepsilon$. Thus, a point $f(x)$ of the range of f exists in each neighborhood of L.

Some examples of functions and limits follow.

Example 7-8 Let g be a function defined on the interval $[0,2]$ by $g(x) = 2$ if $x \ne 1$ and $g(1) = 1$. This function has the limit 1 at each point of $[0,2]$, the point $x = 1$ not excepted. (See Figure 7-5.)

Example 7-9 Let F be defined on $[-1,0) \cup (0,1]$ by $F(x) = x^2 + 1$ if $x < 0$ and $F(x) = 1 - x$ if $x > 0$. We shall show that $\lim_0 F$ exists and has the value 1.

Without loss of generality, let ε be such that $0 < \varepsilon < 1$. It must be shown that $\delta > 0$ exists with the properties

(1) $$|(x^2 + 1) - 1| < \varepsilon \quad \text{or} \quad x^2 < \varepsilon$$

for all x such that $-\delta < x < 0$, and

(2) $$|(1 - x) - 1| < \varepsilon \quad \text{or} \quad x < \varepsilon$$

for all x such that $0 < x < \delta$. Observe that if $\delta = \varepsilon$, then (2) is trivial. Also, if $0 < -x < \delta = \varepsilon < 1$, then $x^2 < \varepsilon^2 < \varepsilon$, so that (1) is true. Consequently, $\lim_0 F = 1$.

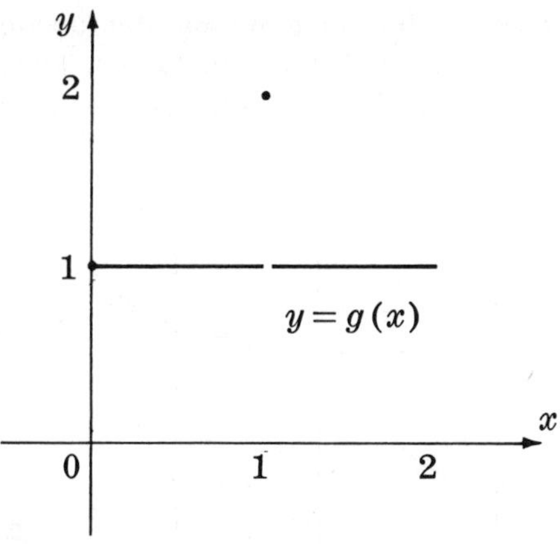

Figure 7-5

Example 7-10 Let f be defined for all values of x by $f(x) = x^2$, and let c be any point. Then we may show that $\lim_c f = c^2$. Observe, first, that

$$|x^2 - c^2| = |x + c|\,|x - c| \leq (|x| + |c|)\,|x - c|.$$

Now let $\varepsilon > 0$ be given, and choose $\delta = \min\{\varepsilon, 1\}$. Then, if x is such that $0 < |x - c| < \delta$, it follows that $|x - c| < \varepsilon$ and $|x - c| < 1$. From the latter, $|x| < |c| + 1$. Hence $|x^2 - c^2| < (2|c| + 1)\varepsilon$, which proves that $\lim_c f = c^2$.

Example 7-11 Define f for all x by $f(x) = x$ if $x \leq 1$ and $f(x) = 2x + 1$ if $x > 1$. We show that f does not have a limit at 1. Suppose $\lim_1 f = L$ to exist. Then $\delta > 0$ exists such that $|f(x) - L| < 1$ for $x \in \mathcal{N}^*(1, \delta)$. In particular, x_1 and x_2 may be found in that neighborhood such that $x_1 < 1 < x_2$, and $|x_1 - L| < 1$ and $|2x_2 + 1 - L| < 1$. These imply, respectively, $L < 1 + x_1 < 2$ and $2 < 2x_2 < L$. We conclude that L does not exist.

The definition of limit of a function at a point is sufficiently general to include functions defined on domains not consisting of intervals.

Example 7-12 Recall the function p of Example 7-5, defined to have the value $p(x) = x/2$ for all rational x and to be undefined for all irrational x. It may be shown to possess the limit $c/2$ at any point c, rational or irrational. For, if $\varepsilon > 0$ is given and if x is any rational number such that $|x - c| < \delta = \varepsilon$, then $|(x/2) - (c/2)| = \frac{1}{2}|x - c| < \frac{1}{2}\delta = \frac{1}{2}\varepsilon$.

Example 7-13 Consider again the function q of Example 7-6, defined to have the value $q(x) = x^2$ at points x of the domain consisting of rational numbers of the form $1/n$, n a positive integer. Since 0 is the only cluster point of the domain, it is the only point at which it is meaningful to speak of limit. To show that q does have a limit at the point 0 and that the value of the limit is 0, we have only to observe that $1/n^2 < \varepsilon$ whenever $1/n < \varepsilon$.

It is of value to describe the meaning of limit of a function at a point in terms of the graphical representation of the function. Let f denote the function sketched in Figure 7-6, and let c be a cluster point of the domain D of f. To investigate the

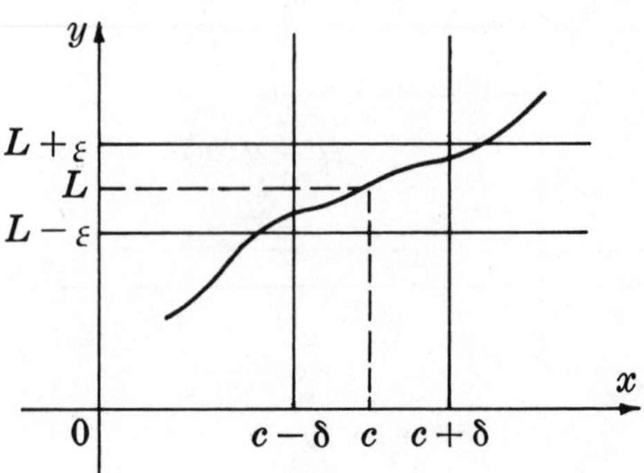

Figure 7-6

possible existence of a limit L for the function f at c, we consider a neighborhood $\mathcal{N}(L,\varepsilon)$. Such a neighborhood of points on the y-axis determines two horizontal lines $y = L - \varepsilon$ and $y = L + \varepsilon$. It is clear that function values $f(x)$ which have the property $|f(x) - L| < \varepsilon$ are depicted by points on the graph between the horizontal lines. In a similar fashion, a deleted neighborhood $\mathcal{N}^*(c,\delta)$ on the x-axis defines two vertical lines $x = c - \delta$ and $x = c + \delta$. Values of x such that $0 < |x - c| < \delta$ then correspond to points on the graph of f (except that associated with c, if such exists) which lie between those vertical lines. The assertion that L is the limit of f at c is then tantamount to the statement that, however close the horizontal lines $y = L - \varepsilon$ and $y = L + \varepsilon$ may be drawn, vertical lines $x = c - \delta$ and $x = c + \delta$ may also be drawn such that all points of the graph of f, except possibly $(c,f(c))$, that lie between the vertical lines also lie between the horizontal lines.

On the other hand, if such construction is not always possible, then f does not possess a limit at c. Consider a function whose graph is that of Figure 7-7. Then,

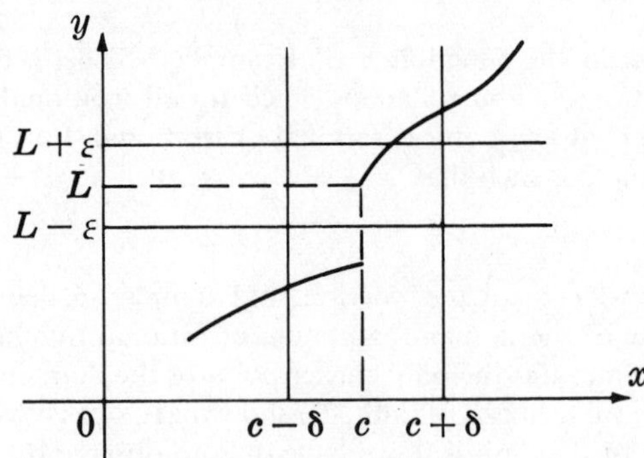

Figure 7-7

if L is any point of the y-axis, it is possible to find horizontal lines for which no corresponding vertical lines exist. Hence f does not have a limit at c.

7-3 Limit Properties

A function f, defined on D, is said to be *bounded* on A ($A \subset D$) if a number M exists such that $|f(x)| \leq M$ for all x in A. It is *bounded above* if $f(x) \leq M$ and *bounded below* if $M \leq f(x)$. Thus, if $A = D$, the bounded character of a function is exactly that of the set of numbers which constitutes its range.

If f is bounded on the nonempty set A, then the set $\{f(x);\ \ x \in A\}$ has a supremum. For convenience, we denote it by $\sup_A f$ or $\sup_A f(x)$ rather than by the more complete notation $\sup\{f(x);\ \ x \in A\}$. If A is the domain of f, we abbreviate further and write simply $\sup f$ or $\sup f(x)$. The symbols $\inf_A f$, $\max_A f$, and $\min_A f$ have similar meanings.

Two basic theorems relate the concepts of boundedness and limit.

Theorem 7-2 If a function f possesses a limit at a point c, then a neighborhood of c exists on which f is bounded.

Proof This is an immediate consequence of the definition of limit of a function at a point ▲

Theorem 7-3 If a function f has a limit L at a point c and if $L \neq 0$, then positive numbers k and δ exist such that $|f(x)| > k$ for all x in $D \cap \mathcal{N}^*(c,\delta)$.

Proof Let $k = \frac{1}{2}|L|$. Then a number $\delta > 0$ exists such that, for all x in $D \cap \mathcal{N}^*(c,\delta)$, the relation $|f(x) - L| < k$ is true. For such x it follows that $|L| - |f(x)| < k$ or $k < |f(x)|$, as desired ▲

The property possessed by f in Theorem 7-3 may be described in words: If f has a nonzero limit at c, then a deleted neighborhood $\mathcal{N}^*(c,\delta)$ exists on which f is *bounded away from zero*.

We next prove a theorem relating limits of functions to limits of combinations of functions.

Theorem 7-4 If f and g are functions possessing limits L_1 and L_2, respectively, at a cluster point c of their common domain D, then (a) $f + g$ has the limit $L_1 + L_2$ at c, (b) fg has the limit L_1L_2 at c, (c) $|f|$ has the limit $|L_1|$ at c, and (d) f/g has the limit L_1/L_2 at c, provided that $L_2 \neq 0$.

Proof Let $\varepsilon > 0$ be given and assume, without loss of generality, that $\varepsilon < 1$. Then a number $\delta > 0$ exists with the properties $|f(x) - L_1| < \varepsilon$ and $|g(x) - L_2| < \varepsilon$ for $x \in D \cap \mathcal{N}^*(c,\delta)$. For such values of x we have

$$|[f(x) + g(x)] - [L_1 + L_2]| \leq |f(x) - L_1| + |g(x) - L_2| < 2\varepsilon,$$

which proves part (a);

$$|f(x)g(x) - L_1L_2| \le |f(x)|\,|g(x) - L_2| + |L_2|\,|f(x) - L_1|$$
$$< (|L_1| + \varepsilon)\varepsilon + |L_2|\,\varepsilon$$
$$< (|L_1| + 1 + |L_2|)\varepsilon,$$

which proves part (b); and

$$||f(x)| - |L_1|| \le |f(x) - L_1| < \varepsilon,$$

which proves part (c).

To prove part (d), it is necessary to show only that $1/g$ has the limit $1/L_2$ at c; for then, by part (b), the function $(f)(1/g) = f/g$ has the limit $(L_1)(1/L_2) = L_1/L_2$. Since $L_2 \ne 0$, it follows by Theorem 7-3 that positive numbers k and δ_1 exist such that $k < |g(x)|$ for all x in $D \cap \mathcal{N}^*(c,\delta_1)$. Let $\delta_2 = \min\{\delta,\delta_1\}$. Then, for all x in $D \cap \mathcal{N}^*(c,\delta_2)$ we have

$$\left|\frac{1}{g(x)} - \frac{1}{L_2}\right| = \frac{|L_2 - g(x)|}{|g(x)|\,|L_2|} < \frac{\varepsilon}{k\,|L_2|} \quad \blacktriangle$$

7-4 Limit of a Function and Limit of a Sequence

Let f be a function defined on a domain D, and let c be a cluster point of D. Denote by $\{x_n\}$ a sequence of points of D, all different from c, which converges to c as a limit. Since to each x_n corresponds a function value $f(x_n)$, the sequence $\{x_n\}$ determines a sequence of function values $\{f(x_n)\}$. It is of interest to investigate the relationship between the existence and value of a limit of the sequence $\{f(x_n)\}$, on the one hand, and the existence and value of a limit of the function f at c, on the other. Theorems 7-5 and 7-6 describe the interrelation completely.

Theorem 7-5 Let f be a function defined on a domain D and possessing a limit L at a point c. Let $\{x_n\}$ $(x_n \ne c)$ denote a sequence of points of D converging to c. Then the sequence $\{f(x_n)\}$ converges to L.

Proof From the definition of limit of f at c we know that, corresponding to any given $\varepsilon > 0$, a number $\delta > 0$ exists such that $|f(x) - L| < \varepsilon$ for all x in $D \cap \mathcal{N}^*(c,\delta)$. From the convergence of $\{x_n\}$ to c it follows that an integer N exists such that x_n is a member of $\mathcal{N}^*(c,\delta)$ for all $n > N$. Therefore $|f(x_n) - L| < \varepsilon$ for all $n > N$ ▲

The direct converse of Theorem 7-5 is not true. That is, if a function f is such that a sequence $\{x_n\}$ $(x_n \ne c)$ converging to c can be found with the property that $\{f(x_n)\}$ converges to a number L, it does not follow that f has the limit L at c or even that f has any limit at all at c. The function Ψ serves to illustrate this. For, if $\{x_n\}$ is any sequence of rational numbers converging to some number c, then $\{\Psi(x_n)\}$ converges to 0. However, it is easily seen that the function Ψ does not possess a limit at c.

In the nature of a converse to Theorem 7-5 we have the following.

Theorem 7-6 Let f be a function defined on D, and let c be a cluster point of D. If, for every sequence $\{x_n\}$ $(x_n \neq c)$ which converges to c, the corresponding sequence $\{f(x_n)\}$ is convergent, it follows that (a) the sequences $\{f(x_n)\}$ have a common limit L, and (b) the function f has the limit L at c.

Proof Let $\{x_n\}$ $(x_n \neq c)$ and $\{x_n'\}$ $(x_n' \neq c)$ be any two sequences converging to c, and denote the limits of $\{f(x_n)\}$ and $\{f(x_n')\}$ by L and L', respectively. Consider the sequence

$$x_1, x_1', x_2, x_2', \ldots, x_n, x_n', \ldots,$$

constructed in an obvious fashion from the elements of $\{x_n\}$ and $\{x_n'\}$. It is clear that this sequence, too, converges to c. Denoting it by $\{x_n''\}$, let L'' be the limit of the corresponding sequence $\{f(x_n'')\}$ of function values. Since $\{f(x_n)\}$ and $\{f(x_n')\}$ are subsequences of $\{f(x_n'')\}$, it follows that $L = L''$ and $L' = L''$. Hence $L = L'$, and we conclude that all sequences $\{f(x_n)\}$ converge to a common limit, which we may denote by L. This proves part (a) of the theorem.

To prove part (b) we make the assumption that it is not true; i.e., that L is not the limit of f at c. This implies that, for some number $\varepsilon > 0$, and for any $\delta > 0$, there exists a number x such that $0 < |x - c| < \delta$ and $|f(x) - L| > \varepsilon$. By assigning to δ a sequence of values of the form $1/n$, where $n = 1, 2, \ldots$, we obtain a sequence of values of x, say $\{x_n\}$, such that $0 < |x_n - c| < 1/n$ and $|f(x_n) - L| > \varepsilon$. Thus, $\{x_n\}$ converges to c, but $\{f(x_n)\}$ does not converge to L. This is a contradiction to the conclusion of part (a) of the theorem and hence denies the assumption that L is not the limit of f at c. Therefore f has the limit L at c▲

The foregoing theorem is useful when it is desired to prove that a given function does not possess a limit at a specified point. This is illustrated by the following example.

Example 7-14 Consider the function f defined for all values of x by $f(x) = \sin(1/x)$ if $x \neq 0$ and $f(0) = 0$. Let sequences $\{x_n\}$ and $\{x_n'\}$ be defined by $x_n = 1/n\pi$ and $x_n' = 2/(1 + 4n)\pi$ for $n = 1, 2, \ldots$. Then, although $\{x_n\}$ and $\{x_n'\}$ both converge to 0, the sequences $\{f(x_n)\}$ and $\{f(x_n')\}$ have the different limits 0 and 1, respectively. Hence f does not have a limit at 0. (See Figure 7-8.)

Theorem 7-6 may also be employed to show that the truth of Theorem 7-4 on limits of sums, products, quotients, and absolute values of functions follows from the comparable theorems on sequences.

7-5 One-sided Limits of Functions

We introduce now the notion of a one-sided limit of a function based on the earlier concept of a one-sided cluster point. Let f be a function defined on a

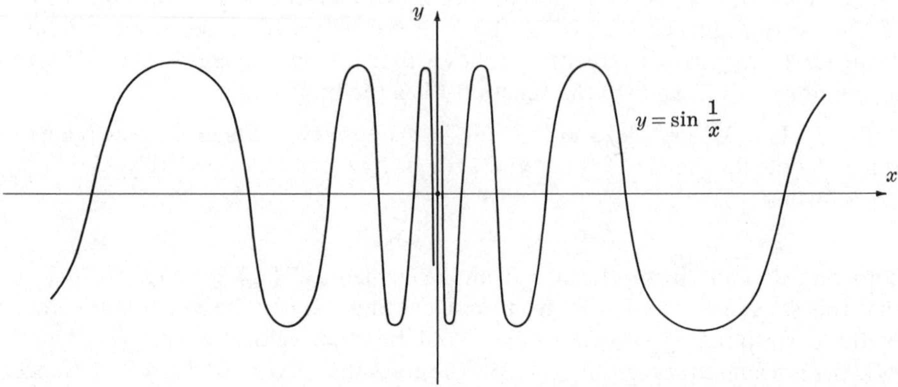

Figure 7-8

domain D, and let c denote a cluster point of D from the right. Then f is said to possess a *limit L from the right at c* if, corresponding to a given $\varepsilon > 0$, there exists a number $\delta > 0$ such that $|f(x) - L| < \varepsilon$ for all x in $D \cap (c, c + \delta)$. In a dual fashion we define the *limit of f from the left at c.* We employ the symbol $\lim_{c+} f$ (or $\lim_{x \to c+} f(x)$ or $f(c+)$) to denote limit of f from the right at c and $\lim_{c-} f$ (or $\lim_{x \to c-} f(x)$ or $f(c-)$) for limit from the left.

All theorems in this chapter remain true if the notion of limit of a function is replaced by limit of a function from the right or left. The intimate connection between limit and limit from right or left is reasonably clear. Suppose f to be a function defined on a domain D. If it possesses limit L from both right and left at a point c, then, corresponding to any $\varepsilon > 0$, a number $\delta > 0$ exists such that $|f(x) - L| < \varepsilon$ for all $x \in D \cap (c - \delta, c)$ and for all $x \in D \cap (c, c + \delta)$. Hence $|f(x) - L| < \varepsilon$ for all $x \in D \cap \mathscr{N}^*(c,\delta)$. Thus the limit at c also exists and has the value L. On the other hand, suppose that c is a cluster point of D from the right and that f possesses a limit L at c. Then, for any $\varepsilon > 0$, there exists $\delta > 0$ such that $|f(x) - L| < \varepsilon$ for $x \in D \cap \mathscr{N}^*(c,\delta)$, and it follows that $\lim_{c+} f$ exists with value L. A similar argument is valid for limit from the left.

It is possible for a function to have a limit at a point c of a domain D and yet not have limits from both sides. Such is the case if c is not a cluster point of D from both left and right.

Example 7-15 If g is defined on the interval $[1,2]$ by $g(x) = 3x$, then it is easily seen that $\lim_{1} g$ and $\lim_{1+} g$ both exist with value 3. However, $\lim_{1-} g$ is not defined since the point 1 is not a cluster point from the left.

Two facts are obvious: first, the existence of $\lim_{c} f$ implies the existence of at least one of the two limits $\lim_{c-} f$ and $\lim_{c+} f$; second, if $\lim_{c-} f$ and $\lim_{c+} f$ both exist and if $\lim_{c-} f \neq \lim_{c+} f$, then $\lim_{c} f$ does not exist.

We conclude the discussion of one-sided limits with some examples.

Example 7-16 Let g be defined for $x \neq 0$ by $g(x) = [1 - e^{-1/x}]^{-1}$. The graph of g appears in Figure 7-9. It may be verified that $\lim_{0-} g = 0$ and $\lim_{0+} g = 1$. The function, consequently, does not possess a limit at 0.

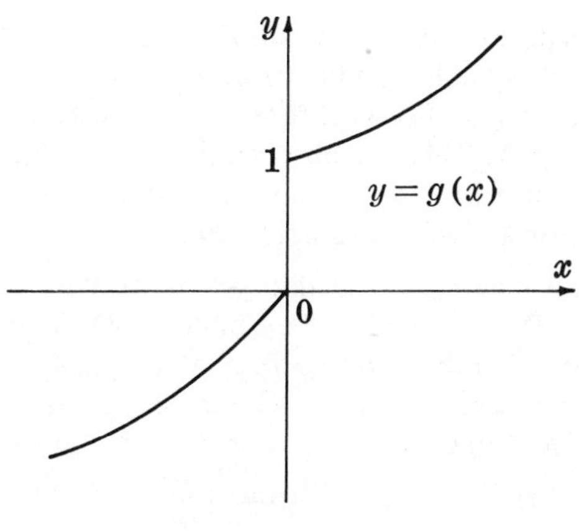

Figure 7-9

Example 7-17 The function f defined for all x as in Example 7-14 has the property that neither $\lim_{0-} f$ nor $\lim_{0+} f$ exists.

7-6 Further Limit Concepts

For a function defined on a subset D of $\mathscr{R}$, the notion of limit may be extended by admitting ∞ and $-\infty$ as possible limit values or as possible members of the closure of D. In effect, the domain and range of the function, although restricted to subsets of $\mathscr{R}$, are to be viewed as subsets of $\mathscr{R}^*$. First, we show how infinite values of c may be included. Let f denote a function defined on D, and suppose ∞ to be a cluster point of D. Then f is said to have a *limit L at ∞* if, corresponding to each positive number ε, there exists a number k with the property $|f(x) - L| < \varepsilon$ for all x in D such that $x > k$. We write $\lim_{\infty} f = L$ and sometimes speak of this as the *limit of f as x increases without bound* or the *limit of f as x approaches infinity.*

In a similar fashion we may define *limit at $-\infty$* or *limit as x decreases without bound* and write $\lim_{-\infty} f = L$.

That limits at ∞ and $-\infty$ are unique and that Theorem 7-4 on sums, products, quotients, and absolute value remains valid if $c = \infty$ or $c = -\infty$ may be ascertained without difficulty.

Example 7-18 If $f(x) = 1/x$ for $x > 0$, then $\lim_{\infty} f = 0$ follows readily from the definition. This may be employed, together with facts concerning sums, products, and quotients to show, for instance, that

$$\lim_{\infty} \frac{3x - x^2}{2x^2 + 1} = \lim_{\infty} \frac{\frac{3}{x} - 1}{2 + \frac{1}{x^2}} = -\frac{1}{2}.$$

Let f be a function defined on the domain D consisting of all positive integers n, and write $f(n) = a_n$. For f to have a limit L at ∞ in accordance with the definition above means the following: For each $\varepsilon > 0$ there exists a number k such that $|a_n - L| < \varepsilon$ for all $n > k$. This, we observe, is precisely the definition of convergence of the sequence $\{a_n\}$ to the limit L. Thus we draw the important conclusion that $\lim \{a_n\}$ is a special instance of $\lim_{\infty} f$.

The notion of limit may be extended in still another way. Let f be a function defined on a domain D in $\mathscr{R}$, and let c be a finite cluster point of D. Then we say that f has *limit* ∞ *at* c if, corresponding to each number k, there exists a positive number δ with the property that $f(x) > k$ for all x in $D \cap \mathscr{N}^*(c,\delta)$. We write $\lim_{c} f = \infty$. Dually, we define *limit* $-\infty$ *at* c and write $\lim_{c} f = -\infty$. Infinite one-sided limits may be defined, too, paralleling those defined in Section 7-5.

Example 7-19 $\lim_{0} 1/x^2 = \infty$. Also, $\lim_{1+} 2^{1/(1-x)} = 0$; $\lim_{1-} 2^{1/(1-x)} = \infty$.

It can be verified that $\lim_{c} f = \infty$ if a deleted neighborhood of c exists on which is positive and $\lim_{c} 1/f = 0$.

The various limit notions considered to this point are special instances of a general limit concept. Let f be a function defined on a domain D in $\mathscr{R}^*$ and having range in $\mathscr{R}^*$. Let c be a cluster point of D. Then f has *limit* L *at* c if, for each neighborhood $\mathscr{N}_L$ of L, there exists a deleted neighborhood $\mathscr{N}_c^*$ of c, with the property that, whenever x is in $D \cap \mathscr{N}_c^*$, then $f(x)$ is in $\mathscr{N}_L$. Notice that in this definition the function f may be defined at ∞ or $-\infty$ and may even take on infinite values. Actually we shall have no occasion to pursue such generalization. We shall be concerned only with functions having domain and range in $\mathscr{R}$. Also, unless otherwise specified, the symbol $\lim_{c} f = L$ will denote a finite limit at a finite point.

Exercises

7-3 Prove that if f has the limit L at c, then a deleted neighborhood exists on which f is bounded away from L.

7-4 Assume that the functions g and h have the same limit L at c. Assume further that a function f exists such that $g(x) \leq f(x) \leq h(x)$ for all x in some $\mathcal{N}^*(c,\varepsilon)$. Prove that $\lim_c f$ exists and equals L.

7-5 Let h and k be functions defined on the interval $[a,b]$. Assume that for each x in (a,b) there is a point ξ such that $a < \xi < x$ and $k(\xi) = h(x)$. Prove that if $\lim_a k$ exists, then so does $\lim_a h$, and the two are equal.

7-6 Let f be a function with domain D and range E. Assume that f has the limit L_1 at c. Let g be a function defined on E, and assume that g has the limit L_2 at L_1. Does the function $h = g \circ f$, defined on D, necessarily have the limit L_2 at c? Support your answer.

7-7 A function f is *increasing* on an interval (a,b) if $f(x_1) \leq f(x_2)$ whenever $a < x_1 < x_2 < b$. Prove that if f is increasing on (a,b), then $f(c-)$ and $f(c+)$ exists at each point c of the interval.

Let f be a bounded function defined on a domain D, and let c be a cluster point of D. For any positive δ, let $h(\delta)$ be the supremum of f on the set $A_\delta = D \cap \mathcal{N}^*(c,\delta)$. Finally let $\overline{\lim}_c f$ denote the infimum of the numbers $h(\delta)$ for $\delta > 0$. The number $\overline{\lim}_c f$ is called the *upper limit* of f at c. Thus $\overline{\lim}_c f = \inf \{\sup_{A_\delta} f;\ \delta > 0\}$. The *lower limit* $\underline{\lim}_c f$ of f at c is defined in a dual manner.

7-8 Prove that $\overline{\lim}_c f$ and $\underline{\lim}_c f$ exist at a cluster point c of the domain of a bounded function f.

7-9 Prove that $\underline{\lim}_c f \leq \overline{\lim}_c f$.

7-10 Give an example of a function f for which $\underline{\lim}_c f < \overline{\lim}_c f$.

7-11 Prove that if $f \leq g$, then $\underline{\lim}_c f \leq \underline{\lim}_c g$ and $\overline{\lim}_c f \leq \overline{\lim}_c g$.

7-12 Prove that $-\overline{\lim}_c f = \underline{\lim}_c (-f)$.

7-13 Prove that $\underline{\lim}_c f + \underline{\lim}_c g \leq \underline{\lim}_c (f+g)$ and $\overline{\lim}_c (f+g) \leq \overline{\lim}_c f + \overline{\lim}_c g$.

7-14 Prove that $\lim_c f$ exists if and only if $\underline{\lim}_c f = \overline{\lim}_c f$. Also prove that if $\lim_c f$ exists, then $\lim_c f = \underline{\lim}_c f = \overline{\lim}_c f$.

8

Continuous Functions

He who would arrive at the appointed end must follow a single road and not wander through many ways.

SENECA (5–65), *Epistulae ad Lucilium*

8-1 Continuity

The concept of *continuous function* distinguishes a certain class of functions of particular interest and importance not only in a purely mathematical study of real functions but also in numerous applications to physics and engineering, where such functions play a dominant role. Intuitively, one associates with the notion of continuity the geometrical concept of an unbroken linear continuum or curve. We shall see, however, that a general analytical formulation of continuity is immediately far less restrictive than that suggested by geometry. Indeed, we shall define continuity in such a way that many continuous functions will actually defy illustration by a curve or graph in a coordinate plane.

A function f, defined on a domain D, is said to be *continuous at a point c of D* if, corresponding to each positive number ε, there exists a positive number δ such that $|f(x) - f(c)| < \varepsilon$ for all x in $D \cap \mathscr{N}(c,\delta)$. A point c at which a function is continuous is called a *continuity point* of the function.

Example 8-1 Consider the function f defined for all real numbers x by $f(x) = x^2$, and let c denote any number. In Example 7-10 it was shown that, corresponding to each $\varepsilon > 0$, the number $\delta = \min\{\varepsilon,1\}$ is such that $|x^2 - c^2| < (2|c| + 1)\varepsilon$ for all x in $\mathscr{N}^*(c,\delta)$. Since the inequality is also true for $x = c$, it follows that it is true for all x in $\mathscr{N}(c,\delta)$. Hence c is a continuity point of f.

Example 8-2 The function f defined for all x by $f(x) = x \sin(1/x)$ for $x \neq 0$ and $f(0) = 0$ is continuous at 0. To see this we have only to observe that, for a given $\varepsilon > 0$, $|x \sin(1/x)| \leq |x| < \varepsilon$ for all x such that $0 < |x| < \delta = \varepsilon$.

If a function f has the property that it is not continuous at a point c of its domain D, then it is said to be *discontinuous at c*, and c is called a *discontinuity* of f. Accordingly, it is not meaningful to speak of a function being continuous or discontinuous at a point not belonging to its domain.

The definition of continuity may be rephrased in two equivalent ways. They will prove to be of value in ensuing discussions.

A function f, defined on a domain D, is continuous at a point c of D if, to any given $\varepsilon > 0$, there corresponds a number $\delta > 0$ with the property that $f(x)$ is a point of the neighborhood $\mathcal{N}(f(c),\varepsilon)$ whenever x is in $D \cap \mathcal{N}(c,\delta)$.

Also, f is continuous at c of the domain D if, given $\varepsilon > 0$, a number $\delta > 0$ exists such that $|f(c+h) - f(c)| < \varepsilon$ for all real numbers h with the properties that $c + h$ is in D and $|h| < \delta$.

A function is said to be *continuous on a set* if it is continuous at each point of the set. In particular, a function is *continuous* if it is continuous at each point of its domain.

Example 8-3 We have seen that the function f defined by $f(x) = x^2$ for all real numbers x is continuous at each point. Hence it is continuous.

Example 8-4 The function h defined by $h(x) = x^2$ on $[-1,0) \cup (0,1) \cup (1,2)$ is continuous since it is continuous at each point of its domain.

We observe a similarity in the definitions of continuity of a function f at a point c, on the one hand, and limit of f at c, on the other. Actually, the difference centers about the existence and value of $f(c)$. While such are important to the notion of continuity, they play no role in the concept of limit. The interrelation between limit and continuity may be described very simply as follows. Let f be a function defined on a domain D, and let c be a point of D which is also a cluster point of D. Then f is continuous at c if and only if f has the limit $f(c)$ at c.

Of course, if c, while a member of D, is not also a cluster point, then it is certainly a continuity point of f, it being a consequence of the definition that a function is continuous at all isolated points of its domain. Thus, it is possible for a function to be continuous at a point yet not possess a limit there. On the other hand, a function may have a limit at a point c without being continuous there. This is possible if c is a cluster point of the domain but does not itself belong to it or if the limit at c is not equal to the function value there.

Example 8-5 Let g denote a function defined by $g(x) = x^2$ on a domain D consisting of all rational numbers of the form $1/n$, where n is a positive integer. Since each point of D is isolated, g is continuous on D, but at no point of D does g possess a limit. On the other hand, g has limit 0 at 0, which is a cluster point of D. But g is not continuous at 0 since it is not defined there.

The following theorem is an immediate consequence of Theorem 7-4 on limits.

Theorem 8-1 If f and g are functions continuous at a point c of a common domain D, then the functions $f+g$, $f-g$, fg, f/g, and $|f|$ are continuous at c.

It is easily shown that a constant function is continuous and that this is the case also for the function defined to have the value x at each point x of its domain. From this and Theorem 8-1 follows the continuity of the *polynomial* function

$$a_0 x^n + a_1 x^{n-1} + \cdots + a_{n-1}x + a_n$$

and of the *rational* function

$$\frac{a_0 x^n + a_1 x^{n-1} + \cdots + a_{n-1}x + a_n}{b_0 x^m + b_1 x^{m-1} + \cdots + b_{m-1}x + b_m}$$

for all values of x which do not cause the denominator to vanish.

Another important fact concerns continuity of a function of a function.

Theorem 8-2 Let f be a function continuous at a point c of its domain D. Denote the range of f by E, and let g be a function defined on E and continuous at the point $f(c)$ of E. Then the function $h = g \circ f$ is continuous at c.

Proof Since g is continuous at $f(c)$ of the domain E, it follows that, corresponding to each given $\varepsilon > 0$, there exists a number $\delta > 0$ with the property that

$$|h(x) - h(c)| = |g(f(x)) - g(f(c))| < \varepsilon$$

for all points $f(x)$ of E such that

$$(1) \qquad |f(x) - f(c)| < \delta.$$

From the continuity of f at c then follows the existence of a positive number δ' such that inequality (1) is true for all points x in $D \cap \mathcal{N}(c,\delta')$. Consequently, corresponding to a given ε, there is a number δ' such that $|h(x) - h(c)| < \varepsilon$ for all x in $D \cap \mathcal{N}(c,\delta')$. Hence h is continuous at c▲

8-2 Discontinuities

Suppose a function f to be discontinuous at a point c of its domain and to have the property that limits from the left and right both exist. The discontinuity is then described as *simple* or *of the first kind.* If it happens that $\lim_{c-} f = \lim_{c+} f$ (in which case $f(c)$ is necessarily different from the common value of those limits), then the discontinuity is said to be *removable*, suggesting that a simple alteration of the value of f at c is enough to create a function continuous at c. On the other hand, if $\lim_{c-} f \neq \lim_{c+} f$, then the discontinuity at c is called a *jump* discontinuity. The number $\lim_{c+} f - \lim_{c-} f$ is called the *jump* of f at c.

Example 8-6 The function f defined by $f(x) = x^2$ for $x \neq 1$ and $f(1) = 3$ has a removable discontinuity at 1. (See Figure 8-1.) The function g defined by $g(x) = x^2$ for all x may be looked upon as having been obtained from f by changing the value of $f(1)$ from 3 to 1. Obviously g is continuous at 1.

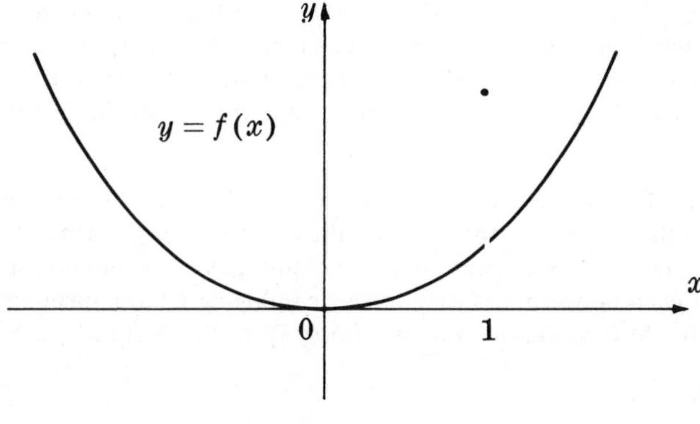

Figure 8-1

Example 8-7 The function h defined by $h(x) = 1$ for $x \leq 0$ and $h(x) = x$ for $x > 0$ possesses a jump discontinuity at 0 since $\lim_{0-} h = 1 \neq \lim_{0+} h = 0$. The value of the jump is -1. (See Figure 8-2.)

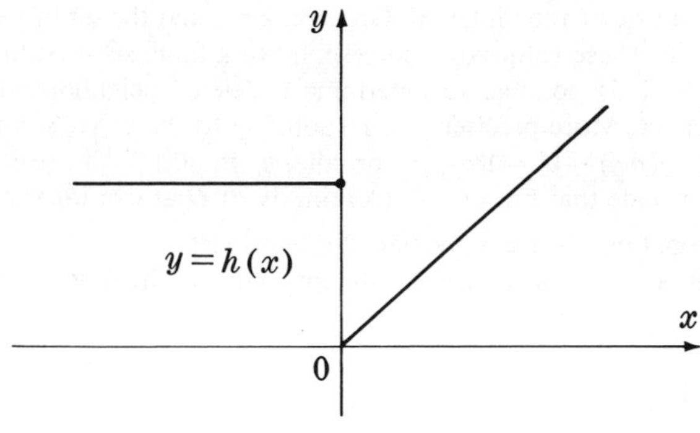

Figure 8-2

Let c be a discontinuity point of f and such that it is a cluster point from both right and left. Then if one or both of the numbers $\lim_{c-} f$ and $\lim_{c+} f$ do not exist, the point c is referred to as a discontinuity *of the second kind.*

Example 8-8 The function F defined by $F(x) = 1/x$ for $x \neq 0$ and $F(0) = 1$ is such that neither $\lim_{0+} F$ nor $\lim_{0-} F$ exists. Hence 0 is a discontinuity of the second kind. (If we consider limits in the extended sense, then in this instance $\lim_{0-} F = -\infty$ and $\lim_{0+} F = \infty$. The point 0 is then a discontinuity point of the first kind with jump $\infty - (-\infty) = \infty$.)

Example 8-9 Consider the function g defined by $g(x) = 0$ for $x \leq 0$ and $g(x) = \sin(1/x)$ for $x > 0$. Here $\lim_{0-} g = 0$, but $\lim_{0+} g$ does not exist. Thus 0 is a discontinuity of the second kind.

Example 8-10 The function Ψ defined earlier is such that every point is a discontinuity of the second kind. For, if c is any number, then for $\varepsilon > 0$ each of the intervals $(c - \varepsilon, c)$ and $(c, c + \varepsilon)$ contains both rational and irrational points with corresponding function values 0 and 1, respectively. Hence, neither $\lim_{c-} \Psi$ nor $\lim_{c+} \Psi$ exists.

Example 8-11 It is possible for a function to be continuous at one point and discontinuous at all others. Such is the case for the function f defined by $f(x) = 0$ if x is rational and $f(x) = x$ if x is irrational. That this possesses discontinuities of the second kind at each nonzero x is easy to see, employing an argument similar to that in Example 8-10. At 0 continuity follows simply from the relation $|f(x) - f(0)| \leq |x|$ for all x.

Example 8-12 An interesting function is that defined on $[0,1]$ by $Q(x) = 1/q$ if $x = p/q$ (p and q relatively prime and $q > 0$) and $Q(x) = 0$ if x is irrational. The function Q is continuous at each irrational point but possesses a discontinuity of the first kind at each rational point. We prove this by showing $\lim_{c} Q$ exists and has the value 0 at each point of the interval.

Let c denote any point of $[0,1]$, and let $\varepsilon > 0$ be given. Then $|Q(x) - 0| = |0 - 0| < \varepsilon$ for any irrational point x in the interval. Observe, also, that the set of values of q such that $\varepsilon \leq 1/q$ is finite. These values of q correspond to a finite set of values of p/q since $p/q \leq 1$. Obviously, it is possible to determine a deleted neighborhood of c which excludes the latter set. More precisely, corresponding to the given ε, a number $\delta > 0$ exists such that $|Q(p/q) - 0| = 1/q < \varepsilon$ for all p/q in $[0,1] \cap \mathcal{N}^*(c,\delta)$. From these inequalities we conclude that $\lim_{c} Q = 0$. Continuity of Q at irrational points and discontinuity at rational points are immediate consequences.

Figure 8-3 is at best an unsatisfactory attempt to depict the function Q graphically.

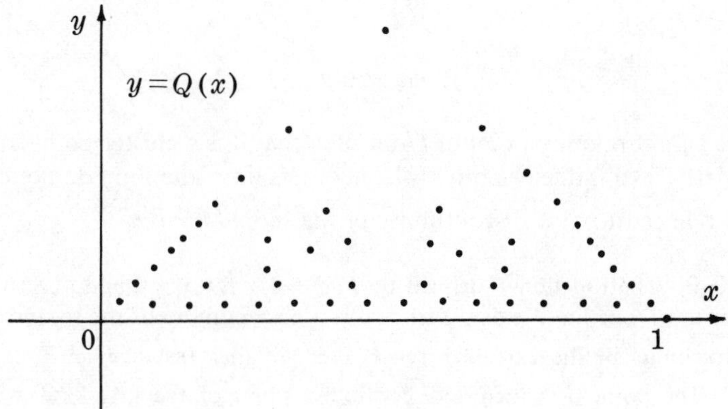

Figure 8-3

The discussion and examples above concern discontinuities at points which are cluster points from both left and right. Certain modifications of terminology are in order for points such as endpoints of intervals. Suppose a point c of a domain of a function f to be such that it is not a cluster point from the right. Then it is meaningless to speak of the limit from the right at c. If f is discontinuous at c, then either $\lim_{c-} f$ does not exist, or it exists with a value different from $f(c)$. If the latter, c is a discontinuity *of the first kind* (and hence removable since the notion of jump is meaningless); and if the former, a discontinuity *of the second kind*. Similar representation applies to a point which is not a cluster point from the left. (If c is neither a cluster point from right nor left, it is isolated and cannot be a discontinuity point.)

Exercises

8-1 Give a "reasonable" definition of *continuous from the right*. Cite an example of a function which is continuous from the right at a certain point but not continuous at that point. Can a function be continuous and yet not be continuous from the right? Explain.

8-2 Let f be a function which is continuous on $\mathscr{R}$ and such that $f(r) = 0$ for each rational point r. Prove that $f(x) = 0$ for any real number x.

8-3 Let f be a function defined and continuous on the domain R of rational numbers. Prove that there exists at most one function g defined and continuous on the set $\mathscr{R}$ of real numbers with the property $g(x) = f(x)$ for each x in $\mathscr{R}$. (The function g may be described as the unique continuous extension of f to the domain $\mathscr{R}$.)

8-4 Let f be defined for $x \neq 0$ by $f(x) = 1/x$. The statement "f is discontinuous at 0" is false because 0 is not in the domain of the function. How can the definition of *discontinuity point* be modified so that the above statement is true? (With your modified definition, is the function of Example 8-4 discontinuous at 4?)

8-5 Let f be a function defined on $\mathscr{R}$ with the property that $f(x + y) = f(x) + f(y)$ for all x and y. Suppose also that f is continuous at at least one point. Prove that a constant a exists such that $f(x) = ax$ for all x.

Two subsets A and B of $\mathscr{R}$ are said to be *topologically equivalent* if a one-to-one mapping f of A onto B exists such that both f and f^{-1} are continuous.

8-6 Prove that the intervals [0,1] and [0,2] are topologically equivalent.

8-7 Prove that $(-1,1)$ and $\mathscr{R}$ are topologically equivalent.

8-8 Prove that if f and g are defined and continuous on a common domain D, then $f \wedge g$ and $f \vee g$ are continuous on D.

8-9 Prove that f is continuous if and only if f^+ and f^- are continuous.

8-10 Define a function f on $\mathscr{R}$ by $f(x) = |a - x|$, where a is a constant. Prove that f is continuous. (That is, the distance of a variable point from a fixed point is a continuous function.)

8-11 Let F denote a closed set of numbers, and let c be a real number in $\mathscr{C}F$. Prove that a continuous function f exists with the property that $f(c) = 0$ and $f(x) = 1$ for all x in F.

8-3 Variation of a Function

If f is a function which is bounded on a subset A of its domain, then the non-negative number $\Omega_A f$ (or $\Omega_A f(x)$) defined by

$$\Omega_A f = \sup_A f - \inf_A f$$

is called the *oscillation of f on A*. Clearly, if $A \subset B$, then $\Omega_A f \leq \Omega_B f$.

Example 8-13

(a) The function f defined by $f(x) = x + 2$ on the set $D = (1,2]$ is such that $\sup_D f = \max_D f = 4$ and $\inf_D f = 3$. Hence $\Omega_D f = 4 - 3 = 1$.

(b) The function g defined by $g(x) = \sin x$ on $[0,2\pi]$ is such that on the subset $A = (\pi/2, 7\pi/6)$ it has supremum 1 and infimum $-\frac{1}{2}$. Hence $\Omega_A f = \frac{3}{2}$.

If a function f is bounded on a set A, the oscillation of f on A is the supremum of the set of numbers $|f(x_1) - f(x_2)|$, where x_1 and x_2 are any two points of A. The reader should verify that this is the case.

Let c be a point of the domain D of a function f, and suppose f to be bounded in some neighborhood $\mathcal{N}(c,\delta')$ of c. With $\delta \leq \delta'$, denote the set $D \cap \mathcal{N}(c,\delta)$ by A_δ. Then the infimum of the set of numbers $\Omega_{A_\delta} f$, with respect to all values of δ, is called the *fluctuation of f at c* and is denoted by $\omega_c f$. (It is clear that $\omega_c f$ is independent of δ'.) The intimate connection between the notions of continuity and fluctuation is apparent in the following theorem and examples.

Theorem 8-3 A function f, defined on D, is continuous at a point c of D if and only if the fluctuation of f at c is zero.

Proof Let $\varepsilon > 0$ be given. Then, if $\omega_c f = 0$, a neighborhood $\mathcal{N}(c,\delta)$ of c exists such that $\Omega_{A_\delta} f < \varepsilon$, where $A_\delta = D \cap \mathcal{N}(c,\delta)$. It follows that if x is any point of A_δ, then $|f(x) - f(c)| < \varepsilon$. Hence f is continuous at c.

On the other hand, suppose f to be continuous at c, and again let $\varepsilon > 0$ be given. Then a neighborhood $\mathcal{N}(c,\delta)$ exists such that $|f(x) - f(c)| < \varepsilon$ for all x in $D \cap \mathcal{N}(c,\delta)$. In particular, if x_1 and x_2 are any two points of $D \cap \mathcal{N}(c,\delta)$, then $|f(x_1) - f(c)| < \varepsilon$ and $|f(x_2) - f(c)| < \varepsilon$, whence $|f(x_1) - f(x_2)| < 2\varepsilon$. Thus, the oscillation of f on $D \cap \mathcal{N}(c,\delta)$ does not exceed 2ε, and it follows that this is also the case for the fluctuation at c; that is, $\omega_c f \leq 2\varepsilon$. Hence $\omega_c f = 0$▲

If a function f is discontinuous at a point c, the fluctuation of f at c serves as a kind of measurement of discontinuity.

Example 8-14 The function f defined by $f(x) = x^2$ for $x \neq 2$ and $f(2) = 1$ has a removable discontinuity at 2. The oscillation of f in a neighborhood $\mathcal{N}(2,\delta)$ of the point 2 is clearly $(2 + \delta)^2 - 1 = \delta^2 + 2\delta + 3$. We have $\omega_2 f = \inf\{\delta^2 + 2\delta + 3;\ \delta > 0\} = 3$.

The fluctuation at a jump discontinuity and the absolute value of the jump are not necessarily one and the same number. The reader may verify that $|\lim_{c+} f - \lim_{c-} f| \le \omega_c f$.

Example 8-15 Consider the function G defined by $G(x) = -1$ for $x < 0$, $G(0) = 7$, and $G(x) = 2$ for $x > 0$. Clearly the jump at 0 is 3 while $\omega_0 G = 8$.

Example 8-16 The function h defined by $h(x) = \sin(1/x)$ for $x \neq 0$ and $h(0) = 0$ has a discontinuity of the second kind at 0. The oscillation in a neighborhood $\mathcal{N}(0,\delta)$ is 2 since $\sin(1/x)$ assumes maximum and minimum values 1 and -1, respectively, in any such neighborhood. Hence $\omega_0 h = 2$.

Discontinuities of the first and second kind differ in an essential way with respect to possible distribution. In Example 8-10 it was seen possible for a function to possess a noncountable set of discontinuities of the second kind. This is not the case for removable or jump discontinuities.

Theorem 8-4 For any function the set of discontinuity points of the first kind is countable.

Proof Let f be a function defined on a domain D, and denote by A_n ($n = 1, 2, \ldots$) the set of points of D at which f has a discontinuity of the first kind with a fluctuation that exceeds $1/n$. Let c be a cluster point of the set A_n for some n. We shall show that c itself is not a point of A_n.

Since c is a cluster point of A_n, it is either a cluster point from the left or from the right. Without loss of generality, suppose the latter to be the case. Let $\varepsilon > 0$ be given. Then the interval $(c, c + \varepsilon)$ contains a point a of A_n, and it, in turn, possesses a neighborhood $\mathcal{N}(a,\varepsilon_1)$ which is entirely contained in $(c, c + \varepsilon)$. Because $\omega_a f > 1/n$ it follows that the oscillation of f on $\mathcal{N}(a,\varepsilon_1)$ exceeds $1/n$, whence points x_1 and x_2 exist in $\mathcal{N}(a,\varepsilon_1)$ such that $|f(x_1) - f(x_2)| > 1/n$. Since x_1 and x_2 are also points of $(c, c + \varepsilon)$ and since ε is arbitrary, it follows that f does not have a limit from the right at c. Thus c is not a discontinuity of the first kind and consequently is not a member of the set A_n.

We conclude that A_n, for each n, is a set of points none of which is a cluster point of the set. As such, the set A_n is a discrete and hence a countable set (see Exercise 4-42). As a consequence, the set $\bigcup_{n=1}^{\infty} A_n$, which is the set of all discontinuity points of the first kind, is a countable set▲

Exercises

8-12 Prove that if f is a function bounded on a set A, the oscillation of f on A is the supremum of the set of numbers $|f(x_1) - f(x_2)|$, where x_1 and x_2 are any two points of A.

8-13 Verify that $|\lim_{c+} f - \lim_{c-} f| \le \omega_c f$.

8-14 Let f be a bounded function defined on a domain D, and let c be a point of D. Define a function A_δ on the set of all positive real numbers δ by $A_\delta = D \cap \mathcal{N}(c,\delta)$. Prove that $\omega_c f = \lim_{0+} \Omega_{A_\delta} f$.

8-15 Let f be a bounded function defined on $[a,b]$, and for each positive integer n define a set D_n by $D_n = \{x; \quad \omega_x f \geq 1/n\}$. Prove that D_n is a closed set.

8-16 Let f be a bounded function defined on $[a,b]$, and let c denote an interior point of the interval with the property that $\omega_c f > 0$. Prove that the oscillation of f on one of the intervals $[a,c]$ and $[c,b]$ exceeds $\frac{1}{2}\omega_c f$.

8-17 Let f be a bounded function continuous on $[a,b]$, and let g be defined on $[a,b]$ by $g(a) = f(a)$ and $g(x) = \max\{f(y); \quad a \leq y \leq x\}$ if $a < x \leq b$. Prove that g is a continuous function.

8-4 Properties of Continuous Functions

In this section we shall discuss various relationships which exist between certain subsets of the domain and the range of a continuous function. Denote the domain of a function by D and its range by E. Then the term *image*, which hitherto has denoted that member $f(x)$ of E which corresponds by the function to the number x of D, we now extend to include sets of points in E. Thus, if A is a subset of the domain D, then the set B of numbers $f(x)$, where x is a member of A, is called the *image* of the set A, and we write $B = f(A)$. Briefly, the image of A by f is $f(A) = \{f(x); \quad x \in A\}$. It is clear that B is a subset of E, and that, in particular, $E = f(D)$.

It is of interest to inquire of a continuous function defined on D to what extent it serves to transfer certain set properties of subsets of D to their images in E. As Examples 8-17 through 8-19 show, we must not expect too much in this vein.

Example 8-17 Let f be the continuous function defined for all x by $f(x) = x^2$. The image of the open set $(0,2)$ is the open set $(0,4)$, but the image of the open set $(-1,1)$ is the set $[0,1)$, which is not open.

Example 8-18 Let g be the continuous function defined by $g(x) = 1/(1+x^2)$ on the domain D consisting of all real numbers. Let A denote the closed (but not bounded) set $\{x; \quad x \geq 3\}$. Then the image of A is the set $g(A) = (0, \frac{1}{10}]$. The latter is bounded but not closed.

Example 8-19 Consider the continuous function h defined by $h(x) = 1/x$ on the bounded (but not closed) domain $(0,1)$. Clearly, the range is the set $\{x; \quad x > 1\}$, which is neither bounded nor closed.

Having seen that openness, closedness, and boundedness need not be preserved by a continuous function, it is perhaps surprising to learn that the image of a compact set is always compact.

Theorem 8-5 The image of a compact set by a continuous function is compact.

Proof Let f be a continuous function defined on a domain which contains a compact set A, and let Γ be an open covering of $f(A)$. We show that some finite subset of Γ covers $f(A)$.

For each x in A the point $f(x)$ is in $f(A)$. Thus there is some member G of Γ such that $f(x) \in G$. Because G is open, there exists a neighborhood of $f(x)$ contained in G, and hence, by the continuity of f, there is a neighborhood $\mathcal{N}$ of x such that the image of $\mathcal{N} \cap A$ is in G.

In this way we associate a neighborhood with each point of A and thereby obtain an open covering of A. The compactness of A then insures that a finite collection of neighborhoods $\mathcal{N}_1, \mathcal{N}_2, \ldots, \mathcal{N}_n$ covers A. Further, for each v $(v = 1, 2, \ldots, n)$ the image C_v of $\mathcal{N}_v \cap A$ is contained in some set G_v of Γ. Since the sets $C_1, C_2, \ldots, C_n$ cover $f(A)$ and since $C_v \subset G_v$, it follows that the sets $G_1, G_2, \ldots, G_n$ also cover $f(A)$. Therefore a finite subcovering exists▲

Theorem 8-5 may be proved in quite another way, employing the characterization of a compact set B as a set with the property that every infinite subset possesses a cluster point belonging to B. It seems instructive to include such a proof as well.

Second Proof of Theorem 8-5 Let B_1 denote an infinite subset of the image set $f(A)$. Then B_1, in turn, contains a countably infinite subset of members $y_1, y_2, \ldots, y_n \ldots$ Corresponding to these members of $f(A)$ are the members of the infinite subset $A_1 = \{x_1, x_2, \ldots, x_n, \ldots\}$ of A with the property that $y_n = f(x_n)$ for all n. Obviously A_1 is bounded since $A_1 \subset A$. Hence A_1 has a cluster point a which, because of the compactness of A, belongs to A. It follows that a sequence $\{x_n'\}$ of members of A_1 exists converging to a and that, since f is continuous, the sequence $\{y_n'\}$, where $y_n' = f(x_n')$, converges to $f(a)$. Hence $f(a)$ is a point of $f(A)$ that is a cluster point of B_1. It follows that $f(A)$ is compact▲

Example 8-20 The function Ψ has the property that every compact set of points is mapped onto one of the sets $\{0\}$, $\{1\}$, or $\{0,1\}$, each of which is compact. It follows that a function is not necessarily continuous if every compact subset of its domain has a compact image.

An important property of continuous functions follows directly from the truth of Theorem 8-5 on the preservation of compact sets.

Theorem 8-6 A continuous function defined on a compact domain attains the values of its infimum and supremum.

Proof We have seen in Chapter 4 that a set which is closed and bounded possesses maximum and minimum values. Since this is exactly the character of the range of the given function, the truth of the theorem follows▲

That the properties of continuity, closedness, and boundedness are essential to the truth of Theorem 8-6 is seen in the next three examples.

Example 8-21 Let g be defined on the closed but unbounded set $\{x;\ x \geq 0\}$ by $g(x) = 1/(x+1)$. Then the range of g has supremum 1 and infimum 0. The value of the supremum is attained by the function at 0, but there is no value of x for which g assumes the value of its infimum.

Example 8-22 Let h be defined on the bounded but not closed domain $(0,1)$ by $h(x) = x(1-x)$. Then $\sup h = \frac{1}{4} = h(\frac{1}{2})$. However, $\inf h = 0$, and for no value of x does $h(x) = 0$.

Example 8-23 Let a discontinuous function f be defined on the compact domain $[-2,2]$ by $f(x) = 0$ for $-2 \leq x \leq -1$, $f(x) = x$ for $-1 < x < 1$, and $f(x) = 0$ for $1 \leq x \leq 2$. Then $\sup f = 1$ and $\inf f = -1$, but neither of these values is assumed by the function.

For the special instance of a function whose domain is an interval we may establish the following additional properties.

Theorem 8-7 If f is continuous on an interval and k is a number lying between $f(x_1)$ and $f(x_2)$, where x_1 and x_2 are points of that interval, then a point c exists between x_1 and x_2 such that $f(c) = k$.

Proof Without loss of generality, suppose that $x_1 < x_2$ and $f(x_1) < k < f(x_2)$. Let C denote the set of all points x in $[x_1,x_2]$ with the property $f(x) \leq k$. Then C is not empty since $f(x_1) < k$, and it is bounded above (by x_2). Hence C possesses a supremum c, and $x_1 \leq c \leq x_2$.

Suppose $f(c) < k$, so that $x_1 \leq c < x_2$. Then, because f is continuous, a neighborhood $\mathcal{N}(c,\varepsilon)$ exists such that $f(x) < k$ for all x in $[x_1,x_2] \cap \mathcal{N}(c,\varepsilon)$. This implies that values of x exist between c and x_2 for which $f(x) < k$, denying the role of c as supremum of C. We conclude $f(c) \geq k$. In a similar fashion, the assumption $f(c) > k$ leads to a contradiction. Hence $f(c) = k$; that is, a point c of $[x_1,x_2]$ exists at which the function assumes the value k▲

Corollary 1 If a function is continuous on a closed and bounded interval, then it assumes all values between its maximum and minimum values on that interval.

Corollary 2 The image of an interval by a continuous function is either a point or an interval.

Corollary 3 If f is continuous on an interval containing points x_1 and x_2 such that $f(x_1) < 0$ and $f(x_2) > 0$, then a point c of (x_1,x_2) exists such that $f(c) = 0$.

It may seem that the above theorem and corollaries have a converse. That is, one is tempted to infer that if a function assumes all values between any two function values, then the function is continuous. That this is not the case can be seen from Example 8-24.

Example 8-24 Let f be defined for all x by $f(x) = \sin(1/x)$ for $x \neq 0$ and $f(0) = 0$. We have seen that this function does not have a limit at 0 and hence is not continuous there. Yet it assumes all values between -1 and 1, and hence between any two function values, in any neighborhood of 0.

Exercises

8-18 Let f be a continuous function defined on a domain D, and let V denote any open set. Prove that an open set U exists such that $x \in U \cap D$ if and only if $f(x) \in V$.

8-19 Suppose f to be a continuous function defined on a domain D, and let T be any closed set. Prove that a closed set S exists such that $x \in S \cap D$ if and only if $f(x) \in T$.

Let f be defined on a domain D, and denote by B any set of points. Then the subset C of D consisting of points x such that $f(x)$ is in B is called the *inverse image* of B by f, and we write $C = f^{-1}(B)$.

8-20 For a given function f and set A, let $B = f(A)$ and $C = f^{-1}(B)$. (a) Prove that $A \subset C$. (b) Find an example in which $A \neq C$.

8-21 The function g is continuous and is defined on an open domain. Prove that the inverse image by g of any open set is open.

8-22 The function f is continuous and is defined on a closed domain. Prove that the inverse image by f of any closed set is closed. In particular, conclude that the set $\{x;\ f(x) = 0\}$ of *zeros* of a continuous function f on a closed domain is closed.

8-23 Cite examples to show that the statements of Exercises 8-21 and 8-22 cease to be true if the hypotheses regarding the domains are omitted.

8-24 Prove Theorem 8-7 by using Exercise 8-21 or 8-22 together with Exercise 4-33.

8-25 Suppose f to be a function defined and continuous on a domain D. Let A be any set and B the image of A. Prove that if $x \in \bar{A} \cap D$, then $f(x) \in \bar{B}$.

8-26 Let f be a continuous function defined on a domain D. Let k be any constant and A any set of numbers. Prove that if $f(x) < k$ for all points x in $A \cap D$, then $f(x) \leq k$ for all points x of $\bar{A} \cap D$.

8-27 If f is a continuous function on a domain D, B any closed set, and C the inverse image of B by f, prove that $\bar{C} \cap D = C$.

8-28 A function f defined on $[0,1]$ has the property that for each real number y the set $\{x;\ f(x) = y\}$ is empty or contains exactly two members. Prove that f is not continuous. Also, construct a function with the stated property.

8-29 Let C be the set of all continuous functions defined on the interval $I = [a,b]$. By the *norm* $\|f\|$ of a function f of C we mean the number $\|f\| = \max\{|f(x)|;\ x \in I\}$. Prove that (a) $\|f+g\| \leq \|f\| + \|g\|$, (b) $\|f\| = 0$ if and only if f is the zero function on I, and (c) $\|kf\| = |k|\ \|f\|$ for any real number k.

8-5 Uniform Continuity

Suppose f to be a function defined and continuous at each point x of a domain D. Then, for a fixed value of x and a prescribed $\varepsilon > 0$, a number $\delta > 0$ exists with the property that $|f(x+h) - f(x)| < \varepsilon$ for all $x + h$ in D such that $|h| < \delta$. It is to be expected, of course, that the determination of δ depends upon the value of ε. Briefly, δ is a function of the given number ε. If ε is assigned a decreasing sequence of values, then we would expect δ to assume a succession of decreasing values also. It is not in general true that a value of δ can be found which will serve for all possible choices of ε.

It is of interest to look at the situation from a different aspect. Having determined a satisfactory value for δ, corresponding to a particular x and a prescribed ε, we propose to consider different values of x without alteration in the given value of ε. The question is: How does variation in x effect the determination of δ? In general, it is the case that for different values of x in D we must expect correspondingly different values of δ. Thus δ is a function of x as well as of ε, and we often write $\delta = \delta(x,\varepsilon)$. We may gain some knowledge of this dependence by considering an example.

Example 8-25 Let f be defined by $f(x) = 1/x$ on $D = (0,1)$. Let x, for the moment, be committed. Then f is continuous at x if, for any given $\varepsilon > 0$, a number $\delta > 0$ can be found such that

(2)
$$\left|\frac{1}{x+h} - \frac{1}{x}\right| < \varepsilon$$

or, what is the same, assuming $\varepsilon < 1$,

(3)
$$-\frac{\varepsilon x^2}{1+\varepsilon x} < h < \frac{\varepsilon x^2}{1-\varepsilon x}$$

for all $x + h$ in D such that $|h| < \delta$. Since $\dfrac{\varepsilon x^2}{1+\varepsilon x} < \dfrac{\varepsilon x^2}{1-\varepsilon x}$, it is easy to see that if $|h| < \delta$, where $\delta = \dfrac{\varepsilon x^2}{1+\varepsilon x}$, then (3) is satisfied; this in turn implies (2). It is important to observe, too, that the value chosen for δ is the largest such that (2) is true for all $|h| < \delta$.

Now let us imagine a fixed value assigned to ε, and observe what happens as x varies on the interval (0,1). It is easy to see that by reducing the value of x, the value of the fraction $\dfrac{\varepsilon x^2}{1+\varepsilon x}$ diminishes also. In fact, by choosing x to be arbitrarily close to 0, we can make $\dfrac{\varepsilon x^2}{1+\varepsilon x}$ as small as we wish. Thus δ is forced to assume ever decreasing values. We conclude that for a given ε no one value of δ will serve for all values of x in D.

Recognizing the possibility that functions may exist which are continuous at all points of a domain D but which, unlike the function in Example 8-25, have the property that δ may be chosen without regard to x, we make the following definition. A function is said to be *uniformly continuous* on a domain D if, corresponding to any given $\varepsilon > 0$, there exists a number $\delta > 0$ with the property that $|f(x+h) - f(x)| < \varepsilon$ for all x and h such that x and $x + h$ are in D and $|h| < \delta$. Equivalently, we can state: A function f is uniformly continuous on a domain D if and only if, corresponding to each $\varepsilon > 0$, a number $\delta > 0$ exists such that $|f(x_1) - f(x_2)| < \varepsilon$ for all pairs of points x_1 and x_2 in D with the property $|x_1 - x_2| < \delta$.

The essence of uniform continuity is the independence of δ and x. Whereas continuity is basically a property of a function *at a point*, uniform continuity is a property of a function *over a domain*. If a function is uniformly continuous over a domain, then certainly it is a continuous function. However, as Example 8-25 indicates, a function may be continuous at each point of its domain without being uniformly continuous.

Example 8-26 Consider the function g defined by $g(x) = 1/x$ on $D = (a,1)$, where $a > 0$. The function differs from the function f of Example 8-25 only in the extent of its domain. This is just enough, however, to make the difference between uniform and nonuniform continuity.

Again, with x fixed and $\varepsilon > 0$ given, we wish to find δ such that (2) is true for all h with the property that $|h| < \delta$. Again we may choose $\delta = \dfrac{\varepsilon x^2}{1 + \varepsilon x}$ and thereby establish the continuity of g over $(a,1)$. But now we can say more. For *all* values of x on the given domain $(a,1)$ the number $\dfrac{\varepsilon a^2}{1 + \varepsilon a}$ is less than $\dfrac{\varepsilon x^2}{1 + \varepsilon x}$. Hence, if we alter our choice of δ to $\delta = \dfrac{\varepsilon a^2}{1 + \varepsilon a}$, then for *all* values of x inequality (2) is valid for all h such that $|h| < \delta$. Hence the function g is uniformly continuous on $(a,1)$.

It has been noted that the nature of the domain of definition of a function is important in distinguishing between continuity and uniform continuity. In the next theorem we see how compactness of a domain resolves the two concepts.

Theorem 8-8 If f is continuous on a compact domain D, then f is uniformly continuous on D.

Proof Let $\varepsilon > 0$ be given, and let c be a point of D. Since f is continuous at c, it follows that a number δ_c exists with the property that $|f(x) - f(c)| < \varepsilon$ for all x in $D \cap \mathcal{N}(c,\delta_c)$. This is the case for each c in D. Thus, employing the number δ_c, we may construct a set of neighborhoods $\mathcal{N}(c,\frac{1}{2}\delta_c)$. This set constitutes an open covering of the compact set D, and hence a finite subset

$$\mathcal{N}(c_1,\tfrac{1}{2}\delta_{c_1}),\ \mathcal{N}(c_2,\tfrac{1}{2}\delta_{c_2}),\ \ldots,\ \mathcal{N}(c_n,\tfrac{1}{2}\delta_{c_n})$$

exists which also covers D.

Now let $\delta = \min\{\frac{1}{2}\delta_{c_1}, \frac{1}{2}\delta_{c_2}, \ldots, \frac{1}{2}\delta_{c_n}\}$, and consider any two points x_1 and x_2 of D with the property that $|x_1 - x_2| < \delta$. Clearly, x_1 belongs to one of the above neighborhoods, say to $\mathcal{N}(c_i,\frac{1}{2}\delta_{c_i})$. Hence $|x_1 - c_i| < \frac{1}{2}\delta_{c_i}$, so that from the continuity of f at c_i follows $|f(x_1) - f(c_i)| < \varepsilon$. Moreover, $|x_2 - c_i| \le |x_2 - x_1| + |x_1 - c_i| < \delta + \frac{1}{2}\delta_{c_i}$, so that $|x_2 - c_i| < \delta_{c_i}$. Again from the continuity of f at c_i follows $|f(x_2) - f(c_i)| < \varepsilon$. But these two inequalities together imply $|f(x_1) - f(x_2)| < 2\varepsilon$ for all x_1 and x_2 in D such that $|x_1 - x_2| < \delta$. Hence f is uniformly continuous▲

Exercises

8-30 Prove that if f is uniformly continuous on a bounded domain D, then f is bounded on D.

8-31 Assume that f is uniformly continuous on D. Prove that f can be extended to $\bar{D}$ in such a way that the extended function is continuous.

A function f, defined on a domain D, is said to be *upper semicontinuous at a point c of D* if, corresponding to each $\varepsilon > 0$, there exists a number $\delta > 0$ such that $f(x) < f(c) + \varepsilon$ for all x in $D \cap \mathcal{N}(c,\delta)$. *Lower semicontinuity* is defined in a dual manner.

8-32 State the definition of *lower semicontinuity* explicitly.

8-33 Prove that if f and g are upper semicontinuous at c, then so is $f+g$.

8-34 Cite an example of functions f and g which are upper semicontinuous at a point c but are such that fg is not upper semicontinuous at c.

8-35 Prove that f is upper semicontinuous at a point c if and only if $-f$ is lower semicontinuous at c.

8-36 Let f be defined by $f(x) = x$ if $x < 0, f(0) = 1$, and $f(x) = \sin(1/x)$ if $x > 0$. Prove that f is upper semicontinuous at each point.

8-37 If a function f, defined on $\mathscr{R}$, is upper semicontinuous at each point and if k denotes any real number, prove that the set $\{x;\quad f(x) \geq k\}$ is closed.

8-38 Prove that if f is a function defined on a closed domain, then the function $\omega_x f$ is upper semicontinuous at each point.

8-39 Let f be a function defined on [0,1] with the property that its set of continuity points is dense in that interval. Let $D_n = \{x;\quad \omega_x f \geq 1/n\}$, where n is a positive integer. Prove (a) that D_n is nowhere dense in [0,1], and (b) that D, the set of points of [0,1] at which f is discontinuous, is of the first category.

8-40 Prove that no function exists on [0,1] which is continuous at each rational point and discontinuous at each irrational point.

8-41 Let f be a continuous function defined on a closed set F. Prove that there exists a continuous function g defined on $\mathscr{R}$ that is an extension of f. Moreover, if f is bounded, g can be constructed so that it attains a maximum equal to that of f. (*Suggestion*: Define g to be a linear function on the open intervals of $\mathscr{C}F$.)

9

Derivatives of Functions

"*I have trod the upward and the downward slope.*"
ROBERT LOUIS STEVENSON (1850–1894), *Songs of Travel*

9-1 Differentiability of a Function

Suppose f to be a function defined on a domain D, and let c be a point of D which is also a cluster point of D. Consider the function F defined at all points of D except c by

$$F(x) = \frac{f(x) - f(c)}{x - c}.$$

Then, if F has a limit at c, we say that the function f is *differentiable at c*. The number $\lim_c F$ is also denoted by the symbol $f'(c)$, that is,

$$f'(c) = \lim_c F = \lim_c \frac{f(x) - f(c)}{x - c},$$

and is called the *derivative of f at c*.

The reasons for the conditions placed on the point c are quite plain. For one thing, c must be a member of D if the symbol $f(c)$ is to have meaning. Also, it will be recalled that the limit of a function (in this case F) is defined at a point c only if c is a cluster point of the domain of the function.

If a function f is differentiable at each point of a subset A of its domain D, then it is said to be *differentiable on A*; and the function f', defined to have the value $f'(x)$ at each point of A, is called the *derivative of f* on A. If f is differentiable on D itself, then we may say simply that it is a *differentiable function*. For convenience, whenever we propose to consider the derivative of a function on a domain D or on a subset A of D we shall assume, unless otherwise stated, that it is meaningful to do so; that is, we assume that D or A is dense in itself.

The definitions above concern two different but related concepts. The derivative of f at c is a number while the derivative of f is a function. Frequently, however, the term *derivative of f* is employed to denote both number and function, and it is left to context to distinguish which is intended.

Example 9-1 Let f be a constant function; that is, $f(x) = k$ for all x in a domain D. Then, for each point c of D we have

$$F(x) = \frac{f(x) - f(c)}{x - c} = \frac{k - k}{x - c} = 0$$

for all $x \neq c$. Hence $f'(c) = 0$. We conclude that f is differentiable on D and that f' is the zero function on D.

Example 9-2 Let g be a function defined on a domain D by $g(x) = x$. Then, for $x \neq c$ we have

$$G(x) = \frac{g(x) - g(c)}{x - c} = \frac{x - c}{x - c} = 1,$$

whence $g'(c) = 1$ for all c in D. The function g' is the unit function on D.

Example 9-3 Consider f defined by $f(x) = x^2$ for all x. If c is any point, we have

$$F(x) = \frac{x^2 - c^2}{x - c} = x + c,$$

whence $\lim_c F = 2c$. Clearly the function f is differentiable at each point of its domain.

Example 9-4 Let g be defined by $g(x) = x^2$ on a domain D consisting of the set of all rational points $1/n$ (n a positive integer) together with the number 0. Then only at 0 is the notion of derivative meaningful. We have

$$g'(0) = \lim_0 \frac{x^2 - 0}{x - 0} = \lim_0 x = 0.$$

Hence g is differentiable at 0.

Example 9-5 Let h be defined by $h(x) = \sqrt{x}$ on the interval $\{x;\ \ x \geq 0\}$. Since this domain is dense in itself, the notion of derivative is meaningful at each point. We have

$$h'(c) = \lim_c \frac{\sqrt{x} - \sqrt{c}}{x - c} = \lim_c \frac{1}{\sqrt{x} + \sqrt{c}}.$$

If $c \neq 0$, we see that $h'(c) = 1/2\sqrt{c}$, but at $c = 0$ no limit exists. The function h, therefore, is differentiable only on the subset $\{x; \quad x > 0\}$ of its domain.

If a function f possesses a derivative f' on a domain D and if f' is itself differentiable, its derivative is denoted by f'' and is called the *second derivative of f*. (Accordingly, for emphasis, the derivative of f is sometimes referred to as the *first derivative of f*.) Clearly this concept may be extended to the general notion of a *derivative of nth order*. Thus, the symbol $f^{(n)}$ denotes the derivative of the function $f^{(n-1)}$ and is called the *nth derivative of the function f*.

For the present we concern ourselves with the (first) derivative of a function. Later we shall have more to say about higher order derivatives.

As may be expected, the concepts of continuity and differentiability are related.

Theorem 9-1 If a function is differentiable at a point, then it is continuous at that point.

Proof Suppose a function f to be differentiable at a point c. Then

$$\lim_{c} \frac{f(x) - f(c)}{x - c} = f'(c)$$

exists and

$$\lim_{c} [f(x) - f(c)] = \lim_{c} \frac{f(x) - f(c)}{x - c} (x - c)$$

$$= \lim_{c} \frac{f(x) - f(c)}{x - c} \lim_{c} (x - c)$$

$$= f'(c) \cdot 0 = 0.$$

Hence $\lim_{c} f(x) = f(c)$, proving continuity of f at c▲

Example 9-6 The function Ψ is not continuous at any point. Therefore it does not possess a derivative at any point.

Although differentiability implies continuity, the converse is not also true. A function may, but need not, be differentiable at a continuity point.

Example 9-7 The function f defined by $f(x) = 0$ if x is rational and $f(x) = x$ if x is irrational was shown to be continuous at 0 and discontinuous elsewhere. To determine whether or not it has a derivative at 0, we consider

$$F(x) = \frac{f(x) - f(0)}{x - 0} = \frac{f(x)}{x}.$$

If x is rational, then $F(x) = x/x = 1$; whereas if x is irrational, we have $F(x) = 0/x = 0$. Since $\lim_{0} F$ does not exist, we conclude that f is not differentiable at 0.

Example 9-8 Consider the function g defined by $g(x) = x^2$ if x is rational and $g(x) = 0$ if x is irrational. Like the function f of Example 9-7, this function is continuous only at 0. We have

$$G(x) = \frac{g(x) - g(0)}{x - 0} = \frac{g(x)}{x}.$$

Hence $G(x) = x^2/x = x$ if x is rational and $G(x) = 0/x = 0$ if x is irrational. Since $\lim_0 G$ exists with value 0, we conclude that g is differentiable at 0 and $g'(0) = 0$.

Example 9-9 Let h be a function defined for all real numbers by $h(x) = |x|$. Then, for any point $c > 0$, let ε be such that $0 < \varepsilon < c$. For all points x with the property $0 < |x - c| < \varepsilon$ we then have

$$H(x) = \frac{h(x) - h(c)}{x - c} = \frac{x - c}{x - c} = 1$$

so that $h'(c) = 1$. Similarly, $h'(c) = -1$ if $c < 0$. Finally, if $c = 0$, then $H(x) = (x - 0)/(x - 0) = 1$ if $x > 0$ and $H(x) = (-x - 0)/(x - 0) = -1$ if $x < 0$, from which we conclude that $\lim_0 H$ does not exist. The function h, therefore, although continuous for all real x, is differentiable only for $c \neq 0$.

While continuity at a point is not sufficient to guarantee differentiability at the point, one wonders if it is possible for a function to be continuous at all points of an interval without a derivative existing at least somewhere on the domain. Intuition, supported by the usual picturization of a function by a graph, suggests that such is not possible, and for some time mathematicians were of this opinion. However, in 1875 Karl Weierstrass gave an example of a function which is defined and continuous for all real x but which has a derivative nowhere. This function will be described at length in Section 12-3.

9-2 Combinations of Differentiable Functions

The differentiability of the sum, product, and quotient of two differentiable functions, as well as that of a differentiable function of a differentiable function, is attested to in Theorems 9-2 through 9-5.

Theorem 9-2 If f and g are functions differentiable at a point c of their common domain, then the function $h = f + g$ is differentiable at c, and $h'(c) = f'(c) + g'(c)$.

Proof The differentiability of f and g at c implies that the functions F and G, given by

$$F(x) = \frac{f(x) - f(c)}{x - c} \quad \text{and} \quad G(x) = \frac{g(x) - g(c)}{x - c}$$

for $x \neq c$, have limits $f'(c)$ and $g'(c)$, respectively, at c. Then

$$H(x) = \frac{h(x) - h(c)}{x - c} = \frac{[f(x) + g(x)] - [f(c) + g(c)]}{x - c}$$

$$= \frac{f(x) - f(c)}{x - c} + \frac{g(x) - g(c)}{x - c} = F(x) + G(x).$$

Hence $H(x)$ possesses a limit at c, and its value is $h'(c) = \lim_c F + \lim_c G = f'(c) + g'(c)$▲

The converse of Theorem 9-2 is not true. For consider functions f and g defined by $f(x) = \Psi(x)$ and $g(x) = -\Psi(x)$. Neither f nor g is differentiable at any point. However, $h = f + g$ is the zero function on the set of all real numbers. It is differentiable everywhere.

Theorem 9-3 If f and g are functions differentiable at a point c of their common domain, then the function $h = fg$ is differentiable at c, and $h'(c) = f(c)g'(c) + g(c)f'(c)$.

Proof Let F, G, and H be functions defined for $x \neq c$ as in the proof of Theorem 9-2. Then

$$H(x) = \frac{f(x)g(x) - f(c)g(c)}{x - c}$$

$$= \frac{f(x)[g(x) - g(c)] + g(c)[f(x) - f(c)]}{x - c}$$

$$= f(x)G(x) + g(c)F(x).$$

Since f is differentiable at c, it is continuous there. Hence $\lim_c f$ exists and equals $f(c)$. From this and from the hypotheses of the theorem it follows that H has a limit at c given by $h'(c) = f(c)g'(c) + g(c)f'(c)$▲

If f and g are differentiable at a point c, and if k_1 and k_2 are any constants, then the function $h = k_1 f + k_2 g$ is differentiable at c and $h'(c) = k_1 f'(c) + k_2 g'(c)$. Thus differentiation is a *linear operation*.

That the converse of Theorem 9-3 is false is seen by considering the functions f and g defined by $f(x) = \Psi(x)$ and $g(x) = 1 - \Psi(x)$. Although neither of these is differentiable anywhere, the product $h = fg$ has the constant value 0 for all x and is therefore differentiable. On the other hand, if $h = kf$ is differentiable for a constant $k \neq 0$, then so is $(1/k)(h) = f$.

Theorem 9-4 If f and g are functions differentiable at a point c of their common domain D, and if $g(c) \neq 0$, then the function $h = f/g$ is differentiable at c, and the value of the derivative there is

$$h'(c) = \frac{g(c)f'(c) - f(c)g'(c)}{[g(c)]^2}.$$

Proof Left as an exercise▲

The general power function defined to have the value ax^n for all x, for all non-negative integers n, and for all real numbers a, has been seen in Examples 9-1 and 9-2 to be differentiable in the special cases $n = 0$ and $n = 1$; that is, a and ax define differentiable functions for all x. By Theorem 9-3 it follows that ax^2, ax^3, and, in general, ax^n are differentiable. Theorem 9-2 then implies differentiability of the polynomial function

$$a_0 x^n + a_1 x^{n-1} + \cdots + a_{n-1} x + a_n$$

for all x, and Theorem 9-4 that of the rational function

$$\frac{a_0 x^n + a_1 x^{n-1} + \cdots + a_{n-1} x + a_n}{b_0 x^m + b_1 x^{m-1} + \cdots + b_{m-1} x + b_m}$$

for all values of x which do not cause the denominator to vanish.

Theorem 9-5 (The Chain Rule) Let f denote a function differentiable at a point x_0 of its domain D. Let g be defined on the range of f, and suppose g to be differentiable at $y_0 = f(x_0)$. Then the function $h = g \circ f$ is differentiable at x_0, and its derivative there is $h'(x_0) = g'(y_0)f'(x_0)$.

Proof Consider the function u defined by $u(x_0) = 0$ and

$$u(x) = \frac{f(x) - f(x_0)}{x - x_0} - f'(x_0) \quad \text{for } x \neq x_0 .$$

Since f is differentiable at x_0, it follows that $\lim_{x_0} u = 0$; and, since this is the value of $u(x_0)$, the function u is continuous at x_0. For all values of x in D, the relation between u and f may then be written

$$f(x) - f(x_0) = [f'(x_0) + u(x)](x - x_0).$$

In a similar fashion a function v may be defined such that v is continuous at y_0, $v(y_0) = 0$, and

$$g(y) - g(y_0) = [g'(y_0) + v(y)](y - y_0).$$

Because f is differentiable at x_0, it is continuous there. Hence, if we write $w(x) = v(f(x))$, then the function w is continuous at x_0. Moreover, $w(x_0) = v(f(x_0)) = 0$.

We now have

$$\begin{aligned} h(x) - h(x_0) &= g(f(x)) - g(f(x_0)) \\ &= [g'(f(x_0)) + v(f(x))][f(x) - f(x_0)] \\ &= [g'(y_0) + w(x)][f'(x_0) + u(x)](x - x_0). \end{aligned}$$

For $x \neq x_0$ this yields

$$\frac{h(x) - h(x_0)}{x - x_0} - g'(y_0)f'(x_0) = w(x)f'(x_0) + u(x)g'(y_0) + w(x)u(x).$$

Since each term of the right member is continuous at x_0 and has the value 0 there, we conclude that the limit of the left member at x_0 exists and equals 0. Hence

$$h'(x_0) = \lim_{x_0} \frac{h(x) - h(x_0)}{x - x_0} = g'(y_0)f'(x_0)$$

as proposed▲

It might seem that Theorem 9-5 could be proved more easily by defining the function H by

$$H(x) = \frac{h(x) - h(x_0)}{x - x_0} = \frac{g(f(x)) - g(f(x_0))}{x - x_0},$$

writing it in the form

$$H(x) = \frac{g(f(x)) - g(f(x_0))}{f(x) - f(x_0)} \cdot \frac{f(x) - f(x_0)}{x - x_0},$$

and applying the rule for limit of a product. This procedure fails, however, since it may be that $f(x) - f(x_0)$ has the value 0 at some point of each neighborhood of x_0, in which case division by $f(x) - f(x_0)$ is not possible. Hence the proof as given above.

If the hypotheses of Theorem 9-5 are satisfied at all points of D, then the function $g \circ f$ is differentiable, and

$$(g \circ f)' = (g' \circ f)f'.$$

Exercises

9-1 Suppose g to be a function defined for all x by $g(x) = 2x + 1$ if x is rational and $g(x) = x^2 - 2x + 5$ if x is irrational. Prove that g is differentiable at 2.

9-2 Let h be a function defined for all x by $h(1)=2$ and $h(x) = (x - 1)/(\sqrt{x} - 1)$ if $x \neq 1$. Show that h is differentiable at 1.

9-3 Can a function be differentiable at each point of (a) a bounded domain, (b) a compact domain, without being uniformly continuous? Defend each answer with an example or an argument.

9-4 Let f be a function differentiable at a point c, and suppose that in each deleted neighborhood of c a point x exists such that $f(x) = f(c)$. Prove that $f'(c) = 0$.

9-5 Let $\{x_n\}$ and $\{y_n\}$ be two sequences converging to c with $x_n \neq y_n$ for all n. Assume that f is defined for all real numbers and that $f'(c)$ exists. Prove that $\left\{\frac{f(y_n) - f(x_n)}{y_n - x_n}\right\}$ converges to $f'(c)$.

9-6 Prove Theorem 9-4.

A function f defined on a domain D is said to satisfy a *Lipschitz condition* at a point c of D if there is a neighborhood $\mathcal{N}(c,\delta)$ and a positive number N such that $|f(x) - f(c)| \leq N|x - c|$ for all $x \in \mathcal{N}(c,\delta) \cap D$.

9-7 Prove that a function which satisfies a Lipschitz condition at c is continuous there but that the converse need not hold.

9-8 Prove that a function which is differentiable at c satisfies a Lipschitz condition there but that the converse need not hold.

9-3 Geometric Significance of Derivative

A geometric interpretation of derivative helps to make the concept more clear. Suppose f to be a function defined on an interval, and let c be an interior point of the interval. From the definition of $f'(c)$, assuming it to exist, there corresponds to a number $\varepsilon > 0$, a number $\delta > 0$ such that, if x is in the neighborhood $\mathcal{N}^*(c,\delta)$, then

$$\left| \frac{f(x) - f(c)}{x - c} - f'(c) \right| < \varepsilon$$

or, what is the same,

$$f'(c) - \varepsilon < \frac{f(x) - f(c)}{x - c} < f'(c) + \varepsilon.$$

From this we conclude that if x is a point of $(c, c + \delta)$, then

(1) $$f(c) + [f'(c) - \varepsilon](x - c) < f(x) < f(c) + [f'(c) + \varepsilon](x - c),$$

and if x is in $(c - \delta, c)$, then

(2) $$f(c) + [f'(c) + \varepsilon](x - c) < f(x) < f(c) + [f'(c) - \varepsilon](x - c).$$

Imagine a graph of the function f to be drawn. Then if lines are drawn through the point $(c,f(c))$, one with slope exceeding $f'(c)$ and one with slope less than $f'(c)$, then what we have shown in (1) and (2) is that a neighborhood $\mathcal{N}(c,\delta)$ of c exists such that the points $(x,f(x))$ lie between the lines for all points x of

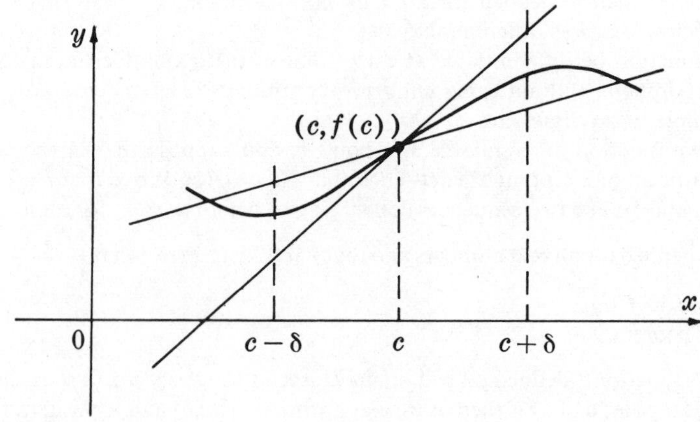

Figure 9-1

the neighborhood. (See Figure 9-1.) The reader may interpret the conclusions in Examples 9-7 and 9-8 in the light of the above geometric analysis.

Assuming an intuitive knowledge of the notion of a tangent line to a curve, we attempt to reconcile this with the notion of derivative. To this end we confine our attention to a certain restricted category of functions, it being clear that many of the functions we have considered so far do not possess graphs which permit construction of tangent lines. To make matters simple, let f be a function which is defined on an interval, and let c be a point of the interval. Then f is said to be *smooth at c* if it is differentiable on some neighborhood of c and if f' is a continuous function on that neighborhood. Furthermore, f is called a *smooth* function if it is smooth at each point of its domain. It is clear that if a function is smooth, it is continuous.

The concept of smoothness over an interval is very nearly in accord with our intuitive notion of a "continuously rolling tangent line." However, the two are not identical. If we regard a *tangent line* to the graph of f and c as a line through the point $(c,f(c))$ having slope $f'(c)$, then a tangent line rolls continuously on the graph of a smooth function. However, the intuitive notion of tangent line goes beyond such definition. It is possible to draw a tangent line to the graph of a function at a point at which the function has no derivative.

Example 9-10 The function g defined for all x by $g(x) = x^{1/3}$ has a vertical tangent at the origin. However, if we attempt to compute the derivative of g at 0, we find that

$$G(x) = \frac{g(x) - g(0)}{x - 0} = \frac{x^{1/3}}{x} = x^{-2/3}$$

does not have a limit at 0. Hence g is not differentiable at 0. (See Figure 9-2.)

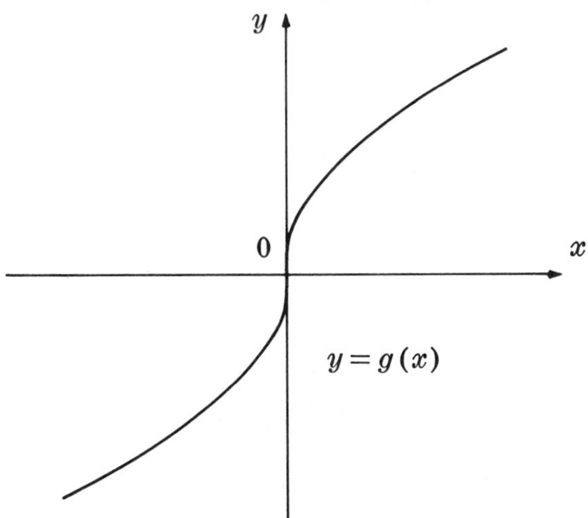

Figure 9-2

In the above example we see that $\lim_{0} G = \infty$. This suggests that we might enlarge our definition of derivative, saying that f has an *infinite derivative at c* in the event that

$$\lim_{c} \frac{f(x) - f(c)}{x - c} = \pm\infty.$$

With this convention the function g of Example 9-10 is smooth at every point.

Even this extended notion of derivative, however, does not reconcile the notions of derivative and tangent line, as we can see from the next example.

Example 9-11 Let h be defined by $h(x) = \sqrt{-x}$ for $x < 0$ and $h(x) = \sqrt{x}$ for $x \geq 0$. Its graph (Figure 9-3) has the property that a tangent line may be drawn at every point, including the origin. To find the derivative of h at 0 we consider

$$H(x) = \frac{h(x) - h(0)}{x - 0} = \frac{h(x)}{x} = \begin{cases} -\dfrac{1}{\sqrt{-x}} & \text{for } x < 0 \\ \dfrac{1}{\sqrt{x}} & \text{for } x > 0. \end{cases}$$

Clearly, $\lim_{0-} H = -\infty$ and $\lim_{0+} H = \infty$. Thus h does not have a derivative at 0, even in the extended sense.

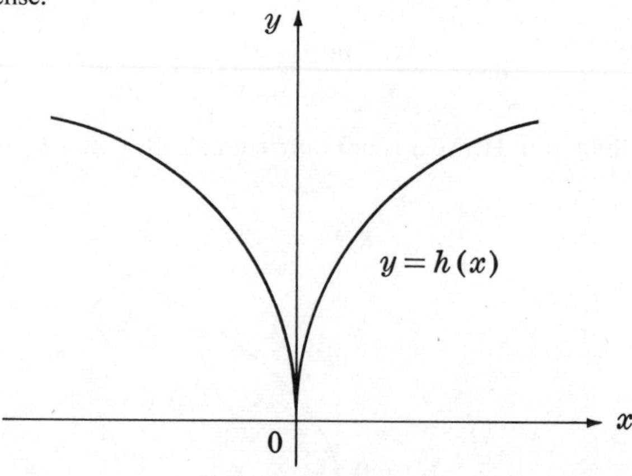

Figure 9-3

In the course of this text *we shall not admit infinite values in the general concept of derivative*. To include such would be to upset, among other things, the role of continuity as a necessary condition for differentiability.

Example 9-12 Let f be defined for all x by $f(x) = x^{1/3}$ for $x \leq 0$ and $f(x) = 1 + x^{1/3}$ for $x > 0$. This function is plainly discontinuous at 0, yet possesses an infinite derivative (if such is allowed) at that point. (See Figure 9-4.)

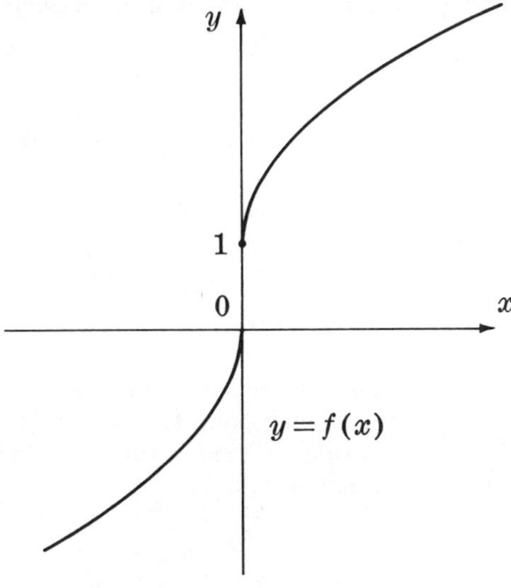

Figure 9-4

9-4 Mean Value Theorems

It is frequently the case that at some special point of the domain of a function a property may be established which reflects, in part, at least, the nature of the function on its full domain. The isolation of such a property is a decided technical advantage; what is true "at a point" is often easier to translate than what is true "over a domain." If a function is defined on an interval $[a,b]$ and if a special property is seen to exist at a point ξ of (a,b), then ξ is termed a *mean value*, and the fact of its existence is referred to as a *mean value theorem*. Certain such principles are basic in connection with differentiation.

Theorem 9-6 Let f be a function differentiable at a point c of a domain D, and suppose $f'(c) > 0$. Then a neighborhood $\mathcal{N}(c,\varepsilon)$ exists such that $f(x) < f(c)$ if $x < c$, and $f(x) > f(c)$ if $x > c$, for all x of $D \cap \mathcal{N}(c,\varepsilon)$. (Dually, if $f'(c) < 0$, then $\mathcal{N}(c,\varepsilon)$ exists such that $f(x) > f(c)$ if $x < c$, and $f(x) < f(c)$ if $x > c$, for all x in $D \cap \mathcal{N}(c,\varepsilon)$.)

Proof From $f'(c) > 0$ it follows that the function defined by $\dfrac{f(x) - f(c)}{x - c}$ is bounded away from 0. Hence a neighborhood $\mathcal{N}(c,\varepsilon)$ of c exists such that $\dfrac{f(x) - f(c)}{x - c} > 0$ for all x in $D \cap \mathcal{N}(c,\varepsilon)$. For such x, the numbers $f(x) - f(c)$ and $x - c$ have the same sign, and the conclusions of the theorem follow. (Similarly for the dual theorem.)▲

The converse of Theorem 9-6 is not true, as may be seen by observing the function defined by $f(x) = x^3$. Here $f(x) < 0$ for $x < 0$, $f(x) > 0$ for $x > 0$, but $f'(0) = 0$. However, from the combined truths of Theorem 9-6 and its dual, we may conclude the following. If $f(x) < f(c)$ for $x < c$, or $f(x) > f(c)$ for $x > c$, then $f'(c) \geq 0$. Also, if $f(x) > f(c)$ for $x < c$, or $f(x) < f(c)$ for $x > c$, then $f'(c) \leq 0$.

A point c is called a *local maximum point* of a function f if a neighborhood $\mathcal{N}(c,\varepsilon)$ exists such that $f(x) \leq f(c)$ for all x which are at once in $\mathcal{N}(c,\varepsilon)$ and in the domain D of the function. Also, c is a *local minimum point* if $\mathcal{N}(c,\varepsilon)$ exists such that $f(x) \geq f(c)$ for all x in $D \cap \mathcal{N}(c,\varepsilon)$.

Example 9-13 The polynomial function defined for all x by $f(x) = 3x^4 - 4x^3 - 12x^2$ may be shown to possess a local maximum point at 0 and local minimum points at -1 and 2. Moreover, the minimum value of the function is attained at 2. There is no maximum since the function is not bounded above.

Example 9-14 The function g defined for all x by $g(x) = \sin(1/x)$ has local maxima at points $2/(4n+1)\pi$ and local minima at $2/(4n-1)\pi$ for $n = 1,2,\ldots$.

A necessary condition for the existence of a local maximum or minimum point concerns the existence and value of a derivative.

Theorem 9-7 Suppose c to be an interior point of the domain D of a function f, and assume f to be differentiable at c. Then $f'(c) = 0$ in the event that c is a local maximum point or a local minimum point.

Proof If $f'(c) > 0$, then by Theorem 9-6 a number $\varepsilon > 0$ exists such that if x is in $(c - \varepsilon, c)$, then $f(x) < f(c)$, and if x is in $(c, c + \varepsilon)$, then $f(x) > f(c)$. Hence c is neither a local maximum nor a local minimum point. A similar result follows if $f'(c) < 0$. We conclude that $f'(c) = 0$▲

The reader may recall the above theorem from a course in calculus in which one frequently encounters the following argument. "Let c be a local maximum point of f, and suppose f' exists and is continuous in some neighborhood of c." (This is, of course, less general than the hypothesis of Theorem 9-7.) "Since c is a local maximum, the function is increasing for $x < c$, so that for such x we have $f'(x) \geq 0$. On the other hand, for $x > c$ the function f is decreasing; so here $f'(x) \leq 0$. Since f' is continuous, it follows that $f'(c) = 0$."

This argument is fallacious. For consider the function f defined by $f(0) = 0$ and $f(x) = -x^4[2 - \sin(1/x)]$ for $x \neq 0$. Since $f(x) < 0$ for all $x \neq 0$, it follows that 0 is a local maximum of f. Further, f' exists and is continuous for all x. However, in any neighborhood of the point 0 there are points to the left of 0 at which $f'(x)$ is negative and points to the right at which $f'(x)$ is positive.

The restriction in Theorem 9-7 that c be an interior point is not essential; the theorem and proof remain valid if c is merely a cluster point from both right and left. However, if c is a one-sided cluster point—for example, an endpoint of an interval—then the situation is somewhat different. If c is a cluster point from the left, then the existence of a local maximum at c implies $f'(c) \geq 0$; also, the existence of a local miminum at c implies $f'(c) \leq 0$. A dual situation prevails if c is a cluster point from the right.

The next few theorems are concerned with domains which are intervals.

Theorem 9-8 (Rolle's Theorem) Let f be a function which is continuous on a closed interval $[a,b]$, differentiable on the open interval (a,b), and such that $f(a) = f(b)$. Then a point ξ exists with the properties $f'(\xi) = 0$ and $a < \xi < b$.

Proof We distinguish two cases. First, suppose f to be a constant function. Then $f(x) = f(a)$ for all x, so that $f'(x) = 0$ at all points of (a,b). Hence any point ξ of (a,b) is such that $f'(\xi) = 0$. Second, suppose that f is not a constant function. Then the continuity of f implies that f attains maximum and minimum values, and, because the function is not constant, these values are different. This implies that either the maximum point or the minimum point lies in the open interval (a,b). Denoting one such point by ξ, it follows from Theorem 9-7 that $f'(\xi) = 0$▲

Example 9-15 The function g defined by $g(x) = 3x - x^3$ on $[-\sqrt{3},\sqrt{3}]$ satisfies the conditions of Rolle's theorem. From $g'(x) = 3 - 3x^2$ follows the existence of two points ξ. Thus $f'(-1) = f'(1) = 0$.

Example 9-16 The function h defined by $h(x) = \sqrt{1 - x^2}$ on the interval $[-1,1]$ satisfies the hypotheses of Rolle's theorem. Note that $h'(x) = -x(1 - x^2)^{-1/2}$ exists only on the open interval $(-1,1)$. The vanishing of h' obviously takes place at 0.

Theorem 9-9 (The Mean Value Theorem) Let f be a function continuous on the interval $[a,b]$ and differentiable on (a,b). Then a point ξ exists such that $a < \xi < b$ and

$$f(b) - f(a) = (b - a)f'(\xi).$$

Proof Let g be a function defined on $[a,b]$ by

$$g(x) = [f(b) - f(a)](x - a) - [f(x) - f(a)](b - a).$$

Clearly, g is continuous on $[a,b]$ and differentiable on (a,b). Moreover, $g(a) = g(b) = 0$. The conditions of Rolle's theorem being satisfied, it follows that ξ exists such that $a < \xi < b$ and

$$g'(\xi) = f(b) - f(a) - (b - a)f'(\xi) = 0,$$

from which the conclusion of the theorem follows▲

Example 9-17 Let f be defined on $[0,1]$ by $f(x) = \sqrt{x}$. Clearly, f is continuous on $[0,1]$, and its derivative $f'(x) = \frac{1}{2}x^{-1/2}$ exists at each point of $(0,1)$. The (unique) mean value may be identified as the point $\xi = \frac{1}{4}$.

The mean value theorem may be stated in a different and frequently useful form: If f is continuous on the closed interval $[x, x+h]$, $h > 0$, and differentiable on $(x, x+h)$, then θ exists such that

$$f(x+h) = f(x) + hf'(x+\theta h) \quad \text{and} \quad 0 < \theta < 1.$$

The remaining theorems in this section are established as consequences of the mean value theorems. Theorem 9-10 will be seen to be of particular importance in connection with Riemann integration.

Theorem 9-10 Suppose f to be continuous on $[a,b]$, differentiable on (a,b), and such that $f'(x) = 0$ for all x in (a,b). Then f is a constant function on $[a,b]$.

Proof Let x be any point such that $a < x \leq b$. Then on the interval $[a,x]$ the hypotheses of the mean value theorem are satisfied. Hence ξ exists such that $a < \xi < x$ and $f(x) - f(a) = (x-a)f'(\xi) = 0$. Since x is arbitrary, it follows that f has the constant value $f(a)$▲

A function f defined on a domain D is said to be *increasing* if $f(x_1) \leq f(x_2)$ for all points x_1 and x_2 of D such that $x_1 < x_2$; also, *strictly increasing* if $f(x_1) < f(x_2)$. It is *decreasing* if $f(x_1) \geq f(x_2)$ for $x_1 < x_2$ and *strictly decreasing* if $f(x_1) > f(x_2)$. A function displaying any one of these four properties is called a *monotone* function; and if it is either strictly increasing or strictly decreasing, it is said to be *strictly monotone*.

It is clear that the functions f and $-f$, if they are monotone, are so in opposite senses; that is, f is increasing if and only if $-f$ is decreasing. Because of this we may fix our attention, without loss of generality, on increasing functions and assume all facts deduced to be true dually for decreasing functions.

Theorem 9-11 Let f be a function continuous on $[a,b]$ and differentiable on (a,b). Then f is increasing on $[a,b]$ if f' is non-negative on (a,b) and strictly increasing if f' is positive on (a,b).

Proof If x_1 and x_2 are any two points on $[a,b]$ such that $x_1 < x_2$, then, by the mean value theorem, a point ξ exists in (x_1,x_2) such that $f(x_2) - f(x_1) = (x_2 - x_1)f'(\xi)$. The truth of the theorem follows from the observation that the right member is non-negative or positive according to whether f' is non-negative or positive▲

Exercises

9-9 Prove that if f is continuous on $[a,b]$ and differentiable on (a,b), then between any two consecutive zeros of f' there occurs at most one zero of f.

9-10 Suppose f to be differentiable on $[a,b]$ and such that $f'(a) = f'(b) = 0$, while for all x in (a,b), $f'(x) \neq 0$. Prove that f has a zero in (a,b) if $f(a)$ and $f(b)$ are one positive and one negative, and no zero if $f(a)$ and $f(b)$ are both positive or both negative.

9-11 Prove that if f is differentiable on $\mathscr{R}$, then at most one zero of f exceeds the greatest zero of f'.

9-12 Prove the following generalization of Rolle's theorem. Let f be differentiable on (a,b) and such that $\lim_{b-} f$ and $\lim_{a+} f$ exist and are equal. Then a number ξ exists with the properties that $a < \xi < b$ and $f'(\xi) = 0$.

9-13 Suppose f to be a function possessing a bounded derivative on $(0,1]$. Prove that the sequence $\{f(1/n)\}$ converges.

9-14 Let f be a function continuous on (a,b) and differentiable on (a,b) except possibly at some point c of that interval. Suppose, too, that $\lim_{c} f'$ exists. Prove that $f'(c)$ exists and equals $\lim_{c} f'$.

9-15 Assume that f has a derivative on (a,b) that is bounded. Use Theorem 7-6 to prove that $f(a+)$ exists.

9-16 Modify the hypothesis of Theorem 9-5 by assuming the functions f and g to have continuous derivatives. Use the mean value theorem to prove the modified theorem.

9-17 Use the mean value theorem to prove that $|\sin x_1 - \sin x_2| \leq |x_1 - x_2|$.

9-18 Let f be a function which is continuous at a point c and differentiable in $\mathscr{N}^*(c,\delta)$ for some δ. Suppose that $f'(x) < 0$ if $c - \delta < x < c$ and $f'(x) > 0$ if $c < x < c + \delta$. Prove that c is a local minimum point of f.

9-19 Cite an example of a function f such that $f'(c) = f''(c) = 0$ but such that c is neither a local minimum point nor a local maximum point of f.

Suppose f to be a function defined on an interval $[a,b]$. Then f is said to be *concave upward* on $[a,b]$ if, for any points x_1, x_2, and x_3 of $[a,b]$ such that $x_1 < x_2 < x_3$, it is the case that

$$\frac{f(x_2) - f(x_1)}{x_2 - x_1} \leq \frac{f(x_3) - f(x_2)}{x_3 - x_2}.$$

It is *concave downward* if the inequality is reversed.

9-20 Prove that if a function is continuous on $[a,b]$, and if f'' is an increasing function on (a,b), then f is concave upward on $[a,b]$.

9-21 Let f be defined on an interval of which c is an interior point. (a) Prove that c is a local minimum point of f if f is concave upward in some neighborhood of c and if $f'(c) = 0$. (Dually, c is a local maximum point if f is concave downward and $f'(c) = 0$.) (b) Conclude that c is a local minimum point of f if $f'(c) = 0$ and if $f''(c)$ is positive over some neighborhood of c. (Dually, c is a local maximum if $f'(c) = 0$ and $f''(c)$ is negative over a neighborhood of c.)

9-22 (a) Prove that $a_1/b_1 \leq a_2/b_2$ if and only if $a_1/b_1 \leq (a_1 + a_2)/(b_1 + b_2)$. (b) Use (a) to prove that a function is concave upward on $[a,b]$ if and only if

$$f(x_2) \leq f(x_1) + \frac{f(x_3) - f(x_1)}{x_3 - x_1}(x_2 - x_1)$$

for any points x_1, x_2, and x_3 of $[a,b]$ such that $x_1 < x_2 < x_3$.

9-5 Continuity of Derivatives

To illustrate certain characteristics of functions which are themselves derivatives of functions, we consider the following examples.

Example 9-18 Let f be defined for all x by $f(x) = \sin(1/x)$ for $x \neq 0$ and $f(0) = 0$. We have seen that this function is not continuous at 0 and hence is not differentiable there. For values of x other than 0, it follows from the chain rule and the usual formula for the derivative of the sine function that $f'(x) = -\frac{1}{x^2}\cos\frac{1}{x}$. It is clear that f' is unbounded in any neighborhood of 0 and therefore has no limit there.

Example 9-19 Let g be defined for all x by $g(x) = x\sin(1/x)$ for $x \neq 0$ and $g(0) = 0$. This function is continuous at 0. However, it is not differentiable at 0 as may be seen by considering the function G, where $G(x) = \frac{g(x) - g(0)}{x - 0} = \sin\frac{1}{x}$ for $x \neq 0$. Clearly G has no limit at 0. Also, we have, for $x \neq 0$, $g'(x) = \sin\frac{1}{x} - \frac{1}{x}\cos\frac{1}{x}$, which is seen to have no limit at 0.

Example 9-20 Let h be defined for all x by $h(x) = x^2 \sin(1/x)$ for $x \neq 0$ and $h(0) = 0$. This function is differentiable as well as continuous at 0. For, from $H(x) = \frac{h(x) - h(0)}{x - 0} = x\sin\frac{1}{x}$ for $x \neq 0$ it is clear that H has limit 0 at 0. Moreover, for $x \neq 0$ we have $h'(x) = 2x\sin\frac{1}{x} - \cos\frac{1}{x}$. Thus h is differentiable for all x. It is to be observed, however, that the function h' is not continuous at 0. Indeed, $\lim_0 h'$ does not exist.

The preceding example shows that it is possible for a function to possess a derivative which is not continuous at all points. While this may not be surprising, what is, perhaps, is that it is not possible for a function which itself is a derivative of a function defined on an interval to have removable or jump discontinuities. This is the import of the following theorem and its corollary.

Theorem 9-12 Let f be a function differentiable on the interval $[a,b]$, and suppose $f'(a) < f'(b)$. Then, if k is any number such that $f'(a) < k < f'(b)$, there exists a number c between a and b for which $f'(c) = k$.

Proof Consider a function g defined on $[a,b]$ by $g(x) = f(x) - kx$. Clearly, g is differentiable and hence continuous on the interval. This implies that g attains a minimum value at some point c of $[a,b]$. From $g'(b) = f'(b) - k$ and the hypothesis of the theorem follows $g'(b) > 0$. Hence for some $\varepsilon > 0$, $g(x) < g(b)$ whenever x is such that $b - \varepsilon < x < b$. This implies that b is not a minimum point of g. In a similar fashion, we deduce from $g'(a) < 0$ that a is not a minimum point of g. Therefore c is an interior point of the interval $[a,b]$. It follows that $g'(c) = 0$, whence $f'(c) = k$ as desired▲

Corollary If f is differentiable on $[a,b]$, then its derivative f' does not have discontinuity points of the first kind on $[a,b]$.

An interesting consequence of Theorem 9-12 is that not every function is the derivative of some other function. For example, there is no function having as derivative the function g defined by $g(x) = 1$ for $x \geq 0$ and $g(x) = -1$ for $x < 0$.

The only function we have cited so far which possesses a discontinuous derviative is that of Example 9-20. Its sole discontinuity is at 0. We show now, by actual construction, that it is possible for a function to be differentiable on an interval and to possess for its derivative a function which is discontinuous at each point of a noncountable set.

Example 9-21 In Chapter 4 it was shown that the Cantor set K is a closed set. Hence its complement with respect to the interval $[0,1]$ is open and therefore is the union of a countable collection of open intervals (a_n,b_n), $n = 1, 2, \ldots$.

Corresponding to each value of n we define a function f_n on (a_n,b_n) by

$$f_n(x) = (x - a_n)^2 \sin \frac{1}{x - a_n}.$$

This function has an infinite set of local maximum points in (a_n,b_n). In particular, there exists a local maximum c_n such that $a_n < c_n < (a_n + b_n)/2$.

Now define a function g_n on (a_n,b_n) by

$$g_n(x) = \begin{cases} f_n(x) & \text{if } a_n < x < c_n \\ f_n(c_n) & \text{if } c_n \leq x \leq a_n + b_n - c_n \\ f_n(a_n + b_n - x) & \text{if } a_n + b_n - c_n < x < b_n. \end{cases}$$

The function $g_n(x)$ is sketched in Figure 9-5. It is symmetric about $(a_n + b_n)/2$ and, as the reader may verify, is differentiable on (a_n,b_n).

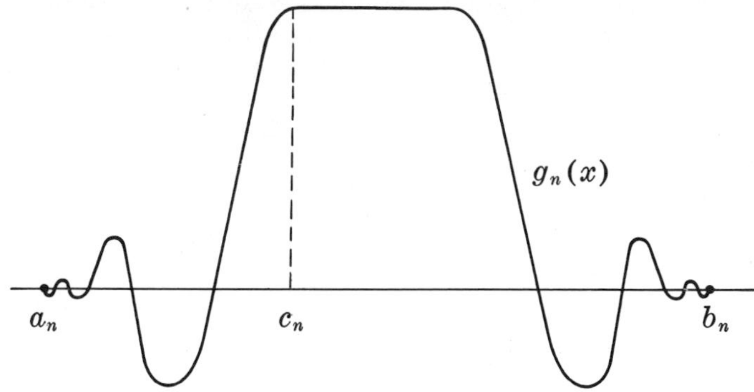

Figure 9-5

Finally, we define the function h on $[0,1]$ by

$$h(x) = \begin{cases} 0 & \text{if } x \text{ is in } K \\ g_n(x) & \text{if } x \text{ is in } (a_n,b_n). \end{cases}$$

Obviously, h is defined in such a way as to be differentiable on each interval (a_n,b_n). To investigate differentiability of h at a point c of the Cantor set K, consider the function H defined by $H(x) = (h(x) - h(c))/(x - c) = h(x)/(x - c)$ for all x other than c in $[0,1]$.

Let $\varepsilon > 0$ be given. Then, following reasoning which proved the function of Example 9-20 to be differentiable at 0, we deduce the existence of a number $\delta > 0$ such that for all n, if $b_n - a_n < \delta$, then

$$\left|\frac{g_n(x)}{x - a_n}\right| < \varepsilon \quad \text{and} \quad \left|\frac{g_n(x)}{x - b_n}\right| < \varepsilon$$

for all x in (a_n,b_n). Let x be a point of $\mathcal{N}^*(c,\delta) \cap [0,1]$. If x is in K, then $h(x) = 0$, whence $H(x) = 0$. If x is not in K, then x is in (a_n,b_n) for some n. Without loss of generality, suppose $c \leq a_n$. Then

$$|H(x)| = \left|\frac{g_n(x)}{x - c}\right| \leq \left|\frac{g_n(x)}{x - a_n}\right| < \varepsilon.$$

Hence $\lim\limits_{c} H = 0$ so that h is differentiable at all points of K. Thus it has been shown that h is a differentiable function on $[0,1]$.

It may now be demonstrated that the derivitive of h is discontinuous at each point of K. Observe that if c is a point of K, then, from the manner in which K is constructed, in each neighborhood of c there exists an interval (a_n,b_n) of the complement of K. For each x such that $a_n < x < c_n$ we then have

$$h_n'(x) = f_n'(x) = 2(x - a_n)\sin\frac{1}{x - a_n} - \cos\frac{1}{x - a_n},$$

from which the oscillation of h' on (a_n,c_n) is seen to be not less than 2. Thus $\omega_c h' \geq 2$, whence h' is not continuous at c.

Exercises

9-23 Consider the function f defined by $f(x) = x^3 \sin(1/x)$ if $x \neq 0$ and $f(0) = 0$. Does $f'(0)$ exist? $f''(0)$?

9-24 Let g be the function defined by $g(x) = x^2 \sin(1/x^2)$ if $x \neq 0$ and $g(0) = 0$. Does $g'(0)$ exist? $g''(0)$?

9-25 Cite an example of a function which is differentiable on a closed and bounded interval but whose derivative is not bounded.

9-26 Prove the corollary to Theorem 9-12.

9-27 Let f be differentiable on (a,b), and suppose that $\lim\limits_{c} f'$ exists at some point c of (a,b). Prove that $\lim\limits_{c} f' = f'(c)$.

9-28 Give an example of a nonconstant function f which satisfies the hypotheses of Theorem 9-5 but which is such that in each deleted neighborhood of x_0 there exists a point x with the property $f(x) = f(x_0)$.

9-6 One-sided Derivatives

At a point c of the domain of a function f it may happen that the function F defined by $F(x) = (f(x) - f(c))/(x - c)$ does not possess a limit but is such that a one-sided limit exists. If $\lim_{c+} F$ exists, we call it the *derivative of f from the right at c*; if $\lim_{c-} F$ exists, it is the *derivative from the left at c*.

Example 9-22 Recall the properties of the function h defined by $h(x) = |x|$ in Example 9-9. It may be seen that h has derivatives from right and left at 0 equal, respectively, to 1 and -1.

It is a direct consequence of facts concerning limits that if both a derivative from the right and a derivative from the left exist at a point and if the two are equal, then a derivative equal to their common value exists at the point. If a point c is a cluster point (of the domain of a function) from one side only—as, for example, an endpoint of a closed interval—and if F has a limit from that side at c, then the notions of derivative and one-sided derivative are one and the same.

An even more general notion of one-sided derivative at a point c may be formulated which does not require that the function concerned be continuous at c or even be defined there. Thus, if the function defined at all points of the domain of f except c by

$$\frac{f(x) - \lim_{c+} f}{x - c}$$

possesses a limit from the right at c, then that limit is called the *generalized derivative of f from the right at c*. A similar definition may be given for a *generalized derivative from the left*.

If a function possesses a derivative from the right at c, then a generalized derivative from the right exists also and is equal to it. That the converse is not true may be seen from the following example.

Example 9-23 Let f be defined for all x by $f(x) = x^2$ if $x \neq 1$ and $f(1) = 3$. Then, although $\lim_{1+} \frac{f(x) - f(1)}{x - 1} = \lim_{1+} \frac{x^2 - 3}{x - 1}$ does not exist, still $\lim_{1+} \frac{f(x) - \lim_{1+} f}{x - 1} = \lim_{1+} \frac{x^2 - 1}{x - 1}$ exists with the value 2.

The notion of a derivative from one side should not be confused with that of a one-sided limit of a derivative. The latter concerns the limit of the derivative f', of a function f, from the right or left at a point c of the domain of f'. These are quite properly denoted by the symbols $\lim_{c+} f'$ and $\lim_{c-} f'$ and may exist where derivatives from the right and left do not. The reader is invited to illustrate with examples.

Exercises

9-29 Prove that if a function has a derivative from the right at c, then it is continuous from the right at c.

9-30 Cite an example of a function f such that for some point c of its domain, $\lim_{c+} f'$ exists but the derivative from the right does not exist.

9-31 Assume that f is differentiable on (a,b). Prove that if $\lim_{a+} f'$ exists, then $\lim_{a+} f$ exists. (Compare Exercise 9-15.)

9-32 Assume that f is differentiable on (a,b). Prove that if $\lim_{a+} f'$ exists, then the generalized derivative from the right at a exists also, and the two are equal.

9-7 Indeterminate Limits

It is frequently the case that one wishes to consider the quotient of two functions that are defined on some common domain and that have the special property that they possess the common value 0 at some point c of that domain. This being the case, the quotient is without meaning at c, and it is of interest to inquire into the limit of the quotient at such a point. Theorems 9-13 through 9-15, which make extensive use of the derivatives of the two functions, are directed to this end.

Theorem 9-13 Let f and g be functions continuous on an interval $[a,b]$, and differentiable in (a,b). Further, suppose $g(a) \neq g(b)$ and that no point of (a,b) exists at which f' and g' both vanish. Then a point ξ of (a,b) exists such that $g'(\xi) \neq 0$ and

$$\frac{f(b) - f(a)}{g(b) - g(a)} = \frac{f'(\xi)}{g'(\xi)}.$$

Proof Let h be a function defined on $[a,b]$ by

$$h(x) = f(x) - f(a) - \frac{f(b) - f(a)}{g(b) - g(a)}[g(x) - g(a)].$$

Clearly, h is continuous on $[a,b]$ and differentiable on (a,b), and $h(a) = h(b) = 0$. Since these constitute the hypotheses of Rolle's theorem, we conclude that a point ξ exists in (a,b) such that $h'(\xi) = 0$. It follows that $f'(\xi) = \dfrac{f(b) - f(a)}{g(b) - g(a)} g'(\xi)$. Now suppose $g'(\xi) = 0$. Then we conclude $f'(\xi) = 0$. This, however, is contrary to hypothesis. Therefore $g'(\xi) \neq 0$, and the truth of the theorem follows▲

Corollary If f and g are continuous on $[a,b]$ and differentiable on (a,b) and if g' does not vanish in (a,b), then a point ξ exists in (a,b) such that $\dfrac{f(b) - f(a)}{g(b) - g(a)} = \dfrac{f'(\xi)}{g'(\xi)}$.

Proof We have only to note that if g' does not vanish in (a,b), then $g(a) \neq g(b)$, since equality would be in contradiction to Rolle's theorem▲

Theorem 9-14 Let f and g be continuous on $[a,b]$ with $f(a) = g(a) = 0$ and differentiable on (a,b) with $g'(x) \neq 0$ for all x in (a,b). Suppose, further, that $\lim_a \frac{f'}{g'}$ exists with value L. Then $\lim_a \frac{f}{g} = L$.

Proof Because the functions f and g satisfy the hypotheses of Theorem 9-13 in any interval $[a,x]$ where $a < x < b$, we conclude the existence in (a,x) of a point ξ such that $f(x)/g(x) = f'(\xi)/g'(\xi)$. The reader should show from the definition of limit of a function that if $\lim_a f'/g' = L$, then, of necessity, $\lim_a f/g = L$. (Compare Exercise 7-5.)▲

The principle expressed in Theorem 9-14, that is, that under proper conditions

$$\lim_a \frac{f}{g} = \lim_a \frac{f'}{g'},$$

is sometimes referred to as *l'Hospital's rule*. Obviously, its application is not restricted to the left end point of an interval. The argument above may be repeated for the endpoint b of $[a,b]$ and, indeed, for any point c of $[a,b]$. If c is in (a,b), we need only consider the function f on the subdomains $[a,c]$ and $[c,b]$ to conclude that

$$\lim_{c+} \frac{f}{g} = \lim_{c+} \frac{f'}{g'} \quad \text{and} \quad \lim_{c-} \frac{f}{g} = \lim_{c-} \frac{f'}{g'}.$$

If it is also the case that f'/g' possesses a limit at c, then, immediately,

$$\lim_c \frac{f}{g} = \lim_c \frac{f'}{g'}.$$

l'Hospital's rule may be shown to apply in the event f and g have infinite limits at a point c.

Theorem 9-15 Let f and g be differentiable functions on (a,b) with $g'(x) \neq 0$ for all x in (a,b). Suppose, too, that $\lim_a g = +\infty$ and that $\lim_a \frac{f'}{g'} = L$. Then $\lim_a \frac{f}{g} = L$.

Proof Let $\varepsilon > 0$ be given, and suppose $\varepsilon < 1$. A point x_2 exists in (a,b) such that, for each x in $(a,x_2]$, $g(x) > 0$ and $|f'(x)/g'(x) - L| < \varepsilon$. Having selected such an x_2, we may find a point x_1 with $a < x_1 < x_2$ and with the property that for all x in (a,x_1) it is the case that $g(x) > g(x_2)/\varepsilon$ and $g(x) > |f(x_2)|/\varepsilon$. The functions f and g satisfy the hypotheses of Theorem 9-13 on $[x,x_2]$ for each x in (a,x_1). Hence a point ξ exists such that $x < \xi < x_2$ and $[f(x_2) - f(x)]/[g(x_2) - g(x)] = f'(\xi)/g'(\xi)$. Multiplying by $[g(x) - g(x_2)]/g(x)$ yields

$$\frac{f(x)}{g(x)} = \frac{f'(\xi)}{g'(\xi)} - \frac{f'(\xi)}{g'(\xi)}\frac{g(x_2)}{g(x)} + \frac{f(x_2)}{g(x)},$$

from which $|f(x)/g(x) - L| \leq \varepsilon + (|L| + \varepsilon)\varepsilon + \varepsilon$ for all x in (a,x_1)▲

The above has significance only if the function f is unbounded. For if f is bounded and g has an infinite limit, then the quotient f/g has limit 0 regardless of differentiability conditions on the functions.

9-8 Differentials

Suppose f to be a function defined on a domain D and differentiable at a point c of D. Let dx denote any nonzero number with the property that $c + dx$ is in D, and represent the product of dx by the derivative of f at c by the symbol df; thus

(3) $$df = f'(c)\, dx.$$

The numbers dx and df are termed *differentials*; dx is called a *differential of* x and df the corresponding *differential of* f. Essentially, they are any two numbers (except for the stated restrictions on dx) such that their quotient is the value of the derivative of f at c. Thus, the frequent usage of df/dx as a notation for the derivative of f is compatible with (3).

Of course, with different points c of D the relative values of dx and df may be expected to differ. However, if c is committed, then df defines a linear function on a domain consisting of all possible values of dx. It is easily seen that $\lim_0 df = 0$, but df is not continuous at 0 since it is undefined there.

Example 9-24 Suppose f defined by $f(x) = x^2$ for all x. Then at $c = 2$ the numbers 12 and 48 are corresponding differentials of x and f. In general, since $f'(2) = 4$, we have $df = 4\, dx$.

The number

(4) $$\Delta f = f(c + dx) - f(c)$$

is called an *increment* of f at c. For a fixed c it, too, defines a function on the set of values of dx. Because f is continuous at c, we have $\lim_0 \Delta f = 0$. Still another function which is of interest is the function η defined on the set of values of dx by

$$\eta = \frac{\Delta f - df}{dx}.$$

With the aid of (4) this may be written in the form

$$\eta = \frac{f(c + dx) - f(c)}{dx} - \frac{df}{dx},$$

and, since the limit at 0 of the function defined by the first term on the right is $f'(c)$, we have $\lim_0 \eta = 0$.

Since $\Delta f = df + \eta\, dx$, not only does $\Delta f - df$ have limit 0, but $(\Delta f - df)/dx$ does as well. Because of this we refer to df as the *principal part* of Δf and note that it is in the nature of a linear approximation to Δf.

Exercises

9-33 Let f and g be differentiable functions on (a,b) such that g' is nonvanishing, $\lim_a g = \infty$ and $\lim_a f'/g'$ and $\lim_a f/g$ exist. Use Theorem 9-14 to prove that $\lim_a f/g = \lim_a f'/g'$.

9-34 Let f and g be defined on $[a,b]$ such that $f(a) = g(a) = 0$. Assume $f'(a)$ and $g'(a)$ exist and that $g'(a) \neq 0$. Prove that $\lim_a f/g$ exists and equals $f'(a)/g'(a)$.

9-35 Cite an example of functions f and g which satisfy the conditions of Exercise 9-34 but which do not satisfy the hypotheses of Theorem 9-14.

9-36 Prove the following generalization of Theorem 9-14. Let f and g be differentiable on (a,b) and such that g' does not vanish on that interval. Assume that $\lim_{a+} f = \lim_{a+} g = 0$. If $\lim_a f'/g'$ exists, then so does $\lim_a f/g$, and the two limits are equal.

9-37 Assume that f and g are differentiable on $\{x;\ x \geq a\}$ and that g' does not vanish on that interval. Also, suppose that $\lim_\infty f = \lim_\infty g = 0$. Prove that if $\lim_\infty f'/g'$ exists, then so does $\lim_\infty f/g$, and the two limits are equal.

9-38 Assume that f and g are differentiable on $\{x;\ x \geq a\}$ and that g' does not vanish there. Also, suppose that $\lim_\infty g = \infty$. Prove that if $\lim_\infty f'/g'$ exists, then so does $\lim_\infty f/g$, and the two limits are equal.

9-39 The function P_n defined by $P_n(x) = \dfrac{1}{2^n(n!)} \dfrac{d^n}{dx^n} (x^2 - 1)^n$ for all x is called the *Legendre polynomial of degree n*. Prove that P_n has n distinct zeros all of which lie in the interval $[-1,1]$.

10
The Riemann Integral

Many a little makes a mickle.

BENJAMIN FRANKLIN (1706–1790), *Poor Richard*

10-1 Introduction

First evidences of the integral calculus are to be found in the exploratory mathematical investigations of the early Greek geometers, who employed what we call today the *method of exhaustion* to give meaning to and to calculate areas of plane figures with circular or parabolic boundaries. Centuries later, pioneers of the calibre of René Descartes (1596–1650), Pierre de Fermat (1608–1665), and Blaise Pascal (1623–1662) set the stage for the genius and inventiveness of Isaac Newton (1646–1727) and Gottfried Wilhelm Leibniz (1646–1716). Subsequent to the independent analytical formulations by these two of the differential and integral calculus, attention became focused primarily on the inverse character of differentiation and integration and on techniques for the evaluation of indefinite and definite integrals. While Leonhard Euler (1707–1783), Carl Friedrich Gauss (1777–1855), and others applied the new ideas to more practical matters, it remained for such giants as Peter Dirichlet (1805–1859), Georg Friedrich Bernhard Riemann (1826–1866), and Augustin Louis Cauchy (1789–1857) to establish the concept of the *definite integral* of a real-valued function of a real variable on a rigorous mathematical basis. Reimann's definition of what is now known as the *Riemann integral* appeared a little over one hundred years ago. Together with Cauchy's extension of the definition to unbounded functions, this, in a sense, was regarded as the culmination of the search for a complete and formal expression of the concept of " limit of a sum." This chapter is devoted to a study of the Riemann integral.

The story does not end here, of course. Toward the close of the last century Thomas J. Stieltjes (1856–1894) introduced a broader concept of integration by replacing a certain linear function germane to the Riemann definition by a function of a more general character. Then early in the twentieth century came the concept of *measure* of a set of real numbers, due in large part to Émile Borel (1871–1938) and Henri Lebesgue (1875–1941), and with it the great contribution of Lebesgue based on a generalization of the partitioning of an interval into sub-intervals. In the chapters to come we shall study in detail the *Riemann-Stieltjes integral*, the *Lebesgue integral*, and an inevitable generalization, the *Lebesgue-Stieltjes integral.*

10-2 Partitions and Riemann Sums

We consider a closed and bounded interval $[a,b]$ and define a *partition* of that interval to be any finite set of real numbers $P = \{x_0, x_1, \ldots, x_n\}$ with the properties $a = x_0 < x_1 < \cdots < x_n = b$. The n closed intervals $[x_0,x_1]$, $[x_1,x_2]$, $\ldots$, $[x_{n-1},x_n]$ determined by P constitute the *segments* of the partition.

We write $\Delta x_\nu = x_\nu - x_{\nu-1}$ for $\nu = 1, 2, \ldots, n$ and call the positive number Δx_ν the *length* of the segment $[x_{\nu-1},x_\nu]$. The greatest of the lengths of the segments of a partition we denote by ΔP, that is, $\Delta P = \max\{\Delta x_\nu; \quad \nu = 1, 2, \ldots, n\}$, and refer to it as the *norm* of P.

A partition P^* is said to be a *refinement* of a partition P or to be *finer* than P if $P \subset P^*$; that is, if every point of P is employed in the construction of P^*. Clearly, $\Delta P^* \leq \Delta P$.

Consider now a bounded function f defined on $[a,b]$. Denote its infimum and supremum on that interval by m and M, respectively. It is clear that if $P = \{x_0,x_1,\ldots,x_n\}$ is any partition of $[a,b]$, then f is bounded on each segment of P. Let $m_\nu = \inf\{f(x); \quad x_{\nu-1} \leq x \leq x_\nu\}$ and $M_\nu = \sup\{f(x); \quad x_{\nu-1} \leq x \leq x_\nu\}$ for $\nu = 1, 2, \ldots, n$. Then form the sums

$$\underline{R}(f,P) = \sum_{\nu=1}^{n} m_\nu \,\Delta x_\nu \quad \text{and} \quad \overline{R}(f,P) = \sum_{\nu=1}^{n} M_\nu \,\Delta x_\nu .$$

The first of these is called the *lower Riemann sum* of f on $[a,b]$ with respect to the partition P; the second is the *upper Riemann sum.* Briefly, we shall refer to these as the *lower* and *upper R-sums* of f with respect to P. Because of the nature of the definitions, these sums are related to each other and to the bounds of f on $[a,b]$ in a very simple manner; thus

$$m(b - a) \leq \underline{R}(f,P) \leq \overline{R}(f,P) \leq M(b - a).$$

Less obvious are the properties of lower and upper R-sums asserted in Theorems 10-1 and 10-2.

Theorem 10-1 If f is a bounded function on $[a,b]$ and if P^* is a refinement of a partition P of $[a,b]$, then $\underline{R}(f,P) \leq \underline{R}(f,P^*)$ and $\overline{R}(f,P^*) \leq \overline{R}(f,P)$.

Proof Let $P = \{x_0, x_1, \ldots, x_n\}$ and suppose P^* to be a partition consisting of all points of P together with a single additional point $\bar{x}$. The latter point necessarily has the property that $x_{\mu-1} < \bar{x} < x_\mu$ for some index μ. Let m_μ denote the infimum of f on the interval $[x_{\mu-1}, x_\mu]$ and m' and m'' the infima of f on the intervals $[x_{\mu-1}, \bar{x}]$ and $[\bar{x}, x_\mu]$, respectively. Certainly $m_\mu \leq m'$ and $m_\mu \leq m''$.

The lower R-sums of f corresponding to P and P^*, respectively, are

$$\underline{R}(f,P) = \sum_{\nu=1}^{n} m_\nu \, \Delta x_\nu$$

and

$$\underline{R}(f,P^*) = \sum_{\nu=1}^{\mu-1} m_\nu \, \Delta x_\nu + m'(\bar{x} - x_{\mu-1}) + m''(x_\mu - \bar{x}) + \sum_{\nu=\mu+1}^{n} m_\nu \, \Delta x_\nu .$$

Therefore

$$\begin{aligned} \underline{R}(f,P^*) - \underline{R}(f,P) &= m'(\bar{x} - x_{\mu-1}) + m''(x_\mu - \bar{x}) - m_\mu(x_\mu - x_{\mu-1}) \\ &= (m' - m_\mu)(\bar{x} - x_{\mu-1}) + (m'' - m_\mu)(x_\mu - \bar{x}). \end{aligned}$$

It follows that $\underline{R}(f,P^*) - \underline{R}(f,P) \geq 0$. In a similar fashion, we can show that $\bar{R}(f,P^*) - \bar{R}(f,P) \leq 0$.

Now suppose P^* to denote *any* refinement of P. Then we may construct a succession of partitions $P_0, P_1, \ldots, P_k$ such that $P_0 = P$ and $P_k = P^*$ and where each is a refinement of the preceding one containing exactly one more point than its predecessor. It follows that

$$\underline{R}(f,P) \leq \underline{R}(f,P_1) \leq \cdots \leq \underline{R}(f,P^*)$$

and

$$\bar{R}(f,P) \geq \bar{R}(f,P_1) \geq \cdots \geq \bar{R}(f,P^*)$$

as desired▲

Thus it is shown that replacement of a partition by a finer partition tends to increase the lower R-sum and to lessen the upper R-sum. We show next that, regardless of the partitions involved, no lower R-sum can exceed any upper R-sum.

Theorem 10-2 If f is a bounded function on $[a,b]$ and if P_1 and P_2 are any partitions of $[a,b]$, then $\underline{R}(f,P_1) \leq \bar{R}(f,P_2)$.

Proof We have only to consider the partition $P = P_1 \cup P_2$. Since P is a refinement of both P_1 and P_2, it follows by Theorem 10-1 that $\underline{R}(f,P_1) \leq \underline{R}(f,P)$ and $\bar{R}(f,P) \leq \bar{R}(f,P_2)$. From $\underline{R}(f,P) \leq \bar{R}(f,P)$, valid for any partition, the truth of the theorem follows▲

10-3 The Riemann Integral

Since the set of all numbers $\underline{R}(f,P)$, with respect to all possible partitions P of $[a,b]$, is bounded above (by $M(b - a)$), a supremum exists, which we call the *lower Riemann integral* or *lower R-integral of f on* $[a,b]$. Similarly, the set of numbers $\overline{R}(f,P)$ is bounded below (by $m(b-a)$) and accordingly possesses an infimum, which we call the *upper Riemann integral* or *upper R-integral of f on* $[a,b]$. We employ the notations

$$\underline{\int_a^b} f = \sup \{\underline{R}(f,P); \quad P \text{ a partition of } [a,b]\}$$

and

$$\overline{\int_a^b} f = \inf \{\overline{R}(f,P); \quad P \text{ a partition on } [a,b]\}.$$

Example 10-1 Consider the function Ψ defined on $[a,b]$. If P is any partition of $[a,b]$, then $\underline{R}(\Psi,P)=0$ and $\overline{R}(\Psi,P)=1$. Hence $\underline{\int_a^b} \Psi = 0$ and $\overline{\int_a^b} \Psi = 1$.

Example 10-2 The function k, defined to have the value k at each point of an interval $[a,b]$, clearly has the property that, whatever the partition P, $\underline{R}(k,P) = \overline{R}(k,P) = k \sum_{\nu=1}^{n} \Delta x_\nu = k(b-a)$. Therefore $\overline{\int_a^b} k = \underline{\int_a^b} k = k(b-a)$.

Examples 10-1 and 10-2 make it plain that the lower and upper R-integrals of a function may, but need not, be equal. It will be recalled (Exercise 4-31) that if A and B are sets of real numbers such that no member of A exceeds any member of B, then $\sup A \leq \inf B$. Since, by Theorem 10-2, the sets of lower and upper R-sums have precisely this character, we conclude that

$$\underline{\int_a^b} f \leq \overline{\int_a^b} f$$

for any bounded function f on $[a,b]$.

We are now in a position to give a definition of the Riemann integral of a bounded function f. Actually, in the course of the chapter we shall state three equivalent definitions and, for convenience, distinguish them by symbols (A), (B), and (C). The first of these may be regarded as the principal definition; we shall discuss it in considerable detail before introducing the others.

Definition (A) If a bounded function f, defined on an interval $[a,b]$, is such that its lower and upper Riemann integrals are equal, then f is said to be *Riemann integrable* or *R-integrable* on $[a,b]$, and the common value of $\underline{\int_a^b} f$ and $\overline{\int_a^b} f$ is further denoted by $\int_a^b f$ and is called the *definite Riemann integral* or *R-integral of f on* $[a,b]$.[1]

That Riemann integrable functions do exist is seen in Example 10-2. There the function having constant value k on $[a,b]$ is seen to have the number $k(b-a)$

[1] The reader should not interpret the employment of the symbol $\int_a^b f$, in preference to the familiar $\int_a^b f(x)\,dx$, as an attempt at novelty. The simple fact is that the latter symbol would be difficult to explain or defend at this point.

as its Riemann integral. On the other hand, the function Ψ of Example 10-1 does not possess a Riemann integral inasmuch as its lower and upper R-integrals are unequal.

A useful necessary and sufficient condition for the existence of a Riemann integral follows. It concerns not the character of the function f but the nature of lower and upper sums.

Theorem 10-3 A bounded function f is R-integrable on $[a,b]$ if and only if, given $\varepsilon > 0$, a partition P exists such that $\overline{R}(f,P) - \underline{R}(f,P) < \varepsilon$.

Proof Partitions P_1 and P_2 of $[a,b]$ exist such that

$$\underline{\int_a^b} f - \underline{R}(f,P_1) < \frac{\varepsilon}{2} \quad \text{and} \quad \overline{R}(f,P_2) - \overline{\int_a^b} f < \frac{\varepsilon}{2}.$$

Let $P = P_1 \cup P_2$. Then, since P is finer than P_1 and P_2, it follows, by Theorem 10-1, that

$$\underline{\int_a^b} f - \underline{R}(f,P) < \frac{\varepsilon}{2} \quad \text{and} \quad \overline{R}(f,P) - \overline{\int_a^b} f < \frac{\varepsilon}{2}.$$

Adding these inequalities and assuming f to be R-integrable leads immediately to $\overline{R}(f,P) - \underline{R}(f,P) < \varepsilon$.

As for the converse, suppose that there exists a partition P such that $\overline{R}(f,P) - \underline{R}(f,P) < \varepsilon$. Then from $\overline{\int_a^b} f \le \overline{R}(f,P)$ and $\underline{\int_a^b} f \ge \underline{R}(f,P)$ follows $\overline{\int_a^b} f - \underline{\int_a^b} f < \varepsilon$. Since $\varepsilon > 0$ is arbitrary, we conclude that $\int_a^b f$ exists▲

It will be of interest to determine necessary and sufficient conditions for Riemann integrability that concern the character of the function f directly and are not simply reflected in the properties of lower and upper R-sums based on it. It is clear that in the foregoing definitions it was essential that f be bounded, for otherwise the set of lower R-sums could have no supremum or the set of upper R-sums an infimum. Boundedness, therefore, is a necessary condition for Riemann integrability. Example 10-1 shows, however, that it is not sufficient. In another vein, it may be shown that continuity of f, although not necessary to the existence of a Riemann integral, is nevertheless sufficient. This we show next.

Theorem 10-4 If f is continuous on $[a,b]$, then $\int_a^b f$ exists.

Proof Since f is uniformly continuous on $[a,b]$, corresponding to any given $\varepsilon > 0$, there exists a positive number δ with the property that $|f(x') - f(x'')| < \varepsilon$ for all x' and x'' of $[a,b]$ such that $|x' - x''| < \delta$. Let P denote a partition of $[a,b]$ with norm $\Delta P < \delta$, and consider a segment $[x_{\nu-1},x_\nu]$ of P. Because of the continuity of f, its infimum and supremum, m_ν and M_ν, on the segment are attained at points ξ_ν and η_ν, respectively, of $[x_{\nu-1},x_\nu]$. Since $|\xi_\nu - \eta_\nu| < \delta$, it follows that $M_\nu - m_\nu < \varepsilon$. This being the case for $\nu = 1, 2, \ldots, n$, we have

$$\overline{R}(f,P) - \underline{R}(f,P) = \sum_{\nu=1}^{n} (M_\nu - m_\nu)\,\Delta x_\nu < \varepsilon(b-a).$$

We conclude from Theorem 10-3 that f is Riemann integrable▲

Frequently it is possible and convenient to demonstrate that a function is Riemann integrable and to actually determine its R-integral simply by considering some countable subset of the set of all partitions of $[a,b]$. For if, as P varies over the members of such a subset, it is the case that $\sup\{\underline{R}(f,P)\} = \inf\{\overline{R}(f,P)\}$, then inclusion of other partitions cannot increase $\sup\{\underline{R}(f,P)\}$ nor decrease $\inf\{\overline{R}(f,P)\}$. It follows that the common value of these bounds is the value of the Riemann integral $\int_a^b f$.

Example 10-3 The function f defined on $[0,1]$ by $f(x) = x^2$ is continuous and hence Riemann integrable. Quite apart from this, however, it is possible by such an argument as that above not only to establish R-integrability but also to determine the value of $\int_0^1 x^2$.

For the set of partitions P_n $(n = 1,2,\ldots)$ defined by $P_n = \{0, 1/n, 2/n, \ldots, n/n\}$ we have

$$\underline{R}(f,P_n) = \sum_{\nu=0}^{n-1} \left(\frac{\nu}{n}\right)^2 \left(\frac{1}{n}\right) = \frac{1}{3}\left(1 - \frac{1}{n}\right)\left(1 - \frac{1}{2n}\right) \leq \underline{\int_0^1} f.$$

Since this is the case for any n, however large, it follows that $\frac{1}{3} \leq \underline{\int_0^1} f$. In a similar fashion we find

$$\overline{R}(f,P_n) = \sum_{\nu=1}^{n} \left(\frac{\nu}{n}\right)^2 \left(\frac{1}{n}\right) = \frac{1}{3}\left(1 + \frac{1}{n}\right)\left(1 + \frac{1}{2n}\right) \geq \overline{\int_0^1} f$$

and conclude $\overline{\int_0^1} f \leq \frac{1}{3}$. From $\underline{\int_0^1} f \leq \overline{\int_0^1} f$ it follows that f is R-integrable and its R-integral is $\frac{1}{3}$.

Exercises

10-1 Prove that a function that is defined and monotone on $[a,b]$ is Riemann integrable on that interval.

10-2 Employ Definition (A) to evaluate the Riemann integral: (a) $\int_a^b x$; (b) $\int_a^b x^k$, where k is any positive integer; (c) $\int_a^b e^x$; (d) $\int_a^b \sin x$.

10-3 Prove that if f is a non-negative and continuous function on $[a,b]$ and if $\int_a^b f = 0$, then f is the zero function on $[a,b]$.

10-4 Let f denote a function that possesses a Riemann integral on $[a,b]$, and suppose g to be a function that is equal to f at all except a finite set of points of $[a,b]$. Prove that g is also Riemann integrable on $[a,b]$ and that $\int_a^b g = \int_a^b f$.

10-5 Suppose f, g, and h are functions defined on $[a,b]$ with the properties $f \leq g \leq h$. Prove that if $\int_a^b f$ and $\int_a^b h$ exist and are equal, then $\int_a^b g$ also exists and equals the first two integrals.

10-6 A function with a constant value on each open segment of some partition of $[a,b]$ is called a *step function* on $[a,b]$. Prove that a function g is Riemann integrable on $[a,b]$ if and only if, corresponding to each $\varepsilon > 0$, there exist step functions f and h such that $f \leq g \leq h$ and $h - f < \varepsilon$ on $[a,b]$.

10-4 Linearity of the Riemann Integral

In the symbol $\int_a^b f$ for the Riemann integral of f, the function f is frequently designated the *integrand*. The two thecrems in this section demonstrate the linear character of a Riemann integral with respect to its integrand.

Theorem 10-5 If f is R-integrable on $[a,b]$ and if k is any constant, then the function kf is R-integrable and, moreover,

$$\int_a^b kf = k\int_a^b f. \tag{1}$$

Proof Let $\varepsilon > 0$ be given. Then from the R-integrability of f follows the existence of a partition P of $[a,b]$ such that

$$\int_a^b f - \varepsilon < \underline{R}(f,P) \quad \text{and} \quad \overline{R}(f,P) < \int_a^b f + \varepsilon. \tag{2}$$

It is immediate, also, from the definitions of lower and upper R-sums that, if k is positive, then for any partition P,

$$\underline{R}(kf,P) = k\underline{R}(f,P) \quad \text{and} \quad \overline{R}(kf,P) = k\overline{R}(f,P). \tag{3}$$

Combining (2) and (3) yields

$$k\int_a^b f - k\varepsilon < \underline{R}(kf,P) \le \underline{\int_a^b} kf$$

and

$$\overline{\int_a^b} kf \le \overline{R}(kf,P) < k\int_a^b f + k\varepsilon.$$

From the arbitrary character of ε then follows

$$k\int_a^b f \le \underline{\int_a^b} kf \le \overline{\int_a^b} kf \le k\int_a^b f, \tag{4}$$

whence the desired conclusion.

If k is negative, the equalities in (3) must be replaced by

$$\underline{R}(kf,P) = k\overline{R}(f,P) \quad \text{and} \quad \overline{R}(kf,P) = k\underline{R}(f,P).$$

These, together with (2), yield

$$k\int_a^b f + k\varepsilon < \underline{R}(kf,P) \le \underline{\int_a^b} kf$$

and

$$\overline{\int_a^b} kf \le \overline{R}(kf,P) < k\int_a^b f - k\varepsilon,$$

from which again follow the relations (4).

Lastly, we observe that if $k = 0$, then the conclusion $\int_a^b kf = k(b - a)$ of Example 10-2 implies that the left member of (1) exists and has the same value, namely 0, as the right member▲

Theorem 10-6 If f_1 and f_2 are R-integrable on $[a,b]$, then the function $f_1 + f_2$ is R-integrable and, moreover,

$$\int_a^b (f_1 + f_2) = \int_a^b f_1 + \int_a^b f_2 .$$

Proof Let $\varepsilon > 0$ be given. Then partitions P_1 and P_2 exist such that

$$\underline{\int_a^b} f_1 - \varepsilon < \underline{R}(f_1,P_1), \quad \overline{R}(f_1,P_1) < \overline{\int_a^b} f_1 + \varepsilon$$

and

$$\underline{\int_a^b} f_2 - \varepsilon < \underline{R}(f_2,P_2), \quad \overline{R}(f_2,P_2) < \overline{\int_a^b} f_2 + \varepsilon.$$

Further, these inequalities remain true if P_1 and P_2 are replaced by the refinement $P = P_1 \cup P_2$.

Now let m_ν' and m_ν'' denote the infima of f_1 and f_2, respectively, on a segment $[x_{\nu-1},x_\nu]$ of the partition P. Since for $\nu = 1, 2, \ldots, n$ it is the case that $m_\nu' + m_\nu'' \le f_1(x) + f_2(x)$ for all x in $[x_{\nu-1},x_\nu]$, it follows that

$$\underline{R}(f_1,P) + \underline{R}(f_2,P) \le \underline{R}(f_1 + f_2, P).$$

A similar argument based on upper bounds leads to

$$\overline{R}(f_1,P) + \overline{R}(f_2,P) \ge \overline{R}(f_1 + f_2, P).$$

By properly combining the above inequalities we obtain

$$\underline{\int_a^b} f_1 + \underline{\int_a^b} f_2 - 2\varepsilon < \underline{R}(f_1 + f_2, P)$$

and

$$\overline{R}(f_1 + f_2, P) < \overline{\int_a^b} f_1 + \overline{\int_a^b} f_2 + 2\varepsilon.$$

From $\underline{R}(f_1 + f_2, P) \le \underline{\int_a^b}(f_1 + f_2)$, $\overline{\int_a^b}(f_1 + f_2) \le \overline{R}(f_1 + f_2, P)$, and the arbitrary character of ε follows

$$\underline{\int_a^b} f_1 + \underline{\int_a^b} f_2 \le \underline{\int_a^b}(f_1 + f_2) \le \overline{\int_a^b}(f_1 + f_2) \le \overline{\int_a^b} f_1 + \overline{\int_a^b} f_2 .$$

The truth of the theorem may then be concluded from the existence of $\int_a^b f_1$ and $\int_a^b f_2$▲

If $\int_a^b$ is regarded as an operational symbol acting upon Riemann integrable functions, then it is clear that it has the character of a *linear operator*. Thus, if f_1 and f_2 are R-integrable on $[a,b]$ and if k_1 and k_2 are any constants, then

$$\int_a^b (k_1 f_1 + k_2 f_2) = k_1 \int_a^b f_1 + k_2 \int_a^b f_2 .$$

10-5 The Interval of Integration

Theorems 10-7 and 10-8 relate the Riemann integrability of a function to the interval on which the function is defined and demonstrate a certain linearity with respect to the interval.

Theorem 10-7 If f is Riemann integrable on $[a,b]$ and if c is such that $a < c < b$, then f is Riemann integrable on $[a,c]$ and on $[c,b]$ and, further,

$$\int_a^c f + \int_c^b f = \int_a^b f.$$

Proof Let $\varepsilon > 0$ be given. Then a partition P of $[a,b]$ exists such that

$$(5) \qquad \int_a^b f - \varepsilon < \underline{R}(f,P) \quad \text{and} \quad \overline{R}(f,P) < \int_a^b f + \varepsilon.$$

Moreover, there is no loss of generality in assuming c to be a point of P since (5) remains true if P is replaced by the refinement $P \cup \{c\}$.

Consider now the intervals $[a,c]$ and $[c,b]$ together with partitions P_1 and P_2 defined by $P_1 = P \cap [a,c]$ and $P_2 = P \cap [c,b]$. Let $\underline{R}_1(f,P_1)$ and $\overline{R}_1(f,P_1)$ denote the lower and upper R-sums of f corresponding to P_1, and $\underline{R}_2(f,P_2)$ and $\overline{R}_2(f,P_2)$ those corresponding to P_2. Because of the construction of P, P_1, and P_2 it follows that

$$(6) \qquad \underline{R}(f,P) = \underline{R}_1(f,P_1) + \underline{R}_2(f,P_2)$$

and

$$(7) \qquad \overline{R}(f,P) = \overline{R}_1(f,P_1) + \overline{R}_2(f,P_2).$$

Combining (6) and (7) with the inequalities of (5) yields

$$\int_a^b f - \varepsilon < \underline{R}_1(f,P_1) + \underline{R}_2(f,P_2) \le \underline{\int_a^c} f + \underline{\int_c^b} f$$

and

$$\overline{\int_a^c} f + \overline{\int_c^b} f \le \overline{R}_1(f,P_1) + \overline{R}_2(f,P_2) < \int_a^b f + \varepsilon.$$

From the arbitrary character of ε follows

$$\int_a^b f \le \underline{\int_a^c} f + \underline{\int_c^b} f \le \overline{\int_a^c} f + \overline{\int_c^b} f \le \int_a^b f \quad \blacktriangle$$

Corollary If f is R-integrable on $[a,b]$ and if $[c,d] \subset [a,b]$, then f is R-integrable on $[c,d]$.

In the nature of a converse we have the following.

Theorem 10-8 If f is R-integrable on each of the intervals $[a,c]$ and $[c,b]$, then it is R-integrable on $[a,b]$.

Proof Let $\varepsilon > 0$ be given. Then, employing the notation of the proof of Theorem 10-7, there exist partitions P_1 of $[a,c]$ and P_2 of $[c,b]$ such that

(8) $$\int_a^c f - \varepsilon < \underline{R}_1(f,P_1), \quad \overline{R}_1(f,P_1) < \int_a^c f + \varepsilon$$

and

(9) $$\int_c^b f - \varepsilon < \underline{R}_2(f,P_2), \quad \overline{R}_2(f,P_2) < \int_c^b f + \varepsilon.$$

Consider the partition $P = P_1 \cup P_2$ of $[a,b]$. Then (6) and (7) of the preceding proof remain valid, and there follows from (8) and (9)

$$\int_a^c f + \int_c^b f - 2\varepsilon < \underline{R}(f,P) \leq \underline{\int_a^b} f$$

and

$$\overline{\int_a^b} f \leq \overline{R}(f,P) < \int_a^c f + \int_c^b f + 2\varepsilon.$$

From the arbitrary character of ε we get

$$\int_a^c f + \int_c^b f \leq \underline{\int_a^b} f \leq \overline{\int_a^b} f \leq \int_a^c f + \int_c^b f \quad \blacktriangle$$

If f is a Riemann integrable function on $[a,b]$, then, for convenience, we introduce the symbol $\int_b^a f$ to denote the negative of the Riemann integral of f or, what is the same, the Riemann integral of $-f$ on $[a,b]$. Thus $\int_b^a f = -\int_a^b f$. Consistently, we also define the symbol $\int_c^c f$ (not a Riemann integral) by $\int_c^c f = 0$ for any point c.

It is left to the reader to verify that the properties of the Riemann integral asserted in Theorems 10-7 and 10-8 remain valid, with some obvious and minor modifications, for the symbol $\int_a^b f$ independent of the relative values of the so-called *limits of integration* a and b. For example, $\int_a^c f + \int_c^b f = \int_a^b f$, whatever the relative values of a, b, and c.

10-6 Further Properties of the Riemann Integral

Here we assert some properties of the R-integral of functions which are subject to certain restraints.

Theorem 10-9 If f is R-integrable and non-negative on $[a,b]$, then $\int_a^b f$ has a non-negative value.

Proof Since $f \geq 0$, it follows that $\underline{R}(f,P) \geq 0$ for any partition P of $[a,b]$ $\blacktriangle$

Corollary 1 If f_1 and f_2 are R-integrable on $[a,b]$ and if $f_1 \leq f_2$, then $\int_a^b f_1 \leq \int_a^b f_2$.

Proof This follows from the theorem with $f = f_2 - f_1$▲

Corollary 2 Suppose f and h to be functions defined on $[a,b]$ such that fh and h are R-integrable. Suppose, too, that $h \geq 0$ and that constants m and M are such that $m \leq f \leq M$. Then $m \int_a^b h \leq \int_a^b fh \leq M \int_a^b h$.

Proof This follows from Corollary 1 upon observing that $fh - mh \geq 0$ and $Mh - fh \geq 0$▲

Corollary 3 If f is R-integrable and non-negative on $[a,d]$, and if b and c are such that $a < b < c < d$, then $\int_b^c f \leq \int_a^d f$.

Proof This follows from the theorem together with the relation $\int_a^d f = \int_a^b f + \int_b^c f + \int_c^d f$, where the integrals on the right exist by virtue of Theorem 10-7▲

Theorem 10-10 If f is R-integrable on $[a,b]$, then so is $|f|$, and

$$\left|\int_a^b f\right| \leq \int_a^b |f|.$$

Proof Let $\varepsilon > 0$ be given. Then, by Theorem 10-3, a partition P of $[a,b]$ exists such that $\overline{R}(f,P) - \underline{R}(f,P) < \varepsilon$. With respect to the segment $[x_{\nu-1}, x_\nu]$ of P, let $m_\nu = \inf f$, $M_\nu = \sup f$, $m_\nu' = \inf |f|$, $M_\nu' = \sup |f|$. It is easily seen that $M_\nu - m_\nu \geq M_\nu' - m_\nu'$ for $\nu = 1, 2, \ldots, n$ and that, as a consequence,

$$\varepsilon > \overline{R}(f,P) - \underline{R}(f,P) \geq \overline{R}(|f|,P) - \underline{R}(|f|,P).$$

From Theorem 10-3 then follows the existence of $\int_a^b |f|$.

Since $|f| \geq f$ and $|f| \geq -f$, we have, by Corollary 1 of Theorem 10-9, $\int_a^b f \leq \int_a^b |f|$ and $-\int_a^b f \leq \int_a^b |f|$. Hence $\left|\int_a^b f\right| \leq \int_a^b |f|$▲

Corollary If M is a bound of an R-integrable function f on $[a,b]$, then

$$\left|\int_a^b f\right| \leq M(b-a).$$

Having established the R-integrability of the sum and difference of R-integrable functions as well as that of a constant multiple and absolute value, we now proceed to show that multiplication is to be included among those operations which preserve integrability.

Theorem 10-11 If f is an R-integrable function on $[a,b]$, then so is the function f^2.

Proof Let $\varepsilon > 0$ be given. Then a partition P of $[a,b]$ exists with the property

$$\overline{R}(f,P) - \underline{R}(f,P) = \sum_{\nu=1}^{n} (M_\nu - m_\nu)\,\Delta x_\nu < \varepsilon,$$

where the M_ν and m_ν are again the suprema and infima of f on the segments of P and are such that $m_\nu \le M_\nu \le M$ for any upper bound M of f.

Suppose f to be a non-negative function on $[a,b]$. Then

$$\begin{aligned}\bar{R}(f^2,P) - \underline{R}(f^2,P) &= \sum_{\nu=1}^{n} M_\nu^2 \,\Delta x_\nu - \sum_{\nu=1}^{n} m_\nu^2 \,\Delta x_\nu \\ &= \sum_{\nu=1}^{n} (M_\nu - m_\nu)(M_\nu + m_\nu)\,\Delta x_\nu \\ &\le 2M \sum_{\nu=1}^{n} (M_\nu - m_\nu)\,\Delta x_\nu < 2M\varepsilon.\end{aligned}$$

Hence f^2 is R-integrable.

If f is not non-negative on $[a,b]$ and is R-integrable, then by Theorem 10-10 the function $|f|$ is R-integrable also. Since $f^2 = |f|^2$, it follows from the above that f^2 is Riemann integrable▲

Theorem 10-12 If f and h are R-integrable functions on $[a,b]$, then so is the product function fh.

Proof Since, by Theorem 10-11, the functions f^2, h^2, and $(f+h)^2$ are R-integrable, so is $\frac{1}{2}[(f+h)^2 - f^2 - h^2] = fh$▲

Exercises

10-7 Cite a counterexample to demonstrate that Riemann integrability of $|f|$ does not imply that of f.

10-8 Establish the *Cauchy-Schwartz inequality*: If f and g are Riemann integrable functions on $[a,b]$, then $[\int_a^b fg]^2 \le [\int_a^b f^2][\int_a^b g^2]$. (*Suggestion:* With $A = \int_a^b f^2$, $B = \int_a^b fg$, and $C = \int_a^b g^2$, prove that $A + 2Bt + Ct^2 \ge 0$ for all real t; use this fact to obtain the desired result.)

10-9 From the Cauchy-Schwartz inequality, deduce the *Minkowski inequality*: If f and g are Riemann integrable functions on $[a,b]$, then $[\int_a^b (f+g)^2]^{1/2} \le [\int_a^b f^2]^{1/2} + [\int_a^b g^2]^{1/2}$.

10-10 Prove that if f is Riemann integrable on $[a,b]$ and if m and M exist such that $0 < m \le f \le M$ on $[a,b]$, then $\int_a^b 1/f$ exists.

10-7 The Fundamental Theorem

In this section we demonstrate the striking relationship between the concepts of differentiation and Riemann integration.

Suppose f to be a function defined on the interval $[a,b]$. Then we shall call a function F a *primitive* of f if F is defined and differentiable on $[a,b]$ and if $F'(x) = f(x)$ for all x in the interval. It is clear, of course, that if a primitive of a function exists, then that primitive is continuous since it is differentiable. It is easily seen, too, that a primitive is not unique, for if F is a primitive of f, then so is any function that differs from F by a constant. Conversely, if F_1 and F_2 are primitives of f, then $(F_1 - F_2)' = F_1' - F_2' = 0$, whence F_1 and F_2 differ by a constant (see Theorem 9-10).

It will be recalled that a differentiable function may have a discontinuous derivative. It follows then that a discontinuous function may possess a primitive.

The following theorem shows the connection between differentiation and Riemann integration.

Theorem 10-13 (The Fundamental Theorem) If f is R-integrable on $[a,b]$ and possesses a primitive F, then

$$\int_a^b f = F(b) - F(a).$$

Proof Let $P = \{x_0,x_1,\ldots,x_n\}$ be any partition of the interval $[a,b]$. Then, by the mean value theorem for derivatives, there exist numbers ξ_ν $(\nu = 1,2,\ldots,n)$ such that $x_{\nu-1} < \xi_\nu < x_\nu$ and

$$F(x_\nu) - F(x_{\nu-1}) = F'(\xi_\nu)\,\Delta x_\nu = f(\xi_\nu)\,\Delta x_\nu.$$

Therefore

$$F(b) - F(a) = \sum_{\nu=1}^{n} [F(x_\nu) - F(x_{\nu-1})] = \sum_{\nu=1}^{n} f(\xi_\nu)\,\Delta x_\nu.$$

Since

$$\underline{R}(f,P) \le \sum_{\nu=1}^{n} f(\xi_\nu)\,\Delta x_\nu \le \overline{R}(f,P),$$

we have

$$\underline{R}(f,P) \le F(b) - F(a) \le \overline{R}(f,P).$$

This being the case for all partitions P of $[a,b]$ we conclude that

$$\underline{\int_a^b} f \le F(b) - F(a) \le \overline{\int_a^b} f.$$

From the hypothesis that f is R-integrable follows the desired result▲

Corollary If a function f possesses a derivative f' which is R-integrable, then

$$\int_a^b f' = f(b) - f(a).$$

It should be noted that in the statement of the fundamental theorem it is not redundant to require that f be R-integrable *and* possess a primitive. It is quite possible for a function to be Riemann integrable and yet not have a primitive. Moreover, a function which has a primitive need not be Riemann integrable. Examples 10-4 and 10-5 attest to the accuracy of these statements.

Example 10-4 Consider the function Q defined on [0,1] by $Q(x) = 1/q$ if $x = p/q$ (p and q relatively prime, $q > 0$) and $Q(x) = 0$ if x is irrational. To show that Q is Riemann integrable, we observe first that for any partition P of [0,1] it is the case that $\underline{R}(Q,P) = 0$. Hence $\underline{\int}_0^1 Q = 0$. Now, let $\varepsilon > 0$ be given. Then (see Example 8-12) only a finite number N of rational numbers exist such that at each the value of $Q(x)$ is or exceeds ε. Consider partitions of [0,1] which are of the form $P = \{0, 1/n, 2/n, \ldots, n/n\}$, where $n > N$. If M_ν is the supremum of Q on the νth segment of such a partition P, then $M_\nu \geq \varepsilon$ for at most N values of ν and $M_\nu \leq 1$ for all ν. Hence

$$\overline{R}(Q,P) = \sum_{\nu=1}^{n} \frac{M_\nu}{n} \leq \frac{1}{n}[N + (n - N)\varepsilon] = \varepsilon + \frac{N}{n}(1 - \varepsilon).$$

Since for each ε this is the case for all $n > N$, it follows that $\inf\{\overline{R}(Q,P)\} \leq \varepsilon$. Because ε is arbitrary, we infer that $\inf\{\overline{R}(Q,P)\} = 0$ or, what is the same, $\overline{\int}_0^1 Q = 0$. We conclude that Q is Riemann integrable and that $\int_0^1 Q = 0$.

Further, by the corollary to Theorem 10-7 and Corollary 3 of Theorem 10-9, it follows that $\int_0^x Q = 0$ for all x in [0,1]. The fundamental theorem then implies that if Q possesses a primitive F, then $F(x) - F(0) = 0$. Thus F has the constant value $F(0)$ on [0,1], from which we conclude that $Q(x) = F'(x) = 0$ for all x in [0,1]. From this contradiction to the definition of Q we conclude that Q is a function which, although Riemann integrable, does not possess a primitive.

Example 10-5 For the function h defined by $h(x) = x^{4/3} \sin(1/x)$ if $x \neq 0$ and $h(0) = 0$ it is easy to show that $h'(x) = \frac{4}{3}x^{1/3} \sin(1/x) - x^{-2/3} \cos(1/x)$ if $x \neq 0$ and $h'(0) = 0$. Since $h'(1/2n\pi) = -(2n\pi)^{2/3}$ for any integer n, it follows that h' is unbounded and hence is not R-integrable on any interval containing 0. However, a primitive, that is, h, exists.

10-8 The Indefinite Riemann Integral

If x_0 is a point of $[a,b]$ and if f is Riemann integrable on $[a,b]$, we call the function φ defined on $[a,b]$ by

$$\varphi(x) = \int_{x_0}^{x} f$$

the *indefinite Riemann integral of f from x_0* or, simply, an *indefinite Riemann integral of f.*

It is clear that $\varphi(x_0) = 0$. Also, that if x_1 and x_2 are any points of $[a,b]$, then the indefinite Riemann integrals from x_1 and x_2 differ by the number $\int_{x_1}^{x_2} f$.

Theorem 10-14 An indefinite Riemann integral of a function f is continuous on the domain $[a,b]$ of f.

Proof First observe that M exists such that $|f(x)| \leq M$ for all points x of $[a,b]$. Now let $\varepsilon > 0$ be given, and suppose x_1 and x_2 to denote any two points of $[a,b]$ such that $|x_2 - x_1| < \varepsilon$. Then, if φ is the indefinite R-integral from x_0, we have

$$|\varphi(x_2) - \varphi(x_1)| = \left|\int_{x_1}^{x_2} f\right| \leq M\,|x_2 - x_1| < M\varepsilon\blacktriangle$$

It cannot also be concluded that an indefinite Riemann integral is differentiable. The following is a counterexample.

Example 10-6 For the function f defined on $[0,2]$ by $f(x) = 1$ if $0 \leq x \leq 1$ and $f(x) = 2$ if $1 < x \leq 2$, we have $\int_0^x f = x$ if $0 \leq x \leq 1$ and $\int_0^x f = 2x - 1$ if $1 < x \leq 2$. The latter defines a function which, although continuous, is not differentiable at $x = 1$.

It is interesting to note, too, that if $\int_{x_0}^x f$ defines a differentiable function on $[a,b]$, it does not necessarily follow that the derivative is the function f. For, consider again the function Q of Example 10-4. It was seen to have the property that $\int_0^x Q = 0$ for all x. The derivative of $\int_0^x Q$ is therefore not the function Q but rather the zero function on $[0,1]$. However, if we confine our attention to continuity points of a function f, then the following affirmative statement can be made.

Theorem 10-15 If f is Riemann integrable on $[a,b]$, φ an indefinite Riemann integral of f, and c a point of $[a,b]$ at which f is continuous, then φ is differentiable at c and $\varphi'(c) = f(c)$.

Proof Let $\varepsilon > 0$ be given and h a nonzero number such that $c + h$ is in $[a,b]$. From $\varphi(x) = \int_{x_0}^x f$ follows

$$\left|\frac{\varphi(c+h) - \varphi(c)}{h} - f(c)\right| = \left|\frac{1}{h}\left\{\int_c^{c+h} f - hf(c)\right\}\right|$$

or, since $hf(c) = f(c)\int_c^{c+h} 1 = \int_c^{c+h} f(c)$,

$$\left|\frac{\varphi(c+h) - \varphi(c)}{h} - f(c)\right| = \left|\frac{1}{h}\int_c^{c+h} [f - f(c)]\right|$$

$$\leq \left|\frac{1}{h}\int_c^{c+h} |f - f(c)|\right|.$$

From the continuity of f at c follows the existence of a number δ such that $|f(x) - f(c)| < \varepsilon$ whenever $|x - c| < \delta$. Hence, if $|h| < \delta$, then

$$\left|\frac{\varphi(c+h) - \varphi(c)}{h} - f(c)\right| < \left|\frac{1}{h}\int_c^{c+h} \varepsilon\right| = \varepsilon$$

from which it follows that $\varphi'(c)$ exists and equals $f(c)\blacktriangle$

Corollary If a function is continuous on a closed interval, then it has a primitive on that interval.

It is also a consequence of the above that if φ is an indefinite Riemann integral of a continuous function f and if F is any primitive of f, then $F = \varphi + C$ for some constant C.

10-9 Mean Value Theorems

It is possible to establish for the Riemann integral, as it was in the case of derivatives, certain principles based upon mean values.

Theorem 10-16 If fh and h are Riemann integrable functions on $[a,b]$, if h is non-negative, and if m and M are such that $m \leq f(x) \leq M$, then a number η exists such that $m \leq \eta \leq M$ and

$$\int_a^b fh = \eta \int_a^b h.$$

Proof Let g be a function defined on $[m,M]$ by $g(x) = x \int_a^b h$. Since it is continuous, it assumes all values between its minimum and maximum values $m \int_a^b h$ and $M \int_a^b h$. Since, by Corollary 2 to Theorem 10-9, we have $m \int_a^b h \leq \int_a^b fh \leq M \int_a^b h$, it follows that η exists in $[m,M]$ such that $g(\eta) = \eta \int_a^b h = \int_a^b fh$▲

Theorem 10-17 (The First Mean Value Theorem) If f is continuous and h Riemann integrable and non-negative on $[a,b]$, then ξ exists such that $a \leq \xi \leq b$ and

$$\int_a^b fh = f(\xi) \int_a^b h.$$

Proof By Theorem 10-16 a number η exists such that $\inf f \leq \eta \leq \sup f$ and $\int_a^b fh = \eta \int_a^b h$. Since f is continuous it assumes all values in the interval $[\min f, \max f]$, in particular, η. Hence ξ exists such that $a \leq \xi \leq b$ and $f(\xi) = \eta$▲

Corollary If f is continuous on $[a,b]$, then a number ξ exists such that $a \leq \xi \leq b$ and

$$\int_a^b f = f(\xi)(b - a).$$

Example 10-7 The mean value theorem finds application in the practical problem of determining approximate values for certain integrals. Consider the so-called *elliptic* Riemann integral $\int_0^{1/2} g$, where g is defined on $[0,\frac{1}{2}]$ by $g(x) = [(1 - x^2)(1 - k^2x^2)]^{-1/2}$ with $|k| < 2$. By identifying g with the function fh of the mean value theorem, with $f(x) = (1 - k^2x^2)^{-1/2}$ and $h(x) = (1 - x^2)^{-1/2}$, there exists a number ξ such that $0 \leq \xi \leq \frac{1}{2}$ and

$$\int_0^{1/2} g = (1 - k^2\xi^2)^{-1/2} \int_0^{1/2} h = (1 - k^2\xi^2)^{-1/2} \arcsin x \Big|_0^{1/2} = \frac{\pi}{6}(1 - k^2\xi^2)^{-1/2}.$$

Finally, since $(\pi/6)(1 - k^2\xi^2)^{-1/2}$ increases on the interval $[0,\frac{1}{2}]$, we have $\pi/6 \leq \int_0^{1/2} g \leq (\pi/6)(1 - \frac{1}{4}k^2)^{-1/2}$.

Theorem 10-18 (The Second Mean Value Theorem) If f is continuous and increasing on $[a,b]$, and if h is Riemann integrable and non-negative, then ξ exists such that $a \le \xi \le b$ and

$$\int_a^b fh = f(a)\int_a^\xi h + f(b)\int_\xi^b h.$$

Proof By the first mean value theorem there exists a number η in $[a,b]$ such that $\int_a^b fh = f(\eta)\int_a^b h$. Also, since f is increasing, we have $f(a) \le f(\eta) \le f(b)$.

Let κ denote a function defined on $[a,b]$ by $\kappa(x) = [f(b) - f(a)]\int_x^b h$. Since $h \ge 0$, it follows that $\int_x^b h$ defines a decreasing function; since $f(b) - f(a) \ge 0$, κ also is decreasing. Further,

$$\kappa(a) = [f(b) - f(a)]\int_a^b h \ge [f(\eta) - f(a)]\int_a^b h \ge 0 = \kappa(b).$$

Because κ is continuous it assumes all values between $\kappa(a)$ and $\kappa(b)$, in particular, $[f(\eta) - f(a)]\int_a^b h$. Hence ξ exists in $[a,b]$ such that

$$\kappa(\xi) = [f(b) - f(a)]\int_\xi^b h = [f(\eta) - f(a)]\int_a^b h.$$

Therefore

$$f(b)\int_\xi^b h - f(a)\int_\xi^b h = \int_a^b fh - f(a)\int_a^b h$$

or

$$\int_a^b fh = f(a)\left(\int_a^b h - \int_\xi^b h\right) + f(b)\int_\xi^b h,$$

which reduces to the desired relation▲

Example 10-8 We may use the second mean value theorem to obtain an upper bound for the integral $\int_a^b [(\sin x)/x]$, where $a > 0$. We have

$$\int_a^b \frac{\sin x}{x} = \frac{1}{a}\int_a^\xi \sin x + \frac{1}{b}\int_\xi^b \sin x$$

for some ξ in $[a,b]$. Therefore, from $|\int_a^\xi \sin x| \le 2$ and $|\int_\xi^b \sin x| \le 2$ (where we have used the property of the sine function that the integral over any interval cannot exceed the integral over one arch) it follows that $\left|\int_a^b \frac{\sin x}{x}\right| \le 2\left(\frac{1}{a} + \frac{1}{b}\right)$.

Exercises

10-11 Prove the following generalization of the fundamental theorem: If f is Riemann integrable on $[a,b]$ and if F is continuous on $[a,b]$ and such that F' exists and equals f at all points of $[a,b]$ except for a finite set, then $\int_a^b f = F(b) - F(a)$. (*Suggestion:* Consider a partition of $[a,b]$ that includes those points for which $F' \ne f$.)

10-12 Prove that if f is Riemann integrable on $[a,b]$ and if G is defined on $[a,b]$ by $G(x) = \int_x^b f$, then $G' = -f$ at each continuity point of f.
10-13 Let f be continuous on $[a,b]$, and suppose u and v to be differentiable functions with ranges in $[a,b]$. Prove that $\left(\int_u^v f\right)' = (f \circ v)v' - (f \circ u)u'$.
10-14 Demonstrate a geometric interpretation of the first mean value theorem in the event that h is the unit function (corollary to Theorem 10-17).

10-10 Properties of the Integrand of a Riemann Integral

In this section we state and prove two theorems that tend to simplify the problem of evaluation of a Riemann integral. The first of these is the formula for *integration by parts*, familiar to the reader from elementary calculus; the second demonstrates the effect of a *change of variable* in the integrand.

Theorem 10-19 (Integration by Parts) If f and h are functions such that f' and h' exist and are Riemann integrable on $[a,b]$, then

$$\int_a^b fh' = f(b)h(b) - f(a)h(a) - \int_a^b f'h.$$

Proof Since f' and h' exist, the functions f and h are continuous and therefore Riemann integrable. It follows that fh' and $f'h$, also, are Riemann integrable. From $(fh)' = fh' + f'h$ it follows that $(fh)'$ is Riemann integrable and that $\int_a^b (fh)' = \int_a^b fh' + \int_a^b f'h$. Furthermore, because of the corollary to the fundamental theorem, we may write $\int_a^b (fh)' = f(b)h(b) - f(a)h(a)$. The desired conclusion follows▲

Theorem 10-20 (Change of Variable) If h is a function having a derivative which is Riemann integrable on an interval $[p,q]$ and if f is defined and continuous on the range of h, then

$$\int_a^b f = \int_p^q (f \circ h)h',$$

where $a = h(p)$ and $b = h(q)$.

Proof Let φ be the indefinite Riemann integral defined on the range of h by $\varphi(x) = \int_a^x f$. From the continuity of f, φ is differentiable and $\varphi' = f$. Consider the function ρ defined on $[p,q]$ by $\rho = \varphi \circ h$. Since φ and h are differentiable, so is ρ and $\rho' = (\varphi' \circ h)h' = (f \circ h)h'$. Because f and h are continuous, so is $f \circ h$. The last is therefore Riemann integrable, and, since, by hypothesis, h' is also, it follows that ρ' possesses a Riemann integral. Hence

$$\int_p^q \rho' = \rho(q) - \rho(p) = \varphi(h(q)) - \varphi(h(p))$$

$$= \varphi(b) - \varphi(a) = \int_a^b f.$$

The desired conclusion follows from $\rho' = (f \circ h)h'$▲

It is not to be inferred from the statement of Theorem 10-20 that the range of h is the interval $[a,b]$. Moreover, no condition is placed on the relative values of a and b. That a may exceed or even equal b is quite consistent with the argument leading to the change of variable formula.

10-11 Zero Sets

Before progressing to further properties of Riemann integrals, it is appropriate that we look more closely at the character of a Riemann integrable function. We have seen that boundedness is necessary but not sufficient for Riemann integrability and that continuity is sufficient but not necessary. With a view to demonstrating a condition which is both necessary and sufficient we introduce at this point the concept of a *zero set* or a *set of measure zero*. In a later chapter we shall find this notion to be encompassed in the broader concept of *measurable set*.

Let I denote the bounded open interval (a,b). Denote its *length* by ΔI, and define it to be the positive number $b - a$.[2] A set A is called a *zero set* if, corresponding to each positive number ε, there exists a sequence $\{I_\nu\}$ of open intervals which covers A and is such that $\sum_{\nu=1}^{\infty} \Delta I_\nu < \varepsilon$.

That any countably infinite set of real numbers is a zero set is virtually immediate. For if $\{x_1,x_2,\ldots\}$ is such a set, then the sequence $\{I_\nu\}$, where $I_\nu = (x_\nu - \varepsilon/4^\nu, x_\nu + \varepsilon/4^\nu)$ for $\nu = 1, 2, \ldots$, is obviously a covering of $\{x_1,x_2,\ldots\}$ and, moreover, $\sum_{\nu=1}^{\infty} \Delta I_\nu = \sum_{\nu=1}^{\infty} 2\varepsilon/4^\nu = 2\varepsilon/3 < \varepsilon$. It is clear, too, that any subset of a zero set is also a zero set. In particular, the empty set is in this category.

We must not leap to the conclusion that every zero set is countable. The following is an example of a noncountable zero set.

Example 10-9 The Cantor set K of Chapter 4 was defined to be the intersection of a collection of closed sets F_n, $n = 0, 1, 2, \ldots$, where for each n the set F_n is the union of 2^n closed intervals, each of length $1/3^n$. Corresponding to any $\varepsilon > 0$, each such interval is clearly contained in an open interval of length $3^{-n} + 2^{-n}\varepsilon$, so that F_n, and hence K, is covered by a finite set of open intervals having total length $2^n(3^{-n} + 2^{-n}\varepsilon) = (2/3)^n + \varepsilon$. For sufficiently large n this is less than 2ε. It follows that the Cantor set K, although noncountable, is a zero set. Historically, it was the first set of this character to be constructed.

We may go even further in revealing the extent of the category of zero sets.

Theorem 10-21 The union of a countable collection of zero sets is a zero set.

Proof Consider $A = \bigcup_{\nu=1}^{\infty} A_\nu$, where each A_ν is a zero set, and let $\varepsilon > 0$ be given. Then, corresponding to each ν, $\nu = 1, 2, \ldots$, there exists a set of open

[2] Earlier the length of a closed interval $[a,b]$ was also defined to be the number $b - a$. Actually, the two definitions are quite compatible, as we shall see in the sequel. It may be pointed out, too, that the discussion which follows, concerning open intervals, is, with minor changes, equally meaningful for closed intervals (and even for intervals which are neither open nor closed).

intervals $I_{\mu\nu}$, $\mu = 1, 2, \ldots$, which covers A_ν and which is such that $\sum_{\mu=1}^{\infty} \Delta I_{\mu\nu} < \varepsilon/4^\nu$. Imagine the countably infinite set of intervals $I_{\mu\nu}$, which clearly covers A, to be arranged in a sequence in any way, and denote the sequence by $\{I_n\}$. Consider the finite set $\{I_1, I_2, \ldots, I_{n'}\}$, where n' is a positive integer. Clearly, integers M and N exist such that each I_n for $n \le n'$ is an interval $I_{\mu\nu}$ with $\mu \le M$ and $\nu \le N$. As a consequence,

$$\sum_{n=1}^{n'} \Delta I_n \le \sum_{\nu=1}^{N} \left(\sum_{\mu=1}^{M} \Delta I_{\mu\nu} \right) \le \sum_{\nu=1}^{N} \left(\sum_{\mu=1}^{\infty} \Delta I_{\mu\nu} \right)$$

$$\le \sum_{\nu=1}^{N} \frac{\varepsilon}{4^\nu} < \sum_{\nu=1}^{\infty} \frac{\varepsilon}{4^\nu} = \frac{\varepsilon}{3}.$$

Since n' is arbitrary, it follows that $\sum_{n=1}^{\infty} \Delta I_n \le \varepsilon/3 < \varepsilon$, whence A is a zero set ▲

10-12 A Necessary and Sufficient Condition for Riemann Integrability

We are now in a position to state a second property essential to the existence of a Riemann integral. Effectively it places a limitation on the set of points at which a Riemann integrable function may be discontinuous.

Theorem 10-22 If a function f is Riemann integrable, then the set of its discontinuity points is a zero set.

Proof Suppose f to be Riemann integrable, and hence bounded on $[a,b]$. Denote by D the set of discontinuity points of f, and, corresponding to each positive integer i, let D_i be the set of points of $[a,b]$ at each of which the fluctuation of f exceeds $1/i$. Then $D = \bigcup_{i=1}^{\infty} D_i$.

Let us assume that D is not a zero set. Then, necessarily, for some integer k the set D_k is not a zero set. Hence a positive number δ exists with the property that any countable open covering $\{I_\nu\}$ of D_k is such that $\sum_{\nu=1}^{\infty} \Delta I_\nu \ge \delta$.

Let $P = \{x_0, x_1, \ldots, x_n\}$ be a partition of $[a,b]$, and suppose $[x_{\nu_1 - 1}, x_{\nu_1}]$, $[x_{\nu_2 - 1}, x_{\nu_2}]$, $\ldots$, $[x_{\nu_\mu - 1}, x_{\nu_\mu}]$ $(\mu \le n)$ to be those segments of P that contain points of D_k. Then $\sum_{j=1}^{\mu} \Delta x_{\nu_j} \ge \delta$. In each of these segments the oscillation of f exceeds $1/k$. Therefore we have $M_{\nu_j} - m_{\nu_j} > 1/k$ for $j = 1, 2, \ldots, \mu$. It follows that $\overline{R}(f,P) - \underline{R}(f,P) > \delta/k$. Since δ/k is independent of P, we conclude from Theorem 10-3 that f is not Riemann integrable. The contradiction implies that D is a zero set ▲

It is quite clear that the possession of a zero set of discontinuities is not sufficient to integrability since unbounded sets exist with this property. Thus boundedness and possession of a zero set of discontinuities are both necessary to the existence of a Riemann integral, but separately neither is sufficient. What we show next is that together these two properties do suffice to assure Riemann integrability.

Theorem 10-23 If f is a bounded function having a zero set of discontinuities on $[a,b]$, then f is Riemann integrable on $[a,b]$.

Proof Let D be the set of discontinuities of f, and suppose $\varepsilon > 0$ to be given. Then a collection $J = \{I_\nu ;\ \ \nu = 1,2,\ldots\}$ of open intervals exists which covers D and is such that $\sum_{\nu=1}^{\infty} \Delta I_\nu < \varepsilon$. Denote by K the collection of all open intervals each such that the oscillation of f on its closure is less than ε. Then $J \cup K$ is a set of open intervals covering $[a,b]$. Since $[a,b]$ is compact, a finite subset of $J \cup K$ exists which also covers $[a,b]$. For convenience, let

$$I_1, I_2, \ldots, I_s, K_1, K_2, \ldots, K_t \tag{10}$$

denote the members of such a collection. Consider the finite set of points of $[a,b]$ which are endpoints of these intervals. Together with the numbers a and b, they constitute a partition $P = \{x_0,x_1,\ldots,x_n\}$ of $[a,b]$. It is clear that each point x_ν is a member of at least one of the intervals of (10).

Denote by symbols $y_0, y_1, \ldots, y_N$ those points x_ν of the partition P each of which is contained in at least one of the intervals $K_1, K_2, \ldots, K_t$; the remaining x_ν denote by $y_{N+1}, \ldots, y_n$. The set of numbers $\{y_0,y_1,\ldots,y_n\}$ is, of course, the same set as $\{x_0,x_1,\ldots,x_n\}$, but it is not necessarily the case that $y_\nu < y_{\nu+1}$ for any ν. Furthermore, the lengths $\Delta x_1, \Delta x_2, \ldots, \Delta x_n$ of the segments of P are appropriately restyled $\Delta y_0, \Delta y_1, \ldots, \Delta y_n$, where Δy_ν denotes the length of the segment having y_ν as its right endpoint. (Note that the inference $\Delta y_\nu = y_\nu - y_{\nu-1}$ is not proper here. Also, that $\Delta y_\nu = 0$ if $y_\nu = x_0$.)

The lower and upper Riemann sums of f on $[a,b]$, which exist by virtue of the hypothesis that f is bounded, may now be expressed in the form $\underline{R}(f,P) = \underline{R}_1 + \underline{R}_2$ and $\overline{R}(f,P) = \overline{R}_1 + \overline{R}_2$, where $\underline{R}_1 = \sum_{\nu=0}^{N} m_\nu \Delta y_\nu$, $\underline{R}_2 = \sum_{\nu=N+1}^{n} m_\nu \Delta y_\nu$, $\overline{R}_1 = \sum_{\nu=0}^{N} M_\nu \Delta y_\nu$, and $\overline{R}_2 = \sum_{\nu=N+1}^{n} M_\nu \Delta y_\nu$. The numbers m_ν and M_ν designate, of course, the infima and suprema, respectively, of f on Δy_ν.

Because of the definitions of the y_ν and the Δy_ν, it follows that $M_\nu - m_\nu < \varepsilon$ for $\nu = 0, 1, \ldots, N$. Hence

$$\overline{R}_1 - \underline{R}_1 = \sum_{\nu=0}^{N} (M_\nu - m_\nu)\, \Delta y_\nu < \varepsilon \sum_{\nu=0}^{N} \Delta y_\nu \le \varepsilon(b - a). \tag{11}$$

On the other hand,

$$\overline{R}_2 - \underline{R}_2 = \sum_{\nu=N+1}^{n} (M_\nu - m_\nu)\, \Delta y_\nu \le (M - m) \sum_{\nu=N+1}^{n} \Delta y_\nu ,$$

where $M = \sup f$ and $m = \inf f$ on $[a,b]$. Since the segments Δy_ν involved in this summation are subsets of closures of intervals of the collection $\{I_1,I_2,\ldots,I_s\}$, it follows that $\sum_{\nu=N+1}^{n} \Delta y_\nu \le \sum_{\nu=1}^{\infty} \Delta I_\nu < \varepsilon$. Therefore

$$\overline{R}_2 - \underline{R}_2 < (M - m)\varepsilon. \tag{12}$$

Finally, we combine (11) and (12) to obtain

$$\overline{R}(f,P) - \underline{R}(f,P) < (b - a + M - m)\varepsilon$$

and conclude from Theorem 10-3 that f is Riemann integrable on $[a,b]$▲

Thus has been established a characterization of Riemann integrable functions. They are those and only those functions which are bounded and whose discontinuities constitute zero sets.

Example 10-10 It is interesting to note that by turning some of the above ideas around we are able to prove the existence of a set of real numbers which is not a zero set. Recall that in Example 10-1 the function Ψ was demonstrated to be not Riemann integrable. Since Ψ is obviously bounded, we may conclude only that the set of its discontinuity points is not a zero set. This set is precisely the interval $[a,b]$.

The next theorem encompasses certain properties of Riemann integrable functions already established. The proof, which follows directly from Theorem 10-23, may be compared with the somewhat less succinct arguments employed earlier.

Theorem 10-24 If f and h are Riemann integrable functions on $[a,b]$, then so are the functions $-f$, $|f|, f+h, fh$; also f/h if $\inf |h| > 0$.

Proof It has been established in Chapter 8 that the functions $-f$ and $|f|$ are discontinuous only at discontinuity points of f. Also, $f+h, fh$, and f/h are not discontinuous at points other than those which are discontinuity points of f or h. Since f and h are Riemann integrable, each possesses a zero set of discontinuity points. It follows that this is also the case for $-f$, $|f|, f+h, fh$, and f/h. Further, boundedness of f and h implies boundedness of $-f$, $|f|$, $f+h$, and fh. The condition $\inf |h| > 0$, moreover, serves to insure that f/h is not unbounded. It follows that $-f$, $|f|, f+h, fh$, and f/h are Riemann integrable▲

The reader may have been struck in the statement of Theorem 10-24 by the absence, among the various combinations listed, of that of a function of a function. The answer, very simply, is this: a Riemann integrable function of a Riemann integrable function does not necessarily define a function which possesses a Riemann integral.

Example 10-11 Consider the function Q of Example 10-4 and the function h defined by $h(x) = 0$ if $x \neq 0$ and $h(0) = 1$. Each of these is Riemann integrable. However, $h \circ Q = \Psi$.

10-13 Alternate Definitions of the Riemann Integral

It is our purpose in this section to state two alternate definitions of the Riemann integral of a function and to establish their equivalence with each other and with the definition, styled Definition (A), with which we have been concerned to this point.

Let $P = \{x_0,x_1,\ldots,x_n\}$ denote a partition of the interval $[a,b]$. Then, by an *intermediate partition* of P we shall understand a set $Q = \{\xi_1,\xi_2,\ldots,\xi_n\}$, of at most n points, with the property that $x_{\nu-1} \le \xi_\nu \le x_\nu$ for $\nu = 1, 2, \ldots, n$. It is consistent with the definition of Q that two consecutive points be identical. Thus, for some value of ν it may be the case that $\xi_{\nu-1} = \xi_\nu = x_{\nu-1}$. Obviously, however, three or more points of Q cannot coincide. Figure 10-1 depicts a partition and an intermediate partition for which $n = 6$.

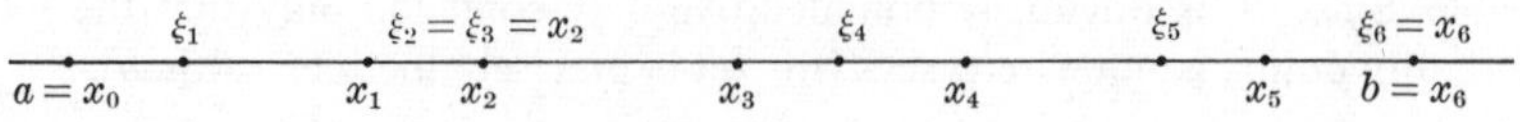

ξ_1 $\quad \xi_2 = \xi_3 = x_2$ $\quad \xi_4$ $\quad \xi_5$ $\quad \xi_6 = x_6$

$a = x_0$ $\quad x_1$ $\quad x_2$ $\quad x_3$ $\quad x_4$ $\quad x_5$ $\quad b = x_6$

Figure 10-1

Suppose f to be a function defined on $[a,b]$, and let P denote a partition of $[a,b]$ and Q an intermediate partition of P. Then we define the *Riemann sum*, or *R-sum*, of f corresponding to P and Q to be the number

$$R(f,P,Q) = \sum_{\nu=1}^{n} f(\xi_\nu)\,\Delta x_\nu .$$

Unlike the case for upper and lower Riemann sums, boundedness of f is not a prerequisite to the existence of a Riemann sum. It is clear, however, that if f is bounded, in which case both lower and upper R-sums exist, then, for a given partition P and any intermediate partition Q of P, we have

$$\underline{R}(f,P) \le R(f,P,Q) \le \overline{R}(f,P).$$

We state now a definition of Riemann integrability based on the notion of a Riemann sum.

Definition (B) A number S is said to be the *Riemann integral* or *R-integral* of the function f on the interval $[a,b]$, and f is said to be *Riemann integrable* on $[a,b]$ if, corresponding to each number $\varepsilon > 0$, there exists a number $\delta > 0$ with the property that $|R(f,P,Q) - S| < \varepsilon$ for all partitions P such that $\Delta P < \delta$ and all intermediate partitions Q of P.

We perceive in this definition a concept resembling that of limit of a function at a point. Indeed, the two notions are sufficiently alike in expression to suggest the use of similar language and notation. Thus, the number S may be described as "the limit of a set of Riemann sums as the norm of the partition P tends to zero." Accordingly, we employ an appropriate limit symbol to denote the R-integral of f, namely

$$\lim_{\Delta P \to 0} R(f,P,Q) = S.$$

It seems proper, however, to forego usage of the notation $\int_a^b f$ for the integral as defined in Definition (B) until such time as we have demonstrated, beyond mere name calling, that such is justified. That the number S, as defined, is unique is left to the reader to establish.

A second alternate definition of the Riemann integral also reflects the notion of limit of a sum. Based upon the idea of a refinement of a partition, it suggests a particularization of Definition (B), but, as we shall see, is no less general.

Definition (C) A number S^* is said to be the *Riemann integral* of a function f on the interval $[a,b]$, and f is styled *Riemann integrable* on $[a,b]$ if, corresponding to each given $\varepsilon > 0$, there exists a partition P_ε with the property that $|R(f,P,Q) - S^*| < \varepsilon$ for each refinement P of P_ε and all intermediate partitions Q of P.

We proceed now to prove the contention that Definitions (A), (B), and (C) are equivalent. Specifically, we shall show that Definition (B) implies Definition (C), that (C) implies (A), and, finally, to complete the cycle, that (A) implies (B).

The first link in this chain is the easiest. We have only to observe that $\varepsilon > 0$ being given, the assertion that S and δ exist such that $|R(f,P,Q) - S| < \varepsilon$ whenever $\Delta P < \delta$, implies that for each partition P with this property all refinements of P have the same property, since the norm of a refinement of P cannot exceed that of P. The following theorem states the conclusion formally.

Theorem 10-25 If a function f, defined on an interval $[a,b]$, possesses a Riemann integral S in the sense of Definition (B), then it possesses a Riemann integral S^* in the sense of Definition (C) and, moreover, $S = S^*$.

In Definition (A) boundedness of f is essential to the statement of the definition. While this is not the case in Definitions (B) or (C), nevertheless it is necessary to the existence of S and S^*. We show next that Definition (C) restricts f to the category of a bounded function. The import of this result is that in connection with all three definitions only bounded functions need be considered.

Theorem 10-26 If f is Riemann integrable on an interval $[a,b]$ in the sense of Definition (C), then it is a bounded function on $[a,b]$.

Proof Let $\varepsilon > 0$ be given. Then, by Definition (C), a partition

$$P_\varepsilon = \{x_0,x_1,\ldots,x_n\}$$

exists such that for any intermediate partitions Q' and Q'' of P_ε it is the case that

$$|R(f,P_\varepsilon,Q') - S^*| < \varepsilon \quad \text{and} \quad |R(f,P_\varepsilon,Q'') - S^*| < \varepsilon.$$

It follows that

$$|R(f,P_\varepsilon,Q'') - R(f,P_\varepsilon,Q')| < 2\varepsilon. \tag{13}$$

Suppose $Q' = \{\xi_1',\xi_2',\ldots,\xi_n'\}$ and let $k = \max\{f(\xi_\nu'); \ \nu = 1,2,\ldots,n\}$. Further, denote by β the minimum of the set $\{\Delta x_\nu; \ \nu = 1,2,\ldots,n\}$.

Assume that f is not bounded above on $[a,b]$. Then a point c of $[a,b]$ exists in some segment, say $[x_{\mu-1},x_\mu]$, of P_ε such that $f(c) > k + 1/\beta$. Suppose Q'' to be obtained from Q' by replacing ξ_μ' by c. Then

$$|R(f,P_\varepsilon,Q'') - R(f,P_\varepsilon,Q')| = [f(c) - f(\xi_\mu')]\,\Delta x_\mu > 1,$$

which is in contradiction to (13). We conclude that f is bounded above. A similar argument shows it is also bounded below▲

Theorem 10-27 If f, defined on $[a,b]$, is Riemann integrable in the sense of Definition (C), then it is Riemann integrable in the sense of Definition (A); moreover, $S^* = \int_a^b f$.

Proof Note first that boundedness of f, which is essential even to the consideration of Definition (A), has been established. Let $\varepsilon > 0$ be given. Then, by Definition (C), a number S^* and a partition P exist such that

$$|R(f,P,Q) - S^*| < \varepsilon \tag{14}$$

for all intermediate partitions Q of P. Clearly, in the interval $[x_{\nu-1},x_\nu]$, for each ν, there exists a point ξ_ν such that $M_\nu < f(\xi_\nu) + \varepsilon$. A set of points ξ_ν defined in this manner constitutes an intermediate partition Q of P, and it is easy to see that

$$\overline{\int_a^b} f \le \overline{R}(f,P) < R(f,P,Q) + \varepsilon(b - a). \tag{15}$$

Combining inequalities (14) and (15) and taking into account the arbitrary nature of ε yields

$$\overline{\int_a^b} f \le S^*.$$

By considering lower Riemann sums we determine the dual inequality

$$S^* \le \underline{\int_a^b} f.$$

Together these imply the existence of $\int_a^b f$ and show its value to be S^*▲

To complete the circuit of implications, it is necessary to make some preliminary remarks. From the definition of the upper Riemann integral $\overline{\int_a^b} f$ of a function f on $[a,b]$ as the infimum of the set of all upper R-sums, it follows that a partition P can be found for which the upper sum $\overline{R}(f,P)$ differs from $\overline{\int_a^b} f$ by as little as one pleases. What is not obvious, however, is whether the difference $\overline{R}(f,P) - \overline{\int_a^b} f$ can be made arbitrarily small for *all* partitions P with norms sufficiently small. That is, with $\varepsilon > 0$ given, is it the case that a number $\delta > 0$ exists with the property that

$$\left|\overline{R}(f,P) - \overline{\int_a^b} f\right| < \varepsilon$$

for all partitions P such that $\Delta P < \delta$? If such should actually be the case we write, consistent with notation introduced earlier,

$$\lim_{\Delta P \to 0} \bar{R}(f,P) = \overline{\int_a^b} f.$$

Similar interpretation should be assigned to the statement

$$\lim_{\Delta P \to 0} \underline{R}(f,P) = \underline{\int_a^b} f.$$

In the following theorem it is demonstrated that the lower and upper Riemann integrals may be achieved in precisely this fashion, that is, as limits of sets of lower and upper sums corresponding to partitions with diminishing norms.

Theorem 10-28 If f is a bounded function on the interval $[a,b]$, then $\lim_{\Delta P \to 0} \bar{R}(f,P) = \overline{\int_a^b} f$ and $\lim_{\Delta P \to 0} \underline{R}(f,P) = \underline{\int_a^b} f$.

Proof From the bounded character of f follows the existence of $\underline{R}(f,P)$ and $\bar{R}(f,P)$ corresponding to any partition of the interval $[a,b]$. Furthermore, M exists such that $|f(x)| \le M$ for all x in $[a,b]$. To simplify the discussion, we introduce a function h by $h(x) = f(x) + M$. Then $0 \le h(x) \le 2M$ throughout $[a,b]$. For any partition P it follows at once that

$$\bar{R}(h,P) = \bar{R}(f,P) + M(b - a). \tag{16}$$

Let $\varepsilon > 0$ be given. Then a partition P' exists such that

$$\bar{R}(f,P') < \overline{\int_a^b} f + \varepsilon. \tag{17}$$

Let $\delta = \varepsilon/n'$, where n' is the number of segments in P', and denote by P any partition such that $\Delta P < \delta$. We distinguish two classes of segments of P; a segment is of type (1) if it contains a point of P' in its interior; otherwise it is of type (2). Observe that each segment of type (2) is contained in some segment of P'.

Consider the sum $\bar{R}(h,P) = \sum_{\nu=1}^{n} M_\nu \, \Delta x_\nu$. Denote by $\bar{A}(h,P)$ the sum of those terms of $\bar{R}(h,P)$ which arise from segments of type (1) and by $\bar{B}(h,P)$ the sum of those arising from type (2) segments. Thus $\bar{R}(h,P) = \bar{A}(h,P) + \bar{B}(h,P)$.

Each term of $\bar{A}(h,P)$ is of the form $M_\nu \, \Delta x_\nu$ with $M_\nu \le 2M$ and $\Delta x_\nu < \delta$. Since there are at most $n' - 1$ such terms, we have $\bar{A}(h,P) < (n' - 1)(2M)\delta < 2M\varepsilon$.

Next, consider the terms $\bar{B}(h,P)$. Each segment of type (2) is a segment of the partition $P \cup P'$. Hence, since h is non-negative, we conclude that $\bar{B}(h,P) \le \bar{R}(h,P \cup P')$. Because $P \cup P'$ is a refinement of P', we have $\bar{R}(h,P \cup P') \le \bar{R}(h,P')$. Therefore $\bar{B}(h,P) \le \bar{R}(h,P')$.

By combining these inequalities we obtain $\bar{R}(h,P) \le \bar{R}(h,P') + 2M\varepsilon$. Further combination with (16) and (17) leads to $\bar{R}(f,P) < \overline{\int_a^b} f + (2M + 1)\varepsilon$, and we conclude that

$$\lim_{\Delta P \to 0} \bar{R}(f,P) = \overline{\int_a^b} f.$$

The proof for lower sums may be carried through in a similar fashion or it may be obtained with less effort by applying the above result to $-f$ and observing that $\bar{R}(-f,P) = -\underline{R}(f,P)$ and $\bar{\int}_a^b -f = -\underline{\int}_a^b f$ ▲

Corollary If f is a bounded function on $[a,b]$, then $\bar{\int}_a^b f = \underline{\int}_a^b f$ if and only if $\lim_{\Delta P \to 0} [\bar{R}(f,P) - \underline{R}(f,P)] = 0$.

The proof that Definition (A) implies Definition (B) now follows quite readily.

Theorem 10-29 If f, defined on $[a,b]$, is Riemann integrable in the sense of Definition (A), then it is Riemann integrable in the sense of Definition (B); moreover, $S = \int_a^b f$.

Proof Let $\varepsilon > 0$ be given. Then, by the corollary to Theorem 10-28, a number $\delta > 0$ exists such that $\bar{R}(f,P) - \underline{R}(f,P) < \varepsilon$ for all partitions P with $\Delta P < \delta$. For such a partition P and any intermediate partition Q we have, as observed earlier, $\underline{R}(f,P) \leq R(f,P,Q) \leq \bar{R}(f,P)$. Hence

$$\left| R(f,P,Q) - \int_a^b f \right| \leq |R(f,P,Q) - \underline{R}(f,P)| + \left| \int_a^b f - \underline{R}(f,P) \right| < 2\varepsilon.$$

Thus it is proved that $\lim_{\Delta P \to 0} R(f,P,Q) = \int_a^b f$, so that f is Riemann integrable in the sense of Definition (B), and $S = \int_a^b f$ ▲

10-14 Notation and Summary

The argument favoring the symbol $\int_a^b f$ for the Riemann integral of a function over the traditional $\int_a^b f(x)\,dx$ is simple and obvious. It says what it has to say and nothing more. It makes explicit reference to the function f, whose Riemann integral it denotes, and to the interval $[a,b]$ on which f is defined. The symbol $\int$, an elongated letter S, is intended to convey the character of a Riemann integral as a limit of a sum, in the spirit of Definitions (B) and (C).

On the other hand, there is certain merit to be found in the conventional and well-known notation $\int_a^b f(x)\,dx$, provided that it is properly interpreted. Pedagogically, the role of the indefinite integral as an inverse of a derivative tends to take precedence over the notion of an integral as a limit of a sum. Accordingly, the symbol $\int_a^b f(x)\,dx$ is often construed to represent an operation to be performed on a function $f(x)$ to yield a new function $F(x)$ whose derivative is $f(x)$ or, expressed otherwise, whose differential is $f(x)\,dx$. The definite integral of f on $[a,b]$ is then obtained as the value of $F(b) - F(a)$.

We have seen, however, that a primitive of $f(x)$ does not always exist; in which case the above construction on the symbol $\int_a^b f(x)\,dx$ is without meaning. If it is desired to denote the Riemann integral of f on $[a,b]$ by $\int_a^b f(x)\,dx$, then an interpretation must be placed on that symbol which does not depend on the concepts of derivative or differential.

Although dx is traditionally referred to as the differential of x in the symbol $\int_a^b f(x)\,dx$, in reality it is not to be regarded as the number described by that name in Chapter 9. Rather, it is not a number at all but an inseparable part of the notation $\int_a^b \cdots dx$ in much the same sense that it contributes to the operational symbol d/dx employed to denote differentiation.

The advantages of the notation $\int_a^b f(x)\,dx$ are several, though perhaps not very striking. For one, suppose, as is the case in numerous applications, that a geometrical or physical interpretation is placed upon a Riemann integral. Then the symbol $\int_a^b f(x)\,dx$ tends to suggest the proper dimensionality of the quantity it represents. For example, the product $f(\xi_\nu)\Delta x_\nu$ occurring in the construction of Riemann sums in Definitions (B) and (C) is frequently regarded as the area of a rectangle of dimensions $f(\xi_\nu)$ and Δx_ν and the integral $\int_a^b f(x)\,dx$ defined to be the area of a certain region of the coordinate plane. The symbol $f(x)\,dx$, rather than f or $f(x)$ alone, because it "looks like" a product, tends to convey the two-dimensional character of the integral.

Another argument favoring the symbol $\int_a^b f(x)\,dx$ lies in the inevitable extension of the notions of the present chapter to functions of several variables. It becomes desirable in such a generalization to distinguish symbolically the so-called variables of integration. This is successfully achieved by the inclusion in integrands of the symbols dx, dy, and so forth.

Not the least argument in support of the notation $\int_a^b f(x)\,dx$ is the fact that the reader has grown accustomed to it and is perhaps dismayed to find in print such symbols as $\int_0^\pi \sin x$ or $\int_0^\pi x^2$ after the frequent and insistent admonitions in an elementary calculus course not to forget the dx in the integrand! Actually, he has seen the dx play a significant role in integration technique (where primitives exist). For example, consider the change of variable formula $\int_a^b f = \int_p^q (f \circ h)h'$. In the alternate notation it becomes

$$\int_a^b f(y)\,dy = \int_p^q f(h(x))h'(x)\,dx,$$

which lends itself more readily to formal manipulation.

A further and more poignant reason than any advanced so far for the employment of the symbol $\int_a^b f(x)\,dx$ in place of, or in addition to, the notation $\int_a^b f$ is one which cannot be stated at this point but will become clear in later chapters after certain more advanced types of integrals have been studied.

In any case, whichever notation the reader may prefer, it will not prove a burden to persist in the employment of the sophisticated and formally correct notation $\int_a^b f$ and, as well, when convenient to do so, to invoke the traditional $\int_a^b f(x)\,dx$.

Exercises

10-15 Prove that a function f is Riemann integrable on $[a,b]$ if and only if the set of Riemann sums is *Cauchy convergent*; that is, corresponding to each $\varepsilon > 0$, there exists a number $\delta > 0$ such that $|R(f,P',Q') - R(f,P'',Q'')| < \varepsilon$ for all partitions P' and P'' for which $\Delta P' < \delta$ and $\Delta P'' < \delta$ and for all intermediate partitions Q' of P' and Q'' of P''.

10-16 Prove that if f and g are Riemann integrable on $[a,b]$, then so are the functions $f \wedge g$ and $f \vee g$.

10-17 (a) Determine the value of $\lim \left\{\frac{1}{n+1} + \frac{1}{n+2} + \cdots + \frac{1}{2n}\right\}$ from the definition of $\int_0^1 \frac{dx}{1+x}$. (b) Evaluate $\lim \left\{n\left(\frac{1}{n^2+1^2} + \frac{1}{n^2+2^2} + \cdots + \frac{1}{2n^2}\right)\right\}$ using the definition of an integral.

If f is Riemann integrable on $[a,b]$ for all $b > a$, we define the *improper Riemann integral of the first kind* by $\int_a^\infty f = \lim_{b\to\infty} \int_a^b f$, provided the limit exists.

10-18 Prove that the improper integral $\int_1^\infty x^{-p}\,dx$ exists with value $1/(p-1)$ if $p > 1$ but that the integral does not exist if $p \leq 1$.

10-19 Prove that if f is continuous and if $\int_a^\infty |f|$ exists, then so does $\int_a^\infty f$.

10-20 Show that if f is defined on each interval $[n, n+1)$ by $f(x) = (-1)^n/n$ for $n = 1, 2, \ldots$, then the integral $\int_1^\infty f$ exists but $\int_1^\infty |f|$ does not.

10-21 Cite an example to show that $\int_a^\infty f$ may exist even if f is not bounded on $\{x;\ \ x \geq a\}$.

10-22 Let f be non-negative and decreasing on $\{x;\ \ x \geq 1\}$. Prove that the improper Riemann integral $\int_1^\infty f$ exists if and only if the series $\sum_1^\infty f(n)$ converges. (This may be viewed as a test for convergence of a series or as a test for the existence of an improper integral.)

10-23 Suppose f to be bounded on $\{x;\ \ x \geq a\}$, and let g be non-negative on $\{x;\ \ x \geq a\}$ and such that $\int_a^\infty g$ exists. Prove that a number η exists such that $\inf f \leq \eta \leq \sup f$ and $\int_a^\infty fg = \eta \int_a^\infty g$.

If f is Riemann integrable on $[c,b]$ for all c such that $a < c < b$, then we define the *improper Riemann integral of the second kind* on $[a,b]$ to be the number $\lim_{c\to a} \int_c^b f$, provided the limit exists.

10-24 Show that if $\lim_{c\to a} \int_c^b f$ exists but that $\int_a^b f$ does not, then f is necessarily unbounded on $[a,b]$.

10-25 Prove that if f is Riemann integrable on $[a,b]$, then the improper Riemann integral of the second kind exists and is equal to the Riemann integral of f on $[a,b]$.

10-26 Prove that $\int_0^1 x^{-p}\,dx$ exists and equals $1/(1-p)$ if $p < 1$ but that the integral does not exist if $p \geq 1$.

11

Sequences of Functions

For men may come and men may go,
But I go on forever.

ALFRED TENNYSON (1808–1892), *The Brook*

11-1 Sequences of Functions

Having considered in some detail in Chapter 5 the concept of a sequence of constants, we now generalize to the notion of a *sequence of functions*. By this we understand a function defined on a set of integers n such that $n \geq n_0$, for some integer n_0, and whose range is a set of real valued functions defined on a common domain D of real numbers. We denote a sequence of functions by $\{f_n\}$ or $f_{n_0}, f_{n_0+1}, \ldots, f_n, \ldots$ and refer to f_n as the nth *element* of the sequence. We remark that, corresponding to each point x of the domain D, such a sequence gives rise to a sequence of constants $\{f_n(x)\}$ or $f_{n_0}(x), f_{n_0+1}(x), \ldots, f_n(x), \ldots$. If convenient, the latter notations are also employed to denote the sequence of functions itself. The integer n_0 will almost always be 0 or 1.

If $\{f_n(x)\}$ is a bounded sequence for each x in D, we describe $\{f_n\}$ as a *bounded sequence*. If, moreover, it is the case that a number M exists such that $|f_n(x)| \leq M$ for all $n \geq n_0$ and all x in D, then the sequence is said to be *uniformly bounded*. The two notions are not the same, of course; a sequence may be bounded yet not uniformly so.

Example 11-1

(a) The sequence $\{f_n\}$ defined on $(0,1]$ by $f_n(x) = 1/(nx)$ for $n = 1, 2, \ldots$ is a bounded sequence but one which is not uniformly bounded.

(b) Since $|\sin nx| \leq 1$ for all $x \in \mathscr{R}$ and $n \in I_+$, the sequence defined by $\{\sin nx\}$ is uniformly bounded on $\mathscr{R}$.

A function sequence $\{f_n\}$ is said to be *convergent to a limit function f* or to *converge to* or *approach f* on a subset D_1 of the common domain of the elements f_n if, corresponding to any prescribed $\varepsilon > 0$ and any given x in D_1, there exists an integer N with the property $|f_n(x) - f(x)| < \varepsilon$ for all $n > N$. We employ the notation $f = \lim \{f_n\}$.

It is important to note that N corresponds, not to the value of ε alone, but to the number pair consisting of ε and x. Thus, generally, N depends for its value upon x as well as upon ε.

Alternately, we may express the notion of convergence of a sequence of functions in terms of the earlier concept of convergence of a sequence of constants: A sequence $\{f_n\}$ converges on a domain D_1 if the sequence of constants $\{f_n(x)\}$ converges for each x in D_1. If $f(x)$ denotes the limit of $\{f_n(x)\}$ for each x, then the values of $f(x)$ define the limit function f of $\{f_n\}$.

Example 11-2 The sequence $\{g_n\}$, defined by $g_n(x) = 1 - 1/nx$ for all x in $(0,1]$ and $n \in I_+$, is sketched in Figure 11-1. Each g_n is unbounded, but for each fixed x, such as x_1 in the figure, the sequence $\{1 - 1/nx\}$ converges to 1. Hence the limit of $\{g_n\}$ is the unit function on $(0,1]$.

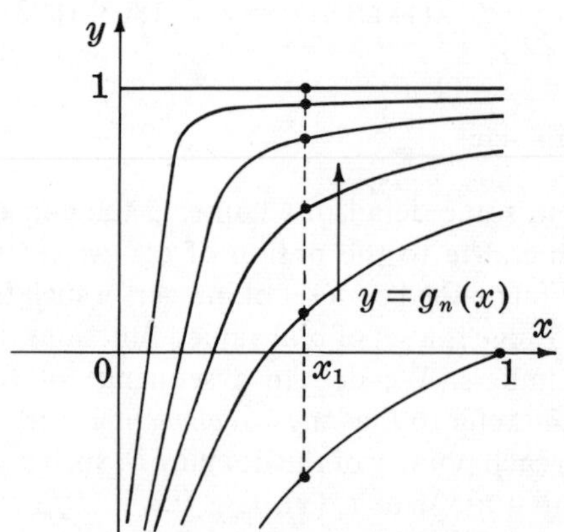

Figure 11-1

Example 11-3 Consider the sequence $\{f_n\}$ defined by $f_n(x) = x^n/(1 + x^{2n})$ for all x in $\mathscr{R}$ and $n = 1, 2, \ldots$. Its domain of convergence comprises all real numbers except -1 since $\{\frac{1}{2}(-1)^n\}$ clearly diverges. Thus the limit function f is given by $f(x) = 0$ for $x \neq \pm 1$, $f(1) = \frac{1}{2}$, and f is undefined at -1.

The previous example makes it plain that a function sequence may converge at some points of the domain of its elements and diverge at others. That set of values of x for which the sequence $\{f_n\}$ converges we shall refer to as the *domain of convergence* of $\{f_n\}$.

Example 11-4

(a) Consider the sequence $\{\cos^{2n} k!\,\pi x\}$, where k is some positive integer and each element is defined on the interval $[0,1]$. If x is a rational point p/q of $[0,1]$ with the property that q is a factor of $k!$, then each element of the sequence has value 1 and the sequence converges to 1. For any other point x of the interval we have $0 \leq \cos k!\,\pi x < 1$, whence, for such x, the sequence converges to 0. It follows that the limit function $g_k(x) = \lim_{n\to\infty} \{\cos^{2n} k!\,\pi x\}$ of the given sequence has the value 0 at each point of $[0,1]$ except for a certain finite set of rational points, determined by the integer k, on which it has the value 1.

(b) Of interest, too, is the sequence $\{g_k\}$ whose elements are the discontinuous limit functions above. If the set of discontinuity points of g_k is denoted by R_k, then clearly $R_k \subset R_{k+1}$ for all k. Further each rational point of $[0,1]$ is a member of R_k for some k. It follows that the sequence $\{g_k\}$ converges on $[0,1]$ to the limit function $1 - \Psi$, having value 1 at each rational point and 0 at each irrational point, a function which we have seen to be discontinuous at each point of the interval $[0,1]$.

It is quite possible for a sequence of discontinuous functions to possess a continuous limit. This is demonstrated in the following example.

Example 11-5 Let f_n for $n = 1, 2, \ldots$ be defined on $[0,1]$ by $f_n(x) = 1$ if $x = p/q$ with $p + q = n$, and $f_n(x) = 0$ otherwise. Observe that as n increases, the number of discontinuity points of f_n increases without bound. Nevertheless, the sequence $\{f_n\}$ converges to the zero function on $[0,1]$.

Since a function sequence may be regarded as a collection of sequences of constants, it is to be expected that much of what has been shown to be the case for the latter is true for the former, either by implication or analogy. We shall not attempt to reiterate for sequences of functions the many truths of Chapter 5 which can be shown to apply. We content ourselves with the following observations concerning combinations of sequences, plus comments necessary to the further development of the present chapter.

Suppose $\{f_n\}$ and $\{g_n\}$ to denote function sequences which converge, respectively, to limit functions f and g, defined on a common domain. Then it is the case that the sequence $\{f_n + g_n\}$, styled the *sum* of $\{f_n\}$ and $\{g_n\}$, converges to $f + g$, the *product* sequence $\{f_n g_n\}$ converges to fg, and $\{|f_n|\}$ has limit $|f|$. Furthermore, if $g_n(x) \neq 0$ and $g(x) \neq 0$ for all n and x, then the *quotient* sequence $\{f_n/g_n\}$ converges to f/g.

Exercises

For each point x in the domain of a sequence $\{f_n\}$, the sequence $\{f_n(x)\}$ has a lower limiting point $\underline{\lim}\, \{f_n(x)\}$, and an upper limiting point $\overline{\lim}\, \{f_n(x)\}$ in $\mathscr{R}^*$. (See Sections 5-10 and 5-11.) We define the *lower limit* $\underline{\lim}\, \{f_n\}$ of $\{f_n\}$ to be the function defined by $\underline{\lim}\, \{f_n(x)\}$ on D; similarly for the *upper limit* $\overline{\lim}\, \{f_n\}$.

11-1 Prove that $\underline{\lim}\,\{f_n\} \le \overline{\lim}\,\{f_n\}$.

11-2 Prove that $-\underline{\lim}\,\{f_n\} = \overline{\lim}\,\{-f_n\}$.

11-3 Prove that (a) $\underline{\lim}\,\{f_n\} + \underline{\lim}\,\{g_n\} \le \underline{\lim}\,\{f_n + g_n\}$, and (b) $\overline{\lim}\,\{f_n + g_n\} \le \overline{\lim}\,\{f_n\} + \overline{\lim}\,\{g_n\}$.

11-4 Prove that the real valued function $\lim\,\{f_n\}$ exists if and only if $\underline{\lim}\,\{f_n\}$ and $\overline{\lim}\,\{f_n\}$ are real valued and equal.

11-2 Uniform Convergence

Concerning sequences of functions, certain questions arise which do not come up in the event of a sequence of constants, questions which concern comparative properties of the functions f_n and f. We have seen in the examples of Section 11-1 instances of sequences of continuous functions which have continuous limits and others with discontinuous limits. Also, sequences of discontinuous functions in some cases converge to continuous limits and in others to discontinuous ones. We have had unbounded sequences with bounded limits and could easily devise bounded sequences with unbounded limits. It is natural, then, to query under what conditions a property common to all elements of a function sequence is transmitted to the limit function. If each element of a sequence is differentiable, what further must be the case if the limit is to have a derivative? Is it possible for a sequence of Riemann integrable functions to have an integrable limit? Etc.

The answers to these questions and others require analysis of the concept of convergence and, in particular, the investigation of a certain modification of that concept, called *uniform convergence*.

In the definition of convergence, as has been pointed out, the choice of N follows from prescribed values of ε and x. If it should be the case in establishing the convergence of a sequence that a value of N can be found which, while dependent upon ε, is independent of x, then we say that the convergence is *uniform* in character. (Observe a likeness to the notions of uniform continuity and uniform boundedness.) A formal statement of the concept follows.

A function sequence $\{f_n\}$ is said to be *uniformly convergent* or to *converge uniformly* to a limit f on a domain D_1 if, corresponding to each $\varepsilon > 0$, there exists an integer N such that $|f_n(x) - f(x)| < \varepsilon$ for all $n > N$ and all x in D_1.

Here the emphasis is on "one N for all x," whereas for ordinary convergence it is "one N for each x." It is immediate, of course, that if a sequence is uniformly convergent, then it is convergent in the simple sense as well. However, sequences exist which, although convergent, are not uniformly so. The following examples will help to bring out the difference between the two concepts.

Example 11-6 Consider the sequence $\{f_n\}$ defined by $f_n(x) = x/n$ for all x in some domain D. Its limit function f is given by $f(x) = 0$ on D.

For the convergence of $\{f_n\}$ to be uniform, it is necessary that, given $\varepsilon > 0$, an integer N exist such that $|x/n| < \varepsilon$ for all $n > N$ and all x in D. This is a condition which cannot be met if D is an unbounded set, say the set $\mathscr{R}$. For, with ε and n given, however large n may be, it is clear that on an unbounded domain a number x exists such that $|x| > n\varepsilon$. Hence, the sequence $\{f_n\}$, although convergent on $\mathscr{R}$, is not uniformly convergent on that domain.

However, if D is a bounded domain, such as the interval $[a,b]$ of Figure 11-2, then a number M exists with the property $|x| \leq M$ for all x in D, and by the choice $N = M/\varepsilon$ we may satisfy the inequality $|x/n| < \varepsilon$ for all $n > N$ and all x in D. It follows that $\{f_n\}$ is uniformly convergent on $[a,b]$ and, indeed, on any bounded domain. It may be noted, too, that the sequence is uniformly bounded on such a domain.

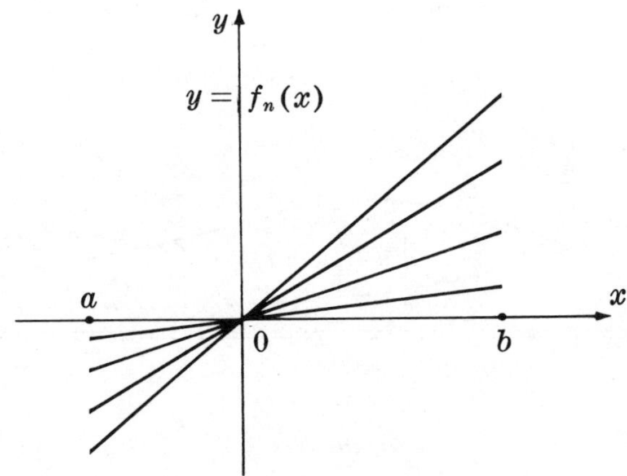

Figure 11-2

Example 11-7 The sequence $\{f_n\}$ depicted in Figure 11-3 and defined by $f_n(x) = 1/nx$ for all x in $(0,1]$ and $n \in I_+$ is a sequence of unbounded functions which for each fixed x converges to 0. Hence $\{f_n\}$ converges to the zero function on $(0,1]$. However, the convergence is not uniform since it is not possible, for all x in $(0,1]$, to make $1/x < n\varepsilon$, however large a value may be chosen for n.

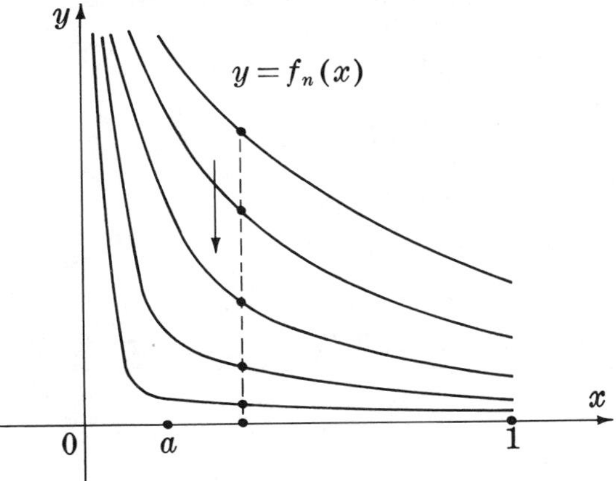

Figure 11-3

If the domain of definition of the elements f_n is revised to $(a,1]$, where $0 < a < 1$, then the modified sequence is both uniformly bounded (by $f_1(a) = 1/a$) and uniformly convergent. To establish the latter, we have only to choose for N (corresponding to a given $\varepsilon > 0$) any integer exceeding $1/\varepsilon$ since $1/nx < \varepsilon/a$ for $n > N$ and all x in $(a,1]$.

Example 11-8 The sequence $\{f_n\}$ defined by $f_n(x) = 2x/(1 + n^2x^2)$ for all $x \geq 0$, and $n = 1, 2, \ldots$ converges to the zero function on its domain. Figure 11-4 demonstrates how, for each fixed x, the sequence of ordinates $\{2x/(1 + n^2x^2)\}$ converges to 0. Moreover, it reveals the uniformly bounded character of the sequence $\{f_n\}$ since for each n the maximum point of the corresponding element is seen to be the point $(1/n,1/n)$; and, accordingly, all points on all curves lie in the range $[0,1]$.

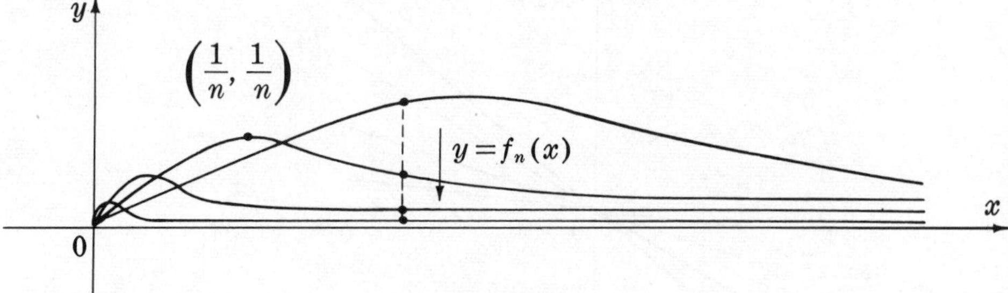

Figure 11-4

We observe that with increasing n, the maximum point or "hump" moves to the left and downward along a path that terminates at the origin. The impact of this is that the sequence is uniformly convergent. To see this we note that $2x/(1 + n^2x^2) \leq 1/n$ for all n and x. As a consequence, corresponding to each $\varepsilon > 0$, if we choose N to be any integer exceeding $1/\varepsilon$, then for $n > N$ we have $2x/(1 + n^2x^2) < \varepsilon$.

Example 11-9 Varying slightly in definition from that of Example 11-8 is the sequence $\{g_n\}$ given by $g_n(x) = 2nx/(1 + n^2x^2)$ for $x \geq 0$ and $n = 1, 2, \ldots$. Like that sequence it is uniformly bounded. However, from Figure 11-5, we see this difference. By in-

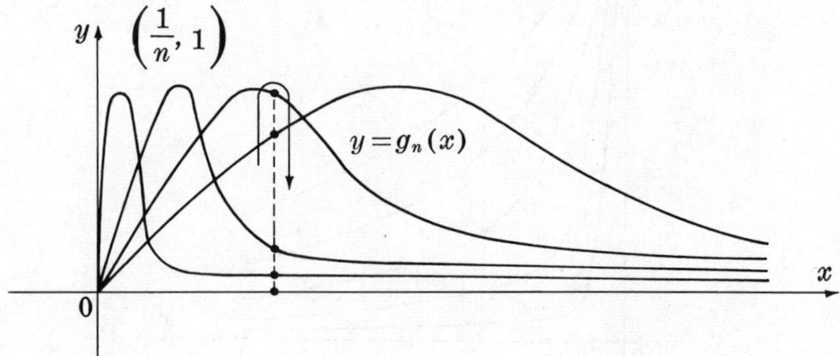

Figure 11-5

creasing n, the hump of f_n does not tend to move downward or disappear. Instead, it persists indefinitely, pressing into a smaller and smaller region against the vertical axis. It should be noted that this does not prevent the sequence from converging to 0 for each x.

It does, however, make for nonuniform convergence since, no matter how small $x > 0$ may be chosen, all elements of the sequence for which $1/n < x$ are such that the hump lies in the interval $(0,x)$. For all such there is a value of x for which $2nx/(1 + n^2x^2) = 1$. Consequently, $2nx/(1 + n^2x^2)$ cannot be made arbitrarily small by a choice of N independent of x. In contrast to Example 11-8, the present sequence is uniformly bounded but not uniformly convergent.

Example 11-10 Contrast with Examples 11-8 and 11-9 the sequence $\{h_n\}$ defined by $h_n(x) = 2n^2x/(1 + n^2x^2)$ for $x \geq 0$ and $n = 1, 2, \ldots$. Here it is the case (see Figure 11-6) that the maximum point of h_n for any n is $(1/n,n)$. Thus the elements of the sequence are bounded functions but the sequence is not uniformly bounded. Although uniform convergence is out of the question for similar but stronger reasons than in Example 11-9, what is surprising perhaps is that in spite of the fact that we may trace a divergent sequence of ordinates, the sequence $\{h_n(x)\}$ converges to 0 for each $x \geq 0$. Thus it is illustrated that, as in Example 11-7, a nonuniformly bounded sequence may converge.

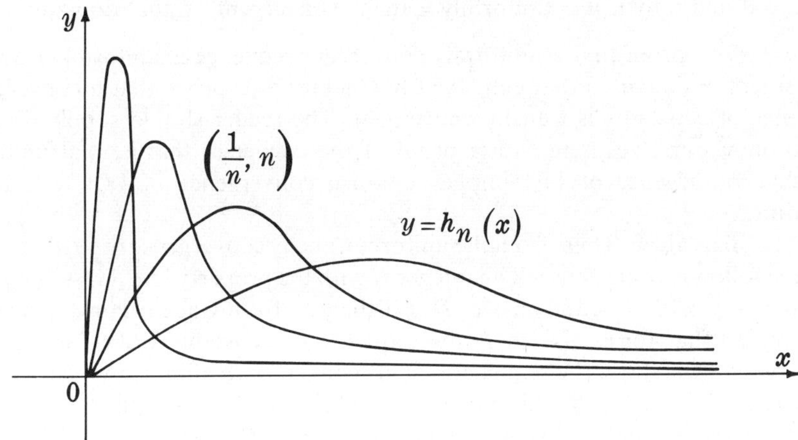

Figure 11-6

The concept of simple convergence, defined to have significance over some domain of convergence, is meaningful at each point of that domain, in the sense that it reduces to a convergent sequence of constants. Uniform convergence, on the other hand, describes a significant property of a function sequence only when considered over a domain comprising more than one point. It tends to compare the nature or "rate" of convergence at the various points of the domain of convergence; asserting that *all* elements of the sequence beyond the Nth differ at *all* points of D from a limit function f by less than a prescribed ε.

11-3 Cauchy Convergence

A sequence of functions $\{f_n\}$ is said to be *Cauchy convergent* on a domain D if, corresponding to each $\varepsilon > 0$ and to each x in D, there exists an integer N with the property that $|f_{n+k}(x) - f_n(x)| < \varepsilon$ for all $n > N$ and $k > 0$.

Essentially this definition iterates for each point x of the domain of $\{f_n\}$ the definition of Cauchy convergence given in Chapter 5 for a sequence of constants. As a consequence, we may conclude that a sequence of functions $\{f_n\}$ converges to a limit function f on a domain D if and only if it is Cauchy convergent at each point of D.

If Cauchy convergence is to be related effectively to the concept of uniform convergence, it, too, must have a uniform character; hence the following definition.

A function sequence $\{f_n\}$ is *uniformly Cauchy convergent* on a domain D if, corresponding to each $\varepsilon > 0$, there is an integer N such that $|f_{n+k}(x) - f_n(x)| < \varepsilon$ for all $n > N$, $k > 0$, and all x in D.

Theorem 11-1 A sequence of functions $\{f_n\}$ is uniformly convergent on a domain if and only if it is uniformly Cauchy convergent on that domain.

Proof The proof that a uniformly convergent sequence is uniformly Cauchy convergent is similar to that employed in Chapter 5 to prove that a convergent sequence of constants is Cauchy convergent. The reader should supply details. We content ourselves here with a proof of the converse; that is, that uniform Cauchy convergence of $\{f_n\}$ implies uniform convergence of $\{f_n\}$ to a limit function f.

Let $\varepsilon > 0$ be given. Then, from the uniform Cauchy convergence of $\{f_n\}$ on a domain D follows the existence of an integer N with the property $|f_{n+k}(x) - f_n(x)| < \varepsilon$ for all $n > N$, all $k > 0$, and all x in D. Further, uniform Cauchy convergence of $\{f_n\}$ implies that for each x in D the sequence of constants $\{f_n(x)\}$ is Cauchy convergent. Therefore, the sequence $\{f_n(x)\}$ converges to a limit $f(x)$. Hence, for any integer n exceeding N and some $k > 0$ we have $|f_{n+k}(x) - f(x)| < \varepsilon$. Combining the above inequalities we conclude that $|f_n(x) - f(x)| < 2\varepsilon$ for $n > N$ (independent of x) and each x in D▲

The equivalence of the concepts of uniform convergence and uniform Cauchy convergence may be employed to prove the next theorem, which states a conclusion that has possibly suggested itself to the reader through the succession of examples of the preceding section.

Theorem 11-2 If $\{f_n\}$ is a uniformly convergent sequence of bounded functions on a domain D, then it is uniformly bounded on D.

Proof From the implied uniform Cauchy convergence of $\{f_n\}$ follows the existence of an integer N such that $|f_{N+1+k}(x)| < 1 + |f_{N+1}(x)|$ for all $k > 0$ and all x in D. Further, because $\{f_n\}$ is bounded, integers M_n, $n = 1, 2, \ldots$, exist such that $|f_n(x)| \leq M_n$ for all x in D. Consequently $|f_{N+1+k}(x)| < 1 + M_{N+1}$ for all $k > 0$ and all x in D. Hence if M denotes the maximum of the set

$$\{M_1, M_2, \ldots, M_N, 1 + M_{N+1}\},$$

then $|f_n(x)| \leq M$ for all $n \in I_+$ and all x in D▲

Example 11-11 Uniform convergence alone, however, is not sufficient to the conclusion of uniform boundedness of a sequence. In order to observe this, we may consider the sequence $\{f_n\}$ defined by $f_n(x) = (1/x) + (x/n)$ for all x in $(0,1]$ and $n \in I_+$. It is clear that for each n the function f_n is unbounded on $(0,1]$. Further, the sequence converges to the limit function f, given by $f(x) = 1/x$. That the convergence is uniform is seen from Example 11-6.

Exercises

11-5 Prove that if a sequence is uniformly convergent on a domain, then it is uniformly convergent on each subset of that domain.

11-6 Prove that if a sequence is uniformly convergent on each of two subsets of its domain, then it is uniformly convergent on the union of those subsets.

11-7 Give a direct statement of the negation of uniform convergence.

11-8 Investigate the following sequences for possible uniform convergence on $[0,1]$: (a) $\{x/(1 + nx)\}$; (b) $\{nxe^{-nx}\}$; (c) $\{nx(1-x)^n\}$; (d) $\{n^2x/(1 + n^3x^2\}$.

11-9 (a) Prove that the sequence $\{xe^{-nx}\}$ is uniformly convergent on $\{x;\ x \geq 0\}$. (b) Prove that $\{nxe^{-nx}\}$ converges uniformly on $\{x;\ x \geq a\}$ for any $a > 0$ but not on $\{x;\ x > 0\}$.

11-10 Cite an example of a sequence of functions on $[0,1]$ that is uniformly convergent on $(0,1)$ but not convergent on $[0,1]$.

11-11 Cite an example to show that a sequence may be convergent on a closed domain, yet not be uniformly convergent there.

11-12 Illustrate by an example that a sequence may be uniformly convergent on a set and be convergent on the closure of the set without being uniformly convergent on the closure.

11-13 Prove that a uniformly convergent sequence of functions is uniformly Cauchy convergent.

11-14 Prove that if a sequence $\{f_n\}$ on a domain D converges uniformly to a bounded limit function, then N and M exist such that $|f_n(x)| < M$ for all $n > N$ and all x in D. Give an example to show why we cannot say "all n" instead of "all $n > N$."

11-15 Prove that if sequences $\{f_n\}$ and $\{g_n\}$ converge uniformly to f and g, respectively, then $\{f_n + g_n\}$ converges uniformly to $f + g$.

11-16 If $f_n(x) = (1/x) + (1/n)$ and $g_n(x) = 1/n$, both on $(0,1)$, prove that $\{f_n\}$ and $\{g_n\}$ are uniformly convergent while $\{f_n g_n\}$ is not.

11-17 Prove that if $\{f_n\}$ and $\{g_n\}$ are uniformly convergent sequences of bounded functions with limits f and g, respectively, then the sequence $\{f_n g_n\}$ is uniformly convergent with limit fg.

11-18 Suppose $\{f_n\}$ to be a uniformly bounded sequence converging uniformly to a limit f on a domain D, and let g be continuous on a domain such that $h = g \circ f$ and $h_n = g \circ f_n$ are defined. Prove that $\{h_n\}$ converges uniformly to h on D.

11-19 Prove that if $\{f_n\}$ is a monotone sequence possessing a uniformly convergent subsequence, then $\{f_n\}$ is uniformly convergent.

11-4 Sequences of Continuous Functions

We are now in a position to answer some questions proposed earlier concerning the transmission to the limit of a function sequence of properties common to the elements of the sequence. Briefly, the answer lies in the concept of uniform convergence. We shall see in the theorems of the next several sections how uniform convergence tends to perpetuate in the limit function the properties of existence of a limit at a point, continuity, differentiability, and Riemann integrability of the elements.

Theorem 11-3 Let $\{f_n\}$ be a uniformly convergent sequence of functions with limit f, and let c denote a cluster point of the common domain D of the f_n. Suppose, too, that for each index n the function f_n has a limit L_n at c. Then the sequence of constants $\{L_n\}$ converges to a limit L, and the function f has limit L at c.

Proof Let $\varepsilon > 0$ be give. Then from the uniform convergence of $\{f_n\}$ follows the existence of N such that $|f_{n+k}(x) - f_n(x)| < \varepsilon$ for all $n > N$, all $k > 0$, and all x in D. Further, since for each such n and k the functions f_n and f_{n+k} have limits L_n and L_{n+k}, respectively, at c, then follows the existence of a number $\delta_{n,k}$ such that $|f_n(x) - L_n| < \varepsilon$ and $|f_{n+k}(x) - L_{n+k}| < \varepsilon$ for all $x \in D \cap \mathcal{N}^*(c,\delta_{n,k})$. Therefore, for any one such x,

$$|L_{n+k} - L_n| \leq |L_{n+k} - f_{n+k}(x)| + |f_{n+k}(x) - f_n(x)| + |f_n(x) - L_n| < 3\varepsilon$$

for $n > N$ and $k > 0$. This testifies to the convergence of $\{L_n\}$ to a limit L.

To demonstrate that f has limit L at c, observe that from the convergence of $\{L_n\}$ to L and from the uniform convergence of $\{f_n\}$ to f follows the existence of an integer N' such that $|L_n - L| < \varepsilon$ and $|f_n(x) - f(x)| < \varepsilon$ for all $n > N'$ and all x in D. Letting n denote an integer which exceeds both N and N', it follows that

$$|f(x) - L| \leq |f(x) - f_n(x)| + |f_n(x) - L_n| + |L_n - L| < 3\varepsilon$$

for all $x \in D \cap \mathcal{N}^*(c,\delta_{n,k})$▲

The import of Theorem 11-3 is essentially this. If a function sequence is uniformly convergent, then the processes of determining a limit at a point and of finding the limit of the sequence may be performed in either order. Thus

$$\lim_c \lim \{f_n\} = \lim \{\lim_c f_n\}.$$

The condition of uniform convergence, although sufficient to the conclusion of the theorem, is not a necessary one. Thus, the formula above may be a valid statement for some nonuniformly convergent sequences.

That the limit of a uniformly convergent sequence of continuous functions is itself continuous follows in part from Theorem 11-3. Suppose $\{f_n\}$ to be uniformly convergent to a limit f and such that the functions f_n, for all n, are continuous at one and the same point c. If c is a cluster point of the domain of convergence D of $\{f_n\}$, then it is immediate from Theorem 11-3 that the limit function is continuous at c. We have only to observe that by the theorem the sequence $\{f_n(c)\}$ converges to the limit of the limit function f at c.

On the other hand, if c is an isolated point of D, then f, simply by virtue of having a value at c, is continuous there. Formally, we state the conclusion as follows.

Theorem 11-4 If $\{f_n\}$ is a uniformly convergent sequence of functions each continuous at a point c of their common domain D, then the limit function f is continuous at c.

It is not difficult to show that, as in the case of limit at a point, the condition of uniform convergence is not necessary to the continuity of the limit function. Nor, indeed, is the hypothesis of continuity of the elements of $\{f_n\}$, whether the sequence be uniformly convergent or not.

If a sequence of polynomials $\{f_n\}$ is uniformly convergent to a limit function f on an interval $[a,b]$, then it follows from Theorem 11-4 that f is continuous. In the theorem that follows, which is due to Weierstrass, the incisive role of polynomials in the study of continuous functions is revealed. It is demonstrated that if f is any continuous function on a closed interval, then a sequence of polynomials exists which converges uniformly to f. Expressed otherwise, any continuous function on a closed interval can be approximated, uniformly, and to any desired degree of accuracy, by a polynomial.

Theorem 11-5 (The Weierstrass Approximation Theorem) If f is a continuous function on a closed interval, then a sequence of polynomials $\{f_n\}$ exists which converges uniformly to f.

Proof Without loss of generality we restrict attention to a function f defined and continuous on the interval [0,1] and consider the sequence of so-called *Bernstein polynomials* $\{f_n\}$ defined on that interval by

$$f_n(x) = \sum_{\nu=0}^{n} \frac{n!}{\nu!(n-\nu)!} f\left(\frac{\nu}{n}\right) x^\nu (1-x)^{n-\nu}$$

for $n = 0, 1, 2, \ldots$. We propose to show that $\{f_n\}$ converges uniformly to the given function f on [0,1].

First, in the interest of simplicity, we introduce the notation

$$p_\nu(x) = \frac{n!}{\nu!(n-\nu)!} x^\nu (1-x)^{n-\nu}$$

and invite the reader to carry out the calculations which establish that

(1) $$\sum_{\nu=0}^{n} p_\nu(x) = 1 \quad \text{and} \quad \sum_{\nu=0}^{n} (\nu - nx)^2 p_\nu(x) = nx(1-x).$$

Further, it should be observed that, for all x in $[0,1]$,

(2) $$x(1-x) = \frac{1}{4} - \left(x - \frac{1}{2}\right)^2 \le \frac{1}{4}.$$

Let $\varepsilon > 0$ be given. Then from the uniform continuity of f follows the existence of $\delta > 0$ with the property that $|f(x_1) - f(x_2)| < \varepsilon$ for all x_1 and x_2 in $[0,1]$ such that $|x_1 - x_2| < \delta$. Further, M exists such that $|f(x)| \le M$ for all x in $[0,1]$.

From the above it follows that, for each x in $[0,1]$,

$$|f_n(x) - f(x)| = \left| \sum_{\nu=0}^{n} f\left(\frac{\nu}{n}\right) p_\nu(x) - f(x) \right|$$

$$= \left| \sum_{\nu=0}^{n} \left[f\left(\frac{\nu}{n}\right) - f(x) \right] p_\nu(x) \right|.$$

Conveniently, we separate the terms of the latter sum according as $|\nu/n - x|$ is less than or not less than δ. Thus, we have

$$|f_n(x) - f(x)| \le \sum_{|\nu/n - x| < \delta} \left| f\left(\frac{\nu}{n}\right) - f(x) \right| p_\nu(x) + \sum_{|\nu/n - x| \ge \delta} \left| f\left(\frac{\nu}{n}\right) - f(x) \right| p_\nu(x),$$

and it follows that

$$|f_n(x) - f(x)| < \varepsilon \sum_{|\nu/n - x| < \delta} p_\nu(x) + 2M \sum_{|\nu/n - x| \ge \delta} p_\nu(x).$$

We observe that

$$\sum_{|\nu/n - x| < \delta} p_\nu(x) \le \sum_{\nu=0}^{n} p_\nu(x) = 1$$

and

$$\sum_{|\nu/n - x| \ge \delta} p_\nu(x) \le \sum_{\nu=0}^{n} \left(\frac{\nu - nx}{n\delta}\right)^2 p_\nu(x)$$

since $|\nu/n - x| \ge \delta$ implies $[(\nu - nx)/n\delta]^2 \ge 1$. Therefore

$$|f_n(x) - f(x)| < \varepsilon + \frac{2M}{n^2\delta^2} \sum_{\nu=0}^{n} (\nu - nx)^2 p_\nu(x)$$

so that, by (1) and (2), $|f_n(x) - f(x)| < \varepsilon + M/2n\delta^2$. This is true for any x in $[0,1]$. Hence for any x and for $n > M/2\varepsilon\delta^2$ we have $|f_n(x) - f(x)| < 2\varepsilon$▲

Exercises

11-20 Cite an example of a nonuniformly convergent sequence of continuous functions having a continuous limit.

11-21 Show by an example that a sequence of discontinuous functions may converge uniformly to a continuous limit.

11-22 Cite an example of a sequence of discontinuous functions that converges nonuniformly to a continuous limit.

11-23 Determine the limit of the sequence $\{f_n\}$, where $f_n(x) = 1$ if $x < 0$, $f_n(x) = 1 - nx$ if $0 \leq x < 1/n$, and $f_n(x) = 0$ if $1/n \leq x$. Prove that the convergence is not uniform.

11-24 Let $\{f_n\}$ be a sequence of continuous functions converging to a limit f on a domain D, and suppose $\{x_n\}$ to denote a sequence of points of D converging to a point c of D. Prove that the convergence of $\{f_n\}$ is nonuniform if the sequence of constants $\{f_n(x_n)\}$ does not converge to $f(c)$.

11-25 Consider again sequences $\{f_n\}$ and $\{x_n\}$ as defined in Exercise 11-24. Prove that if $\{f_n(x_n)\}$ converges to a limit other than $f(c)$, then no subsequence of $\{f_n\}$ exists that is uniformly convergent on D. Illustrate by an example that the conclusion need not follow if we merely assume that $\{f_n(x_n)\}$ does not converge to $f(c)$.

11-26 Use the results of Exercise 11-25 to prove that the sequence $\{f_n\}$ defined on $\{x;\ x \geq 0\}$ by $f_n(x) = 2nx/(1 + n^2x^2)$ is not uniformly convergent on its domain.

11-27 Prove that if a sequence of continuous functions on $[a,b]$ is uniformly convergent on (a,b), then it is uniformly convergent on $[a,b]$.

11-28 Let $\{f_n\}$ be a sequence of continuous functions on a compact domain D with the property that the sequence $\{f_n(x)\}$ is monotone for each x in D. Prove that the sequence is uniformly convergent on D if it converges to a continuous limit.

11-29 Use the sequence $\{f_n\}$ defined for $x > 0$ by $f_n(x) = n^2x$ if $0 < x < 1/n$ and $f_n(x) = 1/x$ if $1/n \leq x$ to demonstrate that the assumption of compactness in Exercise 11-28 is essential to the conclusion.

11-30 Use the Weierstrass approximation theorem to prove that if f is continuous on $[a,b]$ and such that $\int_a^b x^n f(x)\, dx = 0$ for all non-negative integers n, then $\int_a^b f^2(x)\, dx = 0$ and, as a consequence, f is the zero function on $[a,b]$.

11-5 Operations Performed on Sequences

The performance of an operation, such as differentiation or integration, on the elements of a sequence, which tends to transform it into a new and different sequence, may be such that the resultant sequence retains or fails to retain the convergence character of the original sequence. We are principally interested in operations that under suitable conditions preserve such convergence properties and that relate the limits of the original and transformed sequences in a reasonable and significant way. Indeed, we formally define an operation to be applicable *element by element* to a convergent sequence $\{f_n\}$ or we say that the sequence may be *operated on element by element* if the sequence $\{g_n\}$ resulting from the formal performance of such operation converges to a limit g which is otherwise obtainable by direct application of the operation to the limit function f of $\{f_n\}$.

A simple instance is the elementary operation of multiplication of the elements of a sequence by a bounded function. In this vein we have the following theorem, for which the reader is encouraged to supply a proof.

Theorem 11-6 If $\{f_n\}$ is a uniformly convergent sequence of functions with limit f and if h is a bounded function on the domain of $\{f_n\}$, then the sequence $\{hf_n\}$ is uniformly convergent to the limit function hf.

Consider next the effect of the operation of differentiation applied to the elements of a sequence. If $\{f_n\}$ is a sequence of functions differentiable on some common domain, questions arise concerning the derivative sequence $\{f_n'\}$. Does it converge if $\{f_n\}$ does? May it converge if $\{f_n\}$ does not? Further, if both sequences converge, what is the relationship between their limits? What role does uniform convergence play?

Following the definition of element by element operation we shall describe a sequence $\{f_n\}$ as *differentiable element by element* or simply *differentiable* if f_n' exists for each n, if $\{f_n\}$ and $\{f_n'\}$ both converge, and if the limit of $\{f_n'\}$ is the derivative of the limit of $\{f_n\}$.

Example 11-12 The sequence $\{f_n\}$ defined by $\{x^n\}$ on $(-1,1)$ converges to the zero function on that interval. So does the sequence $\{f_n'(x)\} = \{nx^{n-1}\}$. Hence, in this instance, $\lim \{f_n'\}$ exists and equals $(\lim \{f_n\})'$. Consequently, the given sequence is differentiable element by element. Note that the convergence of neither $\{f_n\}$ nor $\{f_n'\}$ is uniform.

Example 11-13 Consider the sequence $\{g_n\}$ defined for all x by $g_n(x) = (\sin nx)/n$. Although it clearly converges uniformly to the zero function on $\mathscr{R}$ and although g_n' exists for each n, the sequence $\{g_n'\}$ is given by $\{\cos nx\}$, which does not converge for all x. It follows, therefore, that uniform convergence of a sequence is not sufficient to element by element differentiability.

Example 11-14 The sequence $\{h_n\}$, defined for all x by $h_n(x) = x + n$, does not converge for any x. However, $h_n'(x) = 1$ for each n and x, whence $\{h_n'\}$ converges uniformly to the unit function on $\mathscr{R}$. Consequently, uniform convergence of $\{h_n'\}$ is not enough to insure differentiability.

Example 11-15 Recall the sequence $\{f_n\}$ defined by $f_n(x) = 2x/(1 + n^2x^2)$ for all $x \geq 0$ and $n \in I_+$ and shown in Example 11-8 to be uniformly convergent to the function $f(x) = 0$ for all $x \geq 0$. Clearly, f_n is differentiable for all n and x, and $f_n'(x) = 2(1 - n^2x^2)/(1 + n^2x^2)^2$. For $x > 0$ the sequence $\{f_n'(x)\}$ converges to 0, but for $x = 0$ we have $f_n'(0) = 2$, whence $\{f_n'(0)\}$ converges to 2. Here, then, it is the case that $\{f_n\}$ and $\{f_n'\}$ both converge. Moreover, the function $f = \lim \{f_n\}$ is differentiable, but its derivative is not identical with the limit of $\{f_n'\}$.

The following is in the nature of a desirable sufficient condition for element by element differentiation of a sequence. Observe how it entails establishing validity of differentiation only after the differentiation has been formally carried out.

Theorem 11-7 Let $\{f_n\}$ be a sequence of functions, each element differentiable on the interval $[a,b]$ and such that $\{f_n(c)\}$ converges for some point c of

$[a,b]$. Further, suppose $\{f_n'\}$ to be uniformly convergent on $[a,b]$. Then the sequence $\{f_n\}$ converges uniformly to a differentiable limit f with the property $f' = \lim \{f_n'\}$.

Proof Let $\varepsilon > 0$ be given. Then N exists such that

(3) $$|f_{n+k}(c) - f_n(c)| < \varepsilon \quad \text{and} \quad |f_{n+k}'(\xi) - f_n'(\xi)| < \varepsilon$$

for all $n > N$, $k > 0$, and all ξ in $[a,b]$.

For any point x of $[a,b]$ we may write

$$f_{n+k}(x) - f_n(x) = \{[f_{n+k}(x) - f_n(x)] - [f_{n+k}(c) - f_n(c)]\} + f_{n+k}(c) - f_n(c)$$

and apply to the function $f_{n+k} - f_n$ the mean value theorem for derivatives. Thus, between x and c exists a number ξ such that

$$\begin{aligned} |f_{n+k}(x) - f_n(x)| &\leq |f_{n+k}'(\xi) - f_n'(\xi)|\, |x - c| + |f_{n+k}(c) - f_n(c)| \\ &\leq (b - a + 1)\varepsilon, \end{aligned}$$

and we conclude that $\{f_n\}$ is uniformly convergent to a limit function f on $[a,b]$.

Let x_0 be any point of $[a,b]$, and consider the functions F_n $(n \in I_+)$ defined on $[a,b]$ by

(4) $$F_n(x) = \begin{cases} \dfrac{f_n(x) - f_n(x_0)}{x - x_0} & \text{if} \quad x \neq x_0 \\ f_n'(x_0) & \text{if} \quad x = x_0 . \end{cases}$$

It is clear from the differentiability of f_n that F_n is continuous at each point of $[a,b]$. Further, by a second application of the mean value theorem we are led to the conclusion that if $x \neq x_0$, then

(5) $$\begin{aligned} |F_{n+k}(x) - F_n(x)| &= \left| \frac{[f_{n+k}(x) - f_n(x)] - [f_{n+k}(x_0) - f_n(x_0)]}{x - x_0} \right| \\ &= |f_{n+k}'(\xi) - f_n'(\xi)| \end{aligned}$$

for some point ξ between x and x_0. By (3) and (5) for $x \neq x_0$ and by (4) for $x = x_0$, we deduce $|F_{n+k}(x) - F_n(x)| < \varepsilon$ for all $n > N$, $k > 0$, and all x in $[a,b]$. Thus the sequence $\{F_n\}$ is uniformly convergent on $[a,b]$. It follows from Theorem 11-4 that its limit function F is continuous on the interval $[a,b]$. Moreover, by (4), F is given by

$$F(x) = \begin{cases} \dfrac{f(x) - f(x_0)}{x - x_0} & \text{if} \quad x \neq x_0 \\ \lim \{f_n'(x_0)\} & \text{if} \quad x = x_0 . \end{cases}$$

From the continuity of F then follows, for each point x_0 of $[a,b]$,

$$f'(x_0) = \lim_{x_0} \frac{f(x) - f(x_0)}{x - x_0} = \lim_{x_0} F = \lim \{f_n'(x_0)\} \blacktriangle$$

A sequence of functions $\{f_n\}$ will be styled *Riemann integrable element by element* on an interval $[a,b]$, or simply *Riemann integrable* on $[a,b]$, if for each n the function f_n is R-integrable on $[a,b]$, if the sequences $\{f_n\}$ and $\{\int_a^b f_n\}$ both converge, and if the limit of the former possesses an R-integral on $[a,b]$ equal to the limit of the latter. The question as to the conditions under which a sequence of R-integrable functions is element by element R-integrable is answered, as in the case of differentiability, by uniform convergence, but in not quite the same way. The theorem which follows is seen to be simpler in nature and more direct than the theorem insuring differentiability.

Theorem 11-8 If $\{f_n\}$ is a uniformly convergent sequence of R-integrable functions on an interval $[a,b]$, then the limit function f is R-integrable and the sequence $\{\int_a^b f_n\}$ converges to $\int_a^b f$.

Proof Given $\varepsilon > 0$ there exists an integer N such that $|f_n(x) - f(x)| < \varepsilon$ for all $n > N$ and all x in $[a,b]$. Hence if P is any partition of $[a,b]$, we have for $n > N$

$$|(\underline{R}f_n,P) - \underline{R}(f,P)| = \left|\sum_\nu (m_\nu^{(n)} - m_\nu)\,\Delta x_\nu\right| \le \varepsilon(b-a),$$

where $m_\nu^{(n)}$ and m_ν denote infima of f_n and f on segments of P. In particular,

$$\underline{R}(f_n,P) \le \underline{R}(f,P) + \varepsilon(b-a).$$

Similarly, we obtain

$$\overline{R}(f,P) \le \overline{R}(f_n,P) + \varepsilon(b-a)$$

and conclude that

$$\overline{R}(f,P) - \underline{R}(f,P) \le \overline{R}(f_n,P) - \underline{R}(f_n,P) + 2\varepsilon(b-a).$$

It follows by Theorem 10-3 that R-integrability of f_n for $n > N$ implies R-integrability of f. Further, we have

$$\left|\int_a^b f_n - \int_a^b f\right| \le \int_a^b |f_n - f| < \varepsilon(b-a)$$

for all $n > N$. Hence $\lim \{\int_a^b f_n\} = \int_a^b f$ ▲

If the hypothesis of uniform convergence is relaxed in the theorem, the conclusion may, but need not, be realized.

Example 11-16 The sequence $\{f_n\}$ of functions defined on $[0,1]$ by $f_n(x) = n$ if $0 < x < 1/n$ and $f_n(x) = 0$ otherwise, converges, but not uniformly, to the zero function on $[0,1]$. Each f_n is R-integrable; indeed $\int_0^1 f_n = 1$ for each n so that $\lim \{\int_0^1 f_n\} = 1$. However $\int_0^1 \lim \{f_n\} = 0$.

Example 11-17 Consider the sequence $\{f_n\}$ of functions defined on $[0,1]$ by $f_n(x) = 0$ if $x = p/q$, where $q < n$ (p and q relatively prime), and $f_n(x) = 1$ otherwise. This sequence converges, but not uniformly, to the function Ψ on $[0,1]$. Although $\int_0^1 f_n$ exists with value 1 for each n, the limit function Ψ is not R-integrable, as we have seen.

Exercises

11-31 (a) Prove that if $\{f_n\}$ is a uniformly convergent sequence of functions with limit f and if h is a bounded function on the domain of $\{f_n\}$, then the sequence $\{hf_n\}$ is uniformly convergent to the limit function hf. (b) Illustrate by examples that neither boundedness of h nor uniform convergence of $\{f_n\}$ is essential to the conclusion of the theorem.

11-32 Cite an example of a function sequence that is element by element differentiable but is not uniformly convergent.

11-33 Show that the sequence $\{nx(1-x)^n\}$ is element by element integrable on $[0,1]$, yet the sequence is not uniformly convergent.

11-34 Show that for the sequence $\{x/(1+nx^2)\}$ the limit of the sequence of derivatives is not equal to the derivative of the limit.

11-35 Show that the sequence $\{1/(1+n^2x^2)\}$ is uniformly convergent on $[0,1]$. Is the sequence element by element integrable?

11-36 Prove that $\{2nxe^{-nx^2}\}$ is not uniformly convergent on $[0,1]$; nor is it element by element integrable.

11-37 Show that although $\{xe^{-nx^2}\}$ converges uniformly on $[-1,1]$, it is not element by element differentiable.

11-38 Investigate possible uniform convergence on $[0,1]$ of the sequence $\{f_n\}$ defined by $f_n(x) = nx/(1+nx)$. Determine the limit function f, and compare $\lim \left\{\int_0^1 f_n\right\}$ with $\int_0^1 f$.

11-39 Let $\{f_n\}$ be a uniformly bounded sequence of continuous functions on $[a,b]$, and suppose f to be a function on $[a,b]$ to which $\{f_n\}$ converges uniformly on $[a,c]$ for each point c of (a,b). Prove that $\lim \left\{\int_a^b f_n\right\} = \int_a^b f$.

11-40 Is $\{e^{-nx}/n\}$ differentiable element by element on $\mathscr{R}$?

11-41 Is $\{n^2x^2/(1+n^2x^2)\}$ uniformly convergent? Can it be integrated element by element?

11-42 Show that $\{(\sin nx)/n\}$ converges uniformly to the zero function on $\mathscr{R}$. On what domain does the sequence of derivatives converge?

11-43 If $f_n(x) = 1/n$ for $0 \leq x \leq n$ and $f_n(x) = 0$ otherwise, prove that $\{f_n\}$ converges uniformly to the zero function on $\{x;\ x \geq 0\}$ and that $\lim \left\{\int_0^\infty f_n\right\} \neq \int_0^\infty \lim \{f_n\}$.

11-44 Prove that if $\{f_n\}$ is a convergent sequence of functions on $[a,b]$ such that $\{f_n'\}$ is uniformly bounded, then the convergence of $\{f_n\}$ is uniform.

11-45 Prove *Arzela's theorem*: If $\{f_n\}$ is a uniformly bounded sequence of Riemann integrable functions converging to a Riemann integrable function on $[a,b]$, then $\lim \left\{\int_a^b f_n\right\} = \int_a^b f$. (*Suggestion*: Let $g_n = |f_n - f|$ and $h_n = \sup \{g_n, g_{n+1}, \ldots\}$. Then $0 \leq g_n \leq h_n$ and $\{h_n\}$ is a decreasing null sequence. Therefore $\left\{\underline{\int}_a^b h_n\right\}$ is decreasing with non-negative limit L. Assume $L > 0$, then partitions P_n exist such that $\underline{R}(h_n, P_n) > L/2$. Let $\varepsilon = \frac{1}{2}L(M+b-a)$, where M is a bound for all h_n. Then $\underline{R}(h_n, P_n) = \sum_1 m_\nu \Delta x_\nu + \sum_2 m_\nu \Delta x_\nu$ where $\sum_1$ is the sum of terms for which $m_\nu > \varepsilon$ and $\sum_2$ the remaining. Deduce from this that $\sum_1 \Delta x_\nu > \varepsilon$. Therefore, the union S_n of intervals contributing to $\sum_1$ is not empty. Show that we may assume $S_{n+1} \subset S_n$, whence $\bigcap_1^\infty S_n \neq \varnothing$. If $x \in \bigcap_1^\infty S_n$, then $h_n(x) > \varepsilon$, which is a contradiction.)

A sequence of functions $\{f_n\}$ is said to be *equicontinuous* on its domain D if, given $\varepsilon > 0$, a number $\delta > 0$ exists with the property that $|f_n(x_1) - f_n(x_2)| < \varepsilon$ for any x_1 and x_2 in D such that $|x_1 - x_2| < \delta$ and for all n.

11-46 Prove the *propagation theorem*: If $\{f_n\}$ is an equicontinuous sequence of functions on a compact domain D and if $\{f_n\}$ converges on a dense subset of D, then $\{f_n\}$ is uniformly convergent on D.

11-47 Prove the *compactness theorem*: If $\{f_n\}$ is a uniformly bounded sequence of equicontinuous functions on a compact domain D, then a subsequence of $\{f_n\}$ exists that is uniformly convergent on D. (In Theorem 13-7 we shall encounter a method of proof, reminiscent of Cantor's diagonal proof, that is applicable in the present problem.)

12
Series of Functions

If a man does not keep pace with his companions, perhaps it is because he hears a different drummer.

HENRY DAVID THOREAU (1817–1862), *Walden*

12-1 Series of Functions

As in the case of series of constants, treated in Chapter 6, a *series of functions*

$$\sum_{\nu=1}^{\infty} f_\nu = \sum f_\nu = f_1 + f_2 + \cdots + f_\nu + \cdots$$

is to be identified with the sequence $\{s_n\}$ of *partial sums*

$$s_n = f_1 + f_2 + \cdots + f_n = \sum_{\nu=1}^{n} f_\nu .$$

It converges on the common domain D of its terms to a *limit function* or *sum* f if, given $\varepsilon > 0$, there exists for each x in D an integer N such that $|s_n(x) - f(x)| < \varepsilon$ for all $n > N$. As before, we shall employ the symbol $\sum f_\nu$ in a dual capacity: to designate the series, whether convergent or divergent, and its sum, if convergent.

Since a series is a sequence, the facts established in Chapter 11 pertain here as well. What remains essentially is the expression of properties of a series with explicit reference to its terms.

Example 12-1 The geometric series $\sum_{\nu=0}^{\infty} x^\nu$ is such that $\sum_{\nu=0}^{n} x^\nu = (1 - x^{n+1})/(1 - x)$ for $x \neq 1$. It follows that $\{\sum_{\nu=0}^{n} x^\nu\}$ converges to $\sum_{\nu=0}^{\infty} x^\nu = 1/(1 - x)$ for $|x| < 1$ and diverges for $|x| \geq 1$.

Example 12-2 For the series

$$\sum f_n = \frac{1}{x^2+1} + \frac{x^2-1}{(x^2+1)(x^4+1)} + \cdots + \frac{x^{2n-2}(x^2-1)}{(x^{2n+2}+1)(x^{2n}+1)} + \cdots$$

we have

$$s_n(x) = \frac{1}{x^2+1} + \left(\frac{x^2}{x^4+1} - \frac{1}{x^2+1}\right) + \cdots + \left(\frac{x^{2n}}{x^{2n+2}+1} - \frac{x^{2n-2}}{x^{2n}+1}\right)$$

$$= \frac{x^{2n}}{x^{2n+2}+1}.$$

The sequence $\{s_n(x)\}$ is easily seen to converge for all x to the limit function with value 0 if $|x| < 1$, 1/2 if $|x| = 1$, and $1/x^2$ if $|x| > 1$.

As with sequences in general, a principal interest in connection with series of functions lies in the implications and nuances of uniform convergence. It is clearly the case that a series of functions $\sum_{\nu=1}^{\infty} f_\nu$ converges uniformly to a sum f on the domain D of the terms f_ν if, corresponding to each $\varepsilon > 0$, there exists an integer N with the property that

$$\left| f(x) - \sum_{\nu=1}^{n} f_\nu(x) \right| < \varepsilon$$

for all $n > N$ and all x in D. Moreover, since a sequence is uniformly convergent if and only if it is uniformly Cauchy convergent, it follows that a necessary and sufficient condition for the uniform convergence of $\sum_{\nu=1}^{\infty} f_\nu$ on D is that to each $\varepsilon > 0$ there corresponds an integer N such that

$$(1) \qquad \left| \sum_{\nu=n+1}^{n+k} f_\nu(x) \right| < \varepsilon$$

for all $n > N$, all $k > 0$, and all x in D.

An immediate consequence of the last condition concerns the *remainder after n terms* of a uniformly convergent series. Thus, from (1) it follows that

$$(2) \qquad \left| \sum_{\nu=n+1}^{\infty} f_\nu(x) \right| \leq \varepsilon$$

for all $n > N$ and all x in D. Conversely, (2) implies uniform convergence of the given series. This affirms that the sequence $\{\sum_{\nu=n+1}^{\infty} f_\nu\}$ is a uniformly convergent null sequence (limit zero for all x) if and only if the series $\sum_{\nu=1}^{\infty} f_\nu$ is uniformly convergent.

We say that a convergent series of functions may be *operated on term by term* if the series resulting from the formal performance of such operation has for its limit the function obtained by performing the operation on the function defined by the given series. For the operations of multiplication by a bounded function, differentiation, and integration, this is clearly the restatement for series of the notion of element by element operation enunciated for sequences in general. This is the case since each of the stated operations, when applied to the terms of a series, is at the same time applied to any finite sum of such terms and, in particular, to the partial sums of the series, and conversely. This is also the case for the properties of limit at a point and continuity: If for each n the function f_n has a limit at c, then so has $\sum_{v=1}^{n} f_v$; if each f_n is continuous, then so is $\sum_{v=1}^{n} f_v$. As a consequence, we may regard Theorems 11-4, 11-7, and 11-8 as truths which concern the sequences of partial sums of series, restate them in terms of the terms of the series, and describe them as term-by-term operations and properties.

Theorem 12-1 If $\sum f_n$ is a uniformly convergent series of continuous functions on a domain D, then the series itself is a continuous function on D.

Example 12-3 The series $\sum f_n$, where $f_n(x) = (1/x^{n-1}) - (1/x^n)$ for $n \in I_+$, converges uniformly on $\{x;\ x > r\}$ for $r > 1$ to the (continuous) unit function on that domain. On $\{x;\ x \geq 1\}$ the series is not uniformly convergent, nor is the limit function continuous.

Theorem 12-2 Let $\sum f_n$ be a series of functions, each differentiable on the interval $[a,b]$ and such that $\sum f_n'$ is uniformly convergent. Suppose, further, that a point c of $[a,b]$ exists such that $\sum f_n(c)$ converges. Then the series $\sum f_n$ converges uniformly to a differentiable limit on $[a,b]$ with the property $(\sum f_n)' = \sum f_n'$.

Example 12-4 The series

$$\sum_{v=1}^{\infty} \frac{\sin vx}{v2^v},$$

defined for all real x, clearly converges to 0 at $x = 0$. Since the series formed of the derivatives of its terms, that is,

$$(3) \qquad \sum_{v=1}^{\infty} \frac{\cos vx}{2^v},$$

is such that

$$\left| \sum_{v=n+1}^{\infty} \frac{\cos vx}{2^v} \right| \leq \sum_{v=n+1}^{\infty} \frac{1}{2^v} = \frac{1}{2^n}$$

for any n, it follows, from the condition relating to (1), that series (3) is uniformly convergent. Hence the given series is differentiable term by term with derivative given by (3).

Theorem 12-3 If $\sum f_n$ is a uniformly convergent series of Riemann integrable functions on an interval $[a,b]$, then the limit function is Riemann integrable and $\sum \int_a^b f_n = \int_a^b \sum f_n$.

Example 12-5 Let x and y be numbers such that $x \in (0,1)$ and $|y| \leq x$. By Example 12-1, the series $\sum_{v=0}^{\infty} (-1)^v y^v$ is seen to converge to $1/(1+y)$. Moreover,

$$\left| \sum_{v=n+1}^{\infty} (-1)^v y^v \right| = \left| \frac{y^{n+1}}{1+y} \right| \leq \frac{x^{n+1}}{1-x}.$$

Since $x^{n+1}/(1-x)$, for any x in $(0,1)$, can be made arbitrarily small by choosing n sufficiently large, it follows that the convergence of $\sum(-1)^\nu y^\nu$ is uniform on the interval $[0,x]$. Therefore the series is term-by-term Riemann integrable, and

$$\sum_{\nu=0}^{\infty}\int_0^x(-1)^\nu y^\nu\,dy=\int_0^x\frac{dy}{1+y},$$

or, what is the same,

$$\sum_{\nu=0}^{\infty}\frac{(-1)^\nu x^{\nu+1}}{\nu+1}=\log(1+x)$$

for all x in $(0,1)$.

Exercises

12-1 Let $\sum f_n$ be a uniformly convergent series of functions with limit f, and let c denote a cluster point of the common domain D of the f_n. Suppose, further, that for each index n the function f_n has limit L_n at c. Prove that the series $\sum L_n$ converges to a limit L and that the function f has limit L at c.

12-2 Use the series $\sum x^n$ on $[-r,r]$, where $r<1$, to illustrate the statement of Exercise 12-1.

12-3 Let f_n be defined on $[0,1]$ by $f_n(x)=1$ if $x=m/n$, where m and n are relatively prime integers, and $f_n(x)=0$ otherwise. Show that $\sum f_n=1-\Psi$. Conclude from Exercise 12-1 that the convergence of $\sum f_n$ is not uniform.

12-4 Let $\sum f_n$ be a series of non-negative continuous functions on a compact domain D. Prove that if $\sum f_n$ converges to a continuous limit function, then the convergence is uniform.

12-5 Use the result of Exercise 12-4 to establish the uniform character of the convergence of $\sum \sin^n x$ on $[0,\pi/2-\varepsilon]$ for any $\varepsilon>0$.

12-6 Prove that if $\sum f_n$ is a uniformly convergent series of functions with limit f and if h is a bounded function on the domain of the f_n, then $\sum hf_n$ converges uniformly to hf.

12-7 Prove that if the series $\sum f_n$ is uniformly convergent, then the sequence $\{f_n\}$ converges uniformly to the zero function.

12-8 Find (a) the derivative for $x>0$, and (b) the integral on $[a,b]$, where $0<a<b$, of the function f defined by $f(x)=\sum e^{-nx}\sin nx$.

12-9 Evaluate: (a) the integral of $\sum_1^\infty(\sin nx)/n^2$ on $[0,\pi]$; (b) the integral of $\sum_1^\infty ne^{-nx}$ on $[1,2]$.

12-10 Find the derivative of (a) $\sum_1^\infty(\sin nx)/n^3$ for all x, and (b) $\sum_1^\infty nx^{-n}$ for $|x|>1$.

12-11 Let $f(x)=\sum x^n/n!$ for all x. Prove that $f'(x)=f(x)$ and that $f(0)=1$. Infer that $f(x)=e^x$.

12-12 Prove that if the *Fourier series*

$$f(x)=\frac{a_0}{2}+\sum_{n=1}^{\infty}(a_n\cos nx+b_n\sin nx)$$

converges uniformly on $[0,2\pi]$, then the values of the constants a_n and b_n are given by the so-called *Euler formulas*

$$a_n=\frac{1}{\pi}\int_0^{2\pi}f(x)\cos nx\,dx\quad\text{and}\quad b_n=\frac{1}{\pi}\int_0^{2\pi}f(x)\sin nx\,dx.$$

12-2 Tests for Uniform Convergence

In order to establish the uniform convergence of a series $\sum f_n$ it is convenient to resort to certain sufficient conditions or tests instead of appealing directly to the definition or to uniform Cauchy convergence. Certain of these follow directly from comparisons made in connection with series of constants in Chapter 6. One, due to Weierstrass, has a rather obvious character and is limited in scope. Others tend to be more penetrating. We shall present certain of these tests here and reserve still others for the reader to develop and illustrate. Aside from application to the determination of possible uniform convergence, they will serve as fertile ground in which to develop an insight into the nature of uniform convergence.

Theorem 12-4 (The Weierstrass M-Test) If $\sum f_n$ is a series of functions and if a convergent series of constants $\sum M_n$ exists such that $|f_n(x)| \le M_n$ for each n and for all x in the common domain D of the f_n, then $\sum f_n$ is absolutely convergent for each x in D and is uniformly convergent on D.

Proof Absolute convergence follows directly from the comparison test for series of constants. To demonstrate uniformity of convergence over D, we have, by hypothesis, $|\sum_{\nu=n+1}^{n+k} f_\nu(x)| \le \sum_{\nu=n+1}^{n+k} M_\nu$ for all x in D and any n and k. Since $\sum_{\nu=1}^{\infty} M_\nu$ is Cauchy convergent, it follows that, $\varepsilon > 0$ being given, N exists such that $\sum_{\nu=n+1}^{n+k} M_\nu < \varepsilon$ for $n > N$ and $k > 0$. Since the choice for N here is quite independent of x, we have $|\sum_{\nu=n+1}^{n+k} f_\nu(x)| < \varepsilon$ for all $n > N$, $k > 0$, and all x in D. This testifies to the uniform Cauchy convergence of $\sum f_n$ and hence to its uniform convergence▲

Example 12-6 The series defined by $\sum_{n=0}^{\infty} 2x/(1 + n^3x^2)$ is such that for all n and all real x, $|2x/(1 + n^3x^2)| \le 1/n^{3/2}$. Since $\sum 1/n^{3/2}$ converges, the given series does so uniformly.

Example 12-7 The series $\sum_{\nu=1}^{\infty} (\sin \nu x)/\nu^p$ where $p > 1$, is such that $|(\sin \nu x)/\nu^p| \le 1/\nu^p$ for any x and is therefore uniformly convergent on $\mathscr{R}$. It follows, incidentally, that the series is Riemann integrable on any closed interval. In particular,

$$\int_0^\pi \sum_{\nu=1}^{\infty} \frac{\sin \nu x}{\nu^p}\, dx = \sum_{\nu=1}^{\infty} \int_0^\pi \frac{\sin \nu x}{\nu^p}\, dx = 2 \sum_{\nu=1}^{\infty} \frac{1}{(2\nu + 1)^{p+1}}.$$

The Weierstrass test for uniform convergence asserts absolute convergence as well. Necessarily, then, if a series is uniformly but not absolutely convergent, the uniformity of its convergence cannot be ascertained from the M-test. Further tests, more incisive in nature and with the capacity to discern the possible uniform convergence of a conditionally convergent series, may be devised. Two such, attributed to Abel and Dirichlet, will be considered next.

Theorem 12-5 (Abel's Test) The series $\sum_{v=1}^{\infty} f_v g_v$ is uniformly convergent on the common domain D of the f_v and g_v if the series $\sum_{v=1}^{\infty} f_v$ is uniformly convergent and the set of functions g_v uniformly bounded on D and if for each x in D the sequence $\{g_v(x)\}$ is monotone.

Proof We have $\sum_{v=1}^{\infty} f_v = s = s_n + t_n$, where $s_n = \sum_{v=1}^{n} f_v$ and $t_n = \sum_{v=n+1}^{\infty} f_v$. We have already seen how uniform convergence of $\{t_n\}$ follows from that of $\sum f_v$. As the reader may verify,

$$f_v g_v = -t_v[g_v - g_{v+1}] + t_{v-1}g_v - t_v g_{v+1}.$$

Summing from $v = n + 1$ to $v = n + k$ yields

$$\sum_{v=n+1}^{n+k} f_v g_v = -\sum_{v=n+1}^{n+k} t_v[g_v - g_{v+1}] + t_n g_{n+1} - t_{n+k} g_{n+k+1},$$

whence

$$\sum_{v=n+1}^{n+k} f_v g_v \leq \sum_{v=n+1}^{n+k} |t_v|\,|g_v - g_{v+1}| + |t_n|\,|g_{n+1}| + |t_{n+k}|\,|g_{n+k+1}|.$$

Let $\varepsilon > 0$ be given. Then, because of the uniform convergence of $\{t_v\}$ to zero on D, there exists an integer N such that $|t_v(x)| < \varepsilon$ for $v > N$ and all x in D. Also, by hypothesis, a number M exists such that $|g_v(x)| \leq M$ for all v and all x in D. It follows that for $v > N$, $k > 0$, and all x in D we have

$$\left|\sum_{v=n+1}^{n+k} f_v(x)g_v(x)\right| < \varepsilon \sum_{v=n+1}^{n+k} |g_v(x) - g_{v+1}(x)| + 2M\varepsilon.$$

Because of the monotonicity of $\{g_v(x)\}$ we find

$$\sum_{v=n+1}^{n+k} |g_v(x) - g_{v+1}(x)| = |g_{n+1}(x) - g_{n+k+1}(x)| \leq 2M.$$

Hence for $v > N$, $k > 0$, and all x in D it is the case that

$$\left|\sum_{v=n+1}^{n+k} f_v(x)g_v(x)\right| < 4M\varepsilon \blacktriangle$$

Example 12-8 The series $\sum_{v=1}^{\infty} (-1)^v/v^x$, defined for $x \geq 1$, may be written in the form $\sum f_v g_v$, where $f_v(x) = (-1)^v/v$ and $g_v(x) = 1/v^{x-1}$. Since $\sum_{v=1}^{\infty} f_v$ is a series of constants, it is uniformly convergent trivially. The functions $g_v(x)$ are uniformly bounded, since $0 < 1/v^{x-1} \leq 1$ for all $x \geq 1$ and $v \in I_+$. Also, for each x the sequence $\{1/v^{x-1}\}$ is monotone. By Abel's test we conclude, therefore, that $\sum_{v=1}^{\infty} (-1)^v/v^x$ is uniformly convergent on the domain $x \geq 1$. (Since the series is only conditionally convergent for $x = 1$, the Weierstrass test proves inadequate in this instance.)

We may draw a further conclusion here. Because of Exercise 12-1 we may find the limit from the right at $x = 1$ of $\sum_{v=1}^{\infty} (-1)^v/v^x$ by finding such limits term by term. We get $\lim_{1+} \sum_{v=1}^{\infty} (-1)^v/v^x = \sum_{v=1}^{\infty} (-1)^v/v$.

Theorem 12-6 (Dirichlet's Test) The series $\sum_{v=1}^{\infty} f_v g_v$ is uniformly convergent on a common domain D of the f_v and g_v if the set of functions s_v ($s_v = f_1 + f_2 + \cdots + f_v$) is uniformly bounded on D, if $\{g_v\}$ is uniformly convergent on D, and if $\{g_v(x)\}$ is a monotone null sequence for each x in D.

Proof Summing the two members of the identity

$$f_\nu g_\nu = s_\nu[g_\nu - g_{\nu+1}] - s_{\nu-1}g_\nu + s_\nu g_{\nu+1}$$

we deduce the inequality

$$\left|\sum_{\nu=n+1}^{n+k} f_\nu g_\nu\right| \le \sum_{\nu=n+1}^{n+k} |s_\nu|\,|g_\nu - g_{\nu+1}| + |s_{n+k}|\,|g_{n+k+1}| + |s_n|\,|g_{n+1}|.$$

Let $\varepsilon > 0$ be given. Then N exists such that $|g_\nu(x)| < \varepsilon$ for $\nu > N$ and all x in D. Also, M exists such that $|s_\nu(x)| \le M$ for all x in D and $\nu \in I_+$. Hence, for $n > N$, $k > 0$, and all x in D,

$$\left|\sum_{\nu=n+1}^{n+k} f_\nu(x)g_\nu(x)\right| < M \sum_{\nu=n+1}^{n+k} |g_\nu(x) - g_{\nu+1}(x)| + 2M\varepsilon.$$

Because $\{g_\nu(x)\}$ is monotone for each x we have

$$\left|\sum_{\nu=n+1}^{n+k} f_\nu(x)g_\nu(x)\right| < M\,|g_{n+1}(x) - g_{n+k+1}(x)| + 2M\varepsilon < 4M\varepsilon\blacktriangle$$

Example 12-9 Consider the series $\sum_{\nu=1}^{\infty} (\sin \nu x)/\nu^p$, where $0 < p \le 1$. (Compare Example 12-7.) We may regard it in the form $\sum_{\nu=1}^{\infty} f_\nu g_\nu$ with $f_\nu(x) = \sin \nu x$ and $g_\nu(x) = 1/\nu^p$ for all x. By employing the formula

$$\sum_{\nu=1}^{n} \sin \nu x = \frac{\sin \dfrac{nx}{2} \sin \dfrac{(n+1)x}{2}}{\sin \dfrac{x}{2}}$$

for $x \ne \pm 2k\pi$ $(k = 0,1,2,\ldots)$, we see that $\left|\sum_{\nu=1}^{n} \sin \nu x\right| \le \csc(\delta/2)$ for any n and on any interval of the form $[\delta, 2\pi - \delta]$, where $\delta > 0$. Thus, the sequence $\left\{\sum_{\nu=1}^{n} \sin \nu x\right\}$ is uniformly bounded on $[\delta, 2\pi - \delta]$. Trivially, the monotone null sequence $\{1/\nu^p\}$ is uniformly convergent. The conditions of the Dirichlet test being met, it follows that the given series is uniformly convergent on $[\delta, 2\pi - \delta]$.

Further tests for uniform convergence—in particular those attributed to du Bois-Reymond and Dedekind—are left for the reader to develop and illustrate. (See Exercises 12-18 and 12-19.)

Exercises

12-13 Let $\sum f_n$ and $\sum g_n$ be such that $|f_n| \le g_n$ on a common domain D. Prove that if $\sum g_n$ is uniformly convergent on D, then so is $\sum f_n$.

12-14 Prove that the following are uniformly convergent on the given interval: (a) $\sum n^2 x^n$, $[-\frac{1}{2},\frac{1}{2}]$; (b) $\sum ne^{-nx}$, $\{x;\ \ x \ge a\}$, $a > 0$; (c) $\sum (\sin nx)/2^n$, $\mathscr{R}$; (d) $\sum a_n x^n/(1 + x^n)$, $[0,1]$ if $\sum a_n$ converges.

12-15 Prove the *alternating series test* for uniform convergence: If $\{f_n\}$ is a decreasing sequence of non-negative functions that converges uniformly to the zero function on a domain D, then $\sum (-1)^n f_n$ is uniformly convergent on D.

12-16 Use the test of Exercise 12-15 to prove that $\sum (-1)^n/(n+x)$ converges uniformly on $\{x;\ x \geq 0\}$. Show also that the series is only conditionally convergent for each x.

12-17 Use the series $\sum f_n$, where $f_n(x) = (-1)^n x^n(1-x)$ on $[0,1]$, to demonstrate that the combination of uniform convergence of $\sum f_n$ and simple convergence of $\sum |f_n|$ is not enough to make $\sum |f_n|$ uniformly convergent.

12-18 Prove the *du Bois-Reymond test*: If $\sum f_n$ and $\sum g_n$ are such that $\sum f_n$ and $\sum |g_n - g_{n+1}|$ are uniformly convergent on a domain D and $\{g_n\}$ is uniformly bounded on D, then the series $\sum f_n g_n$ is uniformly convergent on D. (*Suggestion:* Make use of a summation identity employed in the proof of Abel's test.)

12-19 Prove the *Dedekind test*: If, on a domain D, the sequence of partial sums of $\sum f_n$ is uniformly bounded, the sequence $\{g_n\}$ converges uniformly to the zero function, and the series $\sum |g_n - g_{n+1}|$ is uniformly convergent, then the series $\sum f_n g_n$ is uniformly convergent on D. (*Suggestion:* Make use of summation identities patterned after those in the proof of Abel's test.)

12-20 Establish the important identities

$$\sum_{\nu=1}^{n} \sin \nu x = \frac{\sin \dfrac{nx}{2} \sin \dfrac{(n+1)x}{2}}{\sin \dfrac{x}{2}} \quad \text{and} \quad \sum_{\nu=1}^{n} \cos \nu x = \frac{\sin \dfrac{nx}{2} \cos \dfrac{(n+1)x}{2}}{\sin \dfrac{x}{2}}$$

for $x \neq 2k\pi$. Infer that the partial sums of $\sum \sin nx$ and $\sum \cos nx$ are uniformly bounded on $[h, 2\pi - h]$, where $0 < h < \pi$.

12-21 Apply Dirichlet's test to show that $\sum a_n \sin nx$ and $\sum a_n \cos nx$ converge uniformly on $[h, 2\pi - h]$ if $\{a_n\}$ is any monotone null sequence.

12-22 Use Dedekind's test to establish uniform convergence of $\sum a_n \sin nx$ and $\sum a_n \cos nx$ on $[h, 2\pi - h]$ if $\{a_n\}$ is a null sequence and $\sum |a_n - a_{n+1}|$ converges.

12-23 Prove that if $\{a_n\}$ is a decreasing sequence and $a_n \geq 0$ for all n, then $\{na_n\}$ is a null sequence if and only if $\sum a_n \sin nx$ is uniformly convergent.

12-24 Show that $\sum_1^\infty (\sin nx)/n$ converges on $\mathscr{R}$ but that the series $\sum_1^\infty \cos nx$ of derivatives diverges for each x.

12-25 Show that $\sum_1^\infty (\sin nx)/n^2$ converges uniformly on $\mathscr{R}$ but that the series $\sum_1^\infty (\cos nx)/n$ of derivatives does not converge for all x.

12-26 Prove that $\sum_1^\infty 1/(n^3 + n^4x^2)$ is uniformly convergent on $\mathscr{R}$ and that term by term differentiation is valid.

12-3 A Continuous Nondifferentiable Function

In an earlier chapter was remarked the existence of functions which, while continuous at all points of some domain, fail to be differentiable anywhere. We are now in a position to give an example of such a function. The reader who anticipates a definition of bizarre construction will be pleasantly surprised at the simplicity of a so-called *Weierstrass function* f given, for all real x, by

$$f(x) = \sum_{\nu=0}^{\infty} \left(\frac{2}{3}\right)^{\nu} \cos 15^{\nu}\pi x.$$

Since $|(\frac{2}{3})^\nu \cos 15^\nu \pi x| \le (\frac{2}{3})^\nu$ for all ν and x, it is clear from the M-test that the series defining f is uniformly convergent. As a consequence of the continuity of its terms, it then follows that f is continuous for all x. We have only to show that for no x is f differentiable.

Let c be any fixed point, and denote by $\{x_n\}$ and $\{y_n\}$ any two sequences converging to c with $x_n \neq y_n$ for all n. Consider the sequence

$$\left\{\frac{f(y_n) - f(x_n)}{y_n - x_n}\right\}. \tag{4}$$

It is a consequence of the definition of derivative of f at c that if $f'(c)$ exists, then all sequences of the type (4) converge to $f'(c)$. Hence, if such a sequence can be constructed with the property that it does not converge, it will have been proved that f is not differentiable at c.

To construct a sequence of the form (4), let

$$x_n = \frac{k_n + \frac{1}{2}}{15^n} \quad \text{and} \quad y_n = \frac{k_n - 1}{15^n},$$

where k_n, for each n, is that integer satisfying the condition $k_n - \frac{1}{2} \le 15^n c < k_n + \frac{1}{2}$. It follows directly from the definition of f that

$$\frac{f(y_n) - f(x_n)}{y_n - x_n} = \Sigma_1 + \Sigma_2,$$

where

$$\Sigma_1 = \sum_{\nu=0}^{n-1} \left(\frac{2}{3}\right)^{\nu+1} 15^n \left[\cos \pi\left(\frac{2k_n + 1}{2}\right) 15^{\nu-n} - \cos \pi(k_n - 1)15^{\nu-n}\right]$$

and

$$\Sigma_2 = \sum_{\nu=n}^{\infty} \left(\frac{2}{3}\right)^{\nu+1} 15^n \left[\cos \pi\left(\frac{2k_n + 1}{2}\right) 15^{\nu-n} - \cos \pi(k_n - 1)15^{\nu-n}\right].$$

In the sum Σ_2 observe that $\cos \pi[(2k_n + 1)/2]\, 15^{\nu-n} = 0$ since $\nu \ge n$. Also $\cos \pi(k_n - 1)15^{\nu-n}$ equals 1 or -1 according to whether k_n is odd or even. Hence Σ_2 is a geometric series, and we conclude

$$\frac{f(y_n) - f(x_n)}{y_n - x_n} = \Sigma_1 + 2(-1)^{k_n} 10^n. \tag{5}$$

By the mean value theorem we have

$$\cos \pi\left(\frac{2k_n + 1}{2}\right) 15^{\nu-n} - \cos \pi(k_n - 1)15^{\nu-n} = -\frac{3\pi}{2}\, 15^{\nu-n} \sin \xi,$$

where $\pi(k_n - 1)15^{\nu-n} < \xi < \pi[(2k_n + 1)/2]15^{\nu-n}$. Hence

$$\left|\cos \pi\left(\frac{2k_n + 1}{2}\right) 15^{\nu-n} - \cos \pi(k_n - 1)15^{\nu-n}\right| \le \frac{3\pi}{2}\, 15^{\nu-n}.$$

It follows from this and (5) that

$$\left|\frac{f(y_n) - f(x_n)}{y_n - x_n} - 2(-1)^{k_n}10^n\right| \leq \pi \sum_{\nu=0}^{n-1} 10^\nu < \frac{1}{2} 10^n.$$

This may be written

$$\frac{f(y_n) - f(x_n)}{y_n - x_n} = 2(-1)^{k_n}10^n + \frac{\lambda}{2} 10^n,$$

where $-1 < \lambda < 1$. Rearranging terms on the right yields

$$\frac{f(y_n) - f(x_n)}{y_n - x_n} = (-1)^{k_n}10^n\left[2 + (-1)^{k_n}\frac{\lambda}{2}\right]. \tag{6}$$

Examination of the sequence defined by (6) shows the following to be true. If the sequence $\{k_n\}$ possesses a subsequence of even integers, then the right member increases with n on this subsequence and so is unbounded above. If $\{k_n\}$ does not contain a subsequence of even integers, then it must have a subsequence of odd integers. On such a sequence (6) decreases and is not bounded below. Hence (6) does not possess a limit; from this it follows that f is not differentiable at c. Since c denotes any real number, it follows that f is not differentiable anywhere.

Exercises

12-27 Prove that the function $f = \sum f_n$, where f_n is the "sawtooth" function defined by

$$f_n(x) = \begin{cases} -x & \text{if } -\frac{1}{2}4^{-n} \leq x \leq 0 \\ x & \text{if } 0 < x < \frac{1}{2}4^{-n} \\ f_n(x + 4^{-n}) & \text{for all } x, \end{cases}$$

is continuous for all x and differentiable nowhere.

12-4 Power Series

A special and most important instance of a series of functions is the *power series*

$$\sum_{n=0}^{\infty} a_n(x - x_0)^n = a_0 + a_1(x - x_0) + \cdots + a_n(x - x_0)^n + \cdots,$$

where x_0 and a_n, $n = 0, 1, 2, \ldots$, are constants. Described also as an *expansion* or series *centered at* or *about* x_0, it is clear that for $x = x_0$ the power series converges trivially to the sum a_0. It is entirely possible for a power series to converge at x_0 and only there. The series $\sum_{n=0}^{\infty} n!\,(x - x_0)^n$ has this property; for if $x \neq x_0$, then $\{n!\,(x - x_0)^n\}$ increases without bound. At the other extreme, the series $\sum_{n=0}^{\infty} (x - x_0)^n/n!$ converges for all real x. This is a consequence of the ratio test since $\rho = \lim \{|x - x_0|/(n + 1)\} = 0$ for all x.

We shall see also that power series exist which converge for values of x different from x_0, yet do not converge for all x in $\mathscr{R}$. (Actually, we have already encountered instances of such.) The following theorems tend to establish the domain of convergence of a power series.

Theorem 12-7 If the power series $\sum a_n(x - x_0)^n$ converges for $x = x_1$, then it is absolutely convergent for each x such that $|x - x_0| < |x_1 - x_0|$.

Proof From convergence of $\sum a_n(x_1 - x_0)^n$ follows convergence to zero, and hence boundedness, of the sequence $\{a_n(x_1 - x_0)^n\}$. Thus M exists such that $|a_n(x_1 - x_0)^n| \leq M$ for all n. For any x we have

$$|a_n(x - x_0)^n| = |a_n(x_1 - x_0)^n| \left| \frac{x - x_0}{x_1 - x_0} \right|^n \leq Mr^n,$$

where $r = |(x - x_0)/(x_1 - x_0)|$. In particular, if $|x - x_0| < |x_1 - x_0|$, then $r < 1$, and the series $\sum Mr^n$ converges. By the comparison test it follows that $\sum |a_n(x - x_0)^n|$ converges▲

We conclude from the above that the set of real numbers which constitutes the domain of convergence of $\sum a_n(x - x_0)^n$ is an interval, bounded or unbounded, centered at x_0, or consists of the point x_0 alone.

Theorem 12-8 A power series $\sum a_n(x - x_0)^n$ either converges absolutely for all real x or a non-negative number R exists with the property that the series converges absolutely for all x such that $|x - x_0| < R$ and diverges if $|x - x_0| > R$.

Proof Denote by A the set of all non-negative real numbers r with the property that the series $\sum a_n(x - x_0)^n$ converges for all x such that $|x - x_0| < r$. It is clear that 0 is a member of A and, because of Theorem 12-7, so is r if $0 \leq r < r_1$ and if $r_1 \in A$. Further, if x is a point of divergence of the series and if $|x - x_0| < r$, then $r \notin A$.

In the event that A comprises all non-negative real numbers, then the series $\sum a_n(x - x_0)^n$ converges absolutely for all real x. The alternative is that A is bounded above and hence possesses a supremum R. It is then clear that the series is absolutely convergent if $|x - x_0| < R$ and diverges if $|x - x_0| > R$▲

The number R of Theorem 12-8 is termed the *radius of convergence* of the power series $\sum a_n(x - x_0)^n$. If $R > 0$, we call the interval $(x_0 - R, x_0 + R)$ the *interval of convergence* of the series. It should be noted that Theorem 12-8 says nothing about the endpoints of this interval. We see below that it is possible for one or both of the points $x_0 - R$ and $x_0 + R$ to be points of convergence or for neither to be such. It follows that we distinguish technically between the interval of convergence, which is open, and the domain of convergence, which may include one or two endpoints.

Example 12-10

(a) The series $\sum x^n$ has radius of convergence equal to 1 and converges to $1/(1-x)$ on $(-1,1)$. It does not converge at either endpoint of its interval of convergence.

(b) The series $\sum x^n/n$ also has radius of convergence 1. Since the series $\sum(-1)^n/n$ has been seen to converge and $\sum 1/n$ to diverge, it follows that the domain of convergence of $\sum x^n/n$ is the interval $[-1,1)$.

(c) The series $\sum x^n/n^2$ has interval of convergence $(-1,1)$. Its domain of convergence is $[-1,1]$.

If $R = 0$, the set of numbers x with the property $|x - x_0| < R$ is empty, so the conclusion of absolute convergence in Theorem 12-8 is vacuously satisfied. The series diverges for $|x - x_0| > R$, that is, for $x \neq x_0$, and converges at x_0.

If the series $\sum a_n(x - x_0)^n$ converges for all real x, we say that $\mathscr{R}$ is the interval of convergence. It is also convenient in such event to look upon the series as having an *infinite radius of convergence* and to write, accordingly, $R = \infty$. Reference to a number r such that $0 < r < R$ shall be taken to mean, if $R = \infty$, that r is any positive real number.

Theorem 12-9 If $\sum a_n(x - x_0)^n$ has radius of convergence R and if r is a real number such that $0 < r < R$, then the series converges uniformly on the interval $[x_0 - r, x_0 + r]$.

Proof Let x_1 denote a point of the interval of convergence of $\sum a_n(x - x_0)^n$ such that $r < |x_1 - x_0|$. Let $|x_1 - x_0| = r_1$. Then, since $\sum a_n r_1^n$ converges, a number M exists such that $|a_n r_1^n| \leq M$ for all n. Therefore, if x is any point of the interval $[x_0 - r, x_0 + r]$, we have $|x - x_0| \leq r$, and for all n

$$|a_n(x - x_0)^n| = |a_n r_1^n| \frac{|x - x_0|^n}{r_1^n} \leq M\left(\frac{r}{r_1}\right)^n.$$

Since $r < r_1$, it follows that $\sum M(r/r_1)^n$ is convergent. Consequently, by the Weierstrass test, we conclude that $\sum a_n(x - x_0)^n$ is uniformly convergent on $[x_0 - r, x_0 + r]$▲

Since, as Theorem 12-9 states, a power series converges uniformly on *any* closed interval contained in its open interval of convergence, the reader may wonder if this does not imply uniform convergence on the interval of convergence itself. Actually, this does not follow and, indeed, need not be true. Uniform convergence on $[x_0 - r, x_0 + r]$ implies the existence of an integer N, with its customary significance, which is independent of x on that interval. However, it is clear that N may depend on r and that if r assumes an increasing sequence of values (all less than R), then a sequence of corresponding values of N is generated that need not be bounded, in which case no value of N exists that is independent of x on $(x_0 - R, x_0 + R)$.

Example 12-11 It has already been noted that the series $\sum x^n$ converges on the interval $(-1,1)$ to the limit function $1/(1-x)$. That the convergence is not uniform on that interval is not difficult to see. For suppose it to be so. Then the series is also uniformly

Cauchy convergent, so that, given $\varepsilon > 0$, an integer N exists with the property that $\left|\sum_{v=n+1}^{n+k} x^v\right| < \varepsilon$ for all $n > N$, all $k > 0$, and all x in $(-1,1)$. In particular, this must be the case for $k = 1$; that is, $|x^{n+1}| < \varepsilon$ for all $n > N$ and all x in $(-1,1)$. But this is not possible, since, whatever the magnitude of n, the function defined by $|x^{n+1}|$ on $(-1,1)$ has limit 1 at $x = 1$, whence values of x exist which make $|x^{n+1}|$ arbitrarily close to 1. Thus, uniform convergence of $\sum x^n$, although a fact on $[-1+\delta, 1-\delta]$ for any $\delta > 0$, is not so on $(-1,1)$.

Example 12-12 The series $\sum x^n/n^2$ converges at all points of the interval $[-1,1]$. Moreover, the convergence is uniform on that interval since, by the Weierstrass M-test, we have only to note that $|x^n/n^2| \leq 1/n^2$ for all x in $[-1,1]$ and that $\sum 1/n^2$ converges.

The previous example shows that it is possible for the uniform convergence of a power series to extend to the full interval of convergence and to its closure as well. The suggestion is that if a power series converges at an endpoint of its interval of convergence, then it is uniformly convergent on intervals extending to and including that endpoint. Theorem 12-10, the work of Abel, shows this to be a fact. Because the proof tends to invoke some complicated expressions, it will be convenient to simplify matters by considering a power series of the form $\sum a_n x^n$ with the special property that its radius of convergence is 1. That no loss of generality is entailed in so doing is clear once the student has established that $\sum a_n x^n$ has interval of convergence $(-1,1)$ if and only if $\sum (a_n/R^n)(x - x_0)^n$ has interval of convergence $(x_0 - R, x_0 + R)$.

Theorem 12-10 (Abel's Theorem) If $\sum a_n x^n$ has radius of convergence equal to 1 and if $\sum a_n$ converges, then $\sum a_n x^n$ converges uniformly on the interval $[-1+\delta, 1]$ for $\delta > 0$. Dually, if $\sum (-1)^n a_n$ converges, then $\sum a_n x^n$ is uniformly convergent on $[-1, 1-\delta]$.

Proof Let $\varepsilon > 0$ be given. Then, since $\sum a_n$ converges, an integer N exists such that $\left|\sum_{v=n+1}^{n+k} a_v\right| < \varepsilon$ for all $n > N$ and all $k > 0$. Observe now that, for any n and any $k > 0$, the following is true:

$$\sum_{v=n+1}^{n+k} a_v x^v = a_{n+1}x^{n+1} + \sum_{\mu=n+2}^{n+k} \left(\sum_{v=n+1}^{\mu} a_v - \sum_{v=n+1}^{\mu-1} a_v \right) x^\mu$$

$$= \sum_{\mu=n+1}^{n+k-1} \left[(x^\mu - x^{\mu+1}) \sum_{v=n+1}^{\mu} a_v \right] + x^{n+k} \sum_{v=n+1}^{n+k} a_v .$$

Fixing attention for the moment on the interval $[0,1]$, it is clear that for each x in the interval we have $x^\mu - x^{\mu+1} \geq 0$. Hence we have, for all $n > N$, all $k > 0$, and all x in $[0,1]$,

$$\left| \sum_{v=n+1}^{n+k} a_v x^v \right| < \varepsilon \left| \sum_{\mu=n+1}^{n+k-1} (x^\mu - x^{\mu+1}) + x^{n+k} \right| = \varepsilon x^{n+1} < \varepsilon.$$

Thus it is proved that $\sum a_n x^n$ is uniformly convergent on $[0,1]$. Since we have already seen it to be so on $[-1+\delta, 1-\delta]$ for $\delta > 0$, it follows now that it is uniformly convergent on $[-1+\delta, 1]$. In a similar manner the dual theorem may be proved▲

Corollary If $\sum a_n(x-x_0)^n$ has radius of convergence R and if $\sum a_n R^n$ converges, then for each $\delta > 0$ the given series converges uniformly on $[x_0 - R + \delta, x_0 + R]$. If $\sum (-1)^n a_n R^n$ converges, then $\sum a_n(x-x_0)^n$ is uniformly convergent on $[x_0 - R, x_0 + R - \delta]$.

Example 12-13 The series $\sum x^n/n$ converges on $[-1,1)$ and hence is uniformly convergent on $[-1, 1-\delta]$ for $\delta > 0$. It is interesting to note that, unlike the case for the series $\sum x^n/n^2$, the uniform convergence cannot be demonstrated by an application of the Weierstrass test with the obvious comparison with $\sum 1/n$ since the latter series does not converge.

Exercises

12-28 Show that although $\sum x^n/n!$ converges uniformly on every bounded interval, it does not do so on $\mathscr{R}$.

12-29 Use Abel's theorem to prove that if $\sum a_n x^n$ converges on $(-1,1)$ and if $\sum a_n$ converges, then $\lim_{1-} \sum a_n x^n = \sum a_n$.

12-30 Use the series $\sum (-1)^n x^n$ to demonstrate that the hypothesis "$\sum a_n$ converges" cannot be replaced by "$\lim_{1-} \sum a_n x^n$ exists" in Abel's theorem.

12-31 Prove *Tauber's theorem*, which is in the nature of a converse to Abel's theorem:

If $f(x) = \sum a_n x^n$, $|x| < 1$, is such that $\lim \{na_n\} = 0$ and $\lim_{1-} f = L$, then $\sum a_n$ converges to L. (*Suggestion:* Let $t_n = (1/n) \sum_0^n k\,|a_k|$. Then $\lim \{t_n\} = 0$ by Exercise 5-52. For $\varepsilon > 0$, let N be such that $t_n < \varepsilon$, $n\,|a_n| < \varepsilon$, and $|f(1-1/n) - L| < \varepsilon$ for $n > N$. Write

$$\sum_{k=0}^{n} a_k - L = f(x) - L + \sum_{k=0}^{n} a_k(1-x^k) - \sum_{n+1}^{\infty} a_k x^k.$$

Then use the fact that $1 - x^k \le k(1-x)$ for $0 < x < 1$ to show that $|\sum_0^n a_k - L| < \varepsilon$.)

12-32 Prove that if $a_n \ge 0$ for all n and if $\lim_{1-} \sum_1^\infty a_n x^n = L$, then $\sum a_n = L$.

12-5 The Radius of Convergence

The radius of convergence R of a power series may be determined in several ways. We include here two theorems that will recall to the reader the ratio and root tests for convergence of a series of constants.

Theorem 12-11 If $\sum a_n(x-x_0)^n$ is a series such that $a_n \ne 0$ for all n and for which the number $\rho = \lim \{|a_{n+1}/a_n|\}$, finite or infinite, exists, then the radius of convergence of the series is $R = 1/\rho$ (where $R = 0$ if $\rho = \infty$ and $R = \infty$ if $\rho = 0$).

Proof Recalling the modified ratio test of Chapter 6, we calculate the number

$$\lim \left\{ \left| \frac{a_{n+1}(x-x_0)^{n+1}}{a_n(x-x_0)^n} \right| \right\} = |x-x_0| \lim \left\{ \left| \frac{a_{n+1}}{a_n} \right| \right\} = |x-x_0|\,\rho$$

and conclude, if $0 < \rho < \infty$, that the series converges if $|x - x_0|\rho < 1$ and diverges if $|x - x_0|\rho > 1$; that is, $R = 1/\rho$. Since $|x - x_0|\rho < 1$ for all x if $\rho = 0$, it follows that $\rho = 0$ implies convergence for all real x; that is, $R = \infty$. Finally, suppose the series to converge for $x = x_1 \neq x_0$. Then $|x_1 - x_0|\rho < 1$, or $\rho < 1/|x_1 - x_0|$. It follows that if $\rho = \infty$, then the series diverges for all $x \neq x_0$; that is, $R = 0$▲

Example 12-14 $\sum [(n!)^2/(2n)!](x-1)^n$. Here

$$R = \lim\left\{\frac{(n!)^2(2n+2)!}{(2n)![(n+1)!]^2}\right\} = \lim\left\{\frac{(2n+2)(2n+1)}{(n+1)^2}\right\} = 4.$$

Hence the interval of convergence is $(-3,5)$.

Example 12-15 $\sum x^n/n!$. We see here that

$$\rho = \lim\left\{\left|\frac{n!}{(n+1)!}\right|\right\} = \lim\left\{\frac{1}{n+1}\right\} = 0.$$

Hence the series converges for all x.

Theorem 12-12 For the series $\sum a_n(x - x_0)^n$, let the number $M = \overline{\lim}\,\{\sqrt[n]{|a_n|}\}$ be determined. Then the radius of convergence of the series is $R = 1/M$ (where $R = 0$ if $M = \infty$ and $R = \infty$ if $M = 0$).

Proof Suppose M to be finite, and let x be such that $x \neq x_0$ and $M|x - x_0| < 1$. Then numbers r_1 and r_2 exist with the properties

$$M < r_1 < \frac{1}{|x - x_0|} \quad \text{and} \quad r_1|x - x_0| < r_2 < 1.$$

From the definition of $\overline{\lim}\{\sqrt[n]{|a_n|}\}$ follows the existence of an integer N such that $\sqrt[n]{|a_n|} < r_1$ for $n > N$, whence $\sqrt[n]{|a_n|}\cdot|x - x_0| < r_2$ and $|a_n(x - x_0)^n| < r_2{}^n$ for $n > N$. Since $\sum r_2{}^n$ converges, it follows that the given series $\sum a_n(x - x_0)^n$ is convergent for each x such that $M|x - x_0| < 1$. If $M = 0$, it follows that the series converges for all x; that is, $R = \infty$. Also, if $0 < M < \infty$, then the series converges for all x such that $|x - x_0| < 1/M$; that is, $R \geq 1/M$.

Next, suppose x to be such that $1 < M|x - x_0|$. Then for any N there exists an integer $n' > N$ such that $1/|x - x_0| < \sqrt[n']{|a_{n'}|} < M$, which implies that $1 < |a_n(x - x_0)^n|$ for an infinite set of values of n. Hence for such x the series diverges and we conclude $R \leq 1/M$. Therefore for the case $0 < M < \infty$ we have established $R = 1/M$.

From convergence of $\sum a_n(x - x_0)^n$ at a point $x \neq x_0$ follows $|x - x_0| \leq 1/M$ or $M \leq 1/|x - x_0|$. Hence $M = \infty$ implies divergence for all $x \neq x_0$; that is, $R = 0$▲

Example 12-16 Consider the series $\sum a_n x^n = x/2 + x^2/4 + x^3/4 + x^4/16 + x^5/8 + x^6/64 + \cdots$, where $a_{2n-1} = 1/2^n$ and $a_{2n} = 1/4^n$. Then $\overline{\lim}\,\{\sqrt[n]{|a_n|}\} = 1/\sqrt{2}$, and it follows that the radius of convergence of the given series is $R = \sqrt{2}$.

12-6 The Limit Function of a Power Series

Because of the uniform character of the convergence of a power series we are able to conclude certain properties of the function which it defines, namely, continuity, differentiability, and Riemann integrability.

Theorem 12-13 A power series is a continuous function on its domain of convergence.

Proof Essentially this is a consequence of Theorems 12-1 and 12-10, in which it was demonstrated that uniform convergence extends to subsets of the interval of convergence which include an endpoint of that interval, if it is the case that the series converges at such an endpoint▲

Example 12-17 The series $\sum_{\nu=0}^{\infty}(-1)^{\nu}x^{\nu+1}/(\nu+1)$ converges for $x = 1$ and diverges for $x = -1$, whence we conclude the radius of convergence to be 1. Moreover, by Theorem 12-13, the series is continuous at $x = 1$. In Example 12-5 it was shown that $\sum_{\nu=0}^{\infty}(-1)^{\nu}x^{\nu+1}/(\nu+1) = \log(1+x)$ for $x \in (0,1)$. As a consequence, we may conclude that $\sum_{\nu=0}^{\infty}(-1)^{\nu}/(\nu+1) = \lim_{1} \log(1+x) = \log 2$, obtaining thereby the sum of the alternating harmonic series. Observe that the same result could not have been obtained by integrating $\sum_{\nu=0}^{\infty}(-1)x^{\nu}$ on $[0,1]$ since the series does not converge at $x = 1$.

Theorem 12-14 is a special instance of Theorem 12-3 on term-by-term Riemann integrability of a series. It requires no further proof.

Theorem 12-14 A power series $\sum a_n(x-x_0)^n$ is term-by-term Riemann integrable on any interval $[a,b]$ contained in its domain of convergence.

Preliminary to the statement of a corresponding theorem for differentiability, it is important to compare the interval of convergence of a power series with that of the series formed of the derivatives of its terms.

Theorem 12-15 The power series $\sum_{n=0}^{\infty} a_n(x-x_0)^n$ and $\sum_{n=1}^{\infty} na_n(x-x_0)^{n-1}$ have a common interval of convergence.

Proof If R is the radius of convergence of $\sum a_n(x-x_0)^n$ and if x is any point in the interval of convergence of the series, then a number r exists such that $|x-x_0| < r < R$. By the ratio test, since $|x-x_0|/r < 1$, it may be seen that the series $\sum n|x-x_0|^{n-1}/r^{n-1}$ converges. Therefore, for some number N we have $|n(x-x_0)^{n-1}/r^{n-1}| < 1$ for all $n > N$. It follows that

$$|na_n(x-x_0)^{n-1}| = \frac{1}{r}\left|\frac{n(x-x_0)^{n-1}}{r^{n-1}}\right| |a_n r^n| < \frac{1}{r}|a_n| r^n$$

for $n > N$. But $\sum |a_n| r^n$ converges since $r < R$; therefore $\sum na_n(x-x_0)^{n-1}$ converges for any x in the interval of convergence of $\sum a_n(x-x_0)^n$.

On the other hand, suppose the interval of convergence of $\sum na_n(x - x_0)^{n-1}$ to extend beyond that of $\sum a_n(x - x_0)^n$. Then by Theorem 12-14, for some point x_1 such that $x_1 - x_0 > R$, it is the case that $\sum \int_{x_0}^{x_1} na_n(x - x_0)^{n-1}\, dx$ converges to $\sum a_n(x_1 - x_0)^n$, which is impossible since $\sum a_n(x - x_0)^n$ diverges for $x = x_1$.

We conclude that the series have a common radius of convergence R, which may be positive, zero, or infinite▲

Corollary The series $\sum_{n=0}^{\infty} a_n(x - x_0)^n$ and $\sum_{n=0}^{\infty} a_n(x - x_0)^{n+1}/(n + 1)$ have the same radius of convergence.

It cannot be concluded from Theorem 12-15 that the series $\sum a_n(x - x_0)^n$ and $\sum na_n(x - x_0)^{n-1}$ have the same domain of convergence. It is possible for $\sum a_n(x - x_0)^n$ to converge at an endpoint of its interval of convergence and for $\sum na_n(x - x_0)^{n-1}$ to diverge there.

Example 12-18 The series $\sum x^n/n$ converges on $[-1,1)$. However, the derivative series $\sum x^{n-1}$ diverges at -1.

Term-by-term differentiation, valid on the interval but not on the domain of convergence of a power series, follows from Theorems 12-2 and 12-15.

Theorem 12-16 If $\sum a_n(x - x_0)^n$ has a nonzero radius of convergence, then it is term-by-term differentiable on its interval of convergence.

Expressed otherwise, if f denotes the function defined by $\sum a_n(x - x_0)^n$ on its interval of convergence, then $\sum na_n(x - x_0)^{n-1} = f'(x)$ on that interval.

An obvious but nonetheless significant consequence of the preceding theorems concerns repeated differentiation of a power series.

Theorem 12-17 A power series possesses derivatives of all orders. Each may be obtained by term-by-term differentiation and all converge on a common interval of convergence.

Power series, on their intervals of convergence, lend themselves to algebraic manipulations in much the same manner as polynomials. Consider, for example, two series $\sum a_n(x - x_0)^n$ and $\sum b_n(x - x_0)^n$. Assuming each to have at least a positive radius of convergence, it follows that both converge on the smaller of the two intervals of convergence. From our study of series of constants in Chapter 6 it then follows that at each point x of such interval we may form the sum and Cauchy product of the given series, that is,

$$\sum a_n(x - x_0)^n + \sum b_n(x - x_0)^n = \sum (a_n + b_n)(x - x_0)^n$$

and

$$\sum a_n(x - x_0)^n \sum b_n(x - x_0)^n = \sum_{n=0}^{\infty} \sum_{\nu=0}^{n} a_\nu b_{n-\nu}(x - x_0)^n,$$

and be assured of the convergence of both resultants. The latter conclusion (refer to Chapter 6) makes use of the absolute convergence of the series $\sum a_n(x - x_0)^n$ and $\sum b_n(x - x_0)^n$.

Example 12-19 If x is a point for which the series $\sum_{\nu=0}^{\infty} a_\nu x^\nu$ and $\sum_{\nu=0}^{\infty} x^\nu$ both converge, then

$$\sum_{\nu=0}^{\infty} a_\nu x^\nu \sum_{\nu=0}^{\infty} x^\nu = \sum_{\nu=0}^{\infty} s_\nu x^\nu,$$

where $s_\nu = a_0 + a_1 + \cdots + a_\nu$. There follows the interesting conclusion

$$\sum_{\nu=0}^{\infty} a_\nu x^\nu = (1 - x) \sum_{\nu=0}^{\infty} s_\nu x^\nu.$$

Example 12-20 Suppose the series of constants $\sum a_n$, $\sum b_n$, and $\sum c_n$ to be convergent, where for each n, $c_n = a_0 b_n + a_1 b_{n-1} + \cdots + a_n b_0$. It follows that for $|x| < 1$ each of the series $\sum a_n x^n$, $\sum b_n x^n$, and $\sum c_n x^n$ converges, whence $\sum a_n x^n \sum b_n x^n = \sum c_n x^n$. Therefore, by considering limits at $x = 1$, we obtain

$$\sum a_n \sum b_n = \sum c_n. \tag{7}$$

Thus, convergence of $\sum a_n$, $\sum b_n$, and $\sum c_n$ is sufficient to the conclusion (7). This result should be appended to the collection of facts obtained in Chapter 6 in connection with Cauchy products.

If we consider for the moment the category of all possible series of functions, then it is certainly the case that limit functions do not uniquely declare the series which converge to them. We have seen in examples earlier in the chapter instances where different series of functions converge, and even uniformly, to the same limit. For power series, however, this is not possible. If $\sum a_n(x - x_0)^n$ and $\sum b_n(x - x_0)^n$ are different series, that is, if $a_n \neq b_n$ for at least one value of n, then they converge to different limits, as we shall see below. It follows that the limit of a power series determines uniquely the set of all coefficients a_n. Indeed, we may go even further and assert that a power series is uniquely determined by the values of its limit function in an arbitrarily small neighborhood of the center of the interval of convergence. This is not true of functions in general, of course. As a consequence, power series play a special and exceedingly important role in the analysis of real functions.

As we have seen, a function defined by a power series possesses derivatives of all orders. In the following theorem there is revealed a simple relationship between the coefficients in the series and the values of the derivatives at the center of the interval of convergence.

Theorem 12-18 If $\sum a_n(x - x_0)^n$ has a nonzero radius of convergence and if f denotes the function defined by the series on its interval of convergence, then for each n, $a_n = f^{(n)}(x_0)/n!$.

Proof From $f(x) = \sum_{\nu=0}^{\infty} a_\nu(x - x_0)^\nu$ follows $f(x_0) = a_0$. Further,

$$f'(x) = \sum_{\nu=1}^{\infty} \nu a_\nu(x - x_0)^{\nu-1} = \sum_{\nu=0}^{\infty} (\nu + 1)a_{\nu+1}(x - x_0)^\nu$$

implies $f'(x_0) = a_1$. Quite generally, from

$$f^{(n)}(x) = \sum_{\nu=0}^{\infty} \frac{(\nu + n)!}{\nu!} a_{\nu+n}(x - x_0)^\nu$$

follows $f^{(n)}(x_0) = n!\,a_n$ for all $n \geq 0$ (with the understanding that $f^{(0)} = f$)▲

The unique character of a power series now follows.

Theorem 12-19 If $\sum a_n(x - x_0)^n$ and $\sum b_n(x - x_0)^n$ define the same function f on a neighborhood of x_0, however small, then $a_n = b_n$ for all n; that is, the two series are identical.

Proof If for some $\delta > 0$ it is the case that $f(x) = \sum a_n(x - x_0)^n = \sum b_n(x-x_0)^n$ for all x in $\mathcal{N}(x_0,\delta)$, then, by Theorem 12-18, $a_n = f^{(n)}(x_0)/n!$ and $b_n = f^{(n)}(x_0)/n!$. Hence $a_n = b_n$ for all n▲

Exercises

12-33 If for a sequence $\{a_n\}$ constants m and M exist such that $0 < m \leq |a_n| \leq M$ for all n, show that the series $\sum a_n x^n$ has radius of convergence 1.

12-34 If $\sum a_n x^n$ has radius of convergence R, find the radius of convergence of (a) $\sum a_n{}^2 x^n$, and (b) $\sum a_n x^{2n}$.

12-35 If $\sum a_n x^n$ and $\sum b_n x^n$ have radii of convergence R_a and R_b, respectively, with $R_a < R_b$, what is the radius of convergence of: (a) $\sum (a_n + b_n)x^n$? (b) $\sum a_n b_n x^n$? (c) $\sum (a_n/b_n)x^n$?

12-36 Show that Arctan $x = \sum (-1)^n x^{2n+1}/(2n + 1)$, and use this to evaluate the sum $\sum (-1)^n/(2n + 1)$.

12-37 Show how the sums $\sum nx^n$ and $\sum n^2x^n$ can be obtained from $\sum x^n = 1/(1 - x)$ for $|x| < 1$.

12-38 Show that the series $\sum (n + 1)!\,x^n$ and $\sum n!\,x^n$ have the same domain of convergence, and define the same function on that domain. Comment on this relative to the uniqueness of a power series.

12-7 Elementary Functions Defined by Power Series

Certain of the elementary functions of mathematical analysis may be defined as power series. In this section we shall consider the familiar sine and cosine functions.

We define functions s and c by

(8) $$s(x) = \sum_{n=0}^{\infty} \frac{(-1)^n x^{2n+1}}{(2n + 1)!} \quad \text{and} \quad c(x) = \sum_{n=0}^{\infty} \frac{(-1)^n x^{2n}}{(2n)!}$$

(bypassing the well-known symbols sin x and cos x for the present) and refer to them as the *sine* and *cosine* functions, respectively. Since the power series in (8) may be shown by the ratio test to be convergent for any value of x, we may regard both s and c defined on $\mathscr{R}$.

By direct substitution in the defining series we see that $s(-x) = -s(x)$ and $c(-x) = c(x)$; that is, s is an *odd* function and c an *even* function. Also, $s(0) = 0$ and $c(0) = 1$. Since a power series may be differentiated term by term we find that $s'(x) = c(x)$ and $c'(x) = -s(x)$. Now define a function h on $\mathscr{R}$ by $h = s^2 + c^2$. Then h is differentiable and $h' = 2ss' + 2cc' = 2sc - 2cs = 0$. It follows that h is a constant function on $\mathscr{R}$ and, in particular, is the unit function since $h(0) = s^2(0) + c^2(0) = 1$. Thus $s^2(x) + c^2(x) = 1$ for all x. This in turn implies that for all x, $|s(x)| \leq 1$ and $|c(x)| \leq 1$.

Next, let b denote any real number and define functions f and g by $f(x) = s(x + b) - s(x)c(b) - c(x)s(b)$ and $g(x) = c(x + b) - c(x)c(b) + s(x)s(b)$. Clearly, f and g are differentiable, and a simple calculation reveals that $f' = g$ and $g' = -f$. Thus $[f^2 + g^2]' = 0$, which, together with $f(0) = g(0) = 0$, implies that $f^2 + g^2$ is the zero function on $\mathscr{R}$. Hence f and g vanish identically and

$$\begin{aligned} s(x + b) &= s(x)c(b) + c(x)s(b), \\ c(x + b) &= c(x)c(b) - s(x)s(b). \end{aligned} \tag{9}$$

Further, by letting $b = x$ in (9), we obtain the special results $s(2x) = 2s(x)c(x)$ and $c(2x) = c^2(x) - s^2(x)$.

Returning to the definition of c we find, by substitution in the power series, that $c(2) = 1 - 2 + \frac{2}{3} - \cdots$. Since this is an alternating series, we may conclude that $c(2)$ differs in value from -1 by an amount not exceeding $\frac{2}{3}$. Thus $c(2) < 0$, and since $c(0) > 0$, it follows from continuity that c must vanish between 0 and 2. Moreover, since the zeros of a continuous function constitute a closed set, there must be a least point in the interval $(0,2)$ at which c assumes the value 0. Denote this point by $p/2$. Thus, $0 < p/2 < 2$, $c(p/2) = 0$, and $c(x) > 0$ for $0 \leq x < p/2$. In fact, from the even character of c we can state that $c(x) > 0$ for $-p/2 < x < p/2$. Further, since $s^2(x) + c^2(x) = 1$, with $x = p/2$ it follows that $s(p/2) = 1$ or $s(p/2) = -1$.

Now $c(2x) = c^2(x) - s^2(x) = 2c^2(x) - 1$. Therefore $c(p) = 2c^2(p/2) - 1 = -1$ and $c(2p) = 2c^2(p) - 1 = 1$. Since $s^2(2p) + c^2(2p) = 1$, we also conclude that $s(2p) = 0$. By (9) with $b = 2p$, this implies $s(x + 2p) = s(x)$ for all x. This asserts that s is a *periodic* function on $\mathscr{R}$ with *period* $2p$. In a similar way, or from $c = s'$, we may show that the cosine function has period $2p$, that is, $c(x + 2p) = c(x)$. We leave it to the reader as an exercise to show that neither s nor c has a period less than $2p$.

Since $s' = c$ and c is positive on $(-p/2,p/2)$, it follows that s is strictly increasing on that interval. From $s^2 + c^2 = 1$ we see that $-1 < s(x) < 1$, and since by continuity $s(-p/2) = -1$ and $s(p/2) = 1$, we conclude that s is strictly increasing on the closed interval $[-p/2,p/2]$. In a similar way it follows that c is strictly decreasing on $[0,p]$. As a consequence, there exists a function s^{-1}, inverse to s

and mapping $[-1,1]$ on $[-p/2,p/2]$; and a function c^{-1}, inverse to c, defined on $[-1,1]$ and having range $[0,p]$.

Properties of inverse functions s^{-1} and c^{-1} follow easily from certain facts which appear in later chapters. For one thing, the monotone function s^{-1} is differentiable. Thus, from $s(x) = y$ and $s^{-1}(y) = x$ we may obtain, by the chain rule, $[s^{-1}(y)]'s'(x) = 1$, or, if $s'(x) \neq 0$,

$$[s^{-1}(y)]' = \frac{1}{s'(x)} = \frac{1}{c(x)} = \frac{1}{\sqrt{1 - s^2(x)}} = \frac{1}{\sqrt{1 - y^2}}$$

for $|y| < 1$. Since $1/\sqrt{1 - y^2}$ defines a continuous function, it is Riemann integrable on the interval $[0,y]$ for $0 < y < 1$. Hence

$$s^{-1}(y) = \int_0^y \frac{1}{\sqrt{1 - \xi^2}}\, d\xi.$$

Since the limit of s^{-1} at $y = 1$ is $p/2$, we conclude that $p = 2 \int_0^1 1/\sqrt{1 - \xi^2}\, d\xi$. Thus is derived a formula which yields the value of p. It is left to the reader to determine that p is indeed the familiar rational number π. Henceforth we drop the notations $s(x)$ and $c(x)$ in favor of the more familiar $\sin x$ and $\cos x$, respectively.

Such functions as e^x and $\log x$ can be defined and developed in a similar fashion. These we leave as exercises for the reader.

Exercises

12-39 Prove that neither the function s nor the function c of Section 12-7 has a period less than $2p$. Show that $p = \pi$.

12-40 Prove the Jordan inequality: $2x/\pi \leq \sin x$ for $0 \leq x \leq \pi/2$.

12-41 Let the *exponential function* E be defined for all real x by $E(x) = \sum_0^\infty x^n/n!$. Show that: (a) $E(x_1)E(x_2) = E(x_1 + x_2)$ for any x_1 and x_2; (b) $E(0) = 1$; (c) $E(-x) = 1/E(x)$; (d) $E(x) > 0$ for all x; (e) $E'(x) = E(x)$; (f) $E(x)$ is strictly increasing; (g) $E(x)$ may be written as e^x, where $e = \sum_0^\infty 1/n!$.

12-42 Let the *logarithm function* L be defined as the function inverse to E of Exercise 12-41. (a) Prove that $L(y) = \int_1^y 1/t\, dt$ for $y > 0$. (b) Find a power series that converges to L on some bounded domain.

12-8 Taylor Series

In this section we shall be concerned with the problem inverse to that of determining the limit of a power series, namely, that of finding the power series, if one exists, that converges to a given function. We note, in this connection, that if a power series $\sum a_n(x - x_0)^n$ converges to a function f on an interval $(x_0 - R, x_0 + R)$, then, necessarily, f possesses derivatives of all orders on that interval, and the values of its derivatives at x_0 are related to the coefficients in the series by

$n!a_n = f^{(n)}(x_0)$. Accordingly, if f is a function having derivatives of all orders in an interval centered at a point x_0 and if a power series exists which converges to f on that interval, then that series is necessarily

$$\sum_{\nu=0}^{\infty} \frac{f^{(\nu)}(x_0)}{\nu!}(x - x_0)^{\nu}. \tag{10}$$

When approached from this point of view, the power series which converges to f is referred to as the *Taylor series* or *Taylor expansion* of f *at* or *about* the point x_0. Formally, of course, if f has derivatives of all orders at x_0, then a power series of the form (10) always exists, simply by virtue of our being able to write it down. But it does not follow that it converges to f, or, indeed that it converges at all (except at x_0). Examples 12-21 and 12-22 illustrate these possibilities.

Example 12-21 Let f be the function defined by $f(x) = \sum_{\nu=1}^{\infty}(\cos \nu^2 x)/2^{\nu}$. This series converges, and uniformly so, for all x in $\mathscr{R}$. Further, it is easy to verify that derivatives of all orders exist and that $f^{(2n+1)}(0) = 0$ for $n = 0, 1, 2, \ldots$, also $f^{(2n)}(0) = (-1)^n \sum_{\nu=1}^{\infty} \nu^{4n}/2^{\nu}$. The series of constants in the right member clearly converges for each n. As a consequence, if f is to have a Taylor expansion about $x = 0$, then it is the series $\sum_{n=1}^{\infty} a_{2n} x^{2n}$, where

$$a_{2n} = \frac{(-1)^n \sum_{\nu=1}^{\infty} \frac{\nu^{4n}}{2^{\nu}}}{(2n)!}.$$

To determine the radius of convergence of this series we proceed as follows. First, observe that for each n it is the case that

$$\sum_{\nu=1}^{\infty} \frac{\nu^{4n}}{2^{\nu}} \geq \sum_{\nu=n}^{\infty} \frac{\nu^{4n}}{2^{\nu}} \geq n^{4n} \sum_{\nu=n}^{\infty} \frac{1}{2^{\nu}} = \frac{n^{4n}}{2^{n-1}}.$$

Hence

$$\begin{aligned} |a_{2n}|^{1/2n} &= \left| \frac{(-1)^n \sum_{\nu=1}^{\infty} \frac{\nu^{4n}}{2^{\nu}}}{(2n)!} \right|^{1/2n} \geq \left[\frac{n^{4n}}{(2n)!\,2^{n-1}} \right]^{1/2n} = \frac{n^2}{\sqrt[2n]{(2n)!}\,2^{(n-1)/2n}} \\ &\geq n \cdot \frac{2n}{\sqrt[2n]{(2n)!}} \cdot 2^{(1-3n)/2n}. \end{aligned} \tag{11}$$

Reference to Exercise 5-64 shows the sequence $\{2n/\sqrt[2n]{(2n)!}\}$ to converge to e. Further $\{2^{(1-3n)/2n}\}$ has limit $2^{-3/2}$. Therefore the right member of (11) defines a sequence which is unbounded. By the root test the radius of convergence of the series is zero. We conclude that $\sum a_{2n} x^{2n}$ is not the Taylor series of the function f since it has no interval of convergence.

Example 12-22 Consider the function f defined for all real x by $f(x) = e^{-1/x^2}$ if $x \neq 0$ and $f(0) = 0$. To determine its successive derivatives, we proceed as follows (with some assistance from the reader).

Let $P_0, P_2, \ldots, P_{2n}, \ldots$ denote polynomials defined by the iterative scheme $P_0(x) = 1$ and

$$P_{2n+2}(x) = (2 - 3nx^2)P_{2n}(x) + x^3 P_{2n}'(x)$$

for $n \geq 0$. Then it may be shown by mathematical induction that

$$f^{(n)}(x) = \frac{f(x)P_{2n}(x)}{x^{3n}}$$

for $n = 0, 1, 2, \ldots$ and all $x \neq 0$. Moreover, since P_{2n} is of degree $2n$, it follows from l'Hospital's rule that $\lim_0 f^{(n)}(x)/x = 0$ for $n = 0, 1, 2, \ldots$. From

$$f^{(n)}(0) = \lim_0 \frac{f^{(n-1)}(x) - f^{(n-1)}(0)}{x - 0} = -\lim_0 \frac{f^{(n-1)}(x)}{x}$$

it follows by induction that $f^{(n)}(0) = 0$ for all n. As a consequence, the series

$$\sum_{n=0}^{\infty} \frac{f^{(n)}(x)}{n!} x^n = \sum_{n=0}^{\infty} 0 \cdot x^n$$

converges trivially to 0 for all x in $\mathscr{R}$. Obviously, then, it is not the Taylor series for f since it converges to the value of that function at $x = 0$ only. Therefore f does not possess a Taylor expansion.

Suppose f to be a function possessing derivatives of all orders on an interval centered at a point x_0. Then for any n we may write

$$(12) \qquad f(x) = \sum_{\nu=0}^{n-1} \frac{f^{(\nu)}(x_0)}{\nu!}(x - x_0)^\nu + R_n(x).$$

We call this *Taylor's formula* and refer to the function R_n, for each n, as the *remainder after n terms.* Clearly, the series $\sum_{\nu=0}^{\infty}[f^{(\nu)}(x_0)/\nu!](x - x_0)^\nu$ converges to $f(x)$ if and only if the sequence of functions $\{R_n(x)\}$ is a null sequence for each x. Information concerning the nature of a function which possesses a Taylor series can be gleaned from an analysis of the functions $R_n(x)$.

Theorem 12-20 Let f be defined and have derivatives through order n, for some n, at each point of a neighborhood $\mathscr{N}$ of a point x_0, and let k be a number less than n. Then, corresponding to each x in $\mathscr{N}$, there exists a number θ such that $|\theta| < 1$ and

$$R_n(x) = \frac{f^{(n)}(x_0 + \theta(x - x_0))(1 - \theta)^k}{(n - k)(n - 1)!}(x - x_0)^n.$$

Proof Let $x \neq x_0$ be a fixed point of $\mathscr{N}$, and define a function φ for all points t of $\mathscr{N}$ by

$$\varphi(t) = f(x) - \sum_{\nu=0}^{n-1} \frac{f^{(\nu)}(t)}{\nu!}(x - t)^\nu - c(x - t)^{n-k},$$

where c is a constant (dependent upon x but independent of t) defined by the equation

$$f(x) - \sum_{\nu=0}^{n-1} \frac{f^{(\nu)}(x_0)}{\nu!}(x - x_0)^\nu = c(x - x_0)^{n-k}.$$

The function φ is seen to have the property that $\varphi(x) = \varphi(x_0) = 0$ and to be continuous and differentiable over $\mathcal{N}$ with

$$\varphi'(t) = -\frac{f^{(n)}(t)}{(n-1)!}(x - t)^{n-1} + c(n-k)(x-t)^{n-k-1}. \tag{13}$$

As a consequence, the mean value theorem for derivatives is valid on the interval with endpoints x_0 and x. Accordingly, a number θ exists such that $|\theta| < 1$ and $\varphi'(x_0 + \theta(x - x_0)) = 0$. The value of c follows from (13), and the desired formula is a consequence of $R_n(x) = c(x - x_0)^{n-k}$. That the result is valid also for $x = x_0$ is seen by (12), that is, $R_n(x_0) = 0$▲

Corollary 1 (The Lagrange Form for the Remainder) If f is a function having derivatives through order n on a neighborhood $\mathcal{N}$ of a point x_0, then for any $x \neq x_0$ in $\mathcal{N}$ there is a number ξ lying between x_0 and x such that

$$R_n(x) = \frac{f^{(n)}(\xi)}{n!}(x - x_0)^n.$$

Proof In the result of the theorem let $k = 0$ and $\xi = x_0 + \theta(x - x_0)$▲

Corollary 2 (The Cauchy Form for the Remainder) If f is a function having derivatives through order n on a neighborhood $\mathcal{N}$ of a point x_0, then for any $x \neq x_0$ in $\mathcal{N}$ there is a number ξ lying between x_0 and x such that

$$R_n(x) = \frac{f^{(n)}(\xi)}{(n-1)!}(x - x_0)(x - \xi)^{n-1}.$$

Proof In the result of the theorem let $k = n - 1$ and $\xi = x_0 + \theta(x - x_0)$▲

Example 12-23 (The Binomial Series) For any real number b the function f given by $f(x) = (1 + x)^b$ is defined on a neighborhood of $x = 0$. Since derivatives of all orders exist, given by

$$f^{(n)}(x) = b(b-1)\cdots(b-n+1)(1+x)^{b-n},$$

on such a neighborhood, the Taylor series for f at $x = 0$, if there is one, is

$$\sum_{n=0}^{\infty} \frac{f^{(n)}(0)}{n!}x^n = \sum_{n=0}^{\infty} \binom{b}{n} x^n, \tag{14}$$

where

$$\binom{b}{n} = \frac{b(b-1)\cdots(b-n+1)}{n!}$$

is called the *binomial coefficient*.

If b is a non-negative integer, $\binom{b}{n} = 0$ for $n \geq b + 1$; in this event the series is a polynomial and converges trivially for all real x. We have seen earlier that the sum of this polynomial is $(1 + x)^b$. In present context the same result follows by noting that $R_n(x) = 0$ for all x.

If b is other than a non-negative integer, the sum in (14) is an infinite series, and we apply the ratio test to determine the radius of convergence. We find

$$\rho = \lim \left\{ \left| \frac{\binom{b}{n+1}}{\binom{b}{n}} \right| \right\} = \lim \left\{ \left| \frac{b-n}{n+1} \right| \right\} = 1$$

and conclude that the series converges for $|x| < 1$.

Consider the case $x \in (0,1)$. Using the Lagrange form for the remainder, we have $R_n(x) = \binom{b}{n}(1 + \xi)^{b-n}x^n$, where $0 < \xi < x$. It follows that $(1 + \xi)^{b-n} < 1$ if $n > b$. Hence, for such n, $|R_n(x)| \leq \left|\binom{b}{n}x^n\right|$. Since $\sum_{n=0}^{\infty} \binom{b}{n} x^n$ converges, it follows that $\left\{\binom{b}{n}x^n\right\}$ has limit 0. Hence $\{R_n(x)\}$ is a null sequence for each x in $(0,1)$, which signifies that the series (14) converges to $(1 + x)^b$ on $(0,1)$.

Next consider the interval $(-1,0)$. By the Cauchy form for the remainder $R_n(x)$, a number ξ exists such that $x < \xi < 0$ and

$$\begin{aligned} R_n(x) &= n\binom{b}{n}(1 +)\xi^{b-n}x(x - \xi)^{n-1} \\ &= n\binom{b}{n} \frac{(1 + \xi)^b}{x - \xi} x^{n+1}\left(\frac{x - \xi}{x + x\xi}\right)^n. \end{aligned}$$

From $-1 < x < \xi < 0$ we have $\xi(1 + x) < 0$, whence $x + x\xi < x - \xi < 0$. Therefore $0 < (x - \xi)/(x + x\xi) < 1$, and it follows that

$$|R_n(x)| \leq \left| \frac{(1 + \xi)^b x}{x - \xi} \right| \left| n\binom{b}{n}x^n \right|.$$

The ratio test if applied to $\sum_{n=0}^{\infty} n\binom{b}{n}x^n$ reveals it to converge for $|x| < 1$; hence $\left\{n\binom{b}{n}x^n\right\}$ converges to zero, and we conclude that $\{R_n(x)\}$ is a null sequence for all x in $(-1,0)$. The series (14) thereby converges to $(1 + x)^b$ on that interval.

We have shown then that

$$(1 + x)^b = \sum_{n=0}^{\infty} \binom{b}{n}x^n$$

for all x in $(-1,1)$.

The next two theorems describe a condition on the derivatives of a function f that is both necessary and sufficient to the possession by f of a Taylor series. For convenience, and without loss of generality, attention is fixed on expansions about $x = 0$.

Theorem 12-21 If f is a function possessing derivatives of all orders on an interval $(-\beta,\beta)$ and if a number $M > 0$ exists such that $|f^{(n)}(x)| \leq n!\,M^n$ for all n and for all x in $(-\beta,\beta)$, then the series

$$\sum_{n=0}^{\infty} \frac{f^{(n)}(0)}{n!} x^n \tag{15}$$

converges to f at all points x in $(-\alpha,\alpha)$, where $\alpha = \min\{\beta,1/M\}$.

Proof Let x be a point of $(-\beta,\beta)$. Then, by the Lagrange form for the remainder, a number ξ exists in $(-\beta,\beta)$ such that

$$|R_n(x)| = \frac{|f^{(n)}(\xi)|}{n!} |x|^n.$$

By hypothesis it follows that

$$|R_n(x)| \leq M^n |x|^n = (M\,|x|)^n$$

for all x in $(-\beta,\beta)$. This implies that the sequence $\{R_n(x)\}$ is a null sequence for each x with the property $M\,|x| < 1$, that is, on the interval $(-1/M, 1/M)$. Hence, if $\alpha = \min\{\beta,1/M\}$, we conclude that the series (15) converges to f at all points of $(-\alpha,\alpha)$▲

Theorem 12-22 If $f(x) = \sum_{n=0}^{\infty} a_n x^n$ at all points of an interval $(-\alpha,\alpha)$, then numbers β and M exist such that f has derivatives of all orders on $(-\beta,\beta)$ and $|f^{(n)}(x)| \leq n!\,M^n$ at each point of $(-\beta,\beta)$.

Proof Since $\sum a_n x^n$ converges on $(-\alpha,\alpha)$ we have $\overline{\lim}\,\{\sqrt[n]{|a_n|}\} \leq 1/\alpha$. It follows that a number c exists such that $|a_n| \leq c^n$ for all n, and is such that $1/c < \alpha$. Because $f(x) = \sum a_\nu x^\nu$ on $(-\alpha,\alpha)$, we have

$$f^{(n)}(x) = \sum_{\nu=0}^{\infty} (\nu+1)(\nu+2)\cdots(\nu+n)a_{\nu+n}x^\nu$$

for each x in that interval. We then obtain

$$|f^{(n)}(x)| \leq \sum_{\nu=0}^{\infty} (\nu+1)(\nu+2)\cdots(\nu+n)\,|a_{\nu+n}|\,|x|^\nu$$

$$\leq c^n \sum_{\nu=0}^{\infty} (\nu+1)(\nu+2)\cdots(\nu+n)(c\,|x|)^\nu$$

for $x \in (-\alpha,\alpha)$. The series on the right converges for all nonzero x in $(-1/c,1/c)$ to the nth derivative of $1/(1 - c\,|x|)$, that is, to $n!\,c^n/(1 - c\,|x|)^{n+1}$. Hence, if $0 < \beta < 1/c$, then for all nonzero x in $(-\beta,\beta)$ it is the case that

$$|f^{(n)}(x)| \leq \frac{n!\,c^{2n}}{(1 - c\,|x|)^{n+1}} \leq \frac{n!\,c^{2n}}{(1 - c\beta)^{2n}}$$

or, finally, $|f^{(n)}(x)| \leq n!\,M^n$, where $M = [c/(1 - c\beta)]^2$. That this is true as well for $x = 0$ follows trivially from the continuity of $f^{(n)}(x)$ at $x = 0$▲

Example 12-24 It is apparent that the functions defined in Examples 12-21 and 12-22 do not possess Taylor expansions because somehow they fail to meet the requirements of Theorem 12-21. In the instance of the function of Example 12-21 it follows from (11) and the remarks subsequent to that inequality that $|f^{(2n)}(0)| \geq c(2n)!\,n^{2n}$ for any n, c being a constant. Hence on any interval $(-\beta,\beta)$ no integer M exists for which the inequality of Theorem 12-21 is true.

For the function defined in Example 12-22 we have seen all derivatives to vanish at 0. As a consequence, the remainder after n terms is exactly $f(x)$. Thus, for $x \neq 0$ and all n we have $R_n(x) = e^{-1/x^2}$.

Let β denote any positive number, and for each n let $x_n = 1/\sqrt{n}$. Then N exists such that $x_n \in (-\beta,\beta)$ for all $n > N$. By expressing the remainder $R_n(x)$ in the Lagrange form, it follows that corresponding to each x_n in $(-\beta,\beta)$ there exists a point ξ_n, also in $(-\beta,\beta)$, such that

$$f^{(n)}(\xi_n) = \frac{e^{-1/x_n^2}n!}{x_n^n} = n!\left(\frac{n}{e^2}\right)^{n/2}.$$

Since it is clearly the case that $\sqrt{n/e^2} > M$ for any constant M and sufficiently large n, we conclude that the condition of Theorem 12-21 cannot be satisfied whatever may be the values of β and M.

Exercises

12-43 (a) Find the Taylor series for e^x about $x_0 = 0$. (b) Use the Lagrange form of the remainder to prove that this series converges uniformly on a domain D if and only if D is bounded.

12-44 Starting with the identity $\int_a^x f'(t)\,dt = f(x) - f(a)$, integrate by parts an appropriate number of times to establish the *integral form of the remainder*: If f is a function having derivatives through order n on a neighborhood $\mathcal{N}$ of a point x_0 and if $f^{(n)}$ is Riemann integrable on any interval contained in $\mathcal{N}$, then for all x in $\mathcal{N}$ the remainder after n terms in Taylor's formula is given by

$$R_n(x) = \frac{1}{(n-1)!}\int_{x_0}^{x} (x-t)^{n-1} f^{(n)}(t)\,dt.$$

12-45 Assuming $f^{(n)}$ to be continuous, derive the Lagrange form of the remainder from the integral form. (*Suggestion:* Use the first mean value theorem for integrals.)

12-46 Assuming $f^{(n)}$ to be continuous, derive the Cauchy form of the remainder from the integral form.

12-47 In the Taylor series for a function f about $x_0 = c$, write dx for $x - c$. Show that the approximation of $\Delta f = f(c + dx) - f(c)$ by differentials (compare Section 9-8) amounts to the partial sum of the Taylor series of f through terms of the first degree.

12-48 Prove that any power series is the Taylor series of its sum.

12-49 For each of the functions $\sin x$, $\cos x$, and e^x, prove, by an examination of the remainder in the Taylor series about $x_0 = 0$, that the series does indeed converge to the function.

12-50 Prove that $\sum_1^\infty \left(\frac{x^{2n-1}}{2n-1} - \frac{x^{2n}}{2n}\right)$ is convergent but not uniformly convergent on [0,1].

12-51 Prove that $\sum_1^\infty (-1)^n/\sqrt{n}\,\sin(1 + x/n)$ is uniformly convergent on $\mathscr{R}$.

12-52 Let $\sum_0^\infty a_n x^n$ be defined by the recursion formula $a_{n-2} + ba_{n-1} + ca_n = 0$, where b and c are constants. Show that on its interval of convergence the sum of the series is $[(a_0 b + a_1 c)x + a_0 c]/(x^2 + bx + c)$.

12-53 Establish the following series expansion for the *Fresnel cosine integral*:

$$\int_0^x \cos t^2\, dt = \sum_{n=0}^\infty \frac{(-1)^n x^{4n+1}}{(4n+1)(2n)!}.$$

What is the interval of convergence?

12-54 Find the Taylor series about $x = 0$ for the *incomplete gamma function* $\int_0^x t^{\alpha-1}e^{-t}\,dt$. What ranges of values of x and α are valid?

12-55 Express the *beta function* $B(p,q) = \int_0^1 t^{p-1}(1-t)^{q-1}\,dt$ as a series by using the binomial expansion of $(1-t)^{q-1}$. Use Abel's theorem to justify integrating on [0,1].

12-56 Show that $J_p(x) = \sum_0^\infty \frac{(-1)^n(x/2)^{2n+p}}{n!\,(p+n)!}$ converges for all x, and verify that it satisfies the *Bessel differential equation* $x^2y'' + xy' + (x^2 - p^2)y = 0$. (The quantity $J_p(x)$ is called a *Bessel function* of *order p*.)

13
Monotone Functions and Bounded Variation

Does the road wind up-hill all the way?
Yes, to the very end.

CHRISTINA ROSSETTI (1830–1894), *Up-Hill*

13-1 Monotone Functions

In this chapter we resume the study of properties of monotone functions initiated in Chapter 9. Recall, briefly, that a function f is *monotone increasing*, or simply *increasing*, on its domain if $f(x_1) \leq f(x_2)$ for any two points x_1 and x_2 with the property $x_1 < x_2$. It is *monotone decreasing*, or *decreasing*, if $f(x_1) \geq f(x_2)$. In particular, f is *strictly monotone*, increasing or decreasing, if $f(x_1) < f(x_2)$ or $f(x_1) > f(x_2)$, respectively. Because a function is obviously increasing if and only if its negative is decreasing, properties of decreasing functions are, quite generally, immediate consequences of properties of increasing functions. Accordingly, and since increasing functions (both in the strict and nonstrict senses) will be seen to play an important role in the concept of bounded variation in the present chapter and in the integration theories of later chapters, we shall limit discussion, with a few exceptions, to increasing functions, remarking simply on the dual character of decreasing functions. All functions, unless otherwise stated or apparent from context, will have interval domains, bounded or unbounded.

The class of monotone functions possesses certain elementary properties but rejects others. It is easily seen that if f and g are increasing functions, then so is their sum $f + g$. Furthermore, if f and g are non-negative and increasing, then their product fg is increasing. Trivially, if f is increasing, then $-f$ is decreasing, as is $1/f$ if we add the restriction $f > 0$. However, as the reader may show by examples, the difference of increasing functions may, but need not, increase or decrease; also, that this is the case for a product, if the condition that the factors be non-negative be relaxed. Similar (or dual) statements may be asserted for decreasing functions. Mixed sums, differences, products, and quotients of increasing and decreasing functions may be shown to be monotone in some instances and non-monotone in others.

That the system of monotone functions is not closed with respect to the basic operations may be regarded as a disadvantage. Later, this class of functions will be superseded by the broader and more flexible category of *functions of bounded variation*, which will be seen to be not deficient in this respect.

The relationship between differentiability and monotonicity of a function f was brought out in Chapter 9. There it was shown that if a function f is differentiable over a neighborhood of a point c of the domain of f and if $f'(c) \neq 0$, then f is strictly monotone over some neighborhood of c. It was also seen that if f is differentiable over an interval $[a,b]$ and such that f' does not vanish, then either f' is positive throughout $[a,b]$ and f is a strictly increasing function, or f' is negative at all points of the interval and f is a decreasing function.

Theorems 13-1 through 13-3 show the extent to which continuity and monotonicity are related.

Theorem 13-1 If a function which is continuous on an interval defines a one-to-one correspondence between its domain and range, then it is strictly monotone.

Proof Let f denote a continuous one-to-one mapping of the bounded interval $[a,b]$ onto a range set of real numbers. Then $f(a) < f(b)$ or $f(a) > f(b)$. Suppose the former to be the case, and assume that f is not an increasing function. Then points x_1 and x_2 of $[a,b]$ exist such that $x_1 < x_2$ and $f(x_1) > f(x_2)$. Comparing the necessarily different function values $f(a)$ and $f(x_2)$, we see that if $f(a) < f(x_2)$, then, because of the continuity of f, a number ξ exists such that $a < \xi < x_1$ and $f(\xi) = f(x_2)$; if $f(a) > f(x_2)$, then ξ exists with the properties $x_2 < \xi < b$ and $f(\xi) = f(x_1)$. In either case, there is plainly a denial of the one-to-one character of f, and we conclude that f is strictly increasing. Similar reasoning shows f to be strictly decreasing if $f(a) > f(b)$.

Moreover, if f is defined on an unbounded interval, the same conclusion is forthcoming, since the above argument may be applied to any bounded interval contained in the domain of the function▲

Strict monotonicity implies, of course, that $f(x_1) \neq f(x_2)$ for any distinct points x_1 and x_2 of the domain of f; and from this it follows that a strictly monotone function is necessarily a one-to-one mapping. Thus it is that if a function is strictly monotone, then it is meaningful to speak of the inverse function f^{-1}. Moreover, it is immediate that f^{-1} is strictly increasing or strictly decreasing according as f is. For, from $f(x_1) < f(x_2)$ follows $x_1 < x_2$ or, what is the same, $f^{-1}(f(x_1)) < f^{-1}(f(x_2))$.

What is not quite so obvious, perhaps, is the natural continuity of the inverse function f^{-1}, whether f itself is continuous or not.

Theorem 13-2 If f is a strictly monotone function on an interval, then f^{-1} is a continuous function on the range of f.

Proof Without prejudice to the argument, assume f to be increasing. Let y_0 be a point of the range B of f, and write $x_0 = f^{-1}(y_0)$. Suppose x_0 to be an interior point of the domain of f. Let $\mathcal{N}(x_0,\varepsilon)$ be a neighborhood of x_0 which is entirely contained in the domain of f, and consider the numbers $y_1 = f(x_0 - \varepsilon)$ and $y_2 = f(x_0 + \varepsilon)$. Then, since f is strictly increasing, $y_1 < y_0 < y_2$.

Let $\delta > 0$ be such that $y_1 < y_0 - \delta < y_0 + \delta < y_2$. Then, for any member y of $B \cap \mathcal{N}(y_0,\delta)$ it is the case that $f^{-1}(y) \in \mathcal{N}(x_0,\varepsilon)$. Hence f^{-1} is continuous at y_0. If x_0 is a boundary point, obvious modification of the above argument leads to the same conclusion▲

Note that the conclusion of Theorem 13-2 is not in conflict with that of Theorem 13-1; that is, it is not to be inferred that continuity of f^{-1} implies, by Theorem 13-1, that of f. The reason, very simply, is that the domain R of f^{-1} is not necessarily an interval; it is so only in the event that f is continuous.

Although it is evident that a monotone function need not be continuous, it is not so obvious that, as proved next, the nature of its discontinuities is limited.

Theorem 13-3 Any discontinuity of a monotone function is necessarily of the first kind.

Proof Suppose f to be increasing on an interval, and let c be a cluster point of the domain from the right. Consider the set of function values $f(x)$ corresponding to points x with the property $c < x$. Since it is bounded below, by $f(c)$, an infimum $L \geq f(c)$ exists, and it follows that to each $\varepsilon > 0$ there corresponds a point x_0 such that $c < x_0$ and $L \leq f(x_0) < L + \varepsilon$. Because f is increasing, each point x with the property $c < x < x_0$ is such that $L \leq f(x) < L + \varepsilon$. This implies that f has a limit from the right at c and that its value is L. In like manner we may show that f has a limit from the left at each cluster point from the left. If f is a decreasing function a similar argument leads to a like conclusion▲

Corollary 1 If f is an increasing function and c is an interior point of its domain, then $f(c-) \leq f(c) \leq f(c+)$; if f is decreasing, then $f(c-) \geq f(c) \geq f(c+)$.

Corollary 2 A monotone function has at most a countable set of discontinuities.

Proof An immediate consequence of Theorem 8-4▲

Example 13-1 Let $\sum a_n$ be any convergent series of positive constants, and suppose the set R of all rational numbers to be arranged in a sequence $\{r_n\}$. Then consider a function f defined for all real x by $f(x) = \sum_{r_n \leq x} a_n$, the summation symbol indicating that one is to sum all terms a_n for which the corresponding rational number r_n is less than or at most equal to x.

Since the partial sums of $\sum_{r_n \le x} a_n$ for each x constitute an increasing sequence which is bounded above by $\sum a_n$, it is clear that each sum $\sum_{r_n \le x} a_n$ is a convergent series. It is clear also that f is an increasing function. As for its discontinuities, observe the following. If x is any rational number, say r_m, and if x' is any real number less than x, then $f(x') + a_m \le f(x)$. It follows that $f(x-) + a_m \le f(x)$, whence $f(x-) < f(x)$, since $a_m > 0$. On the other hand (see Corollary 1 to Theorem 13-3), it is the case for any x that $f(x) \le f(x+)$. Therefore $f(x-) < f(x+)$, testifying to the discontinuity of f at each rational point. It is left to the reader to show that f is continuous at each irrational point.

Further insight into the character of a monotone function is gained from the following construction of such a function based on values assigned on a dense set. Suppose $A = \{\alpha_n ;\quad n = 1,2,\ldots\}$ to be a countable and dense subset of $\mathscr{R}$, and let $\{\beta_n\}$ denote a sequence of numbers such that $\beta_n \le \beta_m$ if $\alpha_n < \alpha_m$. Consider a function f defined on $\mathscr{R}$ as follows. If $x \in A$, in which case $x = \alpha_n$ for some n, let $f(x) = \beta_n$; if $x \notin A$, then let $f(x) = \inf\{\beta_n ;\quad \alpha_n > x\}$.

The function f so constructed is easily seen to be increasing on $\mathscr{R}$. For suppose $x_1 < x_2$; then, since A is dense in $\mathscr{R}$, a member α_m of A exists such that $x_1 < \alpha_m < x_2$. As a consequence of the definition of f, we conclude $f(x_1) \le \beta_m \le f(x_2)$.

The above construction is valid as well on any bounded open interval (a,b). Furthermore, if the sequence $\{\beta_n\}$ associated with a dense subset of (a,b) is bounded, then, by appending the definitions $f(a) = \inf\{\beta_n ;\quad n = 1,2,\ldots\}$ and $f(b) = \sup\{\beta_n ;\quad n = 1,2,\ldots\}$, an increasing function is achieved on the closed interval $[a,b]$.

Should the functions defined on the set A by the sequence of values $\{\beta_n\}$ be continuous, it will follow that the function f (on any interval) is continuous and, accordingly, is the only increasing function having the values $\{\beta_n\}$ on A. (Compare Exercise 8-3.) On the other hand, if the function defined by $\{\beta_n\}$ is not continuous, then diversity is possible in the construction of f. Thus, we might define $f(x) = \beta_n$ if $x = \alpha_n$, and $f(x) = \sup\{\beta_n ;\quad \alpha_n < x\}$ if $x \notin A$. Indeed, at any discontinuity point of f which is not in A, any value in the interval $(\sup\{\beta_n ;\ \alpha_n < x\}, \inf\{\beta_n ;\quad x < \alpha_n\})$ is consistent with monotonicity.

Although an increasing function need not be continuous, it can always be normalized by proper adjustment of values on a countable set so as to be continuous from the right (or from the left) at each point. Let f be increasing on an interval, and let g be a function defined for each point x of the interval by $g(x) = f(x+)$. If $x_1 < x_2$, we then have $f(x_1+) \le f(x_2) \le f(x_2+)$, whence $g(x_1) \le g(x_2)$. Thus g is an increasing function on the given interval. Furthermore, let $\varepsilon > 0$ be given and x and x_1 such that $x_1 < x < x_1 + \varepsilon$. Then $g(x_1) \le g(x) = f(x+) \le f(x_1 + \varepsilon)$. Therefore $g(x_1) \le g(x_1+) \le f(x_1+) = g(x_1)$, which implies $g(x_1+) = g(x_1)$; that is, g is continuous from the right at x_1. It is clear that f and g have a common set of discontinuity points and that only at such points, if at all, do their values disagree.

13-2 A Decomposition of an Increasing Function

In this section we propose to show that if f is an increasing function, then it may be decomposed into a sum of two functions j and h, where j is a function having the same discontinuities as f and the same jumps at those discontinuities while h is increasing and continuous everywhere. Although we shall consider a function defined on $\mathscr{R}$, it is clear that a similar decomposition is valid for a function that is increasing on a bounded interval.

Suppose f to be increasing on $\mathscr{R}$. Then for each point t of $\mathscr{R}$ the difference $f(t+)-f(t-)$ is non-negative and, in particular, differs from zero on an at most countable set. Let j denote the function defined on $\mathscr{R}$ by

$$(1)\quad j(x)=\begin{cases} -[f(x+)-f(x)]-\sum\limits_{x<t<0}[f(t+)-f(t-)]-[f(0)-f(0-)] & \text{if } x<0 \\ 0 & \text{if } x=0 \\ [f(0+)-f(0)]+\sum\limits_{0<t<x}[f(t+)-f(t-)]+[f(x)-f(x-)] & \text{if } x>0, \end{cases}$$

where the index t in the summation symbols is to range over the countable collection of jump discontinuities of f. We refer to j as the *jump function* associated with f. Observe that it is constant on any interval that does not contain a discontinuity of f.

A few remarks are necessary to establish the fact that such a function does indeed exist corresponding to any given increasing function f, that is, that the summations appearing in definition (1), if not finite, are convergent infinite sums. Let $t_1, t_2, \ldots$ denote the distinct points of an interval (x', x'') for which $f(t+)-f(t-)>0$ and, for any integer N, consider the finite sum $\sum_{\nu=1}^{N}[f(t_\nu+)-f(t_\nu-)]$. Denote an ordered arrangement of the finite set $\{t_1, t_2, \ldots, t_N\}$ by $\{\tau_1, \tau_2, \ldots, \tau_N\}$, that is, $\tau_1<\tau_2<\cdots<\tau_N$ and, for each ν, $\tau_\nu = t_\mu$ for some μ. Let $a_0, a_1, \ldots, a_N$ be such that $x'<a_0<\tau_1$, $\tau_N<a_N<x''$, and $\tau_\nu<a_\nu<\tau_{\nu+1}$ for $\nu=1,2,\ldots,N-1$. Then $f(\tau_\nu+)-f(\tau_\nu-)\le f(a_\nu)-f(a_{\nu-1})$ for $\nu=1,2,\ldots,N$. It follows that

$$\begin{aligned}\sum_{\nu=1}^{N}[f(t_\nu+)-f(t_\nu-)] &= \sum_{\nu=1}^{N}[f(\tau_\nu+)-f(\tau_\nu-)] \le \sum_{\nu=1}^{N}[f(a_\nu)-f(a_{\nu-1})] \\ &= f(a_N)-f(a_0)\le f(x''-)-f(x'+).\end{aligned}$$

Therefore

$$(2)\qquad \sum_{\nu=1}^{\infty}[f(t_\nu+)-f(t_\nu-)]\le f(x''-)-f(x'+),$$

demonstrating that the sums in (1) converge.

That the jump function j is an increasing function on $\mathscr{R}$ is immediate from its definition. With respect to its continuity, the next theorem shows how its behavior is essentially that of f.

Theorem 13-4 The jump function j of a function f, defined and increasing on $\mathscr{R}$, is continuous at those and only those points at which f is continuous.

Proof Let $\varepsilon > 0$ be given. Then, because f is bounded on $(c, c+1)$, at most a finite set of points t of that interval are such that $f(t) - f(t-) > \varepsilon$. Hence $\delta_1 > 0$ exists with the property that $f(x) - f(x-) < \varepsilon$ for any x in $(c, c+\delta_1)$. Since, as we have seen, the summations involved in the definition of j converge, it follows that $\delta_2 > 0$ exists such that $\sum_{c<t<x}[f(t+) - f(t-)] < \varepsilon$ for any x in $(c, c+\delta_2)$.

Let $\delta = \min\{\delta_1, \delta_2\}$. Then from

$$(3)\quad j(x) - j(c) = f(x) - f(x-) + \sum_{c<t<x}[f(t+) - f(t-)] + [f(c+) - f(c)]$$

follows $f(c+) - f(c) \leq j(x) - j(c) \leq f(c+) - f(c) + 2\varepsilon$. Therefore

$$(4)\qquad j(c+) - j(c) = f(c+) - f(c)$$

with the obvious implication that j is continuous from the right at c if and only if f is. Continuity from the left, and hence continuity, follows by a similar argument▲

Example 13-2 The jump function j corresponding to f, defined by $f(x) = e^x$ if $x < 0$, $f(x) = x + 2$ if $0 \leq x < 1$, $f(1) = \pi$, and $f(x) = 2 + e^x$ if $1 < x$, is seen to be given by $j(x) = -1$ if $x < 0$, $j(x) = 0$ if $0 \leq x < 1$, $j(1) = \pi - 3$, and $j(x) = e - 1$ if $1 < x$. Graphs of the two functions are pictured in Figure 13–1.

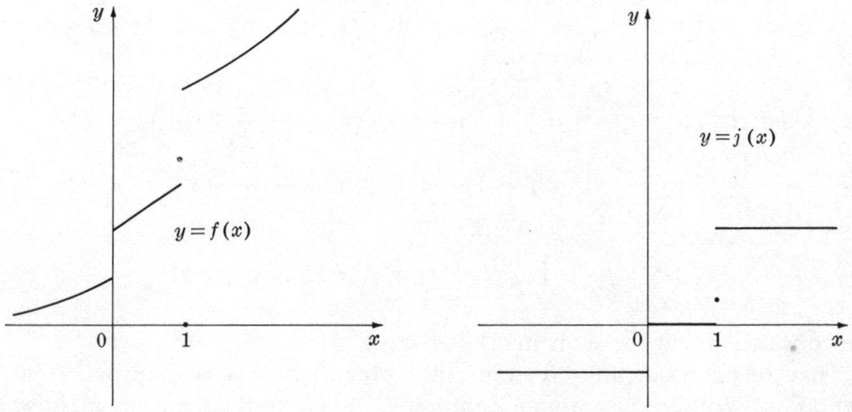

Figure 13-1

We are now in a position to achieve a decomposition of an increasing function into components having the properties stated at the outset of this section. The function j is, of course, the desired jump function. What remains amounts to showing that the difference $f-j$ is an increasing and continuous function.

Theorem 13-5 If f is any increasing function on $\mathscr{R}$, then it may be expressed as a sum of a jump function having the same discontinuities and jumps as f, and an increasing function which is continuous everywhere.

Proof We write $f=(f-j)+j$ and consider the function $h=f-j$. By (2) and (3) we conclude that $h(x') \leq h(x'')$ for $x' < x''$, whence h is increasing. Moreover, by (4),

$$h(c+)-h(c)=[f(c+)-j(c+)]-[f(c)-j(c)]=0.$$

In a similar fashion we determine $h(c-)=h(c)$. Therefore h is continuous at each point c▲

Example 13-3 The continuous function $h=f-j$ corresponding to the function f of Example 13-2 is given by $h(x)=1+e^x$ if $x<0$, $h(x)=x+2$ if $0\leq x\leq 1$, and $h(x)=3-e+e^x$ if $1<x$.

Exercises

13-1 Cite examples of increasing functions f and g such that $f-g$ is (a) increasing, (b) not increasing.

13-2 Cite examples of increasing functions f and g such that fg is (a) increasing, (b) not increasing.

13-3 Prove that if a function is increasing on some neighborhood of each point of (a,b), then it is increasing on (a,b).

13-4 Prove that a function is necessarily monotone on $[a,b]$ if it is continuous and has no local minimum or local maximum at any point of $[a,b]$.

13-5 Prove that the function defined in Example 13-1 is continuous at each irrational point.

13-6 Express the following functions as sums of jump functions and continuous monotone functions: (a) $f(x)=x-1$ for $x<0$, x^2 for $0\leq x<1$, 2 for $x=1$, and $x+4$ for $x>1$. (b) $f(x)=2-x$ for $-3\leq x<-1$, 2 for $-1\leq x<0$, $1-x$ for $0\leq x<1$, and $-x/2$ for $1\leq x\leq 2$.

13-3 Sequences of Monotone Functions

In a later chapter, in connection with an advanced theory of integration, we shall find it necessary to be acquainted with certain facts concerning sequences of functions whose elements are defined and monotone on $\mathscr{R}$. First we shall establish what is in the nature of a *propagation theorem*; that is, that under suitable conditions a sequence of monotone functions which converges to a limit function on a dense subset must also converge at all continuity points of the limit function. Convergence of a sequence to a limit f at the continuity points of f, but not necessarily elsewhere, is sometimes referred to as *weak convergence*.

A consequence of the propagation theorem will be *Helly's compactness theorem* (known also as *Helly's first theorem*), which testifies to the existence of a weakly convergent subsequence of a sequence of monotone functions. In both instances, according to policy, we restrict discussion to increasing functions.

Theorem 13-6 (Propagation Theorem) Let $\{f_n\}$ be a uniformly bounded sequence of functions, each element of which is defined and increasing on the set $\mathscr{R}$ of all real numbers. Suppose that at each point of a set D which is dense in $\mathscr{R}$ the sequence converges to a function f which is also defined and increasing on $\mathscr{R}$. Then $\{f_n\}$ converges to f at each continuity point of the latter function.

Proof Let x be any point of $\mathscr{R}$ and x_1 and x_2 points of D such that $x_1 < x < x_2$, Then for each n

$$f_n(x_1) \le f_n(x) \le f_n(x_2).$$

Hence

$$\lim \{f_n(x_1)\} \le \underline{\lim} \{f_n(x)\} \le \overline{\lim} \{f_n(x)\} \le \lim \{f_n(x_2)\}$$

or

$$f(x_1) \le \underline{\lim} \{f_n(x)\} \le \overline{\lim} \{f_n(x)\} \le f(x_2).$$

This being the case for all x_1 and x_2 as given, it follows that

$$f(x-) \le \underline{\lim} \{f_n(x)\} \le \overline{\lim} \{f_n(x)\} \le f(x+).$$

The assumption of continuity of f at x then implies that $\underline{\lim} \{f_n(x)\} = \overline{\lim} \{f_n(x)\} = f(x)$; that is, $\{f_n(x)\}$ converges to $f(x)$▲

Theorem 13-7 (Helly's Compactness Theorem) Let $\{f_n\}$ be a uniformly bounded sequence of functions each element of which is defined and increasing on $\mathscr{R}$. Then a subsequence of $\{f_n\}$ exists which converges weakly to a bounded and increasing limit function. Moreover, there is no loss in generality in assuming the limit function to be continuous from the right.

Proof Let $A = \{x_k;\ \ k = 1, 2, \ldots\}$ be a countable and dense subset of $\mathscr{R}$. Then the sequence of constants $\{f_n(x_1)\}$ is bounded and therefore possesses a convergent subsequence $\{f_{k_1(n)}(x_1)\}$. Let $f(x_1)$ denote the limit of this subsequence. Then $\{f_{k_1(n)}\}$ converges to $f(x_1)$ at $x = x_1$.

Next consider the sequence $\{f_{k_1(n)}(x_2)\}$. This, too, is bounded, and so possesses a subsequence $\{f_{k_2(n)}(x_2)\}$ converging to a limit $f(x_2)$. Since $\{f_{k_2(n)}\}$ is a subsequence of $\{f_{k_1(n)}\}$, it follows that $\{f_{k_2(n)}\}$ converges to $f(x_1)$ at x_1, as well as to $f(x_2)$ at x_2.

Continuing in this fashion, we find, for each integer j, a subsequence $\{f_{k_j(n)}\}$ of $\{f_{k_{j-1}(n)}\}$ which converges to a limit $f(x_v)$ at each x_v, $v = 1, 2, \ldots, j$.

Finally, consider the sequence $\{f_{k_n(n)}\}$. For each j, all but a finite set of elements of $\{f_{k_n(n)}\}$ are elements of $\{f_{k_j(n)}\}$. Hence $\{f_{k_n(n)}\}$ converges at x_ν for $\nu = 1, 2, \ldots$; that is, it converges on A to a limit function f defined by $f(x_\nu)$ at each point x_ν of A.

(The above procedure for determining a convergent subsequence is known as *Cantor's diagonal process*. The reader will recall an earlier instance in which the diagonal of an array played a significant role.)

The reader should supply the simple argument to show that from the increasing and uniformly bounded character of the functions $f_{k_n(n)}$, it follows that f is necessarily increasing and bounded.

The function f as constructed is defined on A. To extend its definition to $\mathscr{R}$, we employ the construction following Example 13-1. Then, by virtue of the propagation theorem, the proof of the present theorem is complete. The non-unique character of the extension of the definition of f from A to $\mathscr{R}$ accounts for the weak character of the convergence of a subsequence of $\{f_n\}$ to f. By means of the normalization discussed in Section 13-1, the limit function may be taken as continuous from the right at each point▲

Example 13-4 Let $\{r_n\}$ denote a sequence formed of the set of all positive rational numbers. Corresponding to each n let f_n be a function defined by $f_n(x) = 0$ if $x < 0$ and $f_n(x) = 1 - e^{-r_n x}$ if $x \geq 0$. Clearly, $\{f_n\}$ is a uniformly bounded sequence of increasing functions. As a consequence, it possesses a convergent subsequence.

Let $\{r_{k_n}\}$ be a subsequence of $\{r_n\}$ which converges to 0. Then $\{f_{k_n}\}$ converges to the (trivially increasing) zero function on $\mathscr{R}$.

That the constructed subsequence is not unique is easy to see. Let $\{r_{m_n}\}$ be a sequence of positive rational numbers which converges to 1. Then $\{f_{m_n}\}$ converges to the limit function g, where $g(x) = 0$ if $x < 0$ and $g(x) = 1 - e^{-x}$ if $x \geq 0$.

Exercises

13-7 Supply an argument to show that the limit function f of Theorem 13-7 is increasing and bounded.

13-4 Bounded Variation

Let f be a function defined on an interval $[a,b]$, and let $P = \{x_0, x_1, \ldots, x_n\}$ denote a partition of $[a,b]$. We call the non-negative number

$$T_f(P) = \sum_{\nu=1}^{n} |f(x_\nu) - f(x_{\nu-1})|$$

the *variation of f corresponding to P*. It is easy to see that if P' denotes a refinement of a partition P, that is, if $P \subset P'$, then $T_f(P) \leq T_f(P')$.

Consider the set of all variations of f corresponding to the set of all partitions of the interval $[a,b]$. If it is a bounded set, then we say f has *bounded variation* or that it is a *function of bounded variation* on $[a,b]$. We refer to the number

$$T_f^{[a,b]} = \sup \{T_f(P); \quad P \text{ a partition of } [a,b]\}$$

as the *total variation* of f on $[a,b]$.

Example 13-5 If f is a monotone function on $[a,b]$, then for each partition P the variation of f is $T_f(P) = |f(b) - f(a)|$. It follows that f is of bounded variation with $T_f^{[a,b]} = |f(b) - f(a)|$.

Example 13-6 Consider the function f defined on $[0,1]$ by $f(x) = x \sin(1/x)$ if $x \neq 0$ and $f(0) = 0$. For the partition $P_n = \{0, 2/2n\pi, 2/(2n-1)\pi, \ldots, 2/3\pi, 2/2\pi, 1\}$ it can be seen that $T_f(P_n) \geq (4/\pi)[1/3 + 1/5 + \cdots + 1/(2n-1)]$. Since $\{T_f(P_n)\}$ increases without bound, it follows that the given function is not one of bounded variation.

Examples 13-5 and 13-6 show, incidentally, that continuity is neither necessary nor sufficient for a function to be of bounded variation. On the other hand, although boundedness of a function does not insure bounded variation, it is certainly the case that only bounded functions can have this property. The reader should supply a simple proof.

With respect to remarks made earlier, the category of functions of bounded variation, unlike that of monotone functions, lends itself to elementary manipulation.

Theorem 13-8 If f and g are functions possessing bounded variation on an interval $[a,b]$, then so are the functions $f + g$, $f - g$, fg, and $|f|$; also kf if k is any constant. Further, if m exists such that $0 < m \leq |f|$, then $1/f$ has bounded variation.

Proof Left as an exercise▲

Example 13-7 Suppose f to be Riemann integrable on $[a,b]$, and let F be defined by $F(x) = \int_a^x f$. Since f is bounded, $m \leq f$ for some constant m. If $m \geq 0$, then $f \geq 0$, whence F increases and therefore is of bounded variation on $[a,b]$. On the other hand, if $m < 0$, then we write $F(x) = \int_a^x (f - m) + \int_a^x m$ and observe that this is the sum of an increasing and a decreasing function. As such it is again a function of bounded variation.

The next two theorems demonstrate an interrelation among the concepts of differentiation, Riemann integration, and bounded variation. The second provides a simple formula for total variation.

Theorem 13-9 If f is a function which is differentiable and such that $|f'| \leq M$ on $[a,b]$, then it is of bounded variation on $[a,b]$. Moreover, its total variation does not exceed $M(b-a)$.

Proof By applying the mean value theorem for derivatives we have, for each partition P of $[a,b]$,

$$T_f(P) = \sum_{\nu=1}^{n} |f(x_\nu) - f(x_{\nu-1})| = \sum_{\nu=1}^{n} |f'(\xi_\nu)|\,|x_\nu - x_{\nu-1}|,$$

where $x_{\nu-1} < \xi_\nu < x_\nu$ for $\nu = 1, \ldots, n$. Hence $T_f(P) \leq M(b-a)$▲

Theorem 13-10 If f is differentiable and $|f'|$ Riemann integrable on $[a,b]$, then f is of bounded variation on $[a,b]$ and $T_f^{[a,b]} = \int_a^b |f'|$.

Proof Since $|f'|$ is Riemann integrable, it is bounded, and f is of bounded variation by Theorem 13-9. Therefore, given $\varepsilon > 0$, there exists a partition P_1 of $[a,b]$ with the property that

(5) $$|T_f^{[a,b]} - T_f(P_1)| < \varepsilon.$$

It follows also from the R-integrability of $|f'|$ that $\delta > 0$ exists such that

(6) $$\left| R(|f'|,P_2,Q) - \int_a^b |f'| \right| < \varepsilon$$

for any partition P_2 with $\Delta P_2 < \delta$ and any intermediate partition Q of P_2.

Suppose P to be a refinement of the partition P_1 with the property $\Delta P < \delta$. Then (5) and (6) are valid with P replacing P_1 and P_2, respectively. Further, by the mean value theorem for derivatives, an intermediate partition $Q = \{\xi_1, \ldots, \xi_n\}$ of P exists such that

$$T_f(P) = \sum_{\nu=1}^{n} |f(x_\nu) - f(x_{\nu-1})| = \sum_{\nu=1}^{n} |f'(\xi_\nu)|\,\Delta x_\nu = R(|f'|,P,Q).$$

It follows that $T_f^{[a,b]} = \int_a^b |f'|$▲

Example 13-8 Consider the function $\sin x$ on the interval $[0,b]$. Since $|(\sin x)'| = |\cos x|$ is R-integrable on $[0,b]$ for any b, it follows that $\sin x$ is of bounded variation on $[0,b]$ and

$$T_{\sin x}^{[0,b]} = \int_0^b |\cos x| = \begin{cases} \sin b & \text{if } 0 \leq b \leq \frac{\pi}{2} \\ 2 - \sin b & \text{if } \frac{\pi}{2} < b \leq \frac{3\pi}{2} \\ 4 + \sin b & \text{if } \frac{3\pi}{2} < b \leq \frac{5\pi}{2} \\ \cdots & \cdots\cdots\cdots\cdots . \end{cases}$$

Variation of a function on an interval is related in a very simple manner to variation on subintervals. Suppose f to be defined on $[a,b]$, and let c be such that $a < c < b$. If P_1 and P_2 denote partitions of $[a,c]$ and $[c,b]$, respectively, then, directly from the definition, $T_f(P_1) + T_f(P_2) = T_f(P_1 \cup P_2)$. It follows that f has bounded variation on $[a,c]$ and $[c,b]$ if this is the case on $[a,b]$; moreover,

$T_f^{[a,c]} + T_f^{[c,b]} = T_f^{[a,b]}$. Extension to partitions of any number of segments is immediate. If f is of bounded variation on each member of the finite set of intervals $\{[a,a_1],[a_1,a_2],\ldots,[a_{n-1},b]\}$, then and only then is it of bounded variation on $[a,b]$, and the total variation on $[a,b]$ is the sum of the total variation on component intervals.

As a consequence, if f is a piecewise monotone function on an interval $[a,b]$, that is, if it is monotone on each member of a finite set of intervals $\{[a,a_1],\ldots,[a_{n-1},b]\}$, then it is of bounded variation on $[a,b]$. For example, the function $\sin x$ is piecewise monotone on any bounded interval. Hence it has bounded variation. (Refer to Example 13-8.)

Exercises

13-8 Prove that if a function is of bounded variation on an interval, then it is bounded.

13-9 Prove Theorem 13-8.

13-10 Prove that $T_{f+g}^{[a,b]} \leq T_f^{[a,b]} + T_g^{[a,b]}$ if f and g are functions of bounded variation on $[a,b]$.

13-11 Prove that if f and g are functions of bounded variation on $[a,b]$ and if A and B denote bounds of f and g, respectively, then $T_{fg}^{[a,b]} \leq AT_g^{[a,b]} + BT_f^{[a,b]}$.

13-12 Prove that if f is of bounded variation on $[a,b]$ and if $0 < m \leq |f|$, then $T_{1/f}^{[a,b]} \leq (1/m^2)T_f^{[a,b]}$.

13-13 Prove that the function f defined on $[-1,1]$ by $f(0) = 0$ and $f(x) = x^p \sin(1/x^q)$ if $x \neq 0$ is of bounded variation if $0 < q < p$ but not if $0 < p \leq q$.

13-14 Prove that a polynomial is of bounded variation on any closed interval.

13-5 Decomposition of a Function of Bounded Variation

In this section we shall see how a function of bounded variation may be characterized as the difference of two increasing functions. Fundamental properties of functions of bounded variation will then follow from properties of monotone functions.

Suppose f to be of bounded variation on $[a,b]$. Then for each x in $(a,b]$ the total variation $T_f^{[a,x]}$ exists. We denote it also by $t_f(x)$ and, for convenience, define $t_f(a) = 0$. The function t_f, defined on $[a,b]$, we style the *total variation function of f*. It is clear that t_f is an increasing function on $[a,b]$. Moreover, it is bounded since $t_f(x) \leq T_f^{[a,b]}$ for all x.

The function $r_f = t_f - f$ we call the *residual function* of f on $[a,b]$. It is easy to see that it, too, is an increasing function. For if x_1 and x_2 are members of $[a,b]$ and such that $x_1 < x_2$, then the difference $f(x_2) - f(x_1)$ certainly cannot exceed the total variation of f on $[x_1,x_2]$. Thus, $f(x_2) - f(x_1) \leq t_f(x_2) - t_f(x_1)$ or, what is the same,

$$r_f(x_1) = t_f(x_1) - f(x_1) \leq t_f(x_2) - f(x_2) = r_f(x_2).$$

The increasing nature of the total variation function and of the residual function enables us to devise a simple characterization of functions of bounded variation.

Theorem 13-11 A function f is of bounded variation on $[a,b]$ if and only if it is the difference of two increasing functions.

Proof We have already remarked that the difference of two increasing functions is a function of bounded variation. For the converse, we have only to note that if f is of bounded variation, then it may be expressed as the difference of its total variation and residual functions, that is, $f = t_f - r_f$▲

Representation of a function of bounded variation as the difference of increasing functions is not unique, of course, since $f = t_f - r_f = (t_f + g) - (r_f + g)$ is such a representation if g is any increasing function.

Since a monotone function has at most a countable set of discontinuities, all of the first kind, it follows from Theorem 13-11 that this is true as well of a function of bounded variation. As a consequence, if f is of bounded variation on $[a,b]$, then it is Riemann integrable on that interval. The converse is not true since we have seen that not even continuity assures bounded variation. The following provides further insight into the connection between continuity and bounded variation.

Theorem 13-12 A function f which is of bounded variation on an interval $[a,b]$ is continuous at a point c of that interval if and only if its total variation function t_f is continuous at c.

Proof Suppose t_f to be continuous at a point c of $[a,b]$. Then from the inequality $|f(x) - f(c)| \leq |t_f(x) - t_f(c)|$ for all x in $[a,b]$ follows continuity of f at c.

Imagine now that f is continuous at a point c of $[a,b)$. Since $t_f(c+)$ exists, it is the case that, corresponding to any given $\varepsilon > 0$, there exists a number $\delta > 0$ with the properties that $|f(x) - f(c)| < \varepsilon$ and $|t_f(x) - t_f(c+)| < \varepsilon$ for all $x \in [a,b] \cap (c, c+\delta)$. For any such x let $P = \{x_0, x_1, \ldots, x_n\}$ denote a partition of $[c,x]$. Then

$$\begin{aligned} T_f(P) &= |f(x_1) - f(c)| + \sum_{\nu=2}^{n} |f(x_\nu) - f(x_{\nu-1})| \\ &= |f(x_1) - f(c)| + T_f(P'), \end{aligned}$$

where $P' = \{x_1, x_2, \ldots, x_n\}$ is a partition of $[x_1,x]$. Now $T_f(P')$ does not exceed the total variation $t_f(x) - t_f(x_1)$ on $[x_1,x]$. Also, since t_f is increasing, $t_f(x) - t_f(x_1) \leq t_f(x) - t_f(c+)$. Proper combination of the above inequalities yields $T_f(P) < 2\varepsilon$. This being the case for all partitions P of $[c,x]$, we have $t_f(x) - t_f(c) = T_f^{[c,x]} \leq 2\varepsilon$. We conclude that t_f is continuous from the right at c. A similar argument leads to continuity from the left at each point c of $(a,b]$▲

Corollary 1 A function of bounded variation is continuous if and only if its residual function is continuous.

Corollary 2 A continuous function is a function of bounded variation on $[a,b]$ if and only if it is a difference of two continuous increasing functions.

Exercises

Let f be defined on $[a,b]$, and let $P = \{x_0,x_1,\ldots,x_n\}$ be a partition of $[a,b]$. Of the n numbers $f(x_\nu) - f(x_{\nu-1})$, $\nu = 1,\ldots,n$, denote by $M_f(P)$ the sum of those that are non-negative and by $-N_f(P)$ the sum of those that are negative. We then define the *total positive variation* and the *total negative variation* of f on $[a,b]$, respectively, by

$$M_f = \sup\{M_f(P);\quad P \text{ a partition of } [a,b]\}$$
$$N_f = \sup\{N_f(P);\quad P \text{ a partition of } [a,b]\}.$$

13-15 Prove that $T_f(P) = M_f(P) + N_f(P)$ for any partition P. Show that if f is of bounded variation, then M_f and N_f are finite.

13-16 Prove that if f is a function of bounded variation on $[a,b]$, then $T_f = M_f + N_f$.

Let f be defined on $[a,b]$, and denote by $m_f(x)$ and $n_f(x)$ the total positive variation and total negative variation, respectively, of f on $[a,x]$ for $a < x \leq b$. Also, let $m_f(a) = n_f(a) = 0$.

13-17 Show that if f if a function of bounded variation on $[a,b]$, then $f(x) = f(a) + m_f(x) - n_f(x)$. (This is a second representation of a function of bounded variation as a difference of increasing functions.)

13-18 Prove that if f is of bounded variation on $[a,b]$ and if g and h are increasing functions such that $f = g - h$ on $[a,b]$, then $m_f(x_2) - m_f(x_1) < g(x_2) - g(x_1)$ and $n_f(x_2) - n_f(x_1) \leq h(x_2) - h(x_1)$ for $a \leq x_1 \leq x_2 \leq b$. Because of this, we describe the decomposition of f in Exercise 13-17 as *minimal*.

13-19 Let f be defined on $[0,3]$ by $f(x) = x + 1$ if $0 \leq x \leq 1$, $2 - x$ if $1 < x < 2$, and $2x - 4$ if $2 \leq x \leq 3$. Determine the functions t_f, r_f, m_f, and n_f. Compare t_f with m_f and r_f with n_f.

13-20 For any x let $[x]$ denote the unique integer with the property that $x - 1 < [x] \leq x$. Determine the functions t_f, r_f, m_f, and n_f corresponding to the function defined on $[a,b]$ by $f(x) = x - [x]$.

13-21 Prove that if f is differentiable and f' Riemann integrable on $[a,b]$, then $m_f(x) = \int_a^x (f')^+$ and $n_f(x) = \int_a^x (f')^-$.

13-6 Absolute Continuity

A function f defined on a closed interval $[a,b]$ is said to be *absolutely continuous* on that interval if, corresponding to each $\varepsilon > 0$, there exists $\delta > 0$ such that for every finite collection $\{(a_1,b_1), (a_2,b_2),\ldots,(a_n,b_n)\}$ of disjoint intervals whose union is contained in $[a,b]$ and having total length $\sum_{\nu=1}^{n}(b_\nu - a_\nu) < \delta$, it is the case that $\sum_{\nu=1}^{n}|f(b_\nu) - f(a_\nu)| < \varepsilon$.

It is evident that this definition tends to present a concept somewhat more severe than that of simple continuity. Although we shall see shortly that not all functions which are continuous on an interval are also absolutely continuous, yet, with the choice $n = 1$ in the definition, it is apparent that absolute continuity implies uniform continuity of f on $[a,b]$.

Frequently the definition above is stated differently, with the finite collections of disjoint intervals replaced by countably infinite collections and the accompanying finite sums by infinite sums. As the reader may show, however, the two definitions are equivalent.

In a later chapter we shall see an important connection between the concepts of monotonicity and absolute continuity. For the present we simply note some of the basic properties of absolute continuity together with its relationship to the concept of bounded variation.

Theorem 13-13 If f and g are absolutely continuous functions on $[a,b]$, then so are the functions $f + g$, fg and kf for any constant k; also $1/f$ if $0 < m \le f$ for some m.

Proof Left as an exercise▲

Theorem 13-14 If f is absolutely continuous on $[a,b]$, then it is of bounded variation on that interval.

Proof Let $P = \{x_0,x_1,\ldots,x_n\}$ be a partition of $[a,b]$ such that $\Delta P < \delta$, where δ is a constant corresponding to $\varepsilon = 1$ in the definition of absolute continuity. Suppose $P_\nu = \{a_0,a_1,\ldots,a_m\}$ to denote a partition of the segment $[x_{\nu-1},x_\nu]$ of P for $\nu = 1, 2, \ldots, n$. Then

$$\sum_{i=1}^{m} |f(a_i) - f(a_{i-1})| < 1.$$

Thus the variation of f fails to exceed unity on each segment of P. It follows that the variation of f on $[a,b]$ is not greater than n; hence f is of bounded variation▲

Example 13-9 shows that the converse of the above theorem is not true, that, in fact, it is possible for a function to be continuous and to be of bounded variation and yet not be absolutely continuous.

Example 13-9 Recall the Cantor set K defined in Chapter 4 as a subset of the interval $[0,1]$ obtained by deleting certain open intervals. For convenience, we denote the deleted intervals as follows: $G_1{}^1 = (1/3, 2/3)$, $G_2{}^1 = (1/9, 2/9)$, $G_2{}^2 = (7/9, 8/9)$, $G_3{}^1 = (1/27, 2/27)$, $G_3{}^2 = (7/27, 8/27)$, $G_3{}^3 = (19/27, 20/27)$, $G_3{}^4 = (25/27, 26/27)$, and so forth according to an obvious pattern. It is clear that the intervals $G_n{}^m$ for all m and n are disjoint and that their union constitutes the complement of the Cantor set with respect to the interval $[0,1]$. That is, $\bigcup_{m,n} G_n{}^m = [0,1] - K$.

Now, let f denote a function defined on $[0,1] - K$ by

$$f(x) = \frac{2m-1}{2^n} \quad \text{for each } x \in G_n{}^m. \tag{7}$$

It is not difficult to see that f is an increasing function on its domain and, moreover, is continuous, having a constant value on each of the open intervals $G_n{}^m$.

Because of the increasing character of f, limits from both right and left exist at each point of the Cantor set K. Further, these limits are equal, for if x is any point of K, then, corresponding to each integer n, there exists m such that $G_n{}^m \subset (0,x)$ and $G_n^{m+1} \subset (x,1)$. Hence

$$\frac{2m-1}{2^n} \leq \lim_{x-} f \quad \text{and} \quad \lim_{x+} f \leq \frac{2(m+1)-1}{2^n}$$

so that $\lim_{x+} f - \lim_{x-} f \leq 1/2^{n-1}$ for all n.

In a very natural way we extend the definition of f to the domain $\mathscr{R}$ by defining

$$f(x) = \begin{cases} 0 & \text{if } x < 0 \\ \lim_x f & \text{if } x \in K \\ 1 & \text{if } x > 1. \end{cases} \tag{8}$$

Note, in particular, that $f(0) = 0$ and $f(1) = 1$. The function defined by (7) and (8) is continuous and increasing. By virtue of the latter property, it is also a function of bounded variation on [0,1]. We refer to it as the *Cantor function.*

Since, as we have already seen, the Cantor set K is a zero set, corresponding to each $\delta > 0$, there exists a countable collection $\{I_\mu;\quad \mu = 1,2,\ldots\}$ of open intervals such that $K \subset \bigcup_{\mu=1}^{\infty} I_\mu$ and $\sum_{\mu=1}^{\infty} \Delta I_\mu < \delta$. Furthermore, since K is compact, a finite subset of this collection also covers K. Replacing overlapping intervals from this finite set by the single interval they define, we arrive at a finite collection of disjoint open intervals $\{(a_1,b_1), (a_2,b_2), \ldots, (a_r,b_r)\}$ which covers K and is such that

$$\sum_{\nu=1}^{r} (b_\nu - a_\nu) < \delta. \tag{9}$$

If we assume, conveniently, that $b_\nu < a_{\nu+1}$ for $\nu = 1, 2, \ldots, r-1$, then, because the intervals cover K, it follows that for each ν the points b_ν and $a_{\nu+1}$ necessarily lie in the same open interval $G_n{}^m$. Hence $f(b_\nu) = f(a_{\nu+1})$. As a consequence,

$$\sum_{\nu=1}^{r} |f(b_\nu) - f(a_\nu)| = 1. \tag{10}$$

By virtue of (9) and (10) and the arbitrary character of δ, we are led to the conclusion that the Cantor function, although continuous and of bounded variation, is not absolutely continuous.

We may note, incidentally, another interesting feature of the Cantor function. Since f is constant on each of the open intervals $G_n{}^m$, it possesses a zero-valued derivative at each point of $[0,1] - K$. It is, therefore, an instance of a function with a zero derivative everywhere on [0,1] except on a zero set (the Cantor set) and yet is not a constant function on the interval.

13-7 Bounded Variation and Absolute Continuity on Unbounded Intervals

Throughout the preceding sections of this chapter we have been concerned only with functions defined on a bounded interval. At this point we add a few remarks about functions having domain $\mathscr{R}$.

We say that a function f is of *bounded variation on* $\mathscr{R}$ if it is of bounded variation on every bounded interval. It will follow quite easily that certain of the facts established in the case of bounded intervals are valid in the instance of an unbounded domain as well. In particular, if f and g have bounded variation on $\mathscr{R}$, then so have $f+g$, fg, and, if f is bounded away from 0, $1/f$. A monotone function on $\mathscr{R}$ is of bounded variation on $\mathscr{R}$, and from this it follows that the difference of two monotone functions also have this property. It should be noted, however, that although bounded variation on $[a,b]$ implies boundedness on that interval, this limitation does not extend to functions defined on $\mathscr{R}$. Witness the function defined by $f(x) = x$ for all x.

The *total variation function* t_f we define in the case of domain $\mathscr{R}$ by

$$t_f(x) = \begin{cases} -T_f^{[x,0]} & \text{if } x < 0 \\ T_f^{[0,x]} & \text{if } x \geq 0. \end{cases}$$

The *total variation of f on* $\mathscr{R}$ is then the number

$$\lim_{\infty} t_f - \lim_{-\infty} t_f,$$

which may be finite or infinite.

The *residual function* r_f of f is defined by $r_f = t_f - f$. Both t_f and r_f are increasing functions on $\mathscr{R}$ so that, once again, it is the case that a function of bounded variation may be expressed as the difference of increasing functions.

A function defined on $\mathscr{R}$ is said to be *absolutely continuous* if it is absolutely continuous on every bounded interval. Immediately, it is the case that $f+g$, fg, and, if k is a constant, kf are absolutely continuous on $\mathscr{R}$ if f and g, themselves, have this property. Further, if f is absolutely continuous on $\mathscr{R}$, then it has bounded variation on that domain.

Definitions and conclusions similar to those above may be enunciated for domains of the type $\{x;\, x \geq a\}$ and $\{x;\, x \leq a\}$. The reader may supply details.

Exercises

13-22 Prove that the concept of absolute continuity remains unchanged if the finite collection of disjoint intervals in the definition is replaced by a countably infinite collection of disjoint intervals and the accompanying finite sums by infinite sums.

13-23 Prove that the inequality $\sum_{\nu=1}^{n} |f(b_\nu) - f(a_\nu)| < \varepsilon$ in the definition of absolute continuity can be replaced with the weaker inequality $|\sum_{\nu=1}^{n} [f(b_\nu) - f(a_\nu)]| < \varepsilon$ without changing the sense of the definition.

13-24 Prove Theorem 13-13.

13-25 Investigate the possible absolute continuity on $[a,1]$, where $0 < a < 1$, of the function f defined by $f(x) = x \sin(1/x)$.

13-26 If f is absolutely continuous on $[a,1]$ for $0 < a < 1$ and if f is continuous at $x = 0$, does it follow that f is absolutely continuous on $[0,1]$? Explain. Suppose that f is of bounded variation as well on $[0,1]$, does that make a difference?

13-27 Prove that the indefinite Riemann integral of a function on $[a,b]$ is absolutely continuous.

13-28 Let f and g be functions defined on $(0,1]$ by $f(x)=\sqrt{x}$ and $g(x)=x^2|\sin(1/x)|$. Prove that f, g, and $g\circ f$ are absolutely continuous but that $f\circ g$ is not.

13-29 Suppose g to be a function defined on $[a,b]$ and having range $[c,d]$, and let f be defined on $[c,d]$. Prove that if f and g are both absolutely continuous and if g is monotone, then $f\circ g$ is absolutely continuous on $[a,b]$.

A function g satisfies a *uniform Lipschitz condition* on $[a,b]$ if there exists a constant N such that $|g(x_1)-g(x_2)|\leq N|x_1-x_2|$ for all points x_1 and x_2 of $[a,b]$.

13-30 Prove that g is absolutely continuous on $[a,b]$ if it satisfies a uniform Lipschitz condition on that interval.

13-31 Suppose g to be a function defined on $[a,b]$ and having range $[c,d]$, and let f be defined on $[c,d]$. Prove that if g satisfies a uniform Lipschitz condition on $[a,b]$ and if f is absolutely continuous on $[c,d]$, then $f\circ g$ is absolutely continuous on $[a,b]$.

13-32 Show that the definition of bounded variation on $\mathscr{R}$ is not equivalent to the definition on $[a,b]$ with $\mathscr{R}$ simply replacing $[a,b]$. (*Suggestion*: Consider the function $f(x)=x$ on $\mathscr{R}$.)

13-33 Prove that the function f defined by $f(x)=\sqrt{x}$ for $\{x;\ \ 0\leq x\}$ is absolutely continuous on its domain.

14
The Stieltjes Integral

It is always good when a man has two irons in the fire.
FRANCIS BEAUMONT (1584–1616), *The Faithful Friends*

14-1 Generalizing the Riemann Integral

Introductory remarks in Chapter 10 pointed to the Riemann integral as but a first stage in the development of what may be termed the *integral concept*. In the present chapter we embark on a generalization of the Riemann integral that is due to Thomas Jan Stieltjes. This generalization may be regarded as a rather natural one in that it recognizes the central role played by a simple linear function in the Riemann definition; it proposes to replace that function by one of a more general type.

The various definitions of $\int_a^b f$ we have seen to be founded on finite sums of the form $\sum_{\nu=1}^{n} m_\nu \, \Delta x_\nu$, $\sum_{\nu=1}^{n} M_\nu \, \Delta x_\nu$, or $\sum_{\nu=1}^{n} f(\xi_\nu) \, \Delta x_\nu$, in which the numbers $\Delta x_\nu = x_\nu - x_{\nu-1}$, corresponding to segments of a partition, appear as factors in the several terms. If we introduce in the above sums $\Delta g_\nu = g(x_\nu) - g(x_{\nu-1})$ in the place of Δx_ν, where g is a more or less arbitrary function, then a generalization of the Riemann construction is achieved which reduces to the Riemann case in the special instance $g(x) = x$.

In the Stieltjes definition a function or *integrand* f is said to be integrated with respect to or relative to an *integrator* g, and symbols such as $\int_a^b f \, dg$ and $\int_a^b f(x) \, dg(x)$ are employed to denote the integral.

The degree to which the integrator g is arbitrary will be seen to depend on the precise nature of the definition. It will be recalled that in the Riemann case three definitions were formulated and shown ultimately to be equivalent.

Following a similar pattern we shall propose three definitions for the Stieltjes integral paralleling those in the Riemann case. It will not turn out, however, that the three are equivalent or that they impose the same limitations on the functions f and g.

Discussion in the instance of the Riemann integral in Chapter 10 was dominated by Definition (A), based on the notion of lower and upper Riemann sums. In this chapter attention will be focused at the outset on a generalization of Definition (B), based on partitions and intermediate partitions. Ultimately, Definition (C), concerned with refinements of partitions, will be seen to be the most far reaching of the three formulations.

14-2 The Stieltjes Integral

Suppose f and g to denote quite arbitrary functions on an interval $[a,b]$. Let $P = \{x_0,x_1,\ldots,x_n\}$ denote a partition of $[a,b]$ and $Q = \{\xi_1,\xi_2,\ldots,\xi_n\}$ an intermediate partition of P, that is, $x_{\nu-1} \leq \xi_\nu \leq x_\nu$ for $\nu = 1, 2, \ldots, n$. We define the *Stieltjes sum* (or *S-sum*) of f relative to g on $[a,b]$ and corresponding to the partition P and the intermediate partition Q to be the number

$$S(f,g,P,Q) = \sum_{\nu=1}^{n} f(\xi_\nu)\, \Delta g_\nu ,$$

where $\Delta g_\nu = g(x_\nu) - g(x_{\nu-1})$.

The reader will note the similarity between the notion of a Stieltjes sum $S(f,g,P,Q)$ and that of the Riemann sum $R(f,P,Q)$ introduced to support Definition (B) of the Riemann integral. Accordingly, we shall refer to the definition of the Stieltjes integral which follows as Definition (B) also, since it is well to keep the likeness in mind.

Definition (B) (The Stieltjes Integral) A number $\mathop{\$}\nolimits_a^b f\, dg$ is said to be the *Stieltjes integral* (or *S-integral*) *of f relative to g on* $[a,b]$,[1] and f is said to be *Stieltjes integrable* (*S-integrable*) *relative to g on* $[a,b]$ if, corresponding to each $\varepsilon > 0$, there exists a $\delta > 0$ with the property that $|S(f,g,P,Q) - \mathop{\$}\nolimits_a^b f\, dg| < \varepsilon$ for all partitions P such that $\Delta P < \delta$ and all intermediate partitions Q of P.

The symbol $\mathop{\$}\nolimits_a^b f\, dg$, with a superimposed S to denote "Stieltjes" and to avoid any possible confusion with the Riemann integral, will be employed here for the further purpose of emphasizing the conceptual differences between the integral as defined above and those to follow later under the labeling of Definitions (A) and (C).

In keeping with the terminology of Chapter 10, we regard Definition (B) as a "limit type" definition and describe the Stieltjes integral as "the limit of the set of S-sums as the norm ΔP of the partition P tends to zero." Symbolically,

$$\mathop{\$}\nolimits_a^b f\, dg = \lim_{\Delta P \to 0} \sum_{\nu=1}^{n} f(\xi_\nu)\, \Delta g_\nu .$$

[1] The symbol $(S) \int_a^b f\, dg$ is often employed.

As in the Riemann case we append the definitions $\int_b^a f\,dg = -\int_a^b f\,dg$ for $a < b$, and $\int_a^a f\,dg = 0$.

Example 14-1 Suppose f to denote an arbitrary function and $g = k$ a constant function on $[a,b]$. Then, for any partition P and intermediate partition Q of P, we have $\sum_{\nu=1}^{n} f(\xi_\nu)\,\Delta g_\nu = \sum_{\nu=1}^{n} f(\xi_\nu)[g(x_\nu) - g(x_{\nu-1})] = 0$, whence it follows that $\int_a^b f\,dk = 0$. We conclude that any function f on $[a,b]$ is Stieltjes integrable relative to a constant integrator.

Example 14-2 Suppose $f = k$ to be a constant function on $[a,b]$, and let g be any function on that interval. Then, for any partition and intermediate partition, we have $\sum_{\nu=1}^{n} f(\xi_\nu)\,\Delta g_\nu = k \sum_{\nu=1}^{n} [g(x_\nu) - g(x_{\nu-1})]$ whence $\int_a^b k\,dg = k[g(b) - g(a)]$. In particular, if $k = 1$, then we conclude, writing $1\,dg = dg$, $\int_a^b dg = g(b) - g(a)$. There follows, too, the formula $\int_a^b k\,dg = k\int_a^b dg$.

Given functions f and g on $[a,b]$, we say that the set of all S-sums $S(f,g,P,Q)$, corresponding to all pairs of partitions and intermediate partitions on $[a,b]$, is *Cauchy convergent* if, for each $\varepsilon > 0$, there exists $\delta > 0$ with the property that $|S(f,g,P',Q') - S(f,g,P'',Q'')| < \varepsilon$ for all partitions P' and P'' for which $\Delta P' < \delta$ and $\Delta P'' < \delta$, and for all intermediate partitions Q' of P' and Q'' of P''. In the next theorem Cauchy convergence of the Stieltjes sums is seen to be a necessary and sufficient condition for the existence of a Stieltjes integral.

Theorem 14-1 A function f is Stieltjes integrable relative to a function g on $[a,b]$ if and only if the set of Stieltjes sums $S(f,g,P,Q)$ is Cauchy convergent.

Proof It is easy to show that existence of the Stieltjes integral implies Cauchy convergence of the Stieltjes sums. For the converse, suppose the set of S-sums to be Cauchy convergent. Let P_n for $n = 1, 2, \ldots$ denote the partition $\{x_0,x_1,\ldots,x_n\}$ defined by $x_\nu = a + [\nu(b-a)]/n$ and $Q_n = \{\xi_1,\xi_2,\ldots,\xi_n\}$ the associated intermediate partitions given by $\xi_\nu = x_\nu$ for $\nu = 1, 2, \ldots, n$. It is then the case that the sequence $\{S(f,g,P_n,Q_n)\}$ of S-sums is a Cauchy convergent sequence of constants and, as such, possesses a limit L. Therefore, given $\varepsilon > 0$, an integer N exists having the property that

$$|S(f,g,P_n,Q_n) - L| < \varepsilon \tag{1}$$

for all $n > N$. Also, because of the Cauchy convergence of the S-sums, $\delta > 0$ exists such that

$$|S(f,g,P',Q') - S(f,g,P'',Q'')| < \varepsilon$$

for $\Delta P' < \delta$ and $\Delta P'' < \delta$.

Let $\delta' = \min\{\delta, 1/N\}$, and consider only such n as exceed $1/\delta'$. Then P_n and Q_n satisfy (1) and, moreover, $\Delta P_n = 1/n < \delta'$. Denote by P any partition with $\Delta P < \delta'$ and by Q any intermediate partition of P. Then

$$|S(f,g,P,Q) - S(f,g,P_n,Q_n)| < \varepsilon.$$

Combining this with (1) yields $|S(f,g,P,Q) - L| < 2\varepsilon$▲

14-3 Some Conditions Implying Stieltjes Integrability

We see next that for integrands and integrators possessing certain elementary properties the Stieltjes integral exists. Later sections probe more deeply into both necessary and sufficient conditions for integrability.

Theorem 14-2 If f is a continuous function and g an increasing function on $[a,b]$, then f is Stieltjes integrable relative to g on $[a,b]$.

Proof Let $\varepsilon > 0$ be given. Then, from the implied uniform continuity of f, a number $\delta > 0$ exists such that $|f(\xi') - f(\xi'')| < \varepsilon$ whenever $|\xi' - \xi''| < \delta$. Let P' and P'' denote partitions of $[a,b]$ with norms less than δ, and let Q' and Q'' be intermediate to P' and P'', respectively. If $P''' = P' \cup P''$ and if Q''' is intermediate to P''', then it follows that the numbers $|S(f,g,P',Q') - S(f,g,P''',Q''')|$ and $|S(f,g,P'',Q'') - S(f,g,P''',Q''')|$ are both less than $\varepsilon \sum_{\nu=1}^{n} \Delta g_\nu = \varepsilon[g(b) - g(a)]$. Therefore

$$|S(f,g,P',Q') - S(f,g,P'',Q'')| < 2[g(b) - g(a)]\varepsilon;$$

that is, the set of S-sums is Cauchy convergent▲

Example 14-3 Since x is continuous and x^2 is increasing on $[0,1]$, it follows that $\int_0^1 x\,dx^2$ exists. To find its value consider the partition $P_n = \{x_0,x_1,\ldots,x_n\}$, where $x_\nu = \nu/n$, $\nu = 0, 1, \ldots, n$, and $n = 1, 2, \ldots$, and the intermediate partition $Q_n = \{x_1,x_2, \ldots,x_n\}$. Then

$$S(x,x^2,P_n,Q_n) = \sum_{\nu=1}^{n} x_\nu[x_\nu{}^2 - x_{\nu-1}^2] = \frac{1}{n^3}\sum_{\nu=1}^{n} (2\nu^2 - \nu) = \frac{(n+1)(4n-1)}{6n^2}.$$

We conclude

$$\int_0^1 x\,dx^2 = \lim_{n\to\infty} \frac{(n+1)(4n-1)}{6n^2} = \frac{2}{3}.$$

We have seen that a continuous function possesses a Riemann integral. Because a Riemann integral is simply a Stieltjes integral corresponding to the continuous integrator $g(x) = x$, we might be inclined to generalize and to conjecture that a continuous function possesses a Stieltjes integral relative to *any* continuous integrator. It is surprising, perhaps, that this is not the case; that is, continuity of both integrand and integrator is not sufficient to the existence of a Stieltjes integral. Example 14-4 bears out this observation.

Example 14-4 Let f and g be the continuous functions defined on $[0,1]$ by $f(x) = g(x) = \sqrt{x}\sin(1/x)$ if $x \neq 0$ and $f(0) = g(0) = 0$. Consider for any n the partition P_1 and intermediate partition Q_1 defined by

$$P_1 = \left\{0, \frac{1}{2n\pi}, \frac{1}{n}, \frac{2}{n}, \ldots, 1\right\} \quad\text{and}\quad Q_1 = \left\{\frac{1}{2n\pi}, \frac{1}{n}, \frac{2}{n}, \ldots, 1\right\}.$$

Clearly, $\Delta P_1 = 1/n$ and, since $g(0) = g(1/2n\pi) = 0$, we have

$$S(f,g,P_1,Q_1) = \sum_{\nu=1}^{n} f\left(\frac{\nu}{n}\right)\left[g\left(\frac{\nu}{n}\right) - g\left(\frac{\nu-1}{n}\right)\right].$$

Next consider a partition P_2 and corresponding intermediate partition Q_2 obtained through a refinement of P_1 based on a subdivision of the first segment of that partition; thus

$$P_2 = \left\{0, \frac{2}{(8n+1)\pi}, \frac{2}{8n\pi}, \frac{2}{(8n-1)\pi}, \ldots, \frac{2}{(4n+1)\pi}, \frac{2}{4n\pi}, \frac{1}{n}, \frac{2}{n}, \ldots, 1\right\},$$

$$Q_2 = \left\{\frac{2}{(8n+1)\pi}, \frac{2}{8n\pi}, \frac{2}{(8n-1)\pi}, \ldots, \frac{2}{(4n+1)\pi}, \frac{2}{4n\pi}, \frac{1}{n}, \frac{2}{n}, \ldots, 1\right\}.$$

By using the fact that $g(0) = g(2/4n\pi) = g(2/(8n+2)\pi) = 0$, it is easy to verify that for any n

(2)

$$S(f,g,P_2,Q_2) - S(f,g,P_1,Q_1) = \sum_{\nu=0}^{4n+1} f\left(\frac{2}{(4n+\nu)\pi}\right)\left[g\left(\frac{2}{(4n+\nu)\pi}\right) - g\left(\frac{2}{(4n+\nu+1)\pi}\right)\right].$$

(We see here that the difference of the Stieltjes sums is concerned only with terms arising from the refinement of the first segment of P_1. All other terms cancel.)

The right member of (2) is easily simplified and estimated. Since $f(2/2p\pi) = g(2/2p\pi) = 0$ for any integer p and since by definition $f(x) = g(x)$ for any x, we get

$$S(f,g,P_2,Q_2) - S(f,g,P_1,Q_1) = \sum_{p=0}^{2n}\left[f\left(\frac{2}{(4n+2p+1)\pi}\right)\right]^2$$

$$= \frac{2}{\pi}\sum_{p=0}^{2n}\frac{1}{4n+2p+1} \geq \frac{2}{\pi}\frac{2n+1}{8n+1} > \frac{1}{2\pi}$$

for all n. The set of Stieltjes sums, therefore, is not Cauchy convergent, and we must conclude that $\mathfrak{S}_0^1 f\,dg$ does not exist.

Under certain conditions a simple relationship exists between Riemann and Stieltjes integration, apart from the role of the former as a special case of the latter.

Theorem 14-3 If f is a Riemann integrable function on $[a,b]$ and if g is a function possessing a continuous derivative g' on that interval, then $\int_a^b fg'$ and $\mathfrak{S}_a^b f\,dg$ exist and are equal.

Proof Let $\varepsilon > 0$ be given. Since g' is uniformly continuous on $[a,b]$ and since fg' is clearly Riemann integrable, a number $\delta > 0$ exists such that $|g'(\xi) - g'(\xi^*)| < \varepsilon$ and $\left|R(fg',P,Q) - \int_a^b fg'\right| < \varepsilon$ whenever $|\xi - \xi^*| < \delta$ and $\Delta P < \delta$. Consider the Stieltjes sum based on $P = \{x_0,x_1,\ldots,x_n\}$, with $\Delta P < \delta$ and $Q = \{\xi_1,\xi_2,\ldots,\xi_n\}$. Because g is differentiable on the segments $[x_{\nu-1},x_\nu]$

of P, there exists a set of numbers $\{\xi_1{}^*,\xi_2{}^*,\ldots,\xi_n{}^*\}$ such that $\Delta g_\nu = g'(\xi_\nu{}^*)\,\Delta x_\nu$ and $x_{\nu-1} < \xi_\nu{}^* < x_\nu$ for $\nu = 1, 2, \ldots, n$. It follows that

$$|R(fg',P,Q) - S(f,g,P,Q)| = \left|\sum_{\nu=1}^{n} f(\xi_\nu)[g'(\xi_\nu) - g'(\xi_\nu{}^*)]\,\Delta x_\nu\right| < M(b-a)\varepsilon,$$

where M denotes an upper bound of $|f|$. From this it easily follows that $\mathop{\$}_a^b f\,dg$ exists and equals $\int_a^b fg'$ ▲

Example 14-5 The value of the integral of Example 14-3 follows very easily from the previous result. We have $\mathop{\$}_0^1 x\,dx^2 = \int_0^1 2x^2 = \frac{2}{3}$.

14-4 Bilinearity of the Stieltjes Integral

The next two theorems demonstrate a bilinear character of the Stieltjes integral. The first generalizes what was seen to be the case for the Riemann integral, that is, linearity with respect to the integrand. The second, with no counterpart in the Riemann case, testifies to a linear property with respect to the integrator.

Theorem 14-4 If f_1 and f_2 are Stieltjes integrable functions relative to g on $[a,b]$, then so is the function $k_1 f_1 + k_2 f_2$, where k_1 and k_2 are constants. Moreover,

$$\mathop{\$}_a^b (k_1 f_1 + k_2 f_2)\,dg = k_1 \mathop{\$}_a^b f_1\,dg + k_2 \mathop{\$}_a^b f_2\,dg.$$

Proof Left as an exercise ▲

Theorem 14-5 If f is Stieltjes integrable relative to each of the functions g_1 and g_2 on $[a,b]$, then it is Stieltjes integrable relative to $k_1 g_1 + k_2 g_2$, where k_1 and k_2 are constants. Moreover,

$$\mathop{\$}_a^b f\,d(k_1 g_1 + k_2 g_2) = k_1 \mathop{\$}_a^b f\,dg_1 + k_2 \mathop{\$}_a^b f\,dg_2.$$

Proof An immediate consequence of the definition of the Stieltjes integral and the obvious relation

$$S(f, k_1 g_1 + k_2 g_2, P, Q) = k_1 S(f,g_1,P,Q) + k_2 S(f,g_2,P,Q)\blacktriangle$$

Recall that if g is a function of bounded variation on $[a,b]$, then we may write $g = g_1 - g_2$, where g_1 and g_2 are increasing functions. By Theorem 14-2, both $\mathop{\$}_a^b f\,dg_1$ and $\mathop{\$}_a^b f\,dg_2$ exist if f is continuous. It follows by Theorem 14-5 that f is Stieltjes integrable relative to $g_1 - g_2$; that is, $\mathop{\$}_a^b f\,dg$ exists if f is continuous and g of bounded variation.

14-5 The Interval of Integration

The simple linear relationship of Riemann integrals on adjacent intervals to the Riemann integral on the union of those intervals generalizes only in part to the Stieltjes case.

Theorem 14-6 If f is Stieltjes integrable relative to g on $[a,b]$, and if $a < c < b$, then f is Stieltjes integrable relative to g on each of the intervals $[a,c]$ and $[c,b]$. Moreover,

$$\int_a^b f\,dg = \int_a^c f\,dg + \int_c^b f\,dg. \tag{3}$$

Proof Let $\varepsilon > 0$ be given. Then $\delta > 0$ exists such that

$$|S(f,g,P,Q) - S(f,g,P^*,Q^*)| < \varepsilon \tag{4}$$

for any partitions P and P^* of $[a,b]$ for which $\Delta P < \delta$ and $\Delta P^* < \delta$ and for any corresponding intermediate partitions Q and Q^*.

Suppose P_1 and $P_1{}^*$, with $\Delta P_1 < \delta$ and $\Delta P_1{}^* < \delta$, to be partitions of $[a,c]$, and let Q_1 and $Q_1{}^*$ be intermediate to P_1 and $P_1{}^*$, respectively. Furthermore, let P_2 denote a partition of $[c,b]$ such that $\Delta P_2 < \Delta P_1$ and $\Delta P_2 < \Delta P_1{}^*$. Let Q_2 be intermediate to P_2.

Clearly, $P = P_1 \cup P_2$ and $P^* = P_1{}^* \cup P_2$ are partitions of $[a,b]$ with $\Delta P = \Delta P_1$ and $\Delta P^* = \Delta P_1{}^*$. Also, $Q = Q_1 \cup Q_2$ and $Q^* = Q_1{}^* \cup Q_2$ are intermediate to P and P^*, respectively.

It is easy to verify that

$$\begin{aligned} S(f,g,P,Q) &= S(f,g,P_1,Q_1) + S(f,g,P_2,Q_2), \\ S(f,g,P^*,Q^*) &= S(f,g,P_1{}^*,Q_1{}^*) + S(f,g,P_2,Q_2). \end{aligned} \tag{5}$$

Therefore $S(f,g,P,Q) - S(f,g,P^*,Q^*) = S(f,g,P_1,Q_1) - S(f,g,P_1{}^*,Q_1{}^*)$. Since $\Delta P < \delta$ and $\Delta P^* < \delta$, the left member satisfies (4). Hence $|S(f,g,P_1,Q_1) - S(f,g,P_1{}^*,Q_1{}^*)| < \varepsilon$. We conclude that the set of Stieltjes sums of f relative to g on $[a,c]$ is Cauchy convergent; hence $\int_a^c f\,dg$ exists. In a similar fashion $\int_c^b f\,dg$ also exists. Equation (3) is a simple consequence of (5)▲

Corollary If $\int_a^b f\,dg$ exists and if $[c,d] \subset [a,b]$, then $\int_c^d f\,dg$ exists also.

Oddly enough it is not proper to conclude from the existence of the Stieltjes integrals $\int_a^c f\,dg$ and $\int_c^b f\,dg$ that f is Stieltjes integrable relative to g on the full interval $[a,b]$. Thus do we have a departure from the situation for Riemann integrals. The following example is a case in point.

Example 14-6 Suppose f and g to be functions defined on the interval $[0,2]$ by

$$f(x) = \begin{cases} 0 & \text{if } 0 \le x < 1 \\ 1 & \text{if } 1 \le x \le 2, \end{cases} \qquad g(x) = \begin{cases} 0 & \text{if } 0 \le x \le 1 \\ 1 & \text{if } 1 < x \le 2. \end{cases}$$

By Example 14-1, since g is constant on [0,1], we deduce that $\int_0^1 f\,dg$ exists and has the value zero. On [1,2] f is constant. Hence, by Example 14-2, $\int_1^2 f\,dg$ exists with value $g(2)-g(1)=1$.

To investigate Stieltjes integrability of f relative to g on [0,2], let $P=\{x_0,x_1,\ldots,x_n\}$ be any partition of [0,2] which does not include 1; that is, for some μ it is the case that $x_{\mu-1}<1<x_\mu$. As a consequence, $\Delta g_\mu=1$, and $\Delta g_\nu=0$ for $\nu\neq\mu$.

Let $Q=\{\xi_1,\xi_2,\ldots,\xi_n\}$ be intermediate to P and such that $\xi_\nu=x_{\nu-1}$ for $\nu=1,2,\ldots,n$. In particular, $\xi_\mu<1$, so that $f(\xi_\mu)=0$. Therefore, for the Stieltjes sum corresponding to P and Q we have $S(f,g,P,Q)=0$.

Now consider an intermediate partition $Q^*=\{\xi_1{}^*,\xi_2{}^*,\ldots,\xi_n{}^*\}$ of P defined by $\xi_\nu{}^*=x_\nu$ for $\nu=1,2,\ldots,n$. Then $\xi_\mu{}^*>1$, whence $f(\xi_\mu{}^*)=1$ and $S(f,g,P,Q^*)=1$.

It follows that, however small the norm of P, by the above construction we have $|S(f,g,P,Q)-S(f,g,P,Q^*)|=1$, so that the set of Stieltjes sums is not Cauchy convergent. Hence $\int_0^2 f\,dg$ does not exist.

Later we shall prove generally what ostensibly is the case above, that underlying the nonexistence of the integral on an interval is the possession by f and g of a common discontinuity point.

Exercises

14-1 Let g be an increasing function. Prove that f is Stieltjes integrable relative to g if and only if, for each $\varepsilon>0$, there is a $\delta>0$ such that $|S(f,g,P,Q)-S(f,g,P,Q^*)|<\varepsilon$ for any partition P for which $\Delta P<\delta$ and for any intermediate partitions Q and Q^* of P.

14-2 Prove Theorem 14-4.

14-3 Complete the proof of Theorem 14-6 by supplying details that lead to equation (3).

14-6 Integration by Parts

In the present section we develop an expression for a Stieltjes integral that involves an interchange of roles of integrand and integrator. The reader will be reminded of the counterpart styled "integration by parts" in the Riemann case.

Theorem 14-7 If f is Stieltjes integrable relative to g on $[a,b]$, then g is Stieltjes integrable relative to f on that interval, and the respective integrals are related by

$$\int_a^b f\,dg = f(b)g(b) - f(a)g(a) - \int_a^b g\,df. \tag{6}$$

Proof Let $P=\{x_0,x_1,\ldots,x_n\}$ be a partition of $[a,b]$ and $Q=\{\xi_1,\xi_2,\ldots,\xi_n\}$ an intermediate partition of P. Suppose Q^* to denote the set Q with points $\xi_0=a$ and $\xi_{n+1}=b$ adjoined. Then $Q^*=\{\xi_0,\xi_1,\ldots,\xi_{n+1}\}$ is a partition of $[a,b]$. (Technically Q^* is not a partition in the strict sense since some consecutive pairs of points may coincide. In such a case we simply discard repetitions. The reader should verify that the remainder of the proof is indifferent to such adjustment.) Since $\xi_{\nu-1} \leq x_{\nu-1} \leq \xi_\nu$ for $\nu = 1, 2, \ldots, n$, it follows that P is an intermediate partition of Q^*. It is clearly the case that $\Delta Q^* \leq 2\Delta P$.

For the Stieltjes sum of f relative to g corresponding to the partition Q^* and intermediate partition P, it is easy to verify that

$$S(f,g,Q^*,P) = f(b)g(b) - f(a)g(a) - S(g,f,P,Q).$$

Let $\varepsilon > 0$ be given. Then, since $\int_a^b f\,dg$ exists, there is a number $\delta > 0$ such that $|S(f,g,Q^*,P) - \int_a^b f\,dg| < \varepsilon$ for $\Delta Q^* < \delta$. Suppose then that P is such that $\Delta P < \delta/2$. Then $\Delta Q^* < \delta$ and

$$\left| S(g,f,P,Q) - \left[f(b)g(b) - f(a)g(a) - \int_a^b f\,dg \right] \right| < \varepsilon.$$

This establishes at once the existence of $\int_a^b g\,df$ as well as the relation (6) between the integrals▲

Stieltjes integrals of the function x relative to an integrator g are conveniently reduced to a Riemann integral by formula (6).

Example 14-7 $\int_0^{\pi/2} x\,d\sin x = x\sin x\Big|_0^{\pi/2} - \int_0^{\pi/2} \sin x\,dx = \pi/2 - 1.$

A nice result of the "by parts" formula is that it enables us to deduce certain conditions on f and g from others. For example, we have seen that if f is continuous and g of bounded variation on $[a,b]$, then f is Stieltjes integrable relative to g on $[a,b]$. We may conclude from (6) that $\int_a^b f\,dg$ also exists if f is of bounded variation and g is continuous.

14-7 Change of Variable

As in the case of the Riemann integral, the introduction of new variables frequently facilitates evaluation of the integral.

Theorem 14-8 Suppose f to be Stieltjes integrable relative to g on $[a,b]$. Let h denote a function which is strictly increasing and continuous on an interval $[p,q]$ and such that $h(p) = a$ and $h(q) = b$. Imagine F and G to be functions defined on $[p,q]$ by $F = f \circ h$ and $G = g \circ h$. Then F is Stieltjes integrable relative to G on $[p,q]$ and $\int_p^q F\,dG = \int_a^b f\,dg$.

Proof Because of the stated properties of the function h, its inverse h^{-1} exists, and each establishes a one-to-one correspondence between $[a,b]$ and $[p,q]$.

If $P = \{x_0, x_1, \ldots, x_n\}$ is a partition of the interval $[p,q]$, then $P' = \{x_0', x_1', \ldots, x_n'\}$ is a partition of $[a,b]$, where $x_\nu' = h(x_\nu)$. Conversely, to the partition P' of $[a,b]$ corresponds the partition P of $[p,q]$ by virtue of $x_\nu = h^{-1}(x_\nu')$.

Let $\varepsilon > 0$ be given. Then the existence of $\int_a^b f\,dg$ implies that of $\delta' > 0$ with the property

$$(7) \qquad \left| S(f,g,P',Q') - \int_a^b f\,dg \right| < \varepsilon$$

as long as $\Delta P' < \delta'$. Because h is uniformly continuous on $[p,q]$, a number $\delta > 0$ exists with the property that $|h(u) - h(v)| < \delta'$ for all u and v in $[p,q]$ such that $|u - v| < \delta$. Hence if P is a partition of $[p,q]$ with $\Delta P < \delta$, then the corresponding partition P' of $[a,b]$ is such that $\Delta P' < \delta'$.

The Stieltjes sum of F relative to G on $[p,q]$ is

$$S(F,G,P,Q) = \sum_{\nu=1}^{n} F(\xi_\nu)[G(x_\nu) - G(x_{\nu-1})]$$

$$= \sum_{\nu=1}^{n} f(\xi_\nu')[g(x_\nu') - g(x_{\nu-1}')] = S(f,g,P',Q'),$$

where $x_\nu' = h(x_\nu)$ and $\xi_\nu' = h(\xi_\nu)$. We conclude by (7) that $|S(F,G,P,Q) - \int_a^b f\,dg| < \varepsilon$ if $\Delta P < \delta$ ▲

If the function h of Theorem 14-8 were to be discontinuous at a point c of $[a,b]$, then it would be possible for both F and G to be discontinuous at c, in which event the integral $\int_p^q F\,dG$ fails to exist. It is not possible, therefore, to relax the condition of continuity imposed on h in the hypothesis of the theorem.

14-8 Further Conditions Relating to the Existence of a Stieltjes Integral

The suggestion has already been made that if integrand f and integrator g have a discontinuity point in common, then the integral of f relative to g may fail to exist. The intrinsic role played by discontinuities is revealed in the theorems of this section.

Theorem 14-9 If f is Stieltjes integrable relative to g on $[a,b]$, then at each point of the interval either f or g is continuous.

Proof Suppose c to be such that $a < c < b$, and assume that both f and g are discontinuous at c. Consider two cases.

First, suppose $g(c-)$ and $g(c+)$ to exist with a common value. Define $k = |g(c) - g(c+)| = |g(c) - g(c-)| > 0$. Then at each point x of some deleted neighborhood $\mathcal{N}^*(c,\delta)$ of c it is the case that $|g(x) - g(c)| > k/2$. Thus, if P is a partition of $[a,b]$ having c as a member and if $\Delta P < \delta$, then, corresponding to the adjacent segments $[x_{\mu-1},c]$ and $[c,x_{\mu+1}]$ of P, we have

$$(8) \qquad |\Delta g_\mu| > \frac{k}{2} \quad \text{and} \quad |\Delta g_{\mu+1}| > \frac{k}{2}.$$

It follows from its discontinuity that the oscillation of f on at least one of the segments $[x_{\mu-1},c]$ or $[c,x_{\mu+1}]$, say the former, exceeds $\frac{1}{3}\omega_c f$. Therefore, points ξ_μ and $\xi_\mu{}^*$ exist in $[x_{\mu-1},c]$ with the property

(9) $$|f(\xi_\mu) - f(\xi_\mu{}^*)| > \tfrac{1}{3}\omega_c f.$$

Now, let Q denote any intermediate partition of P which includes ξ_μ, and let Q^* be that obtained from Q by replacing ξ_μ by $\xi_\mu{}^*$. Then from (8) and (9) follows

$$|S(f,g,P,Q) - S(f,g,P,Q^*)| = |f(\xi_\mu) - f(\xi_\mu{}^*)|\,|\Delta g_\mu| > \frac{k}{6}\,\omega_c\, f > 0.$$

Since the norm of P is arbitrarily small, we conclude that $\int_a^b f\,dg$ does not exist.

Second, assume the discontinuous behavior of g at c to be other than that treated above. Then $\delta > 0$ being given, it is possible to find $\varepsilon_1 > 0$ and points c_1 and c_2 such that $c - \delta/2 < c_1 < c < c_2 < c + \delta/2$ and $|g(c_2) - g(c_1)| > \varepsilon_1$. If P is a partition of $[a,b]$ with $\Delta P < \delta$ and such that $x_{\mu-1} = c_1$ and $x_\mu = c_2$ for some μ, then $|\Delta g_\mu| > \varepsilon_1$.

Since the oscillation of f on $[x_{\mu-1},x_\mu]$ exceeds $\frac{1}{2}\omega_c f$, it is possible, as in the situation treated first, to find intermediate partitions Q and Q^* differing only in the μth place for which

$$|S(f,g,P,Q) - S(f,g,P,Q^*)| = |f(\xi_\mu) - f(\xi_\mu{}^*)|\,|\Delta g_\mu| > \tfrac{1}{2}\varepsilon_1\omega_c f.$$

Again, since δ is arbitrary, we conclude that the integral does not exist.

With some modifications the above arguments are valid also at the endpoints a and b. The reader should supply details▲

In the event the integrator g is an increasing function, it is possible to state a condition which is both necessary and sufficient to the existence of a Stieltjes integral. Essentially, it concerns a limitation on the variation of g on the set of discontinuity points of the integrand f.

Theorem 14-10 Suppose f to be a bounded function and g an increasing function on $[a,b]$. Let D denote the set of discontinuity points of f. Corresponding to each point x in D consider the set $B_x = \{y;\ \ g(x-) \le y \le g(x+)\}$. Define $B = \bigcup_{x \in D} B_x$. Then f is Stieltjes integrable relative to g on $[a,b]$ if and only if B is a zero set.

Proof Suppose $\int_a^b f\,dg$ to exist. Then, because of Theorem 14-9, g is necessarily continuous at each point of the set D. Hence B_x consists of the single number $g(x)$ for each $x \in D$.

Corresponding to $n = 1, 2, \ldots,$ let $D_n = \{x;\ \ \omega_x f \ge 1/n\}$, so that $D = \bigcup_{n=1}^\infty D_n$. Further, define $C_n = \{g(x);\ \ x \in D_n\}$, so that $B = \bigcup_{n=1}^\infty C_n$.

Let $\varepsilon > 0$ be given. Then a partition P exists such that

$$|S(f,g,P,Q) - S(f,g,P,Q^*)| < \varepsilon$$

for any intermediate partitions Q and Q^* of P. Corresponding to each integer n, let A_n denote the set of all subscripts ν such that the segment $[x_{\nu-1},x_\nu]$ of P contains a point of D_n in its interior. It follows that such a segment contains points ξ_ν and $\xi_\nu{}^*$ with the property $f(\xi_\nu) - f(\xi_\nu{}^*) > 1/2n$. For $\nu \notin A_n$, choose points $\xi_\nu = \xi_\nu{}^*$, thereby defining intermediate partitions of P which differ only at points corresponding to $\nu \in A_n$. Denoting such intermediate partitions by Q and Q^*, we have $|S(f,g,P,Q) - S(f,g,P,Q^*)| > (1/2n) \sum_{\nu \in A_n} \Delta g_\nu$. We conclude that $\sum_{\nu \in A_n} \Delta g_\nu < 2n\varepsilon$.

The finite collection of open intervals $(g(x_{\nu-1}),g(x_\nu))$, corresponding to $\nu \in A_n$, covers the set C_n, with the possible exception of points $g(x)$ such that $x \in P \cap D_n$. Since such exceptions constitute a finite set, they may be covered by a finite collection of open intervals of total length not exceeding ε. It is therefore possible for each value of n to cover all of C_n with a finite collection of open intervals having total length less than $(2n + 1)\varepsilon$. It follows that C_n is a zero set for each n; hence B is a zero set.

For the converse, assume B to be a zero set, and let $\varepsilon > 0$ be given. Denote by $\{I_n\}$ a sequence of open intervals covering B and having total length less than ε.

Note that g is necessarily continuous at each point of D, since if $g(x-) < g(x+)$ for some point x of D, then B_x is an interval, and, as a consequence, B is not a zero set. Thus, the set B_x, corresponding to any point x of D, consists only of the point $g(x)$.

Consider the closed set $D_\varepsilon = \{x; \quad \omega_x f \geq \varepsilon\}$. If c is any point of D_ε, then, because of the continuity of g at c, a number δ_c exists such that $g(x) \in \bigcup_{n=1}^{\infty} I_n$ for $x \in [a,b] \cap \mathcal{N}(c,\delta_c)$. The collection of all neighborhoods $\mathcal{N}(c, \delta_c/2)$ covers D_ε and, as a consequence, a finite set

$$\left\{\mathcal{N}\left(c_1,\frac{\delta_{c_1}}{2}\right),\mathcal{N}\left(c_2,\frac{\delta_{c_2}}{2}\right),\ldots,\mathcal{N}\left(c_s,\frac{\delta_{c_s}}{2}\right)\right\} \tag{10}$$

exists which covers D_ε. Let δ_1 denote the minimum of the numbers $\delta_{c_1}/2$, $\delta_{c_2}/2, \ldots, \delta_{c_s}/2$.

For each $x \in [a,b]$, there is a neighborhood $\mathcal{N}(x,\varepsilon_x)$ with $\Omega_{\mathcal{N}(x,\varepsilon_x)} f \leq \omega_x f + \varepsilon$. Choose such a neighborhood for each x, subject to the additional restriction that $\varepsilon_x < \delta_1$. Then a finite collection

$$\{\mathcal{N}(x_1,\varepsilon_{x_1}),\mathcal{N}(x_2,\varepsilon_{x_2}),\ldots,\mathcal{N}(x_t,\varepsilon_{x_t})\} \tag{11}$$

necessarily covers $[a,b]$. Consider all endpoints of these neighborhoods which lie in $[a,b]$. These, together with a and b, constitute a partition of $[a,b]$. Let δ_2 denote the minimum distance between adjacent members of this partition.

Now, let $\delta = \min\{\delta_1,\delta_2\}$, and consider any partition P of $[a,b]$ with $\Delta P < \delta$ and any Q and Q^* intermediate to P. Since each segment of P has length less than δ_2, it can contain at most one endpoint of the neighborhoods of (11). Thus, each segment of P must lie in at least one of these neighborhoods. We distinguish two types of segments and say a segment is of type I if a neighborhood $\mathcal{N}(x_\mu,\varepsilon_{x_\mu})$ of the collection (11) exists which contains the segment and is such that $x_\mu \notin D_\varepsilon$; otherwise, a segment is of type II.

For a type I segment $[x_{\nu-1},x_\nu]$ we have $\Omega_{[x_{\nu-1},x_\nu]} f \leq \Omega_{\mathcal{N}(x_\mu,\varepsilon_{x_\mu})} f \leq \omega_{x_\mu} f + \varepsilon < 2\varepsilon$. Hence, for points ξ_ν and $\xi_\nu{}^*$ of Q and Q^*, respectively, which lie in $[x_{\nu-1},x_\nu]$, it is the case that

$$(12) \qquad |f(\xi_\nu) - f(\xi_\nu{}^*)| < 2\varepsilon.$$

A type II segment $[x_{\nu-1},x_\nu]$ of P must lie in some neighborhood $\mathcal{N}(x_\mu,\varepsilon_{x_\mu})$ with $x_\mu \in D_\varepsilon$. Hence, by the definition of the covering (10), a point c_j exists with $|x_\mu - c_j| < \frac{1}{2}\delta_{c_j}$. Since $\varepsilon_{x_\mu} < \delta_1 \leq \frac{1}{2}\delta_{c_j}$, it follows that $[x_{\nu-1},x_\nu]$ lies in $\mathcal{N}(c_j,\delta_{c_j})$. Therefore the interval $[g(x_{\nu-1}),g(x_\nu)]$ lies in $\bigcup_{n=1}^\infty I_n$. Since the segments of P do not overlap (except for endpoints), neither can the intervals $[g(x_{\nu-1}),g(x_\nu)]$. Hence their total length does not exceed ε.

Finally, we may write

$$S(f,g,P,Q) - S(f,g,P,Q^*) = \sum_{\text{I}} [f(\xi_\nu) - f(\xi_\nu{}^*)]\, \Delta g_\nu + \sum_{\text{II}} [f(\xi_\nu) - f(\xi_\nu{}^*)]\, \Delta g_\nu,$$

where the sums are taken, as indicated, over type I and type II segments separately. From (12) we obtain

$$\left| \sum_{\text{I}} [f(\xi_\nu) - f(\xi_\nu{}^*)]\, \Delta g_\nu \right| \leq 2\varepsilon[g(b) - g(a)],$$

while, from the preceding paragraph,

$$\left| \sum_{\text{II}} [f(\xi_\nu) - f(\xi_\nu{}^*)]\, \Delta g_\nu \right| \leq 2M\varepsilon,$$

where M denotes an upper bound of $|f|$ on $[a,b]$. Hence

$$|S(f,g,P,Q) - S(f,g,P,Q^*)| < 2[M + g(b) - g(a)]\varepsilon.$$

Therefore $\int_a^b f\, dg$ exists by Exercise 14-1▲

Corollary 1 If f_1 and f_2 are bounded and g is increasing on $[a,b]$ and if $\int_a^b f_1\, dg$ and $\int_a^b f_2\, dg$ exist, then so do $\int_a^b f_1 f_2\, dg$ and $\int_a^b |f_1|\, dg$.

Corollary 2 If f is continuous and g increasing on $[a,b]$, then $\int_a^b f\, dg$ exists. (See Theorem 14-2.)

14-9 Unbounded Integrands

In Example 14-1 it was seen that if an integrator g is constant on $[a,b]$, then $\int_a^b f\, dg$ exists and has the value zero, whether the integrand f be bounded or unbounded. As we shall see below it is indeed necessary, in the event f is unbounded and $\int_a^b f\, dg$ is to exist, for g to have a constant value at least on some subintervals of $[a,b]$. This fact, in turn, effectively implies that there is no loss in generality in assuming, in connection with Stieltjes integration, that all integrands be bounded.

Theorem 14-11 If f is Stieltjes integrable relative to g on $[a,b]$, then a bounded function h exists with the property that $\int_a^x h\,dg = \int_a^x f\,dg$ for all x in $[a,b]$.

Proof Since $\int_a^b f\,dg$ exists, there is a number $\delta > 0$ such that

$$|S(f,g,P,Q') - S(f,g,P,Q'')| < 1 \tag{13}$$

for all partitions P having $\Delta P < \delta$, and any Q' and Q'' intermediate to P.

Let C denote the set of points x of $[a,b]$ each with the property that no neighborhood $\mathcal{N}(x,\varepsilon)$ exists on which f is bounded. Since f is clearly discontinuous at each such point, we conclude by Theorem 14-9, that g is continuous on C.

Suppose $c \in C \cap (a,b)$, and assume g to be non-constant on $\mathcal{N}(c,\delta/2) \cap [a,b]$. Then points s and t exist such that $c - \delta/2 < s < c < t < c + \delta/2$ and $g(s) \neq g(t)$. Since f is not bounded on $[s,t]$, for some point c^* of that interval it is necessarily the case that $f(c^*) > f(c) + (1/|g(s) - g(t)|)$.

Consider a partition P of $[a,b]$ having s and t as consecutive points and such that $\Delta P < \delta$. Let Q' be an intermediate partition of P which contains c^*, and Q'' identical with Q' except that c replaces c^*. Then

$$|S(f,g,P,Q') - S(f,g,P,Q'')| = |f(c^*) - f(c)|\,|g(s) - g(t)| > 1.$$

We are thus led to a contradiction of (13) and conclude therefrom that $\int_a^b f\,dg$ does not exist. This denial of hypothesis forces the conclusion that g has a constant value on $\mathcal{N}(c,\delta/2) \cap [a,b]$. A modification of some of the above statements leads to the same conclusion if $c = a$ or $c = b$.

Consider now the set $D = \bigcup_{x \in C} \mathcal{N}(x,\delta/4)$. Clearly $C \subset D$, and it follows that f is bounded on the compact set $\mathscr{C}D \cap [a,b]$. Let h denote a function defined on $[a,b]$ by

$$h(x) = \begin{cases} f(x) & \text{if } x \in \mathscr{C}D \cap [a,b] \\ 0 & \text{if } x \in D \cap [a,b]. \end{cases}$$

Let x be any point of $(a,b]$, and suppose $P = \{x_0,x_1,\ldots,x_n\}$ to be a partition of the interval $[a,x]$ such that $\Delta P < \delta/4$. Let $Q = \{\xi_1,\xi_2,\ldots,\xi_n\}$ be intermediate to P. Because of the definition of h,

$$\begin{aligned} &S(f,g,P,Q) - S(h,g,P,Q) \\ &= \sum_{\nu=1}^{n} [f(\xi_\nu) - h(\xi_\nu)]\,\Delta g_\nu \\ &= \sum_{\xi_\nu \in \mathscr{C}D \cap [a,b]} [f(\xi_\nu) - h(\xi_\nu)]\,\Delta g_\nu + \sum_{\xi_\nu \in D \cap [a,b]} [f(\xi_\nu) - h(\xi_\nu)]\,\Delta g_\nu \\ &= \sum_{\xi_\nu \in D \cap [a,b]} f(\xi_\nu)\,\Delta g_\nu . \end{aligned}$$

For each $\xi_\nu \in D \cap [a,b]$ there is a point c of C such that $|\xi_\nu - c| < \delta/4$. Since $[x_{\nu-1},x_\nu] \subset \mathcal{N}(\xi_\nu,\delta/4) \subset \mathcal{N}(c,\delta/2)$ and since g is constant on $\mathcal{N}(c,\delta/2)$, it follows that $\Delta g_\nu = 0$, and we conclude that $S(f,g,P,Q) = S(h,g,P,Q)$. This being the case for arbitrarily small ΔP, it follows that $\int_a^x h\,dg$ exists and equals $\int_a^x f\,dg$▲

Exercises

14-4 Find the value of $\int_0^3 f\,dg$ if f is continuous on $[0,3]$ and g is defined by $g(x) = 0$ if $0 \leq x < 2$, $g(2) = k$, and $g(x) = 1$ if $2 < x \leq 3$.

14-5 Show that $\int_0^n f\,d[x] = \sum_{\nu=1}^n f(\nu)$ if f is continuous, where $[x]$ is defined as the unique integer such that $x - 1 < [x] \leq x$.

14-6 Evaluate the following integrals: (a) $\int_0^3 x^2\,d([x] - x)$, (b) $\int_0^4 e^{2x}\,d[x]$, (c) $\int_0^1 x\,dx^2$, (d) $\int_0^{\pi/2} \sin x\,d\cos x$, (e) $\int_{-\pi}^{\pi} x^2\,d|\sin x|$.

14-7 Prove that if f is continuous and g of bounded variation on $[a,b]$, then $|\int_a^b f\,dg| \leq MT_g^{[a,b]}$, where M is a bound of f on $[a,b]$ and $T_g^{[a,b]}$ is the total variation of g on $[a,b]$.

14-8 Prove that if f is continuous and g of bounded variation on $[a,b]$, and if t_g is the total variation function of g on $[a,b]$, then $|\int_a^b f\,dg| \leq \int_a^b |f|\,dt_g$.

14-9 Establish Theorem 14-9 for the endpoints of $[a,b]$.

14-10 Prove that if f possesses a Stieltjes integral relative to each increasing function g on $[a,b]$, then f is continuous. (*Suggestion:* Use Theorem 14-9.)

14-11 Prove that if the assumption in Theorem 14-8 that h be strictly increasing is replaced by the weaker assumption that h is increasing, the conclusion of the theorem is still valid.

14-10 The Generalized Stieltjes Integral

We present next a second definition of Stieltjes integral. Based upon the notion of refinement of a partition, it compares with that definition of the Riemann integral styled Definition (C). Accordingly, we employ the same designation here. Unlike the Riemann situation, however, it will not be the case that Definition (C) is that of a concept exactly equivalent to the one enunciated in the Stieltjes Definition (B). The integral of Definition (C) is styled the *Generalized Stieltjes integral* and the notation $\oint_a^b f\,dg$ employed.[2] The symbol $S(f,g,P,Q)$ appearing below has the same meaning as heretofore.

Definition (C) Suppose f and g to denote functions defined on an interval $[a,b]$. Then a number $\oint_a^b f\,dg$ is said to be the *Generalized Stieltjes integral* (or *GS-integral*) *of f relative to g on* $[a,b]$, and f is said to be *GS-integrable relative to g on* $[a,b]$, if, corresponding to each $\varepsilon > 0$, there exists a partition P_ε with the property that $|S(f,g,P,Q) - \oint_a^b f\,dg| < \varepsilon$ for each refinement P of P_ε and all intermediate partitions Q of P.

As in the Stieltjes case we append the definitions $\oint_b^a f\,dg = -\oint_a^b f\,dg$ for $a < b$, and $\oint_a^a f\,dg = 0$.

[2] The symbol $(GS)\int_a^b f\,dg$ is frequently encountered in the literature.

It is immediate that if a function f is Stieltjes integrable relative to g, then it also possesses a Generalized Stieltjes integral relative to g. That is, $\oint_a^b f\,dg$ exists if $\int_a^b f\,dg$ does and, further, the two have the same value. The converse statement is not true. The less demanding premise of Definition (C), referring as it does to a more limited set of partitions than does Definition (B), permits the existence of integrals which fail to materialize in the ordinary Stieltjes case. This is brought out by the following theorem, which is in obvious contrast with the conclusion of Example 14-6.

Theorem 14-12 If f is GS-integrable relative to g on each of the intervals $[a,c]$ and $[c,b]$, then it is GS-integrable relative to g on $[a,b]$ and, moreover,

$$\oint_a^b f\,dg = \oint_a^c f\,dg + \oint_c^b f\,dg.$$

Proof Let $\varepsilon > 0$ be given. Then partitions P_ε' of $[a,c]$ and P_ε'' of $[c,b]$ exist such that $|S(f,g,P',Q') - \oint_a^c f\,dg| < \varepsilon$ and $|S(f,g,P'',Q'') - \oint_c^b f\,dg| < \varepsilon$ for any refinements P' of P_ε' and P'' of P_ε'' and any intermediate partitions Q' and Q'' of P' and P'', respectively. It follows that $P_\varepsilon = P_\varepsilon' \cup P_\varepsilon''$ is a partition of $[a,b]$. Further, if P is any refinement of P_ε, then the sets $P' = P \cap [a,c]$ and $P'' = P \cap [c,b]$ are refinements of P_ε' and P_ε'', respectively. Also, any Q intermediate to P may be decomposed into intermediate partitions Q' and Q'' of P' and P'', respectively. Finally, it is clear that $S(f,g,P,Q) = S(f,g,P',Q') + S(f,g,P'',Q'')$. Therefore, by combining the above expressions, we obtain

$$\left| S(f,g,P,Q) - \left[\oint_a^c f\,dg + \oint_c^b f\,dg\right]\right| < 2\varepsilon \blacktriangle$$

Properties exhibited for the Stieltjes integral in earlier sections are valid as well for the Generalized Stieltjes integral. In particular, if any two of the three integrals $\oint_a^c f\,dg$, $\oint_c^b f\,dg$, and $\oint_a^b f\,dg$ exist, then so does the third. Moreover, whatever the relative values of a, b, and c, it is the case that the formula of Theorem 14-12 holds.

That the GS-integral has a bilinear character with respect to f and g follows as in the ordinary Stieltjes case. The statement of Cauchy convergence of S-sums based on refinements of partitions, and its relationship with the existence of the GS-integral assumes an obvious form. The theorems concerning integration by parts and change of variable remain valid if $\int_a^b f\,dg$ is replaced by $\oint_a^b f\,dg$. The reader should restate theorems and furnish proofs which reflect the differences in Definitions (B) and (C).

14-11 The Role of Discontinuities in the GS-Integral

In the theorem which follows we note what is essentially the key to the difference between the integrals of Definitions (B) and (C). We have seen that in the ordinary Stieltjes case a discontinuity point common to integrand and integrator is incompatible with the existence of the integral. In the Generalized Stieltjes case common discontinuity points are not ruled out. However, limitations exist on the relative characters of f and g at such points.

Theorem 14-13 If f possesses a Generalized Stieltjes integral relative to g on $[a,b]$, then at no point of $[a,b]$ are f and g both discontinuous from the right or both discontinuous from the left.

Proof Assume each of the functions f and g to be discontinuous from the right at a point c of $[a,b)$. Then for some positive number ε, it is the case that for any $\delta > 0$, points c', ξ, and ξ' exist such that $c \leq \xi < \xi' \leq c' < c + \delta$, and also $|f(\xi') - f(\xi)| > \sqrt{\varepsilon}$ and $|g(c') - g(c)| > \sqrt{\varepsilon}$. Let P_ε be a partition of $[a,b]$ such that $|S(f,g,P,Q) - S(f,g,P,Q')| < \varepsilon$ for $P \supset P_\varepsilon$ and any Q and Q' intermediate to P. Then a refinement P of P_ε may be constructed which has c and c' as members. Let Q be an intermediate partition containing ξ, and suppose Q' to be identical with Q except that ξ' replaces ξ. Then

$$|S(f,g,P,Q) - S(f,g,P,Q')| = |f(\xi') - f(\xi)|\,|g(c') - g(c)| > \varepsilon,$$

which is a contradiction. Therefore f and g cannot both be discontinuous from the right at c. By a dual argument they cannot both be discontinuous from the left at a point c of $(a,b]$▲

Example 14-8 Recall the functions f and g defined in Example 14-6. Because of the common discontinuity of f and g at $x = 1$, the integral $\mathfrak{S}_0^2 f\,dg$ fails to exist. However, since f is discontinuous from the left and g discontinuous from the right at $x = 1$, the door is left open, as it were, for the possible existence of $\oint_0^2 f\,dg$. Indeed, by virtue of the existence of the integrals $\mathfrak{S}_0^1 f\,dg$ and $\mathfrak{S}_1^2 f\,dg$ follows that of $\oint_0^1 f\,dg$ and $\oint_1^2 f\,dg$. Theorem 14-12 then implies that $\oint_0^2 f\,dg$ exists with value 1.

If one revises the definition of g by assigning to it the value 1 at $x = 1$, then both f and g are discontinuous from the left at that point. Accordingly, neither $\oint_0^1 f\,dg$ nor $\oint_0^2 f\,dg$ exists.

If discontinuities of f and g do not provide an obstacle, then simple boundedness of the two functions suffices if the existence of $\oint_a^b f\,dg$ is to imply that of $\mathfrak{S}_a^b f\,dg$. This we see next.

Theorem 14-14 If f and g are bounded functions possessing no common discontinuity points on $[a,b]$, and if f has a Generalized Stieltjes integral relative to g on $[a,b]$, then f is Stieltjes integrable relative to g on that interval.

Proof Let $\varepsilon > 0$ be given. Then a partition $P_\varepsilon = \{x_0{}^*, x_1{}^*, \ldots, x_n{}^*\}$ exists such that

$$\left| S(f,g,P',Q') - \oint_a^b f\,dg \right| < \varepsilon \tag{14}$$

for all $P' \supset P_\varepsilon$. Further, $\delta_1 > 0$ exists with the property that for each ν, $\nu = 1, 2, \ldots, n-1$, and for all x such that $|x - x_\nu| < \delta_1$, it is the case that $|f(x) - f(x_\nu^*)| < \varepsilon/n$ if f is continuous at x_ν^*, and $|g(x) - g(x_\nu^*)| < \varepsilon/n$ if f is discontinuous at x_ν^*.

Denote by δ_2 the minimum distance between points of P_ε and let $\delta = \min\{\delta_1, \frac{1}{2}\delta_2\}$.

Now consider any partition $P = \{x_0, x_1, \ldots, x_m\}$ of $[a,b]$ with $\Delta P < \delta$ and any intermediate partition $Q = \{\xi_1, \xi_2, \ldots, \xi_m\}$. Then for the refinement $P' = P \cup P_\varepsilon$ of P we construct a corresponding intermediate partition Q' as follows. If the segment $[x_{\nu-1}, x_\nu]$ of P contains no point of P_ε in its interior, then it is also a segment of P'; in this event choose ξ_ν to be the member of Q' in the segment. If the segment $[x_{\nu-1}, x_\nu]$ contains a point x^* of P_ε in its interior (because of the choice of δ it can contain at most one such point), then this one segment of P corresponds to two segments $[x_{\nu-1}, x^*]$ and $[x^*, x_\nu]$ of P'. Since ξ_ν must lie in one of these, let it be the member of Q' in that segment; in the remaining segment let Q' be represented by x^*. If $\xi_\nu = x^*$, let it be counted twice. Note that at most $n-1$ segments of P can be of this second type.

Consider now the difference $\Delta S = S(f,g,P,Q) - S(f,g,P',Q')$. By the above construction, those segments of P which are also segments of P' contribute terms which cancel, while if $[x_{\nu-1}, x_\nu]$ of P gives rise to $[x_{\nu-1}, x^*]$ and $[x^*, x_\nu]$ of P', then the contribution of the three segments to the value of ΔS is either

$$f(\xi_\nu)[g(x_\nu) - g(x_{\nu-1})] - \{f(\xi_\nu)[g(x^*) - g(x_{\nu-1})] + f(x^*)[g(x_\nu) - g(x^*)]\}$$
$$= [f(\xi_\nu) - f(x^*)][g(x_\nu) - g(x^*)]$$

or

$$f(\xi_\nu)[g(x_\nu) - g(x_{\nu-1})] - \{f(x^*)[g(x^*) - g(x_{\nu-1})] + f(\xi_\nu)[g(x_\nu) - g(x^*)]\}$$
$$= [f(\xi_\nu) - f(x^*)][g(x^*) - g(x_{\nu-1})]$$

according as ξ_ν is in $[x_{\nu-1}, x^*]$ or $[x^*, x_\nu]$. It follows that, in either case, the contribution to ΔS has an absolute value not exceeding $2M\varepsilon/n$, where M is an upper bound of $|f|$ and $|g|$. Since there are at most $n-1$ such contributions, it follows that $|\Delta S| < 2M\varepsilon$. This, together with (14), yields $|S(f,g,P,Q) - \oint_a^b f\,dg| < (2M+1)\varepsilon$ for all P with $\Delta P < \delta$ ▲

In the following section is established a condition on the functions f and g which is both necessary and sufficient to the existence of the Generalized Stieltjes integral of f relative to g.

14-12 Step Functions as Integrators

A function U_c defined on $\mathscr{R}$ by

$$U_c(x) = \begin{cases} 0 & \text{if } x < c \\ 1 & \text{if } x > c \end{cases} \qquad \text{and} \qquad 0 \le U_c(c) \le 1$$

is referred to as a *unit step function* with *step* or *jump* of magnitude 1 at c. Depending on the value assigned to $U_c(c)$, it is continuous from the left or from the right or neither at c.

As we see in the following theorem, the Generalized Stieltjes integral of any function f relative to a unit step function has a value which depends only on the value of f at c.

Theorem 14-15 If U_c is a unit step function corresponding to $c \in (a,b)$ and if f is any function on $[a,b]$ that does not possess at c a discontinuity of the same character as U_c, then f possesses a Generalized Stieltjes integral relative to U_c on $[a,b]$, and its value is $f(c)$.

Proof Suppose U_c to be continuous from the right at c, that is, $U_c(c) = 1$. Then U_c is discontinuous from the left at c and, in keeping with the hypothesis, the function f is continuous from the left at c. Hence, given $\varepsilon > 0$, there exists $\delta > 0$ such that $|f(x) - f(c)| < \varepsilon$ for all $x \in [a,b] \cap (c - \delta, c)$.

Let P_ε denote a partition of $[a,b]$ having c as a member and such that $\Delta P_\varepsilon < \delta$. If $P = \{x_0,x_1,\ldots,x_n\}$ is a refinement of P_ε with $x_\mu = c$ and if $Q = \{\xi_1,\xi_2,\ldots,\xi_n\}$ is intermediate to P, then $S(f,U_c,P,Q) = f(\xi_\mu)$. From $c - \delta < \xi_\mu \leq c$, it follows that f is integrable in the sense of Definition (C) and $\oint_a^b f\,dU_c = f(c)$.

If U_c is continuous from the left at c, the proof follows a dual pattern. If $0 < U_c(c) < 1$, then f is necessarily continuous at c. The reader should modify the proof accordingly▲

Corollary 1 Let $U_{c_1}, U_{c_2}, \ldots, U_{c_n}$ be unit step functions (of any characters) corresponding to points $c_1, c_2, \ldots, c_n$ of (a,b). Let $g = k_1 U_{c_1} + k_2 U_{c_2} + \cdots + k_n U_{c_n}$, where $k_1, k_2, \ldots, k_n$ are any n constants, and suppose f to be a function on $[a,b]$ possessing different discontinuity properties from g at the points c_ν, $\nu = 1, 2, \ldots, n$. Then f possesses a Generalized Stieltjes integral relative to g on $[a,b]$, and $\oint_a^b f\,dg = \sum_{\nu=1}^n k_\nu f(c_\nu)$.

Corollary 2 Let $\{U_{c_\nu}\}$ be a sequence of unit step functions corresponding to a sequence of distinct points $\{c_\nu\}$ of (a,b). Let $g = \sum_{\nu=1}^\infty k_\nu U_{c_\nu}$, where $\sum_{\nu=1}^\infty k_\nu$ is an absolutely convergent series of constants. (The function g exists by virtue of the uniform boundedness of the U_{c_ν}.) If f is a bounded function on $[a,b]$, and if f and g have no common discontinuities of the same character at the points c_ν, then f has a Generalized Stieltjes integral relative to g on $[a,b]$, and $\oint_a^b f\,dg = \sum_{\nu=1}^\infty k_\nu f(c_\nu)$.

Proof We assume, without loss of generality, that $k_\nu \geq 0$ for all ν. (If this is not the case, the argument below applies separately to the positive and negative parts of $\sum_{\nu=1}^\infty k_\nu$.)

Corresponding to any integer N we write $g = g_N + r_N$, where $g_N(x) = \sum_{\nu=1}^N k_\nu U_{c_\nu}(x)$ and $r_N(x) = \sum_{\nu=N+1}^\infty k_\nu U_{c_\nu}(x)$. Then, by Corollary 1, f is GS-integrable relative to g_N, and $\oint_a^b f\,dg_N = \sum_{\nu=1}^N k_\nu f(c_\nu)$. From the bounded character of f follows absolute convergence of $\sum_{\nu=1}^{\infty} k_\nu f(c_\nu)$ and hence the existence of $\lim_{N\to\infty} \oint_a^b f\, dg_N$. For any $\varepsilon > 0$ let N be chosen so large that

$$\sum_{\nu=N+1}^{\infty} k_\nu < \varepsilon \quad \text{and} \quad \left| \oint_a^b f\, dg_N - \sum_{\nu=1}^{\infty} k_\nu f(c_\nu) \right| < \varepsilon.$$

Suppose a partition P_ε to be such that for any refinement P it is the case that $\left|S(f,g_N,P,Q) - \oint_a^b f\, dg_N\right| < \varepsilon$. Then, if M denotes an upper bound of $|f|$ on $[a,b]$,

$$\begin{aligned} |S(f,g,P,Q) - S(f,g_N,P,Q)| &= |S(f,r_N,P,Q)| \\ &\leq \sum_{\mu=1}^{n} |f(\xi_\mu)|\, \Delta r_{N_\mu} < M\varepsilon, \end{aligned}$$

and it follows that $\left|S(f,g,P,Q) - \sum_{\nu=1}^{\infty} k_\nu f(c_\nu)\right| < (M+2)\varepsilon$▲

The jump function j associated with an increasing function g on $[a,b]$ is defined (see (1) of Chapter 13 for the case of a function increasing on $\mathscr{R}$) by $j(a) = 0$ and

$$j(x) = g(a+) - g(a) + \sum_{a<t<x} [g(t+) - g(t-)] + g(x) - g(x-)$$

if $a < x \leq b$, where in the summation t varies over the set of discontinuities (at most countable) in the stated interval. It is immediate that j may be expressed as a sum of step functions. Thus, for $x \in [a,b]$,

$$\begin{aligned} j(x) = {} & [g(a+) - g(a)]U_a^L(x) + [g(b) - g(b-)]U_b^R(x) \\ & + \sum_{a<t<b} \{[g(t) - g(t-)]U_t^R(x) + [g(t+) - g(t)]U_t^L(x)\}, \end{aligned}$$

where the unit step functions $U_a^L(x)$ and $U_t^L(x)$ are continuous from the left at their discontinuity points, and $U_b^R(x)$ and $U_t^R(x)$ are continuous from the right.

Following is a condition, in the event of an increasing integrator, which is both necessary and sufficient to the existence of a Generalized Stieltjes integral.

Theorem 14-16 If f is bounded and g increasing on $[a,b]$, and if j denotes the jump function of g on $[a,b]$, then $\oint_a^b f\, dg$ exists if and only if f and g have no common discontinuities of the same character and $\mathfrak{f}_a^b f\, d(g-j)$ exists.

Proof If $\oint_a^b f\, dg$ exists, then f and g have no common discontinuities of the same character. It follows from the nature of the jump function j of g that this is also the case for f and j. Hence $\oint_a^b f\, dj$ exists (by virtue of Corollary 2 of Theorem 14-15) and so does $\oint_a^b f\, d(g-j)$, because of the linearity of the GS-integral. Since $g - j$ is continuous on $[a,b]$, the integral $\mathfrak{f}_a^b f\, d(g-j)$ exists as well (Theorem 14-14).

Conversely, if f and g have no common discontinuities on the same side, then neither do f and j. Therefore $\oint_a^b f\, dj$ exists. Moreover, if $\mathfrak{f}_a^b f\, d(g-j)$ exists, then so does $\oint_a^b f\, d(g-j)$. Hence $\oint_a^b f\, d(g-j) + \oint_a^b f\, dj = \oint_a^b f\, dg$ exists▲

Corollary 1 If f_1 and f_2 are bounded and g is increasing on $[a,b]$, and if $\oint_a^b f_1\, dg$ and $\oint_a^b f_2\, dg$ exist, then so do $\oint_a^b f_1 f_2\, dg$ and $\oint_a^b |f_1|\, dg$. (See Corollary 1 to Theorem 14-10.)

Corollary 2 Suppose f to be bounded, g increasing, and j the jump function of g on $[a,b]$. Denote by D the set of discontinuity points of f. Then a necessary and sufficient condition for the existence of $\oint_a^b f\, dg$ is that f and g have no common discontinuities of the same character and that $\bigcup_{x \in D} \{y;\ \lim_{x-}(g-j) \le y \le \lim_{x+}(g-j)\}$ be a zero set.

Exercises

14-12 Prove that the Generalized Stieltjes integral $\oint_a^b f\, dg$ exists if and only if for every $\varepsilon > 0$ there exists a partition P_ε such that for all refinements P' and P'' of P_ε and all intermediate partitions Q' and Q'', it is the case that $|S(f,g,P',Q') - S(f,g,P'',Q'')| < \varepsilon$.

14-13 Prove that if f possesses a Generalized Stieltjes integral relative to g on $[a,b]$, then a bounded function h exists with the property that $\oint_a^x h\, dg = \oint_a^x f\, dg$ for all x in $[a,b]$.

14-14 Restate and prove Theorem 14-7 (integration by parts) for the Generalized Stieltjes integral.

14-15 Restate and prove Theorem 14-8 (change of variable) for the Generalized Stieltjes integral.

14-16 Modify the proof of Theorem 14-15 for the case that $0 < U_c(c) < 1$.

14-13 The Riemann-Stieltjes Integral

We have reserved for the last a definition of an integral of Stieltjes type reminiscent of Definition (A) of the Riemann integral. We present this now, and demonstrate its relationship with the Stieltjes integral of Definition (B) and the Generalized Stieltjes integral (C). We shall refer to it as the *Riemann-Stieltjes integral* and employ the notation $\not\int_a^b f\, dg$.

Suppose f to be a bounded function on $[a,b]$, and let $P = \{x_0, x_1, \ldots, x_n\}$ denote a partition of that interval. As in the Riemann case, we define $m = \inf\{f(x);\ a \le x \le b\}$, $M = \sup\{f(x);\ a \le x \le b\}$, $m_\nu = \inf\{f(x);\ x_{\nu-1} \le x \le x_\nu\}$, and $M_\nu = \sup\{f(x);\ x_{\nu-1} \le x \le x_\nu\}$. Imagine g to be an increasing function defined on $[a,b]$ and let $\Delta g_\nu = g(x_\nu) - g(x_{\nu-1})$, as usual.

We then define the *lower* and *upper Riemann-Stieltjes sums* (or *lower* and *upper RS-sums*) of f with respect to g and corresponding to the partition P, to be, respectively, the numbers

$$\underline{RS}(f,g,P) = \sum_{\nu=1}^{n} m_\nu\, \Delta g_\nu \quad \text{and} \quad \overline{RS}(f,g,P) = \sum_{\nu=1}^{n} M_\nu\, \Delta g_\nu .$$

(Clearly, if $g(x) = x$, these reduce to the lower and upper R-sums defined prior to Definition (A) of the Riemann integral.)

Because Δg_ν is non-negative, it follows, as in the Riemann case, that

$$\underline{RS}(f,g,P) \le \overline{RS}(f,g,P)$$

for any partition P. Further, if P^* denotes a refinement of P, then

$$\underline{RS}(f,g,P) \le \underline{RS}(f,g,P^*) \quad \text{and} \quad \overline{RS}(f,g,P^*) \le \overline{RS}(f,g,P).$$

It follows that if P' and P'' are any two partitions of $[a,b]$, then

$$\underline{RS}(f,g,P') \le \overline{RS}(f,g,P''). \tag{15}$$

As a consequence, the set of all lower RS-sums is bounded above (by any upper RS-sum) and, accordingly, possesses a supremum, which we denote by $\underline{\mathcal{R}\!\int_a^b} f\,dg$ and term the *lower Riemann-Stieltjes integral of f relative to g on* $[a,b]$. Similarly, the set of upper RS-sums possesses an infimum $\overline{\mathcal{R}\!\int_a^b} f\,dg$ called the *upper Riemann-Stieltjes integral of f relative to g on* $[a,b]$. It is clear from (15) that $\underline{\mathcal{R}\!\int_a^b} f\,dg \le \overline{\mathcal{R}\!\int_a^b} f\,dg$.

Definition (A) If f is a bounded function and g an increasing function on $[a,b]$, and if the lower and upper Riemann-Stieltjes integrals of f relative to g are equal, then the common value of these integrals is denoted by $\mathcal{R}\!\int_a^b f\,dg$ and styled the *Riemann-Stieltjes integral* (or *RS-integral*) *of f relative to g on* $[a,b]$; and f is said to be *Riemann-Stieltjes integrable* (or *RS-integrable*) *relative to g on* $[a,b]$. Further, we define $\mathcal{R}\!\int_b^a f\,dg = -\mathcal{R}\!\int_a^b f\,dg$ for $a < b$, and $\mathcal{R}\!\int_a^a f\,dg = 0$.

It is easy to see that if $g(x) = x$, then the lower and upper RS-integrals reduce to the lower and upper R-integrals, respectively; that is,

$$\underline{\mathcal{R}\!\int_a^b} f\,dx = \underline{\int_a^b} f \quad \text{and} \quad \overline{\mathcal{R}\!\int_a^b} f\,dx = \overline{\int_a^b} f.$$

As a consequence, the Riemann-Stieltjes integral $\mathcal{R}\!\int_a^b f\,dx$ of f relative to x is one and the same with the Riemann integral $\int_a^b f$, by virtue of Definition (A) of Chapter 10.

14-14 Comparing Definitions

One notes that, quite apart from conditions which might be necessary to the existence of a Riemann-Stieltjes integral, the very nature of its definition requires that the integrand be bounded and that the integrator be increasing. Such limitations play no role in the statements of Definitions (B) and (C). Indeed, we have seen that the Stieltjes integral and the Generalized Stieltjes integral may exist even if the integrand is unbounded or if the integrator is other than an increasing function. It follows, of course, that a proper comparison of the Riemann-Stieltjes integral with either of the other two must be based on the more severe restrictions of Definition (A). In the following theorem it is shown that under such hypotheses the Riemann-Stieltjes integral and the Generalized Stieltjes integral are, indeed, equivalent.

Theorem 14-17 Suppose f to be bounded and g increasing on $[a,b]$. Then $\mathcal{R}\!\!\!\int_a^b f\,dg$ exists if and only if $\oint_a^b f\,dg$ exists. Further, when both exist they have a common value.

Proof Let $\varepsilon > 0$ be given and assume $\oint_a^b f\,dg$ to exist. Then there is a partition P_ε of $[a,b]$ with the property that

$$\left| S(f,g,P,Q) - \oint_a^b f\,dg \right| < \varepsilon \tag{16}$$

for any refinement $P = \{x_0,x_1,\ldots,x_n\}$ of P_ε and any Q intermediate to P.

For each ν, $\nu = 1, 2, \ldots, n$, denote by M_ν the supremum of f on $[x_{\nu-1},x_\nu]$. Then a set of numbers ξ_ν, $\nu = 1, 2, \ldots, n$, exists such that $\xi_\nu \in [x_{\nu-1},x_\nu]$ and $M_\nu < f(\xi_\nu) + \varepsilon$. For the intermediate partition $Q = \{\xi_1,\xi_2,\ldots,\xi_n\}$ we have

$$\overline{\mathcal{R}\!\!\!\int_a^b} f\,dg \le \overline{RS}(f,g,P) < \sum_{\nu=1}^{n} [f(\xi_\nu) + \varepsilon]\,\Delta g_\nu$$
$$< S(f,g,P,Q) + \varepsilon[g(b) - g(a)].$$

Together with (16) this implies $\overline{\mathcal{R}\!\!\!\int_a^b} f\,dg \le \oint_a^b f\,dg$. In a similar manner one establishes that $\oint_a^b f\,dg \le \underline{\mathcal{R}\!\!\!\int_a^b} f\,dg$ and concludes the existence of $\mathcal{R}\!\!\!\int_a^b f\,dg$ with value $\oint_a^b f\,dg$.

For the converse, suppose $\mathcal{R}\!\!\!\int_a^b f\,dg$ to exist. Then there is a partition P_ε of $[a,b]$ such that

$$\mathcal{R}\!\!\!\int_a^b f\,dg - \varepsilon < \underline{RS}(f,g,P_\varepsilon) \quad \text{and} \quad \overline{RS}(f,g,P_\varepsilon) < \mathcal{R}\!\!\!\int_a^b f\,dg + \varepsilon.$$

For any refinement P of P_ε and any intermediate partition Q of P it is clearly the case that $\underline{RS}(f,g,P) \le S(f,g,P,Q) \le \overline{RS}(f,g,P)$. Consequently,

$$\mathcal{R}\!\!\!\int_a^b f\,dg - \varepsilon < S(f,g,P,Q) < \mathcal{R}\!\!\!\int_a^b f\,dg + \varepsilon,$$

and $\oint_a^b f\,dg$ is seen to exist with value $\mathcal{R}\!\!\!\int_a^b f\,dg$▲

It was pointed out earlier that if an integral exists by Definition (B), then it does also in the sense of Definition (C), but that the converse is not necessarily the case. Now we see that, by virtue of the limitations on integrand and integrator in Definition (A), the Generalized Stieltjes integral may exist even when Definition (A) is devoid of meaning, but that whenever the integral of Definition (A) exists so does that of Definition (C). Further, it is clear that the Riemann-Stieltjes integral may exist when the Stieltjes integral does not.

It would appear that of the three integrals, the Generalized Stieltjes is in fact the most general, and therefore most deserving of attention. Accordingly, we pursue an investigation of properties of the GS-integral, understanding, as we do so, that all conclusions apply to the RS-integral if f is bounded and g increasing, and to the S-integral if f and g are bounded and have no common discontinuities.

In this spirit we also adjust notation and terminology. Henceforth, the symbol $\int_a^b f\,dg$ will be employed for all three integrals. It will always signify the integral of Definition (C) and will also designate those of Definitions (A) and (B) if the functions f and g concur. Also, we shall drop the modifiers Riemann-Stieltjes, Stieltjes, and Generalized Stieltjes and refer simply to the *integral* $\int_a^b f\,dg$. The notation $\int_a^b f\,dg$ itself is sufficient to distinguish it from the Riemann integral $\int_a^b f$.

In much of the literature the symbol $\int_a^b f(x)\,dg(x)$ is in vogue and is referred to variously as the Stieltjes or Riemann-Stieltjes integral, even when it is intended to include the Generalized Stieltjes case. The reader is cautioned not to expect uniform terminology or denotation as he consults one source and then another. In each instance he must refer to definitions.

The notation $\int_a^b f(x)\,dg(x)$ has an appeal, and we shall employ it occasionally. It reduces nicely when $g(x) = x$ to the alternate notation $\int_a^b f(x)\,dx$ introduced for the Riemann integral in Chapter 10.

Exercises

14-17 Prove that the Riemann-Stieltjes integral $\mathcal{R}\!\int_a^b f\,dg$ exists if and only if for every $\varepsilon > 0$, there exists a partition P of $[a,b]$ such that $\overline{RS}(f,g,P) - \underline{RS}(f,g,P) < \varepsilon$.

14-18 Prove that if $a < c < b$, then $\overline{\mathcal{R}\!\int}_a^b f\,dg = \overline{\mathcal{R}\!\int}_a^c f\,dg + \overline{\mathcal{R}\!\int}_c^b f\,dg$.

14-19 Prove (a) $\overline{\mathcal{R}\!\int}_a^b (f+h)\,dg \leq \overline{\mathcal{R}\!\int}_a^b f\,dg + \overline{\mathcal{R}\!\int}_a^b h\,dg$ and (b) $\underline{\mathcal{R}\!\int}_a^b (f+h)\,dg \geq \underline{\mathcal{R}\!\int}_a^b f\,dg + \underline{\mathcal{R}\!\int}_a^b h\,dg$.

14-20 Prove that if f is continuous and g increasing on $[a,b]$, then $\mathcal{R}\!\int_a^b f\,dg$ exists.

14-21 Prove that if g is bounded on $[a,b]$ and if $\mathcal{R}\!\int_a^b f\,dg$ exists with value zero for each bounded increasing function f on $[a,b]$, then g is a constant function.

14-22 Prove that if g is an increasing function on $[a,b]$ and if f is non-negative and integrable relative to g on $[a,b]$, then $\int_a^b f\,dg \geq 0$.

14-23 Prove that if g is increasing and if f and h are integrable relative to g and such that $f \leq h$ on $[a,b]$, then $\int_a^b f\,dg \leq \int_a^b h\,dg$.

14-24 Let g be increasing and f non-negative and integrable relative to g on $[a,b]$. Prove that if $a \leq c \leq d \leq b$, then $\int_c^d f\,dg \leq \int_a^b f\,dg$.

14-25 Let g be increasing and f and h integrable relative to g on $[a,b]$. Prove that $|f|, f^2$, and fh are integrable relative to g on $[a,b]$ and that $|\int_a^b f\,dg| \leq \int_a^b |f|\,dg$.

14-26 Prove that if g is increasing and f and h are such that $m \leq f \leq M$ and $0 \leq h$, and if fh and h are integrable relative to g on $[a,b]$, then (a) $m \int_a^b h\,dg \leq \int_a^b fh\,dg \leq M \int_a^b h\,dg$, and (b) η exists such that $m \leq \eta \leq M$ and $\int_a^b fh\,dg = \eta \int_a^b h\,dg$.

14-27 Prove the *first mean value theorem*: If g is increasing, f continuous, and h non-negative and integrable relative to g on $[a,b]$, then ξ exists such that $a \leq \xi \leq b$ and $\int_a^b fh\,dg = f(\xi) \int_a^b h\,dg$.

14-28 Prove that if g is increasing and f continuous on $[a,b]$, then ξ exists such that $a \leq \xi \leq b$ and $\int_a^b f\,dg = f(\xi)[g(b) - g(a)]$.

14-29 Prove the *second mean value theorem*: If f and g are continuous and increasing, and if h is integrable relative to g on $[a,b]$, then a point ξ of $[a,b]$ exists with the property $\int_a^b fh\,dg = f(a) \int_a^\xi h\,dg + f(b) \int_\xi^b h\,dg$.

14-30 Prove that if f and g are continuous and increasing on $[a,b]$, then a point ξ of $[a,b]$ exists such that $\int_a^b f\,dg = f(a)[g(\xi) - g(a)] + f(b)[g(b) - g(\xi)]$.

14-15 The Indefinite Integral

Let x_0 denote any point of $[a,b]$ and suppose f to be a function which is integrable relative to an increasing integrator g on $[a,b]$. It is then the case that the integral $\int_{x_0}^{x} f\,dg$ exists for each point x of $[a,b]$. Accordingly, the function φ defined on $[a,b]$ by

$$\varphi(x) = \int_{x_0}^{x} f\,dg$$

is called the *indefinite integral, based on* x_0, *of* f *relative to* g *on* $[a,b]$.

Unlike its Riemann counterpart, the indefinite integral need not be a continuous function. For example, if g is discontinuous from the left at a point c of $[a,b]$, and if U_c is a unit step function discontinuous from the right at c, then, as the reader may easily verify, the indefinite integral $\int_a^x U_c\,dg$ is discontinuous at c. We may, however, establish the following theorem on continuity of the indefinite integral.

Theorem 14-18 If g is increasing and is continuous at a point c of $[a,b]$, and if f has an indefinite integral φ relative to g on $[a,b]$, then φ is continuous at c.

Proof From the assumed bounded character of f, i.e., $|f(x)| \le M$, it follows that

$$|\varphi(x) - \varphi(c)| = \left|\int_c^x f\,dg\right| \le M\,|g(x) - g(c)|,$$

whence continuity of φ at c is a consequence of that of g▲

If g is an increasing function which is continuous on $[a,b]$, and if f has an indefinite integral φ relative to g on $[a,b]$, then φ is continuous on $[a,b]$.

Theorem 14-19 If g is increasing and φ the indefinite integral of f relative to g on $[a,b]$, then φ is a function of bounded variation.

Proof Left as an exercise▲

Theorem 14-20 Suppose g to be increasing and f to possess an indefinite integral φ relative to g on $[a,b]$. Further, suppose f to be continuous and g differentiable at a point c of $[a,b]$. Then φ is differentiable at c and $\varphi'(c) = f(c)g'(c)$.

Proof Let x be a point of $[a,b]$ other than c. Then, by Exercise 14-26, a number η_x exists such that $\inf_{[c,x]} f \le \eta_x \le \sup_{[c,x]} f$ and

$$\varphi(x) - \varphi(c) = \int_c^x f\,dg = \eta_x[g(x) - g(c)].$$

Therefore

$$\frac{\varphi(x) - \varphi(c)}{x - c} = \eta_x \frac{g(x) - g(c)}{x - c}$$

and, since $\lim_c \eta_x = f(c)$, because of the continuity of f at c, it follows that $\varphi'(c)$ exists and $\varphi'(c) = f(c)g'(c)$▲

Theorem 14-21 If g is a function which is increasing, φ the indefinite integral of a function h relative to g, and f continuous on $[a,b]$, then f is integrable relative to φ on $[a,b]$, and $\int_a^b f\,d\varphi = \int_a^b fh\,dg$.

Proof Since, by Theorem 14-19, φ is of bounded variation on $[a,b]$, it follows that f is integrable relative to φ on $[a,b]$. Further, if $\varepsilon > 0$ be given, then $\delta > 0$ exists such that $|f(t') - f(t)| < \varepsilon$ for $|t' - t| < \delta$. Let P have norm less than δ. Then from

$$S(f,\varphi,P,Q) = \sum_{\nu=1}^{n} f(\xi_\nu)\,\Delta\varphi_\nu = \sum_{\nu=1}^{n} f(\xi_\nu)\int_{x_{\nu-1}}^{x_\nu} h\,dg$$

and

$$\int_a^b fh\,dg = \sum_{\nu=1}^{n}\int_{x_{\nu-1}}^{x_\nu} fh\,dg$$

follows

$$\left|S(f,\varphi,P,Q) - \int_a^b fh\,dg\right| \le \sum_{\nu=1}^{n}\int_{x_{\nu-1}}^{x_\nu} |f(\xi_\nu) - f(t)|\,|h(t)|\,dg(t)$$
$$< M\varepsilon \sum_{\nu=1}^{n}[g(x_\nu) - g(x_{\nu-1})],$$

M being an upper bound of $|h|$. Since g is increasing the truth of the theorem follows▲

14-16 Sequences and Series

A sequence of Stieltjes integrals may arise by considering a sequence of integrands $\{f_n\}$, a sequence of integrators $\{g_n\}$, or a combination of the two. In the first event we may compare the facts with those in the Riemann case. The theorem below may be compared with Theorem 11-8, and the reader should construct a proof following the pattern in that instance.

Theorem 14-22 If a sequence of functions $\{f_n\}$ converges uniformly to a limit f on an interval $[a,b]$, and if each element f_n is integrable relative to a function g that is of bounded variation on $[a,b]$, then f is integrable relative to g and the sequence $\{\int_a^b f_n\,dg\}$ converges to $\int_a^b f\,dg$.

Corollary If $\sum_{n=1}^{\infty} f_n$ is uniformly convergent on $[a,b]$ and if each f_n is integrable relative to a function g that is of bounded variation on $[a,b]$, then $\sum_{n=1}^{\infty} f_n$ is integrable relative to g and $\sum_{n=1}^{\infty}\int_a^b f_n\,dg = \int_a^b (\sum_{n=1}^{\infty} f_n)\,dg$.

Conditions pertaining to the convergence of sequences of the type $\{\int_a^b f\,dg_n\}$ are not so simply stated as in the above. The following example demonstrates that uniform convergence of $\{g_n\}$ does not suffice to imply the convergence of $\{\int_a^b f\,dg_n.\}$

Example 14-9 The sequence $\{g_n\}$ defined by $g_n(x) = (\sin n^3x)/n$ converges uniformly on any interval since $|g_n(x)| \le 1/n$ for all n and x. Further, it follows from the Weierstrass test that the series f defined by $f(x) = \sum_{\nu=1}^{\infty} (\cos \nu^3 x)/\nu^2$ is uniformly convergent on $\mathscr{R}$. Hence f is a continuous function and, since g_n obviously possesses a continuous derivative for each n, it follows in particular that $\int_0^{2\pi} f\,dg_n$ exists and

$$\int_0^{2\pi} f\,dg_n = \int_0^{2\pi} f g_n' = \int_0^{2\pi} \sum_{\nu=1}^{\infty} \frac{\cos \nu^3 x}{\nu^2} (n^2 \cos n^3 x)$$

$$= n^2 \sum_{\nu=1}^{\infty} \frac{1}{\nu^2} \int_0^{2\pi} \cos \nu^3 x \cos n^3 x = 2\pi$$

for all n. However, since $\{g_n\}$ converges to the zero function on $[0,2\pi]$, we see that $\int_0^{2\pi} f\,d \lim \{g_n\} = 0$.

The following theorem shows how the notion of total variation of a function on an interval may be invoked to provide a sufficient condition for convergence of a sequence $\{\int_a^b f\,dg_n\}$ to the integral of f relative to the limit of $\{g_n\}$. In this respect it should be noted that if the sequence $\{T_{g_n-g}\}$ (T_{g_n-g} denoting the total variation of $g_n - g$ on $[a,b]$) converges to zero, it does not follow that $\{g_n\}$ converges to g, but only that $\lim \{g_n - g\}$ is a constant function on $[a,b]$. Should the g_n and g be normalized, however, by requiring, say, that $g_n(a) = g(a)$ for all n, then from

$$|g_n(x) - g(x)| \le T_{g_n-g} \quad \text{and} \quad \lim \{T_{g_n-g}\} = 0$$

follows the uniform convergence of $\{g_n\}$ to g on $[a,b]$. Actually, no loss of generality is entailed in so doing since, if $c_n = g_n(a) - g(a)$, then integrals relative to g_n and $g_n - c_n$ have the same value. Accordingly, it is convenient, in the statement of the theorem, to incorporate such a normalization.

Theorem 14-23 If $\{g_n\}$ is a sequence of functions of bounded variation on $[a,b]$ such that $\{T_{g_n-g}\}$ is a null sequence, where $g = \lim \{g_n\}$, and if f is a continuous function on $[a,b]$, then the sequence $\{\int_a^b f\,dg_n\}$ converges to $\int_a^b f\,dg$.

Proof Since $g_n - g$, as well as g_n, is of bounded variation on $[a,b]$, it follows that g also is of bounded variation on that interval. Hence $\int_a^b f\,dg$ exists. Further, with M denoting an upper bound of $|f|$ on $[a,b]$, we have

$$\left|\int_a^b f\,dg_n - \int_a^b f\,dg\right| = \left|\int_a^b f\,d(g_n - g)\right| \le M T_{g_n-g}$$

by virtue of the definition of the total variation T_{g_n-g}. From the hypothesis that $\{T_{g_n-g}\}$ converges to zero follows the truth of the theorem▲

Exercises

14-31 Prove Theorem 14-19.

14-32 Prove Theorem 14-23.

14-33 Show that a function f can be integrable relative to a function g of bounded variation without being integrable relative to g_1 and g_2, where g_1 and g_2 are increasing functions such that $g = g_1 - g_2$. (*Suggestion:* Consider the functions given by $f(x) = 0$ if $-1 \leq x \leq 0$, $f(x) = 1$ if $0 < x \leq 1$; $g(x) = 1$ if $-1 \leq x \leq 1$; $g_1(x) = -1$ if $-1 \leq x \leq 0$, $g_1(x) = 1$ if $0 < x \leq 1$; and $g_2 = g_1 - g$.)

14-34 Prove that if f is integrable relative to a function g that is of bounded variation on $[a,b]$, then f is integrable relative to both the total variation function t_g and the residual function r_g of g on $[a,b]$. Further, $\int_a^b f\,dg = \int_a^b f\,dt_g - \int_a^b f\,dr_g$.

14-35 Let φ be defined by $\varphi(x) = \int_a^x f\,dg$, where g is a function of bounded variation on $[a,b]$. Prove (a) φ is of bounded variation on $[a,b]$, (b) φ is continuous at each continuity point of g, (c) $\varphi'(x)$ exists for each x at which f is continuous and $g'(x)$ exists; moreover, $\varphi'(x) = f(x)g'(x)$.

14-36 Suppose $\int_a^b h\,dg$ to exist where g is of bounded variation on $[a,b]$, and let φ be defined by $\varphi(x) = \int_a^x h\,dg$ for x in $[a,b]$. Prove that if f is continuous on $[a,b]$, then $\int_a^b f\,d\varphi = \int_a^b fh\,dg$.

14-37 Prove the Schwarz inequality: If f and h are continuous and g increasing on $[a,b]$, then $|\int_a^b fh\,dg|^2 \leq \int_a^b f^2\,dg \int_a^b h^2\,dg$.

15
Measure of a Set of Real Numbers

'Tis distance lends enchantment to the view,
And robes the mountain in its azure hue.
THOMAS CAMPBELL (1777–1844), *Pleasures of Hope*

15-1 Introduction

In the present chapter we shall be concerned with so-called *measure* properties of sets of real numbers, in contrast to, but intimately related with and supported by, those properties studied in Chapter 4 under the heading of *topology*. In the concept of measure we shall see a generalization of the notion of *length* of an interval, as well as of the concept of a *zero set*, as first defined in Chapter 10. We shall also define and study the category of *measurable functions*, based upon the notion of measure. In the following chapter in connection with the theory of the *Lebesgue integral* this class of functions will be seen to assume a vital role.

The definition of length of a bounded interval, i.e., the number $b - a$ if a and b denote the left and right endpoints, respectively, reflects the intimate association between an interval of real numbers and the geometrical concept of a line segment or continuum of points. In this respect the definition and terminology seem quite natural; however, there is not apparent any suggestion that length is merely a special instance of a broader and more sophisticated principle—one in which a real number is employed to weigh or measure the "amassing" or "accumulation" or "extent" of any set of real numbers.

The problem of generalizing the notion of length of an interval to sets of points which are neither intervals nor unions of intervals has been an intriguing and challenging one. Indeed, it has only been within the last one hundred years that significant and fruitful ideas were promulgated. Initial successes were attained by Hankel in 1882 and by Cantor in 1884. Later, Borel, in 1898, and Lebesgue, in 1902, conceived and developed the notion of measure in a form such as we shall present here.

At the outset, before a definition of measure is given, it is well to remark certain precepts which intuitively appear to be germane to the design of such a concept:

(a) If A is any set and $m(A)$ its measure, then $m(A)$ should be a non-negative real number.

(b) If A_1 and A_2 denote sets such that $A_1 \subset A_2$, then $m(A_1) \leq m(A_2)$. This we may describe as the *increasing* property for measure.

(c) Measure should be *additive*; that is, if A_1 and A_2 are disjoint sets, then $m(A_1 \cup A_2)$ should be the sum of the measures of A_1 and A_2. Further, this characteristic of measure should persist in the event of any finite collection of disjoint sets; and even for infinite collections, if it is meaningful at all to speak of forming sums.

(d) The measure of a set consisting of a single point should be zero, consistent with the "nondimensional" character of a geometric point.

(e) Lastly, the measure of an interval should be its length. This is essential, of course, if measure is to be legitimately an *extension* of length of an interval.

What is proposed is an ambitious undertaking and one should be pleasantly surprised upon the achievement of any degree of success. Complete success, in the sense that to each set of real numbers a measure can be assigned consistent with tenets (a) through (e), is not possible; this, through no deficiency in design but rather that it is inherent in the structure of the real number system to resist such construction.

We shall develop a theory of measure by beginning with the concept of length of a bounded open interval. It will then be an easy step to a measure for unions of disjoint bounded open intervals (bounded open sets). Extension to a measure for nonopen sets is then achieved through a consideration of open coverings together with collections of closed subsets. At the outset the restriction to bounded sets is firm—later we generalize to unbounded sets.

15-2 Lengths of Bounded Open Intervals

It will be our convention in the following development to include the empty set in the general category of *bounded open intervals*. Reference to a bounded open interval I, therefore, will signify either that $I = (a,b)$ for some a and b or that $I = \varnothing$. This, of course, is a departure from past practice.

The *length* of a nonempty open interval (a,b) has been defined to be the number $b - a$. The *length* of the empty set we now define to be zero. Employing the symbol ΔI to denote the length of I we have, therefore,

$$\Delta I = \begin{cases} b - a & \text{if} \quad I = (a,b) \\ 0 & \text{if} \quad I = \varnothing. \end{cases}$$

Length, as defined, is in the nature of a *non-negative valued set function* with a domain consisting of the collection of all bounded open intervals. (Previously we have defined a notion of length for intervals of the types $[a,b)$, $(a,b]$, and $[a,b]$. Such definitions are to be ignored in the present discussion. We shall see later in what sense they prove to be meaningful and consistent.)

Certain attributes of the concept of length, some rather obvious, all important, are enunciated in the theorems below. They tend to set the stage for extension to domains of a broader scope.

Theorem 15-1 If I and I_ν, $\nu = 1, 2, \ldots, n$, are bounded open intervals with the property $I \subset \bigcup_{\nu=1}^{n} I_\nu$, then $\Delta I \leq \sum_{\nu=1}^{n} \Delta I_\nu$.

Proof Since only a finite covering is involved, we regard the truth of this theorem as self-evident and omit a formal inductive proof▲

Corollary If I_1 and I_2 are bounded open intervals such that $I_1 \subset I_2$, then $\Delta I_1 \leq \Delta I_2$.

As a consequence of the last, which compares the length of an interval with that of an interval containing it, we describe length as an *increasing* function on its domain.

In several of the theorems in the present chapter we shall be concerned with an infinite sum of non-negative numbers. If such should diverge, we shall construe the value of the sum to be ∞.

The following theorem, which concerns an infinite covering, appears intuitively sound. Its proof, however, requires the force of the Heine-Borel theorem. It is interesting to note that it was precisely in this connection that Borel found the need to establish the covering principle enunciated in that theorem.

Theorem 15-2 If I and I_ν, $\nu = 1, 2, \ldots$, are bounded open intervals such that $I \subset \bigcup_{\nu=1}^{\infty} I_\nu$, then $\Delta I \leq \sum_{\nu=1}^{\infty} \Delta I_\nu$.

Proof If $I = \varnothing$, the truth of the theorem is obvious; so consider $I = (a,b)$. Let $\varepsilon > 0$ be given. Then the family of intervals consisting of the I_ν for $\nu = 1$, $2, \ldots$, together with the intervals $I_{-1} = (a - \varepsilon, a + \varepsilon)$ and $I_0 = (b - \varepsilon, b + \varepsilon)$, constitutes an open covering of the closed and bounded interval $[a,b]$. Consequently, using the Heine-Borel theorem, a finite subfamily exists consisting of I_{-1}, I_0, and $I_1, I_2, \ldots, I_n$, for some n, which covers $[a,b]$ and hence (a,b). By Theorem 15-1 we conclude

$$\Delta I \leq \sum_{\nu=-1}^{n} \Delta I_\nu = 4\varepsilon + \sum_{\nu=1}^{n} \Delta I_\nu \leq 4\varepsilon + \sum_{\nu=1}^{\infty} \Delta I_\nu .$$

From the arbitrary character of ε then follows $\Delta I \leq \sum_{\nu=1}^{\infty} \Delta I_\nu$▲

Theorem 15-3 If $\{I_\nu;\ \ \nu = 1, 2, \ldots, n\}$ is a finite collection of disjoint bounded open intervals, and if I is a bounded open interval with the property $\bigcup_{\nu=1}^{n} I_\nu \subset I$, then $\sum_{\nu=1}^{n} \Delta I_\nu \leq \Delta I$.

Proof We omit a formal proof since only a finite set of intervals is involved. One should note, of course, the degree to which the disjoint character of the intervals is important to the truth of the theorem▲

Theorem 15-4 If $\{I_\nu\}$ is a countably infinite set of disjoint bounded open intervals, and if I is a bounded open interval with the property $\bigcup_{\nu=1}^{\infty} I_\nu \subset I$, then $\sum_{\nu=1}^{\infty} \Delta I_\nu \leq \Delta I$.

Proof From the hypothesis it follows that $\bigcup_{\nu=1}^{n} I_\nu \subset I$ for any n. Hence, by Theorem 15-3, it is the case that $\sum_{\nu=1}^{n} \Delta I_\nu \leq \Delta I$; that is, the increasing sequence $\{\sum_{\nu=1}^{n} \Delta I_\nu\}$ is bounded by ΔI. As a consequence, the series $\sum_{\nu=1}^{\infty} \Delta I_\nu$ converges to a sum not exceeding ΔI▲

In the sequel we shall employ the symbol $\bigcup_\nu A_\nu$ to denote either the union $\bigcup_{\nu=1}^{n} A_\nu$ of a (nonempty) finite collection of sets or the union $\bigcup_{\nu=1}^{\infty} A_\nu$ of a countably infinite collection of sets. Concommitantly, $\sum_\nu a_\nu$ will signify either a finite sum $\sum_{\nu=1}^{n} a_\nu$ or an infinite sum $\sum_{\nu=1}^{\infty} a_\nu$ of real numbers a_ν. If it is important to distinguish the finite and infinite cases, then the more elaborate symbols committing the range of ν will be used.

Theorem 15-5 If $\{I_\nu\}$ and $\{I_\kappa^*\}$ are countable collections of disjoint bounded open intervals such that $\bigcup_\nu I_\nu \subset \bigcup_\kappa I_\kappa^*$, then $\sum_\nu \Delta I_\nu \leq \sum_\kappa \Delta I_\kappa^*$.

Proof Corresponding to each pair of values of ν and κ the set $I_\nu \cap I_\kappa^*$ is a bounded open interval. Further, for each ν it is the case that $I_\nu = \bigcup_\kappa (I_\nu \cap I_\kappa^*)$; so, by Theorems 15-1 and 15-2, $\Delta I_\nu \leq \sum_\kappa \Delta(I_\nu \cap I_\kappa^*)$. For each κ, the sets $I_\nu \cap I_\kappa^*$ corresponding to different values of ν are disjoint, and $\bigcup_\nu (I_\nu \cap I_\kappa^*) = (\bigcup_\nu I_\nu) \cap I_\kappa^* \subset I_\kappa^*$. Hence, by Theorems 15-3 and 15-4, $\sum_\nu \Delta(I_\nu \cap I_\kappa^*) \leq \Delta I_\kappa^*$. Combining inequalities we get

$$\sum_\nu \Delta I_\nu \leq \sum_\nu \sum_\kappa \Delta(I_\nu \cap I_\kappa^*) = \sum_\kappa \sum_\nu \Delta(I_\nu \cap I_\kappa^*) \leq \sum_\kappa \Delta I_\kappa^* \blacktriangle$$

We have seen that every bounded open set is the union of a unique countable collection of disjoint bounded open intervals—the empty set not excepted, by virtue of its inclusion in the category of open intervals. Thus if G is an open set, then $G = \bigcup_\nu I_\nu$. In view of this it is not unreasonable to refer to the number $\sum_\nu \Delta I_\nu$ as the *length* of G. The sense of Theorem 15-5 then, is that the length function defined on the collection of all open sets is an increasing function; that is, $\Delta G_1 \leq \Delta G_2$ if $G_1 \subset G_2$.

Actually, we shall not regard length of an open set, as defined here, as a stage in the extension of the length concept to noninterval type sets. While the reader may, if he wishes, establish theorems for open sets paralleling those for intervals, we shall proceed, in the next section, to make the jump from bounded open intervals to any bounded sets.

15-3 Outer Measure of a Bounded Set

Suppose E to denote a set which is bounded but without further qualification, and consider the collection $\{\bigcup_\nu I_\nu;\ \ E \subset \bigcup_\nu I_\nu\}$ of open coverings of E. Certainly this collection is not empty since $E \subset (a,b)$ for some a and b. Corresponding to each covering is its length $\sum_\nu \Delta I_\nu$ and, since $\sum_\nu \Delta I_\nu \geq 0$ whatever the intervals I_ν, it follows that the set of all such numbers possesses a finite infimum. We denote it by $\mu^*(E)$ and refer to it as the *exterior* or *outer measure* of the set E. Formally,

$$\mu^*(E) = \inf\left\{\sum_\nu \Delta I_\nu;\quad E \subset \bigcup_\nu I_\nu\right\}.$$

Clearly, $\mu^*(E) \geq 0$ for any bounded set E.

While outer measure of a bounded set is not our ultimate extension of the length concept, it does, indeed, possess attributes essential to such extension, as we see in the following theorems. The first reveals outer measure to be an increasing set function.

Theorem 15-6 If E_1 and E_2 are any bounded sets with the property $E_1 \subset E_2$, then $\mu^*(E_1) \leq \mu^*(E_2)$.

Proof This is an immediate consequence of the observation

$$\left\{\sum_\nu \Delta I_\nu;\quad E_2 \subset \bigcup_\nu I_\nu\right\} \subset \left\{\sum_\nu \Delta I_\nu;\quad E_1 \subset \bigcup_\nu I_\nu\right\} \blacktriangle$$

Theorem 15-7 If $\{E_\nu\}$ is a countable collection of bounded sets with the property that $\bigcup_\nu E_\nu$ is bounded, then $\mu^*(\bigcup_\nu E_\nu) \leq \sum_\nu \mu^*(E_\nu)$.

Proof Let $\varepsilon > 0$ be given. Then, corresponding to each value of ν, there exists a countable collection $\{I_{\nu\kappa}\}$ of bounded open intervals such that $E_\nu \subset \bigcup_\kappa I_{\nu\kappa}$ and $\sum_\kappa \Delta I_{\nu\kappa} < \mu^*(E_\nu) + \varepsilon/2^\nu$. Since $\bigcup_\nu E_\nu \subset \bigcup_\nu(\bigcup_\kappa I_{\nu\kappa})$, we have, from the definition of outer measure, $\mu^*(\bigcup_\nu E_\nu) \leq \sum_\nu \sum_\kappa \Delta I_{\nu\kappa} \leq \varepsilon + \sum_\nu \mu^*(E_\nu)$. Since ε is arbitrary, the theorem is proved$\blacktriangle$

Because of the property of outer measure exhibited in the above theorem, we say that the set function μ^* is *subadditive*.

In the particular instance of a bounded open set the following facts may be observed.

Theorem 15-8 If G is a bounded open set, with $G = \bigcup_\nu I_\nu$, then $\mu^*(G) = \sum_\nu \Delta I_\nu$. (In other words, the length and the outer measure of a bounded open set are equal.)

Proof From the definition of outer measure it is immediately the case that $\mu^*(G) \leq \sum_\nu \Delta I_\nu$. On the other hand, if $\{I_\kappa^*\}$ is an arbitrary countable collection of bounded open intervals such that $G \subset \bigcup_\kappa I_\kappa^*$, then $\bigcup_\nu I_\nu \subset \bigcup_\kappa I_\kappa^*$ and (Theorem 15-5) $\sum_\nu \Delta I_\nu \leq \sum_\kappa \Delta I_\kappa^*$. The definition of $\mu^*(G)$ then implies $\sum_\nu \Delta I_\nu \leq \mu^*(G)\blacktriangle$

It follows, of course, that $\mu^*(I) = \Delta I$ if I is a bounded open interval. Thus we have achieved in outer measure a set function defined on a domain consisting of all bounded sets which is equal to the length function on their common domain.

Theorem 15-9 If $\{G_\nu\}$ is a countable collection of disjoint bounded open sets with the property that $\bigcup_\nu G_\nu$ is bounded, then $\mu^*(\bigcup_\nu G_\nu) = \sum_\nu \mu^*(G_\nu)$.

Proof For each ν we have $\mu^*(G_\nu) = \sum_\kappa \Delta I_{\nu\kappa}$, where the $I_{\nu\kappa}$ are disjoint as both ν and κ vary. Moreover, $\bigcup_\nu G_\nu = \bigcup_\nu \bigcup_\kappa I_{\nu\kappa}$, so that $\mu^*(\bigcup_\nu G_\nu) = \sum_\nu \sum_\kappa \Delta I_{\nu\kappa} = \sum_\nu \mu^*(G_\nu)$▲

Thus, the subadditive outer measure function μ^*, defined on the collection of all bounded sets, is seen to be an *additive* function on the subdomain consisting of all open bounded sets. Below we see that this is also the case in the domain comprising all sets which are both closed and bounded.

Theorem 15-10 If $\{F_\nu\}$ is a countable collection of disjoint bounded closed sets with the property that $\bigcup_\nu F_\nu$ is bounded, then $\mu^*(\bigcup_\nu F_\nu) = \sum_\nu \mu^*(F_\nu)$.

Proof First, consider the case of two disjoint bounded closed sets F_1 and F_2. By Exercise 4-48 open sets G_1 and G_2 exist such that $F_1 \subset G_1$, $F_2 \subset G_2$, and $G_1 \cap G_2 = \varnothing$. Also, if $\varepsilon > 0$ be given, then, by virtue of the definition of outer measure, a countable collection $\{I_\kappa^*\}$ of disjoint open intervals exists with the properties $F_1 \cup F_2 \subset \bigcup_\kappa I_\kappa^*$ and $\sum_\kappa \Delta I_\kappa^* < \mu^*(F_1 \cup F_2) + \varepsilon$. Let $G_1' = G_1 \cap (\bigcup_\kappa I_\kappa^*)$ and $G_2' = G_2 \cap (\bigcup_\kappa I_\kappa^*)$. Then G_1' and G_2' are disjoint open sets such that $F_1 \subset G_1'$, $F_2 \subset G_2'$, and $G_1' \cup G_2' \subset \bigcup_\kappa I_\kappa^*$. Therefore

$$\begin{aligned}\mu^*(F_1) + \mu^*(F_2) &\le \mu^*(G_1') + \mu^*(G_2') = \mu^*(G_1' \cup G_2')\\ &\le \sum_\kappa \Delta I_\kappa^* < \mu^*(F_1 \cup F_2) + \varepsilon.\end{aligned}$$

Since ε is arbitrary, we conclude

$$\mu^*(F_1) + \mu^*(F_2) \le \mu^*(F_1 \cup F_2).$$

It is left to the reader to show, by means of mathematical induction, that what is true of two disjoint sets is true of any finite collection of disjoint sets, i.e.,

$$\sum_{\nu=1}^{n} \mu^*(F_\nu) \le \mu^*\left(\bigcup_{\nu=1}^{n} F_\nu\right).$$

Lastly, let $\{F_\nu;\ \nu = 1, 2, \ldots\}$ be a countably infinite collection of disjoint closed and bounded sets. Since, for any n we have $\bigcup_{\nu=1}^n F_\nu \subset \bigcup_{\nu=1}^\infty F_\nu$, it follows that $\sum_{\nu=1}^n \mu^*(F_\nu) \le \mu^*(\bigcup_{\nu=1}^n F_\nu) \le \mu^*(\bigcup_{\nu=1}^\infty F_\nu)$. Therefore $\sum_{\nu=1}^\infty \mu^*(F_\nu)$ converges to a sum not exceeding $\mu^*(\bigcup_{\nu=1}^\infty F_\nu)$. Thus, quite generally,

$$\sum_\nu \mu^*(F_\nu) \le \mu^*\left(\bigcup_\nu F_\nu\right).$$

The desired conclusion follows from this and Theorem 15-7▲

Certain conclusions consistent with the tenets of Section 15-1 are worthy of mention. If E is a set whose only member is a real number x, then $\mu^*(E) = 0$. This follows from the observation that $E \subset \mathcal{N}(x,\varepsilon)$ and $\mu^*(E) \leq 2\varepsilon$ for arbitrary ε. We have achieved, therefore, in outer measure, a concept recognizing the dimensionless character of a single point. From Theorem 15-10 we conclude also that any bounded countable set has outer measure equal to zero.

Since sets exist having outer measure zero and since $\varnothing$ is a subset of any such, it follows, from Theorem 15-6, that $\mu^*(\varnothing) = 0$.

If $I = (a,b)$, then $\mu^*(I) = b - a$. This is true, of course, since I is an open set. It is also the case that the intervals $[a,b)$, $(a,b]$, and $[a,b]$ have outer measure $b - a$. For example, from $[a,b] \subset (a,b) \cup \{a\} \cup \{b\}$ follows

$$\mu^*([a,b]) \leq \mu^*((a,b)) + \mu^*(\{a\}) + \mu^*(\{b\}) \leq b - a + 0 + 0.$$

On the other hand, from $(a,b) \subset [a,b]$ we have $\mu^*((a,b)) \leq \mu^*([a,b])$. Hence $\mu^*([a,b]) = b - a$.

The outer measure of the set of irrational numbers in the interval $[a,b]$ is $b - a$. For if C is the set of irrationals and D the set of rationals in $[a,b]$, then $[a,b] \subset C \cup D$, whence $\mu^*([a,b]) \leq \mu^*(C) + \mu^*(D) = \mu^*(C)$. On the other hand, since $C \subset [a,b]$, we have $\mu^*(C) \leq \mu^*([a,b])$. It follows that $\mu^*(C) = \mu^*([a,b]) = b - a$.

The outer measure function μ^*, defined on the collection of all bounded sets of real numbers, fails to fulfill the criteria for a measure function as enunciated at the outset of discussion. In particular, it does not possess the much desired additive property, i.e., that for disjoint sets A and B it should follow that $\mu^*(A) + \mu^*(B) = \mu^*(A \cup B)$. Counterexamples pointing to the violation of this principle will be introduced later. However, by properly restricting the domain of μ^*, additivity may be effected. This is the intent of the following sections.

Exercises

15-1 Supply details of proofs of Theorems 15-1 and 15-3.

15-2 Prove by induction that $\sum_{\nu=1}^{n} \mu^*(F_\nu) \leq \mu^*\left(\bigcup_{\nu=1}^{n} F_\nu\right)$ if the sets F_ν are closed, bounded, and pairwise disjoint.

15-3 Prove that if A is any set and if B is such that $\mu^*(B) = 0$, then $\mu^*(A \cup B) = \mu^*(A)$.

15-4 Prove that if E_1 and E_2 are such that $\bar{E}_1 \cap \bar{E}_2 = \varnothing$, then $\mu^*(E_1 \cup E_2) = \mu^*(E_1) + \mu^*(E_2)$.

15-4 Inner Measure of a Bounded Set

Suppose E to denote any bounded set and consider the collection of all closed subsets of E. That this collection is not empty is clear since the empty set $\varnothing$ obviously belongs to it. Also nonempty is the set of real numbers which are outer measures of the members of this collection. Because this set of numbers is bounded above (by $\mu^*(E)$), it possesses a finite supremum. This we denote by $\mu_*(E)$ and call the *interior* or *inner measure* of E. Formally,

$$\mu_*(E) = \sup \{\mu^*(F); \quad F \text{ closed and } F \subset E\}.$$

Every bounded set possesses a non-negative inner measure.

It is not difficult to conjecture the simple and important relationship between outer and inner measure stated in the following theorem, based, as these concepts are, one upon covering sets and the other on covered sets.

Theorem 15-11 If E is any bounded set, then $\mu_*(E) \leq \mu^*(E)$.

Proof For all closed sets F such that $F \subset E$ it is the case that $\mu^*(F) \leq \mu^*(E)$. Hence the supremum $\mu_*(E)$ of the set of all numbers $\mu^*(F)$ cannot exceed $\mu^*(E)$▲

An immediate consequence of the last theorem is that a set having outer measure zero also has inner measure zero. In particular, the inner measure of a countable set is zero, as is that of the empty set.

The increasing character of outer measure is shared by the inner measure function.

Theorem 15-12 If E_1 and E_2 are bounded sets such that $E_1 \subset E_2$, then $\mu_*(E_1) \leq \mu_*(E_2)$.

Proof This follows very simply from the observation that any closed subset of E_1 is a subset of E_2 as well▲

Contrast the following property of inner measure with the subadditive character of outer measure.

Theorem 15-13 If $\{E_\nu\}$ is a countable collection of disjoint bounded sets such that $\bigcup_\nu E_\nu$ is bounded, then $\sum_\nu \mu_*(E_\nu) \leq \mu_*(\bigcup_\nu E_\nu)$.

Proof Let $\varepsilon > 0$ be given. Then, corresponding to each ν, there exists a closed set F_ν such that $F_\nu \subset E_\nu$ and $\mu_*(E_\nu) - \varepsilon/2^\nu < \mu^*(F_\nu)$. Further, for any n the sets F_ν, $\nu = 1, 2, \ldots, n$, are disjoint and have the property $\bigcup_{\nu=1}^n F_\nu \subset \bigcup_\nu E_\nu$. Since $\bigcup_{\nu=1}^n F_\nu$ is closed, we have, by the definition of inner measure, $\mu^*(\bigcup_{\nu=1}^n F_\nu) \leq \mu_*(\bigcup_\nu E_\nu)$. Hence, using Theorem 15-10,

$$\sum_{\nu=1}^{n} \left(\mu_*(E_\nu) - \frac{\varepsilon}{2^\nu}\right) < \sum_{\nu=1}^{n} \mu^*(F_\nu) = \mu^*\left(\bigcup_{\nu=1}^{n} F_\nu\right) \leq \mu_*\left(\bigcup_{\nu} E_\nu\right)$$

or $\sum_{\nu=1}^n \mu_*(E_\nu) < \varepsilon + \mu_*(\bigcup_\nu E_\nu)$. This being the case for any n, we have

$$\sum_{\nu} \mu_*(E_\nu) \leq \varepsilon + \mu_*\left(\bigcup_{\nu} E_\nu\right)▲$$

The next results demonstrate that certain sets frequently encountered possess equivalued inner and outer measures.

Theorem 15-14 If F is a bounded closed set, then $\mu_*(F) = \mu^*(F)$.

Proof From $F \subset F$ and the definition of inner measure follows $\mu^*(F) \leq \mu_*(F)$. Together with Theorem 15-11 this implies the desired equality▲

Because of Theorem 15-14 we have $\mu_*([a,b]) = \mu^*([a,b]) = b - a$. As for (a,b), we see that from $[a+\varepsilon, b-\varepsilon] \subset (a,b)$ for any $\varepsilon > 0$, we have $b - a - 2\varepsilon = \mu_*([a+\varepsilon, b-\varepsilon]) \leq \mu_*((a,b))$, whence $b - a \leq \mu_*((a,b))$. Since $\mu_*((a,b)) \leq \mu^*((a,b)) = b - a$, we conclude $\mu_*((a,b)) = \mu^*((a,b)) = b - a$. Similar arguments lead to $\mu_*([a,b)) = \mu^*([a,b)) = \mu_*((a,b]) = \mu^*((a,b]) = b - a$.

Theorem 15-15 If G is a bounded open set, then $\mu_*(G) = \mu^*(G)$.

Proof Since $G = \bigcup_\nu I_\nu$ and since, by the foregoing remarks, $\mu_*(I_\nu) = \mu^*(I_\nu) = \Delta I_\nu$, we conclude, using Theorems 15-8 and 15-13, that

$$\mu^*(G) = \sum_\nu \Delta I_\nu = \sum_\nu \mu_*(I_\nu) \leq \mu_*\left(\bigcup_\nu I_\nu\right) = \mu_*(G).$$

It follows that $\mu^*(G) = \mu_*(G)$▲

15-5 Measure of a Bounded Set

Associated with each bounded set E is its outer measure $\mu^*(E)$ and its inner measure $\mu_*(E)$, and these stand in the relationship $\mu_*(E) \leq \mu^*(E)$. Moreover, as has just been proved, the two have one and the same value if E is a bounded open set, a bounded closed set, or a bounded interval. Further, if E has outer measure zero, then its inner measure is zero too. It follows, in particular, that bounded countable sets, such as the set of rationals of the form $1/n$ or the dense set of all rationals in any bounded interval, have equal inner and outer measures.

If the outer measure and inner measure of a set E are equal, we call their common value the *measure* (or *Lebesgue measure*) of E and denote it by $\mu(E)$; the set E itself we speak of as being *measurable* (or *Lebesgue measurable*). Measure is our ultimate concept. While, as noted above, bounded open sets, bounded closed sets, bounded intervals, and bounded countable sets are all measurable, the notion of measure is even more far-reaching. We shall encounter measurable sets of real numbers with quite intricate constructions. It will turn out, however, that sets do exist which are *nonmeasurable*; that is, there are sets with the property $\mu_*(E) < \mu^*(E)$. An example of such will be given later, after we have fully explored the nature of measurable sets and have shown, in particular, that in the concept of measure we have achieved the properties enunciated as desirable in Section 15-1.

A simple but useful characterization of the measure of a set recognizes its role as infimum and supremum of measures of open and closed sets, respectively. Its proof is left to the reader.

Theorem 15-16 A bounded set E is measurable if and only if, corresponding to any given $\varepsilon > 0$, there exists a bounded open set G and a bounded closed set F with the properties $F \subset E \subset G$ and $\mu(G) - \mu(F) < \varepsilon$.

The following two theorems express properties of the measure function which are quite immediate consequences of properties of outer and inner measure. The first testifies to the increasing character of the measure function and the second to its additivity.

Theorem 15-17 If E_1 and E_2 are bounded measurable sets such that $E_1 \subset E_2$, then $\mu(E_1) \leq \mu(E_2)$.

Proof A reiteration of Theorem 15-6 for measurable sets▲

Corollary Any subset of a bounded set of measure zero is a set of measure zero.

Reference to sets of measure zero at this point prompts recollection of the notion of a zero set, defined in Chapter 10. With the notation of the present chapter, a set E was called a zero set if, corresponding to each $\varepsilon > 0$, an open covering $\bigcup_\nu I_\nu$ of E exists with the property that $\sum_\nu \Delta I_\nu < \varepsilon$. One sees in this definition a simple reiteration of the definition of outer measure of a set in the special instance in which that outer measure has the value zero. Accordingly, a zero set and a set of measure zero are one and the same concept.

If a statement of fact relative to a set E is valid at all points of E except possibly for a subset having measure zero, we shall refer to the statement as being true *almost everywhere* on E. This phraseology is prompted by the numerous instances in which an hypothesis which leads to a certain conclusion on a set may be violated on a set of measure zero without significantly altering the conclusion. For example, we have seen that continuity of a bounded function on a closed interval $[a,b]$ implies Riemann integrability of the function; so does continuity almost everywhere on $[a,b]$.

Theorem 15-18 If $\{E_\nu\}$ is a countable collection of disjoint bounded measurable sets with the property that $\bigcup_\nu E_\nu$ is bounded, then $\bigcup_\nu E_\nu$ is measurable, and $\mu(\bigcup_\nu E_\nu) = \sum_\nu \mu(E_\nu)$.

Proof From Theorems 15-7 and 15-13 we have

$$\mu^*\left(\bigcup_\nu E_\nu\right) \leq \sum_\nu \mu^*(E_\nu) = \sum_\nu \mu_*(E_\nu) \leq \mu_*\left(\bigcup_\nu E_\nu\right),$$

from which the existence of $\mu(\bigcup_\nu E_\nu)$ and the desired equality follow▲

If F is any bounded closed set and I a bounded open interval, then the set difference $I - F$ is an open set, and hence measurable. Also, if G is a bounded open set, then $\bar{I} - G$ is closed, and therefore measurable. Additivity implies, of course, that

$$\mu(F) + \mu(I - F) = \mu(I) \quad \text{and} \quad \mu(G) + \mu(\bar{I} - G) = \mu(I)$$

if $F \subset I$ and $G \subset \bar{I}$.

The following theorems are in the nature of a generalization to sets which are measurable, but not necessarily open or closed. Note that Theorem 15-19 is a special instance of Theorem 15-20 and is used in the proof of that theorem.

Theorem 15-19 If E is a measurable subset of a bounded open interval I, then $\bar{I} - E$ is measurable.

Proof Suppose $\varepsilon > 0$ to be given and G and F bounded open and closed sets, respectively, such that $F \subset E \subset G \subset I$ and $\mu(G) - \mu(F) < \varepsilon$. Then, since $\bar{I} - G \subset \bar{I} - E$ and $I - E \subset I - F$, we have $\mu(\bar{I} - G) \leq \mu_*(\bar{I} - E)$ and $\mu^*(I - E) \leq \mu(I - F)$. Now $(\bar{I} - E) - (I - E)$ is a finite set; call it A. Therefore $\mu^*(\bar{I} - E) \leq \mu^*(I - E) + \mu^*(A) \leq \mu(I - F)$. By combining inequalities we get

$$\mu^*(\bar{I} - E) - \mu_*(\bar{I} - E) \leq \mu(I - F) - \mu(\bar{I} - G) \leq \mu(G) - \mu(F) < \varepsilon$$

and conclude that the set $\bar{I} - E$ is measurable▲

Theorem 15-20 If E_1 and E_2 are bounded measurable sets such that $E_1 \subset E_2$, then $E_2 - E_1$ is measurable, and $\mu(E_2 - E_1) = \mu(E_2) - \mu(E_1)$.

Proof Let I and J be open intervals such that $E_2 \subset I$ and $\bar{I} \subset J$, and let $E = E_2 - E_1$ and $A = \bar{J} - J$. By Theorem 15-19 we know $\bar{I} - E_2$ to be measurable. Since $E_1, \bar{I} - E_2$, and $J - \bar{I}$ are disjoint, it follows from Theorem 15-18 that $E_1 \cup (\bar{I} - E_2) \cup (J - \bar{I}) = J - E$ is measurable. Again by Theorem 15-19, $\bar{J} - (J - E) = E \cup A$ is measurable.

Suppose that F is a closed subset of $E \cup A$. Then $F_1 = F \cap \bar{I}$ is a closed subset of E, and $\mu(F_1) = \mu(F)$. From this it follows that $\mu_*(E) \geq \mu_*(E \cup A) = \mu(E \cup A)$. But $\mu_*(E) \leq \mu^*(E \cup A) = \mu(E \cup A)$ implies that E is a measurable set. The desired inequality is an immediate consequence of the disjoint character of E_1 and $E_2 - E_1$▲

As a consequence of the above theorem we conclude that the set of all irrational numbers in a bounded interval (a,b) has a measure which is equal to the measure $b - a$ of the interval itself. This is true since the rational points of (a,b) constitute a set having measure zero. It should be noted that until Theorem 15-20 we were not in a position to evaluate the measure of the irrationals in (a,b). Additivity of the measure function along with the measure of the interval and the measure of the rational points in the interval do not suffice. It must first be established that the set of irrationals actually possesses a measure; only then may its value be asserted. We note, incidentally, that the set of irrationals in (a,b) is our first cited instance of a set which has a positive measure but is not a union of intervals.

The remaining theorems in this section are concerned with the establishment of measurability of unions and intersections of measurable sets, together with statements testifying to the relationships between the measures of the sets. A relation of some interest follows directly if it is assumed that E_1, E_2, $E_1 \cup E_2$, and $E_1 \cap E_2$ are all bounded measurable sets. If, for convenience, we introduce sets $A = E_1 - (E_1 \cap E_2)$ and $B = E_2 - (E_1 \cap E_2)$, then it is clear that A, B, and $E_1 \cap E_2$ are disjoint and that, as a consequence, $\mu(A) + \mu(E_1 \cap E_2) = \mu(E_1)$ and $\mu(B) + \mu(E_1 \cap E_2) = \mu(E_2)$. The identity $A \cup B \cup (E_1 \cap E_2) = E_1 \cup E_2$ being used in conjunction with these, we are led to

$$\mu(E_1 \cap E_2) + \mu(E_1 \cup E_2) = \mu(E_1) + \mu(E_2).$$

As we see in the following theorem, the same result ensues even if less is assumed, that is, if it is not assumed to begin with that $E_1 \cup E_2$ and $E_1 \cap E_2$ are measurable.

Theorem 15-21 If E_1 and E_2 are bounded measurable sets, then so are $E_1 \cap E_2$ and $E_1 \cup E_2$. Moreover, $\mu(E_1 \cap E_2) + \mu(E_1 \cup E_2) = \mu(E_1) + \mu(E_2)$.

Proof Let $\varepsilon > 0$ be given and suppose closed sets F_1 and F_2 and open sets G_1 and G_2 to be such that $F_1 \subset E_1 \subset G_1$, $\mu(G_1) - \mu(F_1) < \varepsilon$, $F_2 \subset E_2 \subset G_2$, and $\mu(G_2) - \mu(F_2) < \varepsilon$. Since $F_1 \cap F_2 \subset E_1 \cap E_2 \subset G_1 \cap G_2$ and $F_1 \cup F_2 \subset E_1 \cup E_2 \subset G_1 \cup G_2$, we have

$$\mu(F_1 \cap F_2) \leq \mu_*(E_1 \cap E_2) \leq \mu^*(E_1 \cap E_2) \leq \mu(G_1 \cap G_2),$$

and

$$\mu(F_1 \cup F_2) \leq \mu_*(E_1 \cup E_2) \leq \mu^*(E_1 \cup E_2) \leq \mu(G_1 \cup G_2).$$

It follows that

$$\begin{aligned}\mu(F_1 \cap F_2) + \mu(F_1 \cup F_2) &\leq \mu_*(E_1 \cap E_2) + \mu_*(E_1 \cup E_2)\\ &\leq \mu(G_1 \cap G_2) + \mu(G_1 \cup G_2).\end{aligned}$$

Since F_1, F_2, $F_1 \cap F_2$, $F_1 \cup F_2$, G_1, G_2, $G_1 \cap G_2$, and $G_1 \cup G_2$ are all measurable, it follows from remarks preliminary to the theorem that

$$\begin{aligned}\mu(F_1) + \mu(F_2) &\leq \mu_*(E_1 \cap E_2) + \mu_*(E_1 \cup E_2)\\ &\leq \mu(G_1) + \mu(G_2) \leq \mu(F_1) + \mu(F_2) + 2\varepsilon.\end{aligned}$$

Similarly

$$\mu(F_1) + \mu(F_2) \leqslant \mu^*(E_1 \cap E_2) + \mu^*(E_1 \cup E_2) \leqslant \mu(F_1) + \mu(F_2) + 2\varepsilon$$

so that

$$\mu^*(E_1 \cap E_2) + \mu^*(E_1 \cup E_2) - \mu_*(E_1 \cap E_2) - \mu_*(E_1 \cup E_2) \leqslant 2\varepsilon.$$

Since ε is arbitrary, it follows that

$$\mu_*(E_1 \cap E_2) + \mu_*(E_1 \cup E_2) = \mu^*(E_1 \cap E_2) + \mu^*(E_1 \cup E_2).$$

Clearly, if $\mu_*(E_1 \cap E_2) < \mu^*(E_1 \cap E_2)$, then $\mu_*(E_1 \cup E_2) > \mu^*(E_1 \cup E_2)$, which is absurd. Hence $\mu_*(E_1 \cap E_2) = \mu^*(E_1 \cap E_2)$ and, correspondingly, $\mu_*(E_1 \cup E_2) = \mu^*(E_1 \cup E_2)$▲

A simple induction proof will extend the facts of the above theorem to any finite unions and intersections. The reader should supply details. Further generalization to countably infinite collections is to be found among the facts established in the theorems that follow. These are especially concerned with certain types of sequences of measurable sets and the accompanying sequences of measures.

Theorem 15-22 If $\{E_\nu\}$ is a sequence of measurable sets such that $E_\nu \subset E_{\nu+1}$ for all ν, and if $\bigcup_{\nu=1}^{\infty} E_\nu$ is bounded, then $\bigcup_{\nu=1}^{\infty} E_\nu$ is measurable and $\lim \{\mu(E_\nu)\} = \mu(\bigcup_{\nu=1}^{\infty} E_\nu)$.

Proof Let $A_1 = E_1$ and, for $\nu > 1$, $A_\nu = E_\nu - E_{\nu-1}$. It is evident that $E_\nu = \bigcup_{\kappa=1}^{\nu} A_\kappa$ for each ν and that the sets $\bigcup_{\nu=1}^{\infty} A_\nu$ and $\bigcup_{\nu=1}^{\infty} E_\nu$ are one and the same. Further, the sets A_ν are disjoint, each A_ν is measurable, and, since $\bigcup_{\nu=1}^{\infty} E_\nu$ is bounded, it, too, is measurable. Therefore

$$\mu\left(\bigcup_{\nu=1}^{\infty} E_\nu\right) = \mu\left(\bigcup_{\kappa=1}^{\infty} A_\kappa\right) = \sum_{\kappa=1}^{\infty} \mu(A_\kappa)$$

$$= \lim \left\{\sum_{\kappa=1}^{\nu} \mu(A_\kappa)\right\} = \lim \left\{\sum_{\kappa=1}^{\nu} \mu(E_\kappa - E_{\kappa-1})\right\}$$

where, for convenience, $E_0 = \varnothing$. By Theorem 15-20

$$\mu\left(\bigcup_{\nu=1}^{\infty} E_\nu\right) = \lim \left\{\sum_{\kappa=1}^{\nu} [\mu(E_\kappa) - \mu(E_{\kappa-1})]\right\} = \lim \{\mu(E_\nu)\}▲$$

Theorem 15-23 If E_ν is a nested sequence of bounded measurable sets, then $\bigcap_{\nu=1}^{\infty} E_\nu$ is measurable and $\lim \{\mu(E_\nu)\} = \mu(\bigcap_{\nu=1}^{\infty} E_\nu)$.

Proof It is easily verified that $E_1 - \bigcap_{\nu=1}^{\infty} E_\nu = \bigcup_{\nu=1}^{\infty} (E_1 - E_\nu)$. For each ν we know $E_1 - E_\nu$ to be measurable and, because of the nested character of $\{E_\nu\}$, $E_1 - E_\nu \subset E_1 - E_{\nu+1}$. Hence, by the preceding theorem, $E_1 - \bigcap_{\nu=1}^{\infty} E_\nu$ is measurable, whence, so is $\bigcap_{\nu=1}^{\infty} E_\nu$. Further, by the same theorem,

$$\mu\left(E_1 - \bigcap_{\nu=1}^{\infty} E_\nu\right) = \mu\left(\bigcup_{\nu=1}^{\infty} (E_1 - E_\nu)\right) = \lim \{\mu(E_1 - E_\nu)\}$$

$$= \lim \{\mu(E_1) - \mu(E_\nu)\} = \mu(E_1) - \lim \{\mu(E_\nu)\}.$$

Therefore

$$\mu\left(\bigcap_{\nu=1}^{\infty} E_\nu\right) = \mu(E_1) - \mu\left(E_1 - \bigcap_{\nu=1}^{\infty} E_\nu\right) = \lim \{\mu(E_\nu)\}▲$$

Finally, we consider two theorems testifying to the measurability of the union and intersection of countably infinite collections of measurable sets.

Theorem 15-24 If $\{E_\nu\}$ is a countably infinite collection of measurable sets with the property that $\bigcup_{\nu=1}^{\infty} E_\nu$ is bounded, then $\bigcup_{\nu=1}^{\infty} E_\nu$ is measurable and $\mu(\bigcup_{\nu=1}^{\infty} E_\nu) \leq \sum_{\nu=1}^{\infty} \mu(E_\nu)$.

Proof Let $B_1 = E_1$ and, for $\nu > 1$, $B_\nu = E_\nu - \bigcup_{\kappa=1}^{\nu-1} E_\kappa$. Then for all ν the set B_ν is measurable, by virtue of Theorem 15-20 and Exercise 15-6. Further, since the B_ν are disjoint, it follows, by Theorem 15-18 that $\bigcup_{\nu=1}^{\infty} B_\nu$ is measurable. Thus $\bigcup_{\nu=1}^{\infty} E_\nu$ is measurable since $\bigcup_{\nu=1}^{\infty} E_\nu = \bigcup_{\nu=1}^{\infty} B_\nu$. The desired inequality is a consequence of Theorem 15-7▲

Theorem 15-25 If $\{E_\nu\}$ is a countably infinite collection of bounded measurable sets, then the set $\bigcap_{\nu=1}^{\infty} E_\nu$ is measurable.

Proof Let $C_\nu = \bigcap_{\kappa=1}^{\nu} E_\nu$. Then, by Exercise 15-6, C_ν is measurable for each ν. Also, $C_{\nu+1} \subset C_\nu$, so that, by Theorem 15-23, $\bigcap_{\nu=1}^{\infty} C_\nu$ is measurable. The truth of the theorem follows from the observation $\bigcap_{\nu=1}^{\infty} E_\nu = \bigcap_{\nu=1}^{\infty} C_\nu$▲

The following result expresses a connection between outer and inner measures which is frequently employed as a definition of inner measure.

Theorem 15-26 If E is a set such that $E \subset I$, I being a bounded interval, then $\mu_*(E) + \mu^*(I - E) = \mu(I)$.

Proof Corresponding to each $\varepsilon > 0$ is a measurable set F such that $F \subset E$ and $\mu_*(E) < \mu(F) + \varepsilon$. From $I - E \subset I - F$ follows $\mu^*(I - E) \leq \mu(I - F)$ and

$$\mu_*(E) + \mu^*(I - E) < \mu(I - F) + \mu(F) + \varepsilon = \mu(I) + \varepsilon.$$

Hence $\mu_*(E) + \mu^*(I - E) \leq \mu(I)$.

Also, to each $\varepsilon > 0$ corresponds a measurable set G such that $I - E \subset G \subset I$ and $\mu(G) < \mu^*(I - E) + \varepsilon$. Since $I - G \subset E$, we have $\mu(I - G) \leq \mu_*(E)$ and

$$\mu(G) + \mu(I - G) < \mu_*(E) + \mu^*(I - E) + \varepsilon.$$

Hence $\mu(I) \leq \mu_*(E) + \mu^*(I - E)$▲

The next theorem serves to express measurability of sets in terms of outer measures. Later we shall employ the relationship in the definition of a general measure concept.

Theorem 15-27 A bounded set E is measurable if and only if

$$\mu^*(A) = \mu^*(A \cap E) + \mu^*(A \cap \mathscr{C}E)$$

for each bounded subset A of $\mathscr{R}$.

Proof Let E be measurable and consider $A = (A \cap E) \cup (A \cap \mathscr{C}E)$. For any $\varepsilon > 0$ a measurable set G exists with $A \subset G$ and $\mu(G) < \mu^*(A) + \varepsilon$. It follows that

$$\mu^*(A \cap E) + \mu^*(A \cap \mathscr{C}E) \leq \mu(G \cap E) + \mu(G \cap \mathscr{C}E) = \mu(G) < \mu^*(A) + \varepsilon$$

or

$$\mu^*(A \cap E) + \mu^*(A \cap \mathscr{C}E) \leq \mu^*(A).$$

The "only if" part of the theorem follows from this and the fact that

$$\mu^*(A) \leq \mu^*(A \cap E) + \mu^*(A \cap \mathscr{C}E).$$

Suppose now that E is bounded and that $\mu^*(A) = \mu^*(A \cap E) + \mu^*(A \cap \mathscr{C}E)$ for any bounded set A. Then if $A = I$, where $E \subset I$, we have

$$\mu(I) = \mu^*(E) + \mu^*(I \cap \mathscr{C}E).$$

Together with Theorem 15-26, this implies $\mu^*(E) = \mu_*(E)$▲

Exercises

15-5 Prove Theorem 15-16.

15-6 Prove by induction that the union and intersection of a finite collection of bounded measurable sets is measurable.

15-7 Prove that if A is any bounded set, then a measurable set E exists such that $A \subset E$ and $\mu^*(A) = \mu(E)$.

15-8 Prove that if $A_\nu \subset A_{\nu+1}$ for $\nu \in I_+$ and if $\bigcup_\nu A_\nu$ is bounded, then $\mu^*(\bigcup_\nu A_\nu) = \lim \{\mu^*(A_\nu)\}$.

15-9 Prove that a bounded set E is measurable if and only if, corresponding to each $\varepsilon > 0$, there exists a closed set F such that $F \subset E$ and $\mu^*(E - F) < \varepsilon$.

15-10 Prove that a bounded set E is measurable if and only if, corresponding to each $\varepsilon > 0$, there exists an open set G such that $E \subset G$ and $\mu^*(G - E) < \varepsilon$.

A forerunner of the concept of measure of a set was the notion of *Jordan content*, which was based on unions of finite collections of intervals rather than on countable collections. Thus the *exterior Jordan content* $c^*(E)$ of a bounded set E is defined by $c^*(E) = \inf \{\sum_{\nu=1}^{n} \Delta I_\nu;\quad E \subset \bigcup_{\nu=1}^{n} I_\nu\}$.

15-11 Prove that if E is a finite set, then $c^*(E) = 0$.

15-12 Prove that the set of rational numbers in $[a,b]$ has exterior Jordan content $b - a$.

15-13 Give a definition of *interior Jordan content* $c_*(E)$ paralleling that of inner measure.

15-14 Give a definition of *Jordan content* patterned after that of measure. Prove that if a bounded set E has Jordan content, then it is measurable, but that the converse need not be the case.

15-15 Prove that the Cantor set has Jordan content zero.

15-16 Prove that the union of a countably infinite collection of sets each possessing Jordan content need not have Jordan content.

15-6 Measure of Unbounded Sets

The extension of the measure concept to sets which are not bounded is accomplished quite easily through sequences of bounded sets. Thus, an unbounded set E will be described as *measurable* (or *Lebesgue measurable*) if for each integer n the bounded set $E_n = E \cap (-n,n)$ is measurable. The sequence $\{\mu(E_n)\}$ is clearly increasing. If it converges, its limit is defined to be the *measure* $\mu(E)$ of E; if it diverges, we say that E has *infinite measure* and write $\mu(E) = \infty$. It is convenient in this respect to think in terms of the extended number system $\mathscr{R}^*$. Then in all instances, if E is measurable, the sequence $\{\mu(E_n)\}$ converges to a limit, finite or infinite.

Example 15-1

(a) The set $E = \bigcup_{\nu=1}^{\infty} I_\nu$, where $I_\nu = (\nu - 1/2 - 1/2^\nu, \nu - 1/2 + 1/2^\nu)$ is clearly unbounded. Since for any n we have $E \cap (-n,n) = \bigcup_{\nu=1}^{n} I_\nu$, it follows that E is measurable and that

$$\mu(E) = \lim \left\{\mu\left(\bigcup_{\nu=1}^{n} I_\nu\right)\right\} = \lim \left\{\sum_{\nu=1}^{n} \frac{1}{2^{\nu-1}}\right\} = 2.$$

(b) The set $E = \bigcup_{\nu=1}^{\infty} I_\nu$, where $I_\nu = (\nu - 1/2 - 1/2\nu, \nu - 1/2 + 1/2\nu)$ is also unbounded and measurable. For its measure we have

$$\mu(E) = \lim \left\{\mu\left(\bigcup_{\nu=1}^{n} I_\nu\right)\right\} = \lim \left\{\sum_{\nu=1}^{n} \frac{1}{\nu}\right\} = \infty.$$

(c) The set $\mathscr{R}$ of all real numbers has infinite measure.
(d) The set of all irrational numbers has infinite measure.
(e) The set R of all rational numbers has measure zero.

Measure having been defined for two types of sets, bounded and unbounded, a question arises as to the compatibility of the two definitions. Thus if E is *any* set and if the definition of measure of an unbounded set is applied to it, does the definition effectively reduce to a statement valid in the event E is bounded? Some simple observations show that this is indeed the case. First, a bounded set E is measurable if and only if for each n the set $E_n = E \cap (-n,n)$ is measurable. It is certainly clear that if $E \cap (-n,n)$ is measurable for each n and if E is bounded, then for a sufficiently large n, say N, we have $E \cap (-N,N) = E$, whence E is measurable.

On the other hand, if E is measurable, then so is $E \cap (-n,n)$ for each n, since $(-n,n)$ is measurable. It is easily seen that if E is bounded and measurable, then $\mu(E) = \lim \{\mu(E \cap (-n,n))\}$.

Having extended the domain of definition of the measure function to include unbounded as well as bounded sets and augmented its range to contain the infinite number ∞, it is well to review the many facts established in the case of bounded sets to see if they remain valid with or without modification in the instance of unbounded sets.

With respect to outer and inner measures we have defined no counterparts for unbounded sets. Therefore the statements of the first four sections are without generalization. Nevertheless it is possible to state a result comparable to that of Theorem 15-16 concerning an open covering set and a closed covered set. It will be convenient to delay this until later; the generalization appears in Theorem 15-34.

Theorems 15-17 through 15-25 extend to the case of unbounded sets with little alteration. Omitting proofs we restate them free of the hypotheses of boundedness.

Theorem 15-28 If E_1 and E_2 are measurable sets such that $E_1 \subset E_2$, then $\mu(E_1) \leq \mu(E_2)$.

Corollary Any subset of a set of measure zero is a set of measure zero.

In connection with the latter statement it should be noted that it is entirely possible for an unbounded set to have zero measure. The set of all integers is such a set—so is the set R of all rational numbers.

Theorem 15-29 If $\{E_\nu\}$ is a countable collection of disjoint measurable sets, then $\bigcup_\nu E_\nu$ is measurable, and $\mu(\bigcup_\nu E_\nu) = \sum_\nu \mu(E_\nu)$.

Theorem 15-30 If E_1 and E_2 are measurable sets such that $E_1 \subset E_2$, then $E_2 - E_1$ is measurable.

Corollary If E is a measurable set, then $\mathscr{C}E$ is a measurable set.

Facts enunciated in Theorems 15-21, 15-24, and 15-25 may be incorporated conveniently into a single statement.

Theorem 15-31 If $\{E_\nu\}$ is a countable collection of measurable sets, then $\bigcap_\nu E_\nu$ and $\bigcup_\nu E_\nu$ are measurable. Further, $\mu(\bigcup_\nu E_\nu) \leq \sum_\nu \mu(E_\nu)$.

Theorem 15-22 on increasing sequences of measurable sets remains essentially intact.

Theorem 15-32 If $\{E_\nu\}$ is a sequence of measurable sets such that $E_\nu \subset E_{\nu+1}$ for all ν, then $\bigcup_{\nu=1}^{\infty} E_\nu$ is measurable and $\mu(\bigcup_{\nu=1}^{\infty} E_\nu) = \lim \{\mu(E_\nu)\}$.

In Theorem 15-23 on nested sequences the condition that the elements be bounded may be relaxed to a degree. It is essential that, while E_1 may be unbounded, it should have a finite measure; otherwise the theorem fails. Witness the nested sequence $\{E_\nu\}$ where $E_\nu = \{x; \quad x \geq \nu\}$. Clearly, $\mu(E_\nu) = \infty$ for all ν, yet $\mu(\bigcap_{\nu=1}^{\infty} E_\nu) = \mu(\varnothing) = 0$. The proper statement of the extended theorem follows.

Theorem 15-33 If $\{E_\nu\}$ is a nested sequence of measurable sets such that $\mu(E_1)$ is finite, then $\bigcap_{\nu=1}^{\infty} E_\nu$ is measurable and $\mu(\bigcap_{\nu=1}^{\infty} E_\nu) = \lim \{\mu(E_\nu)\}$.

Next we prove a generalization of Theorem 15-16. Observe that no reference is made to the difference of measures of two sets since such a difference is not defined in the event both measures are infinite.

Theorem 15-34 A set E is measurable if and only if, corresponding to each $\varepsilon > 0$, there exists an open set G and a closed set F with the properties $F \subset E \subset G$ and $\mu(G - F) < \varepsilon$.

Proof Suppose E to be measurable and $\varepsilon > 0$ to be given. Corresponding to each positive integer n define $A_n = \{x; \quad n - 1 \leq |x| < n\}$ and let $E_n = E \cap A_n$. Then each E_n is measurable and $E = \bigcup_{n=1}^{\infty} E_n$.

By Theorem 15-16 there exist, corresponding to each n, an open set G_n and a closed set F_n such that $F_n \subset E_n \subset G_n$ and $\mu(G_n - F_n) < \varepsilon/2^n$. Clearly $G = \bigcup_{n=1}^{\infty} G_n$ is open; further, employing the fact that $F_n \subset A_n$, it is not difficult to see that $F = \bigcup_{n=1}^{\infty} F_n$ is a closed set.

Since $G - F \subset \bigcup_{n=1}^{\infty}(G_n - F_n)$, it follows that $\mu(G - F) \leq \sum_{n=1}^{\infty} \mu(G_n - F_n) \leq \varepsilon$. Moreover, $F \subset E \subset G$. This establishes one part of the theorem.

For the converse, corresponding to a positive integer n, let $E_n = E \cap (-n,n)$. Let $\varepsilon > 0$ be given, along with an open set G and a closed set F such that $F \subset E \subset G$ and $\mu(G - F) < \varepsilon$. Letting $F_n = F \cap (-n,n)$ and $G_n = G \cap (-n,n)$, it is clear that $G_n - F_n \subset G - F$, whence $\mu(G_n - F_n) < \varepsilon$ or, since G_n and F_n have finite measures, $\mu(G_n) < \mu(F_n) + \varepsilon$. From $F_n \subset E_n \subset G_n$ then follows

$$\mu(F_n) \leq \mu_*(E_n) \leq \mu^*(E_n) \leq \mu(G_n) < \mu(F_n) + \varepsilon.$$

It follows that $\mu_*(E_n) = \mu^*(E_n)$ for each n. Hence E is measurable▲

15-7 Postlude

Measure having been formally defined, some remarks in the nature of summary and appraisal are in order in the light of the proposed principles enunciated in Section 15-1.

Measure is a set function defined on a domain whose members are measurable sets, bounded and unbounded, and whose range is the set of all non-negative real numbers together with ∞. It is important to note that measure is not committed to a domain consisting of all possible subsets of $\mathscr{R}$, that its very definition as the common value of outer and inner measure leaves the way open for the possible existence of nonmeasurable sets. We shall cite an example of such a set in a later section.

Measure is additive; the union of any countable collection of disjoint measurable sets is itself measurable, and its measure is the sum of the measures of the constituent sets. From additivity follows the increasing character of measure; if $A \subset B$ and if A and B are both measurable, then the measure of A does not exceed that of B.

The measure of a bounded interval of any of the four types (a,b), $[a,b]$, $(a,b]$, and $[a,b)$ is $b-a$. Thus measure is an extension of the length function.

The measure of a single real number is zero. This is consistent, as desired, with the familiar intuitive notion of a dimensionless geometric point. Because of additivity all countable sets also have measure zero. In particular, as has been pointed out, this is the case for the set of all rationals in any bounded or unbounded interval; this in spite of the apparently significant contribution by such a dense set to the substance of the interval.

An immediate consequence is that the set of irrational numbers in $[a,b]$ has measure equal to that of the interval itself. Additivity, together with measurability of the disjoint sets of rationals and irrationals, brings this about.

It is not to be construed that the nonzero measure of the set of irrationals in $[a,b]$ is in conflict with additivity on the grounds that, after all, the set of irrationals in $[a,b]$ constitutes the union of sets consisting of single points and thereby has zero measure. Such argument is invalid, of course, since additivity is meaningful only in the event of *countable* unions. Since one cannot add the measures of a noncountable collection of sets, even if those measures are all zeros, it is not a contradiction that the measure of their union exists and has a nonzero value. Actually, it is just this characteristic of noncountable sets that makes the notion of measure work at all. The age-old question of how dimensionless points can add up to a line segment has its answer in this very feature. It is not that the points *add* to form a segment but rather that they *cannot be added* that does the trick!

It must be noted that while the measure of a countable set is zero, it does not follow that this is an attribute reserved for countable sets alone. Noncountable sets exist with measure zero; recall the Cantor perfect set shown in Chapter 10 to be a noncountable zero set. (In the terminology of the present chapter a zero set is defined as a set having measure zero.) Thus measure is not simply a device to distinguish countable and noncountable sets.

If two sets are measurable, then so are their union and intersection; this extends, in fact, to any countable collection. Moreover, the measure of a countable union does not exceed the sum of the measures of the component sets —and equals that sum if the sets are disjoint.

Lastly, recall that if A and B are measurable sets then so are their differences $A-B$ and $B-A$. Moreover, because of additivity, $\mu(A)+\mu(B-A)=\mu(B)$ if $A \subset B$.

Exercises

15-17 Show that if $A \subset B$ and if $\mu(B)=0$, then A is measurable and $\mu(A)=0$. (This *complete* character of measure will be discussed further in Chapter 17 in connection with the general measure concept.)

15-18 Let E be a set of positive finite measure. Prove that an appropriate portion of E may be approximated by an interval in the sense that for any $k \in [0,1)$ there exists an open interval I with the property $\mu^*(E \cap I) \geq k\mu(I)$.

15-8 Invariance of Measure Under a Translation

If c is any real number and A any set of real numbers, then the set denoted by $A + c$ and defined by

$$A + c = \{x + c; \quad x \in A\}$$

is said to have been obtained through a *translation* or *displacement* of the set A through a distance c. For example, if A is the interval (a,b), then the *translated set* $A + c$ is the interval $(a + c, b + c)$.

One fact is immediate, namely, the measures of a bounded open interval I and the interval $I + c$ obtained through a translation are one and the same. Formally, $\mu(I + c) = (b + c) - (a + c) = b - a = \mu(I)$. One finds it easy to believe that this is the case for measurable sets in general. A simple argument follows.

If G is a bounded open set so that $G = \bigcup_\nu I_\nu$, where the I_ν are disjoint, then $G + c = \bigcup_\nu (I_\nu + c)$ and, since the translated intervals are clearly disjoint, $\mu(G + c) = \sum_\nu \mu(I_\nu + c)$. From the invariance of measure of an interval under translation we conclude $\mu(G + c) = \sum_\nu \mu(I_\nu) = \mu(G)$. Thus, measure of a bounded open set is preserved in translation.

If F is a bounded closed set and I an open interval with the property $F \subset I$, then $F + c \subset I + c$. Further, since $G = I - F$ is open, it follows that $G + c$ is open and therefore measurable. Clearly, $\mu(F + c) = \mu(I + c) - \mu(G + c) = \mu(I) - \mu(G) = \mu(F)$; so invariance of measure under translation is extended to bound closed sets.

Finally, suppose E to denote any bounded measurable set. Then, corresponding to each $\varepsilon > 0$, bounded sets F and G, closed and open, respectively, exist, such that $F \subset E \subset G$ and $\mu(G) - \mu(F) < \varepsilon$. From $F + c \subset E + c \subset G + c$ and $\mu(G + c) - \mu(F + c) = \mu(G) - \mu(F) < \varepsilon$, we conclude that $E + c$ is measurable. Further, $\mu(F) = \mu(F + c) \leq \mu(E + c) \leq \mu(G + c) = \mu(G)$. From the arbitrary character of ε follows $\mu(E + c) = \mu(E)$.

Thus it is established that measure of a bounded measurable set is invariant under a simple translation. That this is the case as well for unbounded sets is a direct consequence of the definition of measure of such a set.

The reader may verify for himself that the arguments above apply with equal force to the concepts of outer and inner measure of a bounded set, so that these functions, too, are invariant in a translation.

15-9 A Nonmeasurable Set

If E is a bounded set, it has an outer measure and an inner measure. If these are equal, the set is measurable, if not, it is nonmeasurable. In the present section it is our purpose to demonstrate that nonmeasurable sets do, in fact, exist.

One is not surprised to learn that there are no simple nonmeasurable sets. All possess a complexity that virtually precludes anything in the nature of visualization. Nevertheless, the construction necessary to the definition of such a set and the accompanying arguments in defense of such construction are surprisingly easy to follow. Actually, it is our intention here, as a pedagogical device, to give merely a step by step outline of the design of a particular nonmeasurable set of numbers, leaving to the student the relatively simple task of supplying the few necessary proofs.

Suppose A to denote a bounded set of positive measure. It will, of course, be a noncountable set and for some $a > 0$ will be a subset of $[-a,a]$. With each point x of A let there be associated the set A_x of all members ξ of A which differ from x by a rational number. Formally,

$$A_x = \{\xi;\quad \xi = x + r, r \text{ rational}, \xi \in A\}.$$

The sets A_x have properties enunciated in the following theorem—proofs to be provided by the reader.

Theorem 15-35

(a) $A = \bigcup_{x \in A} A_x$.

(b) If $\xi \in A_x$, with $\xi = x + r$, then $-2a \leq r \leq 2a$.

(c) Each set A_x is countable.

(d) If x is rational, then A_x is the set of all rational points of A.

(e) If x is irrational, then each member of A_x is an irrational number.

(f) If x_1 and x_2 are points of A with the property that $x_1 - x_2$ is a rational number, then $A_{x_1} = A_{x_2}$.

(g) If x_1 and x_2 are points of A such that $x_1 - x_2$ is irrational, then A_{x_1} and A_{x_2} are disjoint.

(h) The collection of distinct sets A_x is noncountable.

It is thus established that in the manner above described the set A is decomposed into a noncountable collection of disjoint sets, which may be regarded as equivalence classes of the members of A; each class is countable, one is the set of all rational numbers in A, and the remaining consist of irrational numbers only.

Now, in any fashion whatever, let there be selected from each member of this collection of disjoint sets A_x a representative member, and denote by B the set of all such representatives. It is clear that B is a noncountable set and that for each x the set $B \cap A_x$ consists of a single point. It is our purpose to show that B is nonmeasurable.

To this end suppose the set of all rational numbers in the interval $[-2a,2a]$ to be arranged in some arbitrary manner to form a sequence $r_1, r_2, \ldots, r_\nu, \ldots$. Corresponding to each element r_ν of this sequence, consider the translated set $B + r_\nu = \{x + r_\nu;\quad x \in B\}$. The countable collection of sets $B + r_\nu$, $\nu = 1, 2, \ldots$, has the following properties.

Theorem 15-36

(a) $B + r_\nu$ is noncountable for each ν.

(b) $B + r_\nu \subset [-3a,3a]$ for each ν.

(c) $(B + r_\nu) \cap (B + r_\kappa) = \varnothing$ if $\nu \neq \kappa$.

Thus the collection of translated sets $B + r_\nu$, derived from the set B, is a countable collection of disjoint noncountable sets. Denote the union of the sets $B + r_\nu$ by C. Then the following may be proved.

Theorem 15-37 The set C covers the set A; that is, $A \subset \bigcup_{\nu=1}^{\infty} (B + r_\nu)$.

We are now in a position to assert the main thesis.

Theorem 15-38 The set B is nonmeasurable.

Proof Assume B to be a measurable set. Then conclude the following in sequence.

(a) $B + r_\nu$ is measurable and $\mu(B + r_\nu) = \mu(B)$.
(b) C is measurable and $\mu(C) \leq 6a$.
(c) $\mu(B) = 0$.
(d) $\mu(C) = 0$.
(e) $\mu(A) = 0$.
(f) B is a nonmeasurable set▲

The construction of the nonmeasurable set B above invokes the so-called *axiom of choice* at that point where single members are "selected in any fashion whatever" from the sets A_x. It would be desirable, perhaps, to be more precise, that is, to say which member is to be selected from A_x, which from A_y, and, quite generally, to specify a function, defined on the noncountable domain of sets A_x, which will declare which number is to be selected from each set A_x. The concept of such a function, however, leads to questions which today remain unanswered. Accordingly, we assert an axiom of choice which means essentially that we assume the right to contemplate a selection of numbers one each from a noncountable collection of nonempty sets. The problem of constructing a nonmeasurable set without recourse to the axiom of choice is without a solution as of this writing.

Our interest in nonmeasurable sets is limited to the fact of their existence. We shall not be concerned with a detailed examination of their properties. However, we may note some simple facts. If A is a nonmeasurable subset of a measurable set B, then $B - A$ is nonmeasurable also, for otherwise would be in contradiction to the fact established in Theorem 15-30. If A and B both fail to possess measure $A \cup B$ may be measurable or nonmeasurable; similarly for $A \cap B$. The reader may construct examples. We have seen, too, that increasing and decreasing sequences of measurable sets have measurable limits. It follows that a nonmeasurable set cannot be constructed as a limit of such a sequence of measurable sets.

15-10 Modified Cantor Sets

To gain further insight into the measure concept, we consider some variations on the theme of the now familiar Cantor set. Numerous references have been made in the course of the text to the latter set, obtained, as it was, through successive deletions from the interval $[0,1]$ of open subintervals of lengths $1/3$, $1/3^2, \ldots$, because of its rather remarkable properties. It was introduced originally as an instance of a nowhere dense perfect set, and has lately been cited as a noncountable set having zero measure. It is of interest to observe how, by some slight variations in design, we are able to construct modifications of the Cantor set with preassigned nonzero measures.

Suppose from the middle of the interval $F_0 = [0,1]$ is deleted a subinterval of length $\lambda/3$, where λ is any number such that $0 < \lambda < 1$. From each of the components of the residue $F_1 = [0, (3 - \lambda)/6] \cup [(3 + \lambda)/6, 1]$ let there be removed an interval of length $\lambda/3^2$, and continue in this manner. The nested sequence $\{F_n\}$ thus constructed is such that F_{n+1} follows from F_n by deletion of an open interval of length $\lambda/3^{n+1}$ from each of the 2^n components of F_n. The set $F = \bigcap_{n=1}^{\infty} F_n$ is noncountable, nowhere dense, and perfect. Since the total of the lengths of open intervals deleted in its construction is $\lambda \sum_{n=0}^{\infty} 2^n/3^{n+1} = \lambda$, it follows that the measure of F is $1 - \lambda$. Accordingly, $\mu(F)$ may be made arbitrarily close to 1. In view of the nowhere dense character of F, this is, perhaps, surprising.

To pursue matters further, observe that if a modified Cantor set F is constructed as above, then the set $[0,1] - F$, being open, is a union of disjoint open intervals. Following the construction of Example 9-21, in which the Cantor set itself was employed, we may design a function h with the following properties. It possesses a derivative at each point of $[0,1]$ and the derivative function h' is bounded. Moreover, h' is discontinuous at each point of F. Unlike the function of the earlier example, however, the present one is such that its discontinuities constitute a set of positive measure. Accordingly, h' does not possess a Riemann integral. It serves, therefore, as another instance (see Example 10-5) of a function which possesses a primitive, yet is not Riemann integrable.

15-11 A Covering Theorem

Let E be any set and $\mathscr{I}$ a collection of closed intervals with the property that, corresponding to any $\varepsilon > 0$ and to any point x of E, there exists a member J of $\mathscr{I}$ such that $x \in J$ and $\mu(J) < \varepsilon$. Expressed otherwise, suppose $\mathscr{I}$ to be such that to each point x of E there corresponds a sequence $\{J_\nu\}$ of members of $\mathscr{I}$ such that $x \in J_\nu$ for all ν and $\{\mu(J_\nu)\}$ is a null sequence. The collection $\mathscr{I}$ is a highly specialized type of covering of the set E. We call it a *Vitali covering* of E or say that E is *covered in the sense of Vitali* by $\mathscr{I}$. It is our purpose below to demonstrate a fact that will be used repeatedly in Chapter 16, namely that from such a covering it is always possible to determine a countable subcollection whose members are disjoint and such that the union of those members differs from the set E itself by a set whose measure is zero. Upon the set E are placed no qualifications at all; even measurability is not assumed.

Theorem 15-39 (Vitali) Let E be any set and suppose $\mathscr{I}$ to denote a collection of closed sets which covers E in the sense of Vitali. Then a countable subcollection $\{J_\nu\}$ of disjoint members of $\mathscr{I}$ exists with the property $\mu(E - \bigcup_\nu J_\nu) = 0$.

Proof If E is a bounded set and if I is an open interval such that $E \subset I$, then members of $\mathscr{I}$ which are not subsets of I may be discarded and the remaining subset of $\mathscr{I}$ is still a Vitali covering of E. For, if $x \in E$, then for some $\varepsilon > 0$ it is the case that $\mathscr{N}(x,2\varepsilon) \subset I$. Hence a member J of $\mathscr{I}$ exists such that $x \in J$, $\mu(J) < \varepsilon$, and $J \subset I$. It involves no loss of generality, therefore, to assume all members of $\mathscr{I}$ to be subsets of I. This we do as we proceed to prove the theorem for the case of a bounded set E.

Select from $\mathscr{I}$ disjoint sets $J_1, J_2, \ldots, J_\kappa$. If $E - \bigcup_{\nu=1}^{\kappa} J_\nu$ is measurable (which it need not be since E is not assumed to be measurable) and if $\mu(E - \bigcup_{\nu=1}^{\kappa} J_\nu) = 0$, the conclusion of the theorem has been attained. If such is not the case, that is, if $\mu^*(E - \bigcup_{\nu=1}^{\kappa} J_\nu) \neq 0$, let x denote a point of the set $E - \bigcup_{\nu=1}^{\kappa} J_\nu$. Since $\bigcup_{\nu=1}^{\kappa} J_\nu$ is closed, a neighborhood $\mathscr{N}(x,\varepsilon)$ of x exists which is contained in $\mathscr{C} \bigcup_{\nu=1}^{\kappa} J_\nu$. By virtue of the nature of a Vitali covering, a member J of $\mathscr{I}$ exists such that J is a subset of $\mathscr{N}(x,\varepsilon)$ and, as a consequence, is disjoint from $\bigcup_{\nu=1}^{\kappa} J_\nu$.

Consider the set of all those members of $\mathscr{I}$ which are disjoint from $\bigcup_{\nu=1}^{\kappa} J_\nu$ and let ε_κ denote the supremum of the set of measures of such intervals. Suppose $J_{\kappa+1}$ to be any member of $\mathscr{I}$ which is disjoint from $\bigcup_{\nu=1}^{\kappa} J_\nu$ and such that

$$(1) \qquad \mu(J_{\kappa+1}) > \frac{\varepsilon_\kappa}{2}.$$

Regard the set $E - \bigcup_{\nu=1}^{\kappa+1} J_\nu$. If $\mu^*(E - \bigcup_{\nu=1}^{\kappa+1} J_\nu) \neq 0$, repeat the above construction to obtain a set $E - \bigcup_{\nu=1}^{\kappa+2} J_\nu$, etc. Either at some stage it is the case that the desired zero outer measure has been achieved, or a sequence $\{J_\nu\}$ of disjoint members of $\mathscr{I}$ is generated. If the latter is the case, we consider the set $E_0 = E - \bigcup_{\nu=1}^{\infty} J_\nu$ and proceed to show that $\mu(E_0) = 0$.

To this end, let $\{K_\nu\}$ denote a sequence of closed intervals (not necessarily from $\mathscr{I}$) such that for each ν the element K_ν has the same midpoint as J_ν, and $\mu(K_\nu) = 5\mu(J_\nu)$. Clearly, for each ν, if $x' \in J_\nu$ and $x'' \in \mathscr{C}K_\nu$, then

$$(2) \qquad |x' - x''| \geq 2\mu(J_\nu) > \varepsilon_{\nu-1},$$

by (1). Further, since the sets J_ν involved in the definition of E_0 are disjoint and all are subsets of I, it follows that $\sum_{\nu=1}^{\infty} \mu(K_\nu) = 5\sum_{\nu=1}^{\infty} \mu(J_\nu) \leq 5\mu(I)$, that is, the series $\sum_{\nu=1}^{\infty} \mu(K_\nu)$ converges. This implies that $\{\mu(K_\nu)\}$ is a null sequence and, therefore, that $\lim \{\mu(J_\nu)\} = 0$ since $J_\nu \subset K_\nu$. This, in turn, forces the conclusion, by (1), that $\lim \{\varepsilon_\nu\} = 0$.

Suppose now that $\mu^*(E_0) > 0$. Because of the convergence of $\sum_{\nu=1}^{\infty} \mu(K_\nu)$, for some integer n_1, we have $\mu(\bigcup_{\nu=n_1}^{\infty} K_\nu) \leq \sum_{\nu=n_1}^{\infty} \mu(K_\nu) < \mu^*(E_0)$. Hence a point x_1 of E_0 exists such that $x_1 \notin \bigcup_{\nu=n_1}^{\infty} K_\nu$ and, from the definition of E_0, $x_1 \notin \bigcup_{\nu=1}^{n_1} J_\nu$.

Following the argument of an earlier paragraph, we conclude the existence of an interval J' of $\mathscr{I}$ such that $x_1 \in J'$ and

$$J' \cap \bigcup_{\nu=1}^{n_1} J_\nu = \varnothing. \tag{3}$$

Moreover,

$$x_1 \in J' \cap \mathscr{C}K_\nu \tag{4}$$

for all $\nu \geq n_1$.

Consider the possibility $J' \cap J_\nu = \varnothing$ for all ν. Then, by the definition of ε_ν, we have $\mu(J') \leq \varepsilon_\nu$ for all ν; but this is impossible since $\{\varepsilon_\nu\}$ is a null sequence. It follows that a smallest integer n_2 exists with the property $J' \cap J_{n_2} \neq \varnothing$. This being the case,

$$\mu(J') \leq \varepsilon_{n_2 - 1} \tag{5}$$

and $n_1 < n_2$ by (3). Let $x_2 \in J' \cap J_{n_2}$. By (4) we get $x_1 \in J' \cap \mathscr{C}K_{n_2}$, whence $|x_2 - x_1| \leq \varepsilon_{n_2 - 1}$ by (5). This contradiction to (2) denies the assumption $\mu^*(E_0) \neq 0$. We conclude that the sequence $\{J_\nu\}$, as constructed, is such that $\mu(E - \bigcup_{\nu=1}^{\infty} J_\nu) = 0$. This proves the theorem in the case where E is bounded.

Suppose now that E is unbounded, and again let $\mathscr{I}$ denote a Vitali covering. Corresponding to each positive integer n, consider the set $E_n = E \cap (-n,n)$. Because of the construction in the proof of the theorem for a bounded set, we have, for any κ,

$$\mu^*\left(E_\kappa - \bigcup_{\nu=1}^{n_\kappa} J_\nu\right) < \frac{1}{\kappa}, \tag{6}$$

where $\{J_\nu;\ \nu = 1,2,\ldots,n_\kappa\}$ is some set of disjoint members of $\mathscr{I}$ and n_κ is chosen so that (6) holds. It is then the case that, for any κ, those members of $\mathscr{I}$ that are contained in the open set $(-\kappa - 1, \kappa + 1) - \bigcup_{\nu=1}^{n_\kappa} J_\nu$ constitute a Vitali covering of the set $E_{\kappa+1} - \bigcup_{\nu=1}^{n_\kappa} J_\nu$. As a result, disjoint members $J_{n_\kappa+1}, \ldots, J_{n_{\kappa+1}}$ of $\mathscr{I}$ exist which are disjoint from the intervals $J_1, \ldots, J_{n_\kappa}$ and such that

$$\mu^*\left(\left(E_{\kappa+1} - \bigcup_{\nu=1}^{n_\kappa} J_\nu\right) - \bigcup_{\nu=n_\kappa+1}^{n_{\kappa+1}} J_\nu\right) < \frac{1}{\kappa+1}$$

or, what is the same,

$$\mu^*\left(E_{\kappa+1} - \bigcup_{\nu=1}^{n_{\kappa+1}} J_\nu\right) < \frac{1}{\kappa+1}. \tag{7}$$

In this fashion, a sequence $\{J_\nu\}$ of disjoint intervals from $\mathscr{I}$ is constructed. From (7) we conclude $\mu^*(E_\kappa - \bigcup_{\nu=1}^{\infty} J_\nu) < 1/\kappa$ for all κ and, since $E_\kappa - \bigcup_{\nu=1}^{\infty} J_\nu \subset E_{\kappa+1} - \bigcup_{\nu=1}^{\infty} J_\nu$, we have $\mu^*(E_\kappa - \bigcup_{\nu=1}^{\infty} J_\nu) < 1/n$ for all $n > \kappa$. Therefore $E_\kappa - \bigcup_{\nu=1}^{\infty} J_\nu$ is measurable with measure zero for each κ. Further

$$\mu\left(E - \bigcup_{\nu=1}^{\infty} J_\nu\right) = \mu\left(\bigcup_{\kappa=1}^{\infty}\left(E_\kappa - \bigcup_{\nu=1}^{\infty} J_\nu\right)\right) = \lim\left\{\mu\left(E_\kappa - \bigcup_{\nu=1}^{\infty} J_\nu\right)\right\} = 0 \blacktriangle$$

As an application of the Vitali theorem we next prove a generalization of the fact that a countable union of intervals is a measurable set.

Theorem 15-40 The union of any collection of intervals is a measurable set.

Proof Let Ω be any collection of intervals and let $B = \bigcup_{A \in \Omega} A$. Denote by $\mathscr{J}$ the collection of all closed intervals J such that $J \subset A$ for at least one member A of Ω. Then $\mathscr{J}$ is a Vitali covering of B, and it follows that a countable collection of members J_ν of $\mathscr{J}$ has the property that $\mu(B - \bigcup_\nu J_\nu) = 0$. Since $J_\nu \subset B$ for each ν, we have $B = (\bigcup_\nu J_\nu) \cup (B - \bigcup_\nu J_\nu)$. Measurability of B follows from that of $\bigcup_\nu J_\nu$ and $B - \bigcup_\nu J_\nu$.▲

15-12 Measurable Functions

A real valued function g defined on a subset S of $\mathscr{R}$ is said to be *measurable*, or *Lebesgue measurable*, on S if, corresponding to each real number k, the set of points $\{x;\ x \in S, g(x) \leq k\}$ is a measurable set (of either finite or infinite measure).

This new classification of function, so intimately concerned with the concept of measure, will play a role of central importance in the theory of the Lebesgue integral. It will assume, in this connection, a place comparable to that of functions of bounded variation in the instance of the Stieltjes integral and of the class of functions which are bounded and continuous almost everywhere in the Riemann case.

Example 15-2 Consider the function g defined on $\mathscr{R}$ by $g(x) = x + 5$ if $x < -1$, $g(x) = 2$ if $-1 \leq x < 0$, and $g(x) = x^2$ if $0 \leq x$. That g is a measurable function on $\mathscr{R}$ may be seen by identifying the set $\{x;\ g(x) \leq k\}$ for certain significant ranges of values of k. We have

$$\{x;\ g(x) \leq k\} = \begin{cases} (-\infty, k-5] & \text{if } k < 0 \\ (-\infty, -5] \cup \{0\} & \text{if } k = 0 \\ (-\infty, k-5] \cup [0, \sqrt{k}] & \text{if } 0 < k < 2 \\ (-\infty, k-5] \cup [-1, \sqrt{k}] & \text{if } 2 \leq k < 4 \\ (-\infty, \sqrt{k}] & \text{if } 4 \leq k. \end{cases}$$

Clearly, all sets defined are measurable.

Example 15-3 Consider the familiar Ψ function defined on $[0,1]$. For $k \geq 1$ the set $\{x;\ \Psi(x) \leq k\}$ is the interval $[0,1]$; for $0 \leq k < 1$ it is the set of rational numbers in $[0,1]$; for $k < 0$ it is the empty set. Since each of these is a measurable set, the function Ψ is measurable.

Example 15-4 Consider the *characteristic funtion* K_S of a set S on $\mathscr{R}$ defined by

$$K_S(x) = \begin{cases} 1 & \text{if } x \in S \\ 0 & \text{if } x \notin S. \end{cases}$$

Since $\{x;\ \ x \in \mathscr{R}, K_S(x) \le k\}$ is $\varnothing$, $\mathscr{C}S$, or $\mathscr{R}$ according as $k < 0$, $0 \le k < 1$, or $1 \le k$, it follows that K_S is a measurable function if S is a measurable set. On the other hand, if K_S is not a measurable function, then S is a nonmeasurable set.

Example 15-5 Suppose f to denote a monotone function on $\mathscr{R}$. Then for any k the set $\{x;\ \ f(x) \le k\}$ is clearly an interval. It follows that any monotone function is measurable.

Immediate consequences of the notion of measurable function provide alternate definitions for that concept. First, one observes that the domain S of a measurable function is necessarily a measurable set. For if g is measurable and bounded (by M) on S, then by definition $\{x;\ \ x \in S, g(x) \le M\}$, which is the set S itself, is measurable. If g is measurable and unbounded, then the sequence of sets $\{x;\ \ x \in S, g(x) \le n\}$ has S as its measurable limit.

Certain other facts are virtually immediate. If g is measurable on a set S and if S_1 is a measurable subset of S, then g is measurable on S_1. This follows simply from

$$\{x;\ \ x \in S_1, g(x) \le k\} = S_1 \cap \{x;\ \ x \in S, g(x) \le k\}.$$

Further, if g is defined and measurable on each of two sets S_1 and S_2, then it is measurable on each of the sets $S_1 \cap S_2$ and $S_1 \cup S_2$. As the reader may verify, these statements extend readily to any countable unions and intersections.

The following theorem points to possible alternate definitions of measurability of a function.

Theorem 15-41 Each of the following conditions is necessary and sufficient to measurability of a function g on a set S:

(a) $\{x;\ \ x \in S, g(x) > k\}$ is measurable for each k;
(b) $\{x;\ \ x \in S, g(x) \ge k\}$ is measurable for each k;
(c) $\{x;\ \ x \in S, g(x) < k\}$ is measurable for each k.

Proof To prove the theorem it suffices to show that measurability implies (a), which, in turn, implies (b), which implies (c), which implies measurability.

Suppose g to be measurable on S. Then S and $\{x;\ \ x \in S, g(x) \le k\}$ are measurable sets. It follows that $S - \{x;\ \ x \in S, g(x) \le k\}$, that is, the set $\{x;\ \ x \in S, g(x) > k\}$, is measurable.

Next, assuming (a), it follows that for each positive integer n and each real k the set $\{x;\ \ x \in S, g(x) > k - 1/n\}$ is measurable. In turn, $\{x;\ \ x \in S, g(x) \ge k\}$ $= \bigcap_{n=1}^{\infty} \{x;\ \ x \in S, g(x) > k - 1/n\}$ is a measurable set.

From (b) it follows that $\{x;\ \ x \in S, g(x) < k\} = S - \{x;\ \ x \in S, g(x) \ge k\}$ is also measurable.

Finally, from the measurability of the set $\{x;\ \ x \in S, g(x) < k\}$ follows that of $\{x;\ \ x \in S, g(x) < k + 1/n\}$ for each positive integer n. This implies that $\bigcap_{n=1}^{\infty} \{x;\ \ x \in S, g(x) < k + 1/n\}$, that is, the set $\{x;\ \ x \in S, g(x) \le k\}$, is a measurable set, and therefore, that g is a measurable function on S▲

Corollary If g is measurable on S, then for any a and b, the sets $\{x;\ \ x \in S, a < g(x) < b\}$, $\{x;\ \ x \in S, a \leq g(x) \leq b\}$, $\{x;\ \ x \in S, a < g(x) \leq b\}$, and $\{x;\ \ x \in S, a \leq g(x) < b\}$ are measurable sets.

Incidentally, it follows from the measurability of the sets $\{x;\ \ x \in S, g(x) \leq k\}$ and $\{x;\ \ x \in S, g(x) < k\}$ that the set $\{x;\ \ x \in S, g(x) = k\}$ is measurable. Thus, in particular, if g is a measurable function, its zeros constitute a measurable set.

The next two theorems explore the connection between continuity and measurability.

Theorem 15-42 A continuous function on a measurable domain is measurable.

Proof Suppose S to be a measurable set and f a continuous function on S, and let k be any number. Consider the set $A = \{x;\ \ x \in S, f(x) < k\}$. For each $x \in A$, let $\mathcal{N}_x$ be a neighborhood of x such that $f(y) < k$ for each $y \in S \cap \mathcal{N}_x$. Let $S_1 = \bigcup_{x \in A} \mathcal{N}_x$. Then the open set S_1 is measurable. Since $A = S_1 \cap S$, it follows that A is a measurable set and hence f is a measurable function▲

If f is a Riemann integrable function on $[a,b]$, then the set D of its discontinuity points has measure zero. It follows that f is measurable on D. Since it is continuous on $[a,b] - D$, it is also measurable on that set. We conclude that a Riemann integrable function is measurable on its domain.

Theorem 15-43 If g is a measurable function on a domain S and f a continuous function defined on $\mathcal{R}$, then the function $f \circ g$ is measurable on S.

Proof Because of the continuity of f, the set $\{x;\ \ f(x) < k\}$ is open whatever the value of k. (See Exercise 8-21.) Hence $\{x;\ \ f(x) < k\} = \bigcup_\nu I_\nu$, where the I_ν are open, disjoint intervals. Since g is measurable on S, it follows by the corollary to Theorem 15-41 that the set $\{x;\ \ x \in S, g(x) \in I_\nu\}$ is measurable for each ν. From the identity

$$\{x;\ \ x \in S, f(g(x)) < k\} = \bigcup_\nu \{x;\ \ x \in S, g(x) \in I_\nu\}$$

then follows measurability of $f \circ g$ on S▲

Elementary properties of measurable functions are established in the following theorems.

Theorem 15-44 If f and g are measurable functions on a common domain S, then the set $\{x;\ \ x \in S, f(x) < g(x)\}$ is measurable.

Proof Consider, for each rational number r, the set defined by

$$\begin{aligned} B_r &= \{x;\ \ x \in S, f(x) < r < g(x)\} \\ &= \{x;\ \ x \in S, f(x) < r\} \cap \{x;\ \ x \in S, g(x) > r\}. \end{aligned}$$

Since it is the intersection of measurable sets, it is a measurable set. It follows then that the union $\bigcup_r B_r$ of all the sets B_r is measurable, too. However,

$$\bigcup_r B_r = \{x;\ \ x \in S, f(x) < g(x)\}▲$$

Theorem 15-45 If g is a measurable function on a domain S, and if c is any constant, then $g + c$, cg, g^2, and $|g|$ are measurable on S. In particular, $-g$ is measurable.

Proof This is an immediate consequence of Theorem 15-43, in which the continuous function f is defined on $\mathcal{R}$ in the respective instances by $x + c$, cx, x^2, and $|x|$▲

We have seen that a function of bounded variation is expressible as the sum of two monotone functions. Since monotone functions are measurable, we conclude that functions of bounded variation are measurable also.

Theorem 15-46 If f and g are measurable functions on a domain S, then $f + g$ and fg are measurable on S.

Proof By Theorem 15-45, the function $k - g$ is measurable for each real number k. Therefore $\{x;\ \ x \in S, f(x) < k - g(x)\}$, that is, the set $\{x;\ \ x \in S, f(x) + g(x) < k\}$ is measurable. Measurablility of fg is established by the identity $fg = \frac{1}{4}[(f+g)^2 - (f-g)^2]$▲

Theorem 15-47 If f and g are defined on a common domain S such that $f = g$ almost everywhere and g is measurable, then f is measurable.

Proof Let $S' = \{x;\ \ f(x) = g(x)\}$ and $S'' = \{x;\ \ f(x) \neq g(x)\}$. Then $S = S' \cup S''$ and $\mu(S'') = 0$. Let $A = \{x;\ \ f(x) \leq k\}$. Since $A \cap S'' \subset S''$, it follows that $A \cap S''$ has measure zero. Also, $A \cap S' = \{x;\ \ g(x) \leq k\} \cap S'$, whence $A \cap S'$ is measurable. Measurability of f follows from $A = (A \cap S') \cup (A \cap S'')$▲

Theorem 15-48 If $\{g_n\}$ is a uniformly bounded sequence of measurable functions on a domain S, then the functions $\sup\{g_n\}$ and $\inf\{g_n\}$ are measurable on S.

Proof For each k we have

$$\{x;\ \ \sup\{g_n(x)\} \leq k\} = \bigcap_{n=1}^{\infty} \{x;\ \ g_n(x) \leq k\},$$

and, since each set appearing on the right is measurable, it follows that $\sup\{g_n\}$ is a measurable function. Measurability of $\inf\{g_n\}$ is implied by $\inf\{g_n\} = -\sup\{-g_n\}$▲

Corollary 1 If $\{g_n\}$ is a uniformly bounded sequence of measurable functions on a domain S, then the functions $\overline{\lim}\{g_n\}$ and $\underline{\lim}\{g_n\}$ are measurable on S. In particular, if the sequence converges, then $\lim\{g_n\}$ is a measurable function.

Proof This follows from $\overline{\lim}\{g_n\} = \inf\limits_{m} \sup\limits_{n \geq m}\{g_n\}$ and $\underline{\lim}\{g_n\} = \sup\limits_{m} \inf\limits_{n \geq m}\{g_n\}$▲

Corollary 2 If f and g are measurable functions on S, then so are $f \vee g$, $f \wedge g$, g^+, and g^-.

Proof In the theorem let $g_1 = f$ and $g_n = g$ for $n > 1$. Then $f \vee g = \sup\{g_n\}$ and $f \wedge g = \inf\{g_n\}$, whence $f \vee g$ and $f \wedge g$ are measurable functions. As a consequence, so are $g^+ = g \vee 0$ and $g^- = (-g) \vee 0$, where 0 is the zero function on S▲

What is surprising, perhaps, is that a measurable function of a measurable function need not be measurable! Indeed, it is even the case that a measurable function of a continuous function may be nonmeasurable (contrast with Theorem 15-43).

Example 15-6 Consider the Cantor set K and the increasing and continuous Cantor function f defined on [0,1] as in Example 13-9. Let g be defined on [0,1] by $g(x) = f(x) + x$. It is clear that g is strictly increasing and continuous on [0,1]. Hence the inverse function g^{-1} exists, with domain [0,2] and range [0,1]. Moreover, g^{-1} is continuous, and therefore measurable, on its domain.

Suppose L to denote the image of the Cantor set K by the function g. Then $L \subset [0,2]$. Since $[0,1] - K$ is an open set, it is a union of disjoint intervals (a_ν, b_ν). On each such interval the Cantor function f has a constant value, say c_ν. Therefore the image of (a_ν, b_ν) by g is the open interval $(a_\nu + c_\nu, b_\nu + c_\nu)$, and this interval has the same measure as (a_ν, b_ν). Since the image intervals are disjoint, it follows that the set $[0,2] - L$ has the same measure as $[0,1] - K$. Hence $\mu(L) = 1$. This conclusion, in itself, is rather striking. It cites the continuous function g as one which maps a set of measure zero (the Cantor set) into an image set having nonzero measure.

Now, suppose B to denote a nonmeasurable subset of L; for example, such a set as constructed in Section 15-9. Denote by h the characteristic function of B. Then h is a nonmeasurable function.

Consider the function φ defined on [0,1] by $\varphi = h \circ g$, and having range consisting simply of the numbers 0 and 1. If x is such that $\varphi(x) = 1$, then $g(x) \in B$, whence $x \in K$ since $B \subset L$. Therefore $\{x;\ \ \varphi(x) = 1\} \subset K$ and, since K has measure zero, so does the set $\{x;\ \ \varphi(x) = 1\}$. As a consequence, the function φ differs from the zero function on [0,1] only on a set of measure zero. Accordingly, by Theorem 15-47, φ is a measurable function on [0,1]. Therefore the function $h = \varphi \circ g^{-1}$ is an instance of a nonmeasurable function that is a measurable function of a continuous function.

15-13 Further Properties of Measurable Functions

As we have seen the sequence $\{x^n\}$, with each element defined on [0,1), is convergent, but not uniformly so, to the zero function on [0,1). However, for any number α such that $0 < \alpha < 1$, the sequence is uniformly convergent on $[0,\alpha)$. Thus it is that, in this instance at least, uniform convergence is attainable on a domain whose length is arbitrarily close to that of the original interval. In Egoroff's theorem, which follows, this principle is extended in quite a general way to sequences of measurable functions. Application of this result will be found in Chapter 16.

Theorem 15-49 (Egoroff) Let g and $\{g_n\}$ be defined on a set E of finite measure. Suppose each of the functions g_n to be measurable and the sequence $\{g_n\}$ to converge to g almost everywhere on E. Then, corresponding to each $\varepsilon > 0$, there exists a subset E_ε of E such that $\mu(E - E_\varepsilon) < \varepsilon$ and $\{g_n\}$ converges uniformly to g on E_ε.

Proof Let $\varepsilon > 0$ be given, and, corresponding to positive integers m and n, define measurable sets $E_{m,n}$ by $E_{m,n} = \bigcap_{\nu=n}^{\infty} \{x;\ |g_\nu(x) - g(x)| < 1/m\}$. If A denotes that subset of E on which $\{g_n\}$ converges to g, then $A \subset \bigcup_{n=1}^{\infty} E_{m,n} \subset E$. Since $\mu(A) = \mu(E)$ by hypothesis, we have $\mu(\bigcup_{n=1}^{\infty} E_{m,n}) = \mu(E)$ for each m. Since $E_{m,n} \subset E_{m,n+1}$ for all n, Theorem 15-32 implies

$$\lim_{n\to\infty} \mu(E - E_{m,n}) = \mu(E) - \lim_{n\to\infty} \mu(E_{m,n})$$
$$= \mu(E) - \mu\left(\bigcup_{n=1}^{\infty} E_{m,n}\right) = 0.$$

Consequently, corresponding to each m, there exists an integer n_m such that $\mu(E - E_{m,n_m}) < \varepsilon/2^m$.

Let $E_\varepsilon = \bigcap_{m=1}^{\infty} E_{m,n_m}$. Then E_ε is a measurable set and

$$\mu(E - E_\varepsilon) = \mu\left(\bigcup_{m=1}^{\infty} (E - E_{m,n_m})\right)$$
$$\leq \sum_{m=1}^{\infty} \mu(E - E_{m,n_m}) < \sum_{m=1}^{\infty} \frac{\varepsilon}{2^m} = \varepsilon.$$

Further, for any m and for all $n > n_m$, we have $|g_n(x) - g(x)| < 1/m$ for all x in E_{m,n_m}. Since $E_\varepsilon \subset E_{m,n_m}$ for any m, it follows that this inequality is true for all x in E_ε. Hence the convergence of $\{g_n\}$ to g on E_ε is uniform▲

Let E and E_1 be any sets, bounded or unbounded, with the property $E_1 \subset E$. Then we define the *characteristic function* $K_{E_1}^E$ *of* E_1 *on the domain* E by

$$K_{E_1}^E(x) = \begin{cases} 0 & \text{if } x \in E - E_1 \\ 1 & \text{if } x \in E_1 \end{cases}$$

for all x in E. If $E = \mathscr{R}$, we write simply K_{E_1}.

Now, suppose $E_1, E_2, \ldots, E_n$ to be a finite collection of disjoint measurable subsets of a measurable set E. Then the function f defined on E by

$$f(x) = \sum_{\nu=1}^{n} a_\nu K_{E_\nu}^E(x),$$

where $a_1, a_2, \ldots, a_n$ are any real numbers, is called a *simple function*.

It is clear that a simple function is measurable and bounded and that its range is finite. We propose to show that any measurable function may be approximated by simple functions in much the same manner as continuous functions have been seen to be approximated by polynomials.

Theorem 15-50 If f is a measurable function defined on a set E, then a sequence of simple functions exists which converges to f at all points of E. In particular, if f is non-negative, then an increasing sequence of non-negative simple functions can be found having this property.

Proof Suppose, first, that f is non-negative on E and consider a finite collection of subsets of E defined by

$$E_{n,\nu} = \left\{x;\quad x \in E, \frac{\nu - 1}{2^n} \leq f(x) < \frac{\nu}{2^n}\right\}$$

for $\nu = 1, 2, \ldots, 2^{2n}$, and

$$E_{n,1+2^{2n}} = \{x;\quad x \in E, f(x) \geq 2^n\}.$$

It is easily verified that the sequence $\{f_n\}$ of simple functions

$$f_n = \sum_{\nu=1}^{1+2^{2n}} \left(\frac{\nu - 1}{2^n}\right) K^E_{E_{n,\nu}}$$

is such that $0 \leq f_n \leq f_{n+1}$ and, for any x in E and any $\varepsilon > 0$, there exists an integer n such that $0 \leq f(x) - f_n(x) < \varepsilon$. Thus, in the case of a non-negative function f, the sequence $\{f_n\}$ increases and converges to f. In the event f is any measurable function on E, we have only to apply the above to the resolution $f = f^+ + (-f^-)$▲

We have seen that if a function is continuous except at a single point of its domain, then deletion of the discontinuity point yields a domain on which the function is continuous. Further, the deleted domain has a measure equal to that of the original. This principle may be generalized to the class of measurable functions, revealing thereby an intimacy between measurability of functions and continuity. It is shown below that, whatever the measure of the set of discontinuities of a measurable function, a reduction of the domain by a set of arbitrarily small measure is possible which will yield a new domain on which the given function is continuous. The idea here is not unlike that expressed in Egoroff's theorem where convergence is specialized to uniform convergence by deletion of a set of arbitrarily small measure.

Theorem 15-51 (Lusin) If f is a measurable function on a domain E, then corresponding to each $\varepsilon > 0$, there exists a closed subset F_ε of E such that $\mu(E - F_\varepsilon) < \varepsilon$ and f is continuous on F_ε.

Proof First, suppose f to be a simple function, that is, $f = \sum_{\nu=1}^m a_\nu K^E_{E_\nu}$, the sets E_ν being measurable and disjoint. Further, write $E_{m+1} = E - \bigcup_{\nu=1}^m E_\nu$ and let $\varepsilon > 0$ be given. Then, corresponding to each E_ν, $\nu = 1, 2, \ldots, m+1$, a closed set T_ν exists such that $T_\nu \subset E_\nu$ and $\mu(E_\nu - T_\nu) < \varepsilon$. Let $T = \bigcup_{\nu=1}^{m+1} T_\nu$. Then T is closed, $T \subset E$, and $\mu(E - T) = \sum_{\nu=1}^{m+1} \mu(E_\nu - T_\nu) < (m+1)\varepsilon$. The sets T_ν, $\nu = 1, 2, \ldots, m+1$, being disjoint and the function f having a constant value on each T_ν, it follows that f is continuous on the reduced domain T.

Next, let f be any measurable function defined on E. Then, by the preceding theorem, a sequence $\{f_n\}$ of simple functions exists which converges to f. By the argument of the preceding paragraph it is then the case that, $\varepsilon > 0$ being given, for each n there exists a measurable set A_n such that f_n is continuous on A_n and $\mu(E - A_n) < \varepsilon/2^n$. Let $A = \bigcap_{n=1}^{\infty} A_n$. Then each f_n is continuous on A, and A is measurable, with the property $\mu(E - A) = \mu(\bigcup_{n=1}^{\infty} (E - A_n)) < \sum_{n=1}^{\infty} (\varepsilon/2^n) = \varepsilon$.

Corresponding to each positive integer p, let $B_p = A \cap \{x;\quad p - 1 \le |x| < p\}$. Clearly, each B_p is measurable and $A = \bigcup_{p=1}^{\infty} B_p$. By Egoroff's theorem, corresponding to each p, there exists a measurable set C_p such that $C_p \subset B_p$, $\mu(B_p - C_p) < \varepsilon/2^p$, and $\{f_n\}$ converges uniformly on C_p. Further, closed sets F_p exist with the properties $F_p \subset C_p$ and $\mu(C_p - F_p) < \varepsilon/2^p$. Therefore $F_p \subset B_p$, $\mu(B_p - F_p) < 2\varepsilon/2^p$, each F_p is closed, and $\{f_n\}$ converges uniformly on F_p. From the last we are able to conclude that the function f is continuous on each F_p.

Let $F = \bigcup_{p=1}^{\infty} F_p$. Then F is closed. (See Exercise 4-20.) Moreover, f is continuous on F because of the nature of the construction of F. (The reader should note that what is true in the instance of $\bigcup_{p=1}^{\infty} F_p$ is not true of unions of closed sets in general.) Finally, we see that

$$\mu(E - F) = \mu(E - A) + \mu(A - F) < \varepsilon + \mu\left(\bigcup_{p=1}^{\infty} B_p - \bigcup_{p=1}^{\infty} F_p\right)$$

$$< \varepsilon + \mu\left(\bigcup_{p=1}^{\infty} (B_p - F_p)\right) = \varepsilon + \sum_{p=1}^{\infty} \mu(B_p - F_p)$$

$$< \varepsilon + \sum_{p=1}^{\infty} \frac{2\varepsilon}{2^p} = 3\varepsilon \blacktriangle$$

Theorem 15-52 If f is a measurable function on a domain E, then, corresponding to each $\varepsilon > 0$, there exists a continuous function g defined on $\mathscr{R}$ with the property that $\mu\{x;\quad x \in E, f(x) \ne g(x)\} < \varepsilon$. Further, if $|f(x)| \le M$ on E, then g may be chosen so that $|g(x)| \le M$ on $\mathscr{R}$.

Proof By Lusin's theorem a closed set F_ε exists such that $F_\varepsilon \subset E$, $\mu(E - F_\varepsilon) < \varepsilon$, and f is continuous on F_ε. By Exercise 8-41, a function g exists on $\mathscr{R}$ which is continuous on $\mathscr{R}$ and identical with f on F_ε. Hence $\{x;\quad x \in E, f(x) \ne g(x)\} \subset E - F_\varepsilon$. The set $\{x;\quad x \in E, f(x) \ne g(x)\}$ is measurable since f and g are measurable functions. Obviously, it has a measure which is less than ε. The bounded character of g follows very simply from its construction$\blacktriangle$

The principle stated in Lusin's theorem may be formulated in a manner closely reminiscent of the Weierstrass approximation theorem.

Theorem 15-53 (Frechet) If f is a measurable function on a domain E, then a sequence $\{g_\nu\}$ of continuous functions exists on $\mathscr{R}$ such that $\lim \{g_\nu\} = f$ almost everywhere in E.

Proof By the preceding theorem it is the case that for each ν a continuous function g_ν exists such that the set $D_\nu = \{x;\quad x \in E, f(x) \ne g_\nu(x)\}$ has measure not exceeding $1/2^\nu$. Hence for each n the set $\bigcup_{\nu=n}^{\infty} D_\nu$ is measurable and $\mu(\bigcup_{\nu=n}^{\infty} D_\nu) \leq \sum_{\nu=n}^{\infty} \mu(D_\nu) < 1/2^{n-1}$. Let $D = \bigcap_{n=1}^{\infty} \bigcup_{\nu=n}^{\infty} D_\nu$. Then D is measurable and, since $\{\bigcup_{\nu=n}^{\infty} D_\nu\}$ is a nested sequence, $\mu(D) = \lim \{\mu(\bigcup_{\nu=n}^{\infty} D_\nu)\} = 0$.

If $x \in E \cap \mathscr{C}D$, then for some n we have $x \notin \bigcup_{\nu=n}^{\infty} D_\nu$, whence $x \notin D_\nu$ for all $\nu \geq n$. This implies that $g_\nu(x) = f(x)$ for all $\nu \geq n$. Hence $\lim \{g_\nu(x)\} = f(x)$. The truth of the theorem follows from $\mu(E \cap \mathscr{C}D) = \mu(E)$▲

Exercises

15-19 Prove Theorems 15-35 through 15-38.

15-20 Let f be a continuous function on a domain E. Let $\varepsilon > 0$ be given and denote by $\mathscr{I}$ the collection of closed intervals I each with the property that $|f(x_1) - f(x_2)| < \varepsilon$ if x_1 and x_2 are members of $I \cap E$. Prove that $\mathscr{I}$ is a Vitali covering of E.

15-21 Prove that any measurable function is measurable on each measurable subset of its domain.

15-22 Prove that if a function is measurable on each of a countable collection of sets, then it is measurable on their union.

15-23 Prove that if g is a measurable function and f an increasing function on the range of g, then $f \circ g$ is measurable.

15-24 Use the sequence $\{g_n\}$ of measurable functions defined on $E = \{x;\ x \geq 0\}$ by $g_n(x) = 0$ if $0 \leq x \leq n$ and $g_n(x) = 1$ if $n < x$ to prove that the hypothesis that E be of finite measure cannot be relaxed in Egoroff's theorem.

16

The Lebesgue Integral

Oh, the little more, and how much it is!

ROBERT BROWNING (1812–1889), *By the Fireside*

16-1 Partitions

Let S denote a set of real numbers, bounded or unbounded, having finite measure. By a *Lebesgue partition* of S we mean any finite collection $\{S_\nu;\ \nu = 1,2,\ldots,n\}$ of disjoint measurable subsets of S having the property that $\bigcup_{\nu=1}^{n} S_\nu = S$.

A comparison with the concept of Riemann partition introduced in Chapter 10 reveals the notion of Lebesgue partition to be different in two respects. For one thing, a Lebesgue partition is meaningful for a category of sets considerably broader than the collection of all closed and bounded intervals; for another, the term *partition* itself is now employed to designate, not a set of subdivision points, but rather the subsets themselves into which the primary set is divided. Suppose, for example, that $P = \{x_0,x_1,\ldots,x_n\}$ is a Riemann partition of an interval $[a,b]$. Then examples of Lebesgue partitions of the same interval are such collections as $\{[x_0,x_1),[x_1,x_2),\ldots,[x_{n-1},x_n]\}$, $\{[x_0,x_1],(x_1,x_2],(x_2,x_3],\ldots,(x_{n-2},x_{n-1}],(x_{n-1},x_n]\}$, and $\{P,(x_0,x_1),(x_1,x_2),\ldots,(x_{n-1},x_n)\}$.

In the present chapter our concern is with Lebesgue partitions. Accordingly, the term partition, when employed alone, is to be understood in this sense. For brevity, we shall employ the notation $\{S_\nu\}$ for a partition of a set S.

We style a partition $\{S_\kappa'\}$ a *refinement* of a partition $\{S_\nu\}$ or say that $\{S_\kappa'\}$ is *finer* than $\{S_\nu\}$ if each member S_κ' of $\{S_\kappa'\}$ is a subset of some member S_ν of $\{S_\nu\}$. By *refining* a partition we mean, quite simply, its replacement by a finer partition.

Example 16-1

(a) If S_r and S_i denote the sets of rational and irrational numbers, respectively, in an interval $[a,b]$, then together these constitute a partition of $[a,b]$.

(b) If K is the Cantor set, then K and $[0,1] - K$ constitute a partition of $[0,1]$.

(c) Let K_r and K_i denote the sets of rational and irrational numbers, respectively, in the Cantor set K. Then $\{S_\nu\}$, where $S_1 = K_r$ and $S_2 = K_i$, is a partition of K. In turn, the collection $\{K_r, K_i, [0,1] - K\}$ is a refinement of the partition of $[0,1]$ of the preceding example.

(d) Let $S = \bigcup_{\nu=1}^{\infty} I_\nu$, where $I_\nu = (\nu - 1/2^\nu, \nu + 1/2^\nu)$. Then for any k the collection $\{I_1, I_2, \ldots, I_k, \bigcup_{\nu=k+1}^{\infty} I_\nu\}$ is a partition of S. Refinements may be achieved by increasing k.

16-2 The Lebesgue Integral of a Bounded Function on a Domain of Finite Measure

Suppose g to be a bounded function defined on a set S of finite measure and let $\{S_\nu\}$ be a partition of S consisting of n subsets. Denote by M and m the supremum and infimum, respectively, of g on S, and by M_ν and m_ν, $\nu = 1, 2, \ldots, n$, similar extrema on S_ν. We define *lower* and *upper Lebesgue sums* of g on, or with respect to, the partition $\{S_\nu\}$ by

$$\underline{L}(g,\{S_\nu\}) = \sum_{\nu=1}^{n} m_\nu \mu(S_\nu) \quad \text{and} \quad \overline{L}(g,\{S_\nu\}) = \sum_{\nu=1}^{n} M_\nu \mu(S_\nu),$$

respectively.

It is immediate from $\bigcup_{\nu=1}^{n} S_\nu = S$ that

$$m\mu(S) \le \underline{L}(g,\{S_\nu\}) \le \overline{L}(g,\{S_\nu\}) \le M\mu(S).$$

Also, if $\{S_\kappa'\}$ denotes a refinement of $\{S_\nu\}$, then

$$\underline{L}(g,\{S_\kappa'\}) \ge \underline{L}(g,\{S_\nu\}) \quad \text{and} \quad \overline{L}(g,\{S_\kappa'\}) \le \overline{L}(g,\{S_\nu\});$$

that is, refining a partition has the effect of increasing the lower Lebesgue sum and decreasing the upper sum. Further, it is the case that lower sums can never exceed upper sums; that is, if $\{S_\nu\}$ and $\{S_\kappa'\}$ denote *any* two partitions, neither of which is necessarily a refinement of the other, then

$$\underline{L}(g,\{S_\nu\}) \le \overline{L}(g,\{S_\kappa'\}).$$

Consider now the set of all lower Lebesgue sums of a function g corresponding to all possible partitions $\{S_\nu\}$ of a set S. Since such a set is not empty and is bounded above (by any upper sum), a supremum exists. We refer to it as the *lower Lebesgue integral of g on S* and write

$$\underline{\mathbf{L}\!\!\int_S} g = \sup\{\underline{L}(g,\{S_\nu\})\}.$$

In a similar fashion, the infimum of the set of upper sums of g on S defines the *upper Lebesgue integral of g on S*; thus

$$\overline{\fint}_S g = \inf\{\overline{L}(g,\{S_\nu\})\}.$$

An immediate consequence of the dominance of all lower sums by all upper sums is that of the lower integral by the upper; that is, $\underline{\fint}_S g \leq \overline{\fint}_S g$.

It is clear that every bounded function defined on a set of finite measure possesses both a lower and an upper Lebesgue integral. If a function g, defined on S, has the further property that its upper and lower integrals have one and the same value, then g is said to be *Lebesgue integrable on S* or to possess a *Lebesgue integral*

$$\fint_S g = \underline{\fint}_S g = \overline{\fint}_S g.$$

If the set S is delimited to closed intervals $[a,b]$ and if partitions $\{S_\nu\}$ are restricted to collections of abutting intervals, then the Lebesgue integral $\fint_{[a,b]} g$ of g on $[a,b]$ is precisely the Riemann integral $\int_a^b g$. It is clear, then, that the concept of Lebesgue integral is a generalization of that of Riemann, based on a less restrictive type of domain and an extended notion of partition. It is immediate that if $S = [a,b]$, then lower and upper Riemann and Lebesgue integrals all exist, and are such that

$$\underline{\int_a^b} g \leq \underline{\fint}_{[a,b]} g \leq \overline{\fint}_{[a,b]} g \leq \overline{\int_a^b} g.$$

As a consequence, a bounded function on an interval $[a,b]$ may possibly be Lebesgue integrable without being Riemann integrable. On the other hand, if the Riemann integral exists, then so does the Lebesgue, and the two are equal.

Example 16-2 We have seen that the function Ψ fails to be Riemann integrable on $[a,b]$ because its discontinuities do not constitute a set of measure zero. Or look at it this way. Any partition which consists of abutting intervals is such that in each of these intervals, however small, there are both rational and irrational numbers. Therefore all lower sums have the value zero while all upper sums equal $b - a$. Zero and $b - a$, consequently, are the respective values of the lower and upper Riemann integrals of Ψ. Since they differ, Ψ is not Riemann integrable. On the other hand, partitions of the Lebesgue type give rise to more extensive collections of lower and upper sums. Consider the partition $\{S_\nu\}$ of $[a,b]$ consisting of the sets S_r and S_i of rational and irrational points, respectively, of $[a,b]$. We see that $\underline{L}(\Psi,\{S_\nu\}) = 0 \cdot \mu(S_r) + 1 \cdot \mu(S_i) = 0 \cdot 0 + 1(b - a) = b - a$ and $\overline{L}(\Psi,\{S_\nu\}) = 0 \cdot \mu(S_r) + 1 \cdot \mu(S_i) = b - a$. Hence

$$b - a = \underline{L}(\Psi,\{S_\nu\}) \leq \underline{\fint}_{[a,b]} \Psi \leq \overline{\fint}_{[a,b]} \Psi \leq \overline{L}(\Psi,\{S_\nu\}) = b - a$$

forces the conclusion that Ψ is Lebesgue integrable on $[a,b]$ and $\fint_{[a,b]} \Psi = b - a$.

We shall find it convenient, if g is Lebesgue integrable on an interval $[a,b]$, to denote the integral by $\fint_a^b g$. If the Riemann integral exists as well, then, of course, $\fint_a^b g = \int_a^b g$.

The more general the concept of integral the more difficult does it become to cite examples of functions which are not integrable. While functions do certainly exist which cannot be integrated in the Lebesgue sense, they are of sufficient complexity to preclude a reasonable example at this juncture. In a later section we shall see that a non-Lebesgue integrable function is necessarily nonmeasurable.

16-3 Elementary Properties of the Lebesgue Integral

Some of the facts and theorems which follow tend to describe properties of the Lebesgue integral which are reminiscent of those of the Riemann integral. Frequently, supporting arguments and proofs are simple modifications of those employed in the Riemann case. In such instances, to avoid repetition, it is left to the reader to supply details.

Theorem 16-1 If g_1 and g_2 are bounded Lebesgue integrable functions on a set S of finite measure, and if k is any constant, then the functions kg_1, $g_1 + g_2$, $|g_1|$, and $g_1 g_2$ are all Lebesgue integrable on S. Further, $\fint_S kg_1 = k\fint_S g_1$ and $\fint_S (g_1 + g_2) = \fint_S g_1 + \fint_S g_2$.

Proof See Theorems 10-5, 10-6, 10-10, 10-11, and 10-12▲

Theorem 16-2 Suppose g to be a bounded function on a set S of finite measure. Then g is Lebesgue integrable on S if and only if the functions g^+ and g^- are Lebesgue integrable on S.

Proof This follows directly from the preceding theorem upon making the identifications $g^+ = \frac{1}{2}(|g| + g)$ and $g^- = \frac{1}{2}(|g| - g)$▲

Corollary If g_1 and g_2 are bounded Lebesgue integrable functions on a set S of finite measure, then so are the functions $g_1 \vee g_2$ and $g_1 \wedge g_2$.

Proof We have only to observe that $g_1 \vee g_2 = g_1 + (g_2 - g_1)^+$ and $g_1 \wedge g_2 = g_1 - (g_2 - g_1)^-$▲

The following two theorems have to do with the resolution of the domain of an integrable function into disjoint components.

Theorem 16-3 Let S_1 and S_2 be disjoint sets of finite measure and let g be a bounded function defined on $S_1 \cup S_2$. Then g possesses a Lebesgue integral on S_1 and S_2 if and only if g is Lebesgue integrable on $S_1 \cup S_2$; moreover, $\fint_{S_1} g + \fint_{S_2} g = \fint_{S_1 \cup S_2} g$.

Proof The arguments employed in the proof of Theorems 10-7 and 10-8 are valid here with certain obvious modifications with respect to partitions▲

Theorem 16-4 If $\{S_\nu\}$ is a nonempty countable collection of disjoint measurable sets such that $S = \bigcup_\nu S_\nu$ has finite measure, and if g is bounded and Lebesgue integrable on S, then g is Lebesgue integrable on S_ν for each ν, and $⨏_S g = \sum_\nu ⨏_{S_\nu} g$.

Proof We shall assume the case for a finite collection $\{S_\nu\}$ to follow readily from Theorem 16-3 by induction. Consider then an infinite collection $\{S_\nu\}$. Theorem 16-3 also implies that $⨏_{S_\nu} g$ exists for each ν. If, for any n, we let $R_n = \bigcup_{\nu=n+1}^{\infty} S_\nu$, then, again, by Theorem 16-3, $⨏_{R_n} g$ exists. As a consequence, we have $⨏_S g = \sum_{\nu=1}^{n} ⨏_{S_\nu} g + ⨏_{R_n} g$.

The sequence $\{R_n\}$ is decreasing, and $\bigcap_{n=1}^{\infty} R_n = \varnothing$. Hence $\lim \{\mu(R_n)\} = 0$. Thus, corresponding to each $\varepsilon > 0$, N exists such that $\mu(R_n) < \varepsilon$ for $n > N$. If M denotes an upper bound of $|g|$, we then have $|⨏_S g - \sum_{\nu=1}^{n} ⨏_{S_\nu} g| < \varepsilon M$ for $n > N$; that is, $\sum_{\nu=1}^{\infty} ⨏_{S_\nu} g$ converges to $⨏_S g$▲

From the definition of Lebesgue integral it follows that any bounded function whose domain has measure zero is Lebesgue integrable on its domain and the value of its integral is zero. To see this one has only to evaluate lower and upper sums corresponding to the partition whose only member is the domain itself. An immediate consequence of this fact and of the theorems above concerns the decomposition of the domain of a function into two sets, one of which has measure zero. Suppose g to be Lebesgue integrable on S and let $S = S' \cup S''$ where S' and S'' are disjoint and $\mu(S'') = 0$. Then $⨏_S g = ⨏_{S'} g + ⨏_{S''} g = ⨏_{S'} g$. It is evident, therefore, that with respect to the value of the integral of a Lebesgue integrable function, knowledge of the function on its entire domain is not essential. It is sufficient that it be known almost everywhere, i.e., at all point of its domain except for a subset of measure zero. For example, if g is bounded on a domain S and has a constant value k almost everywhere, then $⨏_S g = k\mu(S)$. Also, if f and g are bounded Lebesgue integrable functions on a common domain S and if $f = g$ almost everywhere on S, then $⨏_S f = ⨏_S g$.

Example 16-3

(a) The function Ψ has the value 1 almost everywhere on $[a,b]$. Hence $⨏_{[a,b]} \Psi = b - a$.

(b) Let g be defined by $g(x) = x$ on $S = [0,1] - K$, where K is the Cantor set. Since K is a zero set we have $⨏_S g = \int_0^1 x = \frac{1}{2}$.

It is clear that if g is bounded and Lebesgue integrable on S and non-negative almost everywhere, then $⨏_S g \geq 0$. As a consequence, if g_1 and g_2 are bounded and integrable on S, and if $g_1 \leq g_2$ almost everywhere, then $⨏_S g_1 \leq ⨏_S g_2$.

If g_1 and g_2 are bounded functions on S with the property that $g_1 g_2$ and g_2 are Lebesgue integrable, and if, further, $g_2 \geq 0$ and $m \leq g_1 \leq M$ almost everywhere on S, then $m ⨏_S g_2 \leq ⨏_S g_1 g_2 \leq M ⨏_S g_2$. We conclude, by assuming g_2 to be the unit function on S, that if g is such that $m \leq g \leq M$ almost everywhere, then $m\mu(S) \leq ⨏_S g \leq M\mu(S)$.

The next two theorems are proved using essentially the same arguments as in the Riemann case.

Theorem 16-5 If g is bounded and Lebesgue integrable on S of finite measure, then $|\mathfrak{L}\int_S g| \le \mathfrak{L}\int_S |g|$.

Proof Left as an exercise▲

Theorem 16-6 If $\{g_\nu\}$ is a uniformly convergent sequence of Lebesgue integrable functions on a domain S of finite measure, then $g = \lim \{g_\nu\}$ is Lebesgue integrable and

$$\lim \left\{ \mathfrak{L}\int_S g_\nu \right\} = \mathfrak{L}\int_S g. \tag{1}$$

Proof Left as an exercise▲

Example 16-4 If uniform convergence is relaxed in the last theorem the conclusion need not follow. Consider the nonuniformly convergent sequence $\{g_\nu\}$ defined on [0,1] by $g_\nu(x) = \nu$ if $0 < x < 1/\nu$ and $g_\nu(x) = 0$ otherwise. Clearly, $\mathfrak{L}\int_0^1 g_\nu = 1$ for all ν, while $\{g_\nu\}$ converges to the zero function on [0,1]. Hence $\mathfrak{L}\int_0^1 \lim \{g_\nu\} = 0 \ne \lim \{\mathfrak{L}\int_0^1 g_\nu\}$.

On the other hand, if $g_\nu(x) = \nu$ on $(0, 1/\nu^2)$ and 0 otherwise, then (1) holds even though $\{g_\nu\}$ is not uniformly convergent.

One observes in Theorem 16-6 a reiteration for Lebesgue integrals of the important theorem permitting interchange of limit and integration operations in the Riemann case. As we shall see, however, in Theorems 16-9, 16-23, and 16-25, such interchange is possible in the case of Lebesgue integration under conditions far less restrictive than uniform convergence. This feature, among others, tends to make the Lebesgue integral a more useful concept than the Riemann integral.

Exercises

16-1 Use the function Ψ on $[a,b]$ to show that a definition of a Lebesgue type integral following the pattern of Definition (B) of the Riemann integral of Chapter 10 would not be equivalent to the definition of the Lebesgue integral as given in Section 16-2.

16-2 Show that a definition of the Lebesgue integral equivalent to that of Section 16-2 can be patterned after Definition (C) of the Riemann integral.

16-3 Prove that a bounded function g on a domain S of finite measure is Lebesgue integrable if and only if, for any $\varepsilon > 0$, a partition $\{S_\nu\}$ exists with the property $\bar{L}(g,\{S_\nu\}) \le \underline{L}(g,\{S_\nu\}) + \varepsilon$.

16-4 Supply details of proof of Theorem 16-1.

16-5 Supply details of proof of Theorem 16-3.

16-6 Prove Theorem 16-5.

16-7 Prove Theorem 16-6.

16-4 Integrability and Measurability

In the present section we shall show that of the class of all bounded functions defined on a domain S of finite measure, those and only those are Lebesgue integrable which are measurable on S. Integrability follows from measurability quite directly as we see in the theorem below. Certain preliminary facts precede establishment of the converse.

Theorem 16-7 If g is a bounded and measurable function on a domain S having finite measure, then g is Lebesgue integrable on S.

Proof Let M be such that $|g(x)| < M$ for each x in S. Corresponding to each integer n, define points $y_\nu = -M + 2\nu M/n$, $\nu = 0, 1, \ldots, n$. Then $y_\nu - y_{\nu-1} = 2M/n$ and $-M = y_0 < y_1 < \cdots < y_n = M$. Consider the sets

$$S_\nu = \{x;\quad x \in S, y_{\nu-1} \le g(x) < y_\nu\}$$

for $\nu = 1, 2, \ldots, n$. They are clearly disjoint and have the property that their union is S. Further, each is a measurable set since g, by hypothesis, is a measurable function. Therefore, the S_ν constitute a Lebesgue partition of S, whereupon

$$\bar{L}(g,\{S_\nu\}) - \underline{L}(g,\{S_\nu\}) \le \sum_{\nu=1}^{n} \frac{2M}{n}\,\mu(S_\nu) = \frac{2M}{n}\,\mu(S).$$

Since n may be arbitrarily large, it follows that g is a Lebesgue integrable function▲

Theorem 16-8 If g is a bounded and measurable function on a domain S of finite measure and has the properties $⨏_S g = 0$ and $g \ge 0$ on S, then $g = 0$ almost everywhere on S.

Proof From the hypothesis it follows that the set $S_n = \{x;\quad x \in S, g(x) \ge 1/n\}$ is measurable for each $n = 1, 2, \ldots,$ that g is Lebesgue integrable on S_n, and that $⨏_{S_n} g \ge (1/n)\mu(S_n)$. Further, the condition $g \ge 0$ on S implies $⨏_{S-S_n} g \ge 0$. Therefore

$$0 = ⨏_S g = ⨏_{S-S_n} g + ⨏_{S_n} g \ge \frac{1}{n}\,\mu(S_n) \ge 0,$$

whence $\mu(S_n) = 0$. Since $\{S_n\}$ is an increasing sequence, it is the case that $\mu(\bigcup_{n=1}^{\infty} S_n) = \lim\,\{\mu(S_n)\}$. Hence the set $\bigcup_{n=1}^{\infty} S_n = \{x;\quad g(x) > 0, x \in S\}$ has measure zero or, what is the same, $g = 0$ almost everywhere▲

Corollary If g and h are bounded measurable functions on a domain S of finite measure, and if $g \le h$ and $⨏_S g = ⨏_S h$, then $g = h$ almost everywhere on S.

The following theorem is an improvement on Theorem 16-6 since uniform boundedness is less stringent than uniform convergence. Note that Theorem 16-6 is employed in the proof and that Egoroff's theorem provides the link with uniform convergence.

Theorem 16-9 If $\{g_\nu\}$ is a uniformly bounded sequence of measurable functions on a domain S of finite measure, and if $\{g_\nu\}$ converges to a limit g almost everywhere on S, then g is Lebesgue integrable and $\lim \{\oint_S g_\nu\} = \oint_S g$.

Proof Suppose $|g_\nu(x)| \leq M$ for all ν and all x in S. If $\lim \{g_\nu(x)\}$ fails to exist for some point x, define $g(x) = 0$. Then g is a measurable function (Corollary 1, Theorem 15-48) and $|g(x)| \leq M$ for all x in S. Hence $\oint_S g$ exists, by virtue of an earlier theorem.

Let $\varepsilon > 0$ be given. Then, by Egoroff's theorem, there exists a measurable set S_0 with $S_0 \subset S$, $\mu(S - S_0) < \varepsilon$, and such that $\{g_\nu\}$ converges uniformly to g on S_0. Therefore $\lim \{\oint_{S_0} g_\nu\} = \oint_{S_0} g$ (Theorem 16-6). As a consequence, for some N, if $\nu > N$, then $|\oint_{S_0} g_\nu - \oint_{S_0} g| < \varepsilon$. Also $|\oint_{S-S_0} g_\nu| \leq M\mu(S - S_0) \leq M\varepsilon$ and $|\oint_{S-S_0} g| \leq M\varepsilon$. Hence, for $\nu > N$,

$$\left|\oint_S g_\nu - \oint_S g\right| \leq \left|\oint_{S_0} g_\nu - \oint_{S_0} g\right| + \left|\oint_{S-S_0} g_\nu\right| + \left|\oint_{S-S_0} g\right| \leq (2M + 1)\varepsilon \blacktriangle$$

Corollary If $\sum_{\nu=1}^{\infty} g_\nu$ is a series of measurable functions that converges almost everywhere on a domain of finite measure S, and if the partial sums of the series are uniformly bounded, then $\oint_S \sum_{\nu=1}^{\infty} g_\nu$ exists and equals $\sum_{\nu=1}^{\infty} \oint_S g_\nu$.

We are now in a position to prove the converse to Theorem 16-7 and thereby establish the equivalence of integrability and measurability for bounded functions on domains of finite measure.

Theorem 16-10 If g is a bounded Lebesgue integrable function on a domain S of finite measure, then g is a measurable function.

Proof To each positive integer n there corresponds a partition $\{S_\nu\}$ of S such that

$$\oint_S g - \frac{1}{n} < \underline{L}(g,\{S_\nu\}) \leq \overline{L}(g,\{S_\nu\}) < \oint_S g + \frac{1}{n}.$$

As usual, let m_ν and M_ν denote infima and suprema of g on S_ν for $\nu = 1, 2, \ldots, n$. Define functions $f_n{}^*$ and $h_n{}^*$ on S by $f_n{}^*(x) = m_\nu$ and $h_n{}^*(x) = M_\nu$ for x in S_ν. Then $f_n{}^*$ and $h_n{}^*$ are bounded and measurable (simple) functions on S; as a consequence, each is Lebesgue integrable. Further, $\oint_S f_n{}^* = \underline{L}(g,\{S_\nu\})$ and $\oint_S h_n{}^* = \overline{L}(g,\{S_\nu\})$, and $f_n{}^* \leq g \leq h_n{}^*$.

Now consider the functions $f_n = f_1{}^* \vee f_2{}^* \vee \cdots \vee f_n{}^*$ and $h_n = h_1{}^* \wedge h_2{}^* \wedge \cdots \wedge h_n{}^*$. Clearly, each is measurable and $f_n{}^* \leq f_n \leq g \leq h_n \leq h_n{}^*$ for each n. Moreover,

$$\oint_S (h_n - f_n) \leq \oint_S (h_n{}^* - f_n{}^*) < \frac{2}{n}. \tag{2}$$

Let $f = \lim \{f_n\}$, which exists everywhere on S since $\{f_n\}$ is increasing and is bounded above (by g). Similarly, let $h = \lim \{h_n\}$. Then $f \le g \le h$ and both f and h are Lebesgue integrable with $\oint_S f = \lim \{\oint_S f_n\}$ and $\oint_S h = \lim \{\oint_S h_n\}$, by Theorem 16-9. From (2) follows the conclusion $\oint_S f = \oint_S h$, and the corollary to Theorem 16-8 implies that $f = h$ almost everywhere on S. Hence $f = g = h$ almost everywhere on S and we conclude, from Theorem 15-47, that g is a measurable function▲

16-5 Extension of the Definition

We next undertake to extend the definition of the Lebesgue integral of a function g on a domain S by relaxing conditions heretofore placed on the domain and range of g. In the present section we shall require no more of the domain of g than that it be measurable; it may be bounded or unbounded and have either finite or infinite measure. Further, the range of g may be bounded or unbounded. In effect the concepts referred to as improper integrals in the Riemann case are treated together.

Suppose to begin with that g is a non-negative function defined on a measurable domain S. Corresponding to each positive real number b let there be a function $g^{[b]}$ defined on S by

$$g^{[b]}(x) = \begin{cases} g(x) & \text{if } x \in S \text{ and } g(x) \le b \\ 0 & \text{if } x \in S \text{ and } g(x) > b. \end{cases}$$

It is clear that each such function is bounded and that $b_1 \le b_2$ implies $g^{[b_1]} \le g^{[b_2]}$.

If, for each b, the function $g^{[b]}$ is Lebesgue integrable on the domain $S \cap [-b,b]$, then the integral $\oint_{S \cap [-b,b]} g^{[b]}$ defines an increasing function on the domain consisting of all positive real numbers. The limit of this function as b increases without bound exists, being either finite or infinite. We define it to be the *Lebesgue integral of g on S*, that is,

$$\oint_S g = \lim_{b \to \infty} \oint_{S \cap [-b,b]} g^{[b]}. \tag{3}$$

If S has finite measure and if g is a bounded function possessing a Lebesgue integral on S, then g is measurable on S and it follows that $\oint_{S \cap [-b,b]} g^{[b]}$ exists for each b. Moreover, the limit of the latter function is clearly the number $\oint_S g$. Thus it is that the definition (3) truly extends the integral defined in Section 16-2.

Unlike the convention in the case of the Riemann integral, we now acknowledge ∞ as a possible value for a Lebesgue integral. That this is convenient becomes clear as we develop further properties of the integral. To distinguish finite and infinite cases it is common to describe a function g as *summable on S* if $\oint_S g$ exists and is finite.

If g is any function on a measurable domain S, it may be represented as the difference of non-negative functions, i.e., $g = g^+ - g^-$. For g^+ and g^- the notion of integrability already has meaning. We now define g itself to be *Lebesgue integrable on* S if both g^+ and g^- are integrable on S and if one, at least, of these functions is summable; further

$$\fint_S g = \fint_S g^+ - \fint_S g^-.$$

(It is clear that if neither g^+ nor g^- is summable, then it is meaningless to speak of the difference of their integrals.) We add that g is *summable* if its integral is finite.

Following are theorems asserting facts concerning the extended Lebesgue integral which tend to repeat or to generalize properties already established in the case of a bounded function on a domain of finite measure. It is clear that in some connections summability and integrability must be treated separately since the latter does not imply the former.

Theorem 16-11 If g is Lebesgue integrable on a domain S having measure zero, then $\fint_S g = 0$.

Proof Left as an exercise▲

Theorem 16-12 If g is integrable (summable) on S and if S_1 is a measurable subset of S, then g is integrable (summable) on S_1.

Proof Left as an exercise▲

Theorem 16-13 If f and g are integrable on a common domain S and if $f \le g$ almost everywhere on S, then $\fint_S f \le \fint_S g$. Further, if f is integrable on S and if $f = g$ almost everywhere, then g is integrable on S and $\fint_S f = \fint_S g$.

Proof Left as an exercise▲

The next four theorems concern measurability, integrability, and summability.

Theorem 16-14 If g is Lebesgue integrable on a measurable domain S, then g is measurable on S.

Proof Suppose first that g is non-negative on S. Then for each b the function $g^{[b]}$ is integrable and hence measurable on $S \cap [-b,b]$. Since $g = \lim_{b\to\infty} g^{[b]}$, it follows that g is measurable on $S \cap [-b,b]$ for any b, hence is measurable on S. If g is not non-negative on S we may write $g = g^+ - g^-$ and draw the same conclusion▲

Theorem 16-15 If g is Lebesgue integrable on a measurable domain S, then g is summable on S if and only if $|g|$ is summable on S.

Proof If g is summable on S, it follows that g^+ and g^- both have this property. By $\fint_S |g| = \fint_S g^+ + \fint_S g^-$ we conclude summability of $|g|$. Conversely, if $\fint_S |g|$ is finite, then so are $\fint_S g^+$ and $\fint_S g^-$▲

Theorem 16-16 If g is non-negative and measurable on a measurable domain S, then g is Lebesgue integrable on S.

Proof For each b the function $g^{[b]}$ is measurable on $S \cap [-b,b]$, whence $\oint_{S \cap [-b,b]} g^{[b]}$ exists▲

Theorem 16-17 If f is measurable on a measurable domain S, and if a summable function g exists with the property $|f| \leq g$, then f is summable on S.

Proof From $f^+ \leq g$ follows $\oint_S f^+ \leq \oint_S g$, whence f^+ is summable. Similarly, $f^- \leq g$ implies summability of f^-. Hence f is summable of S▲

Some observations concerning $|g|$ are in order. One sees that if g is measurable on S, then, whether g be integrable or not, $|g|$, being non-negative, is integrable. Moreover, the function $|g|$ may be summable even if g is not a measurable function—note that this is not in conflict with Theorem 16-15 since g is assumed integrable there. One need only consider the function g defined to have value 1 on some non-measurable subset A of $[0,1]$ and -1 on $[0,1] - A$.

The equivalence of integrability and measurability in the event of a bounded function on a domain of finite measure (Theorems 16-7 and 16-10) requires modification in the light of generalizations of the definition of integral. Summarizing conclusions established above, we see that if g is any integrable function, then it is certainly measurable. On the other hand, if it is measurable and bounded and if its domain has finite measure, then it is integrable, as it is if it is measurable and non-negative on any measurable domain. Lastly, if g is measurable and is dominated by a summable function, then it is integrable (summable).

Theorem 16-18 If g is summable on each of a finite collection of disjoint measurable sets $S_1, S_2, \ldots, S_n$, then g is summable on $S = \bigcup_{\nu=1}^{n} S_\nu$ and $\oint_S g = \sum_{\nu=1}^{n} \oint_{S_\nu} g$. (If g is non-negative, then the hypothesis of summability may be replaced by integrability.)

Proof Left as an exercise▲

The remaining theorems in the present section attest to a linearity of the Lebesgue integral.

Theorem 16-19 If g is summable on a domain S and c any (finite) constant, then cg is integrable on S and $\oint_S cg = c\oint_S g$.

Proof First assume g to be non-negative on S. The case $c = 0$ being trivial, consider $c > 0$. For any b it may be verified that $cg^{[b]} = (cg)^{[cb]}$, whence

(4) $$c \oint_{S \cap [-cb,cb]} g^{[b]} = \oint_{S \cap [-cb,cb]} cg^{[b]} = \oint_{S \cap [-cb,cb]} (cg)^{[cb]}.$$

By definition we have

$$\lim_{b\to\infty} \oint_{S \cap [-cb,cb]} (cg)^{[cb]} = \lim_{b'\to\infty} \oint_{S \cap [-b',b']} (cg)^{[b']} = \oint_S cg.$$

Suppose $c \geq 1$. Then $S \cap [-b,b] \subset S \cap [-cb,cb]$ so that, by (4),

$$c \oint_{S \cap [-b,b]} g^{[b]} \leq c \oint_{S \cap [-cb,cb]} g^{[b]} \leq \oint_{S \cap [-cb,cb]} (cg)^{[cb]} \leq \oint_S cg,$$

and it follows that $c\oint_S g \leq \oint_S cg$. Also, $(cg)^{[b]} \leq cg^{[cb]}$. Therefore

$$\oint_{S \cap [b,b]} (cg)^{[b]} \leq \oint_{S \cap [-cb,cb]} cg^{[cb]} = c\oint_{S \cap [-cb,cb]} g^{[cb]} \leq c\oint_S g$$

from which $\oint_S cg \leq c\oint_S g$. As a consequence, $\oint_S cg = c\oint_S g$ in the case $c \geq 1$.

If $0 < c < 1$, let $b = 1/c$. Then cg is non-negative and measurable, whence $b\oint_S cg = \oint_S bcg = \oint_S g$ or

$$\oint_S cg = \frac{1}{b}\oint_S g = c\oint_S g.$$

The desired conclusion having been established in the event g is non-negative and $c \geq 0$, let g now be regarded as arbitrary. Then for $c \geq 0$ we have $(cg)^+ = cg^+$ and $(cg)^- = cg^-$, whence

$$\oint_S cg = \oint_S (cg)^+ - \oint_S (cg)^- = \oint_S cg^+ - \oint_S cg^-$$

$$= c\oint_S g^+ - c\oint_S g^- = c\oint_S g.$$

If $c < 0$, then $(cg)^+ = (-cg)^-$ and $(cg)^- = (-cg)^+$. Therefore

$$\oint_S cg = \oint_S (cg)^+ - \oint_S (cg)^- = \oint_S (-cg)^- - \oint_S (-cg)^+$$

$$= -\left[\oint_S (-cg)^+ - \oint_S (-cg)^-\right]$$

$$= -\oint_S (-cg) = -(-c)\oint_S g = c\oint_S g \blacktriangle$$

Theorem 16-20 If f and g are non-negative and integrable on a measurable domain S, then $f + g$ is integrable on S and $\oint_S (f + g) = \oint_S f + \oint_S g$.

Proof Since $f^{[b]} + g^{[b]} \leq f + g$, we conclude, using Theorems 16-1 and 16-13, that

$$\oint_{S \cap [-b,b]} f^{[b]} + \oint_{S \cap [-b,b]} g^{[b]} = \oint_{S \cap [-b,b]} (f^{[b]} + g^{[b]}) \leq \oint_S (f + g),$$

and this being the case for any b, it follows that

(5) $$\oint_S f + \oint_S g \leq \oint_S (f + g).$$

On the other hand, it is the case that $(f+g)^{[b]} \le f^{[b]} + g^{[b]}$, whence

$$\oint_{S \cap [-b,b]} (f+g)^{[b]} \le \oint_{S \cap [-b,b]} (f^{[b]} + g^{[b]})$$

$$= \oint_{S \cap [-b,b]} f^{[b]} + \oint_{S \cap [-b,b]} g^{[b]}$$

$$\le \oint_S f + \oint_S g.$$

It follows that

$$\oint_S (f+g) \le \oint_S f + \oint_S g$$

which, together with (5), establishes the theorem▲

Theorem 16-21 If f and g are summable functions on a domain S, then $f+g$ is summable on S, and

$$\oint_S (f+g) = \oint_S f + \oint_S g. \tag{6}$$

Proof The functions $|f|$ and $|g|$ are summable and hence so is $|f|+|g|$ (Theorems 16-15 and 16-20). From $|f+g| \le |f| + |g|$, it follows (Theorem 16-17) that $f+g$ is summable. To establish (6) consider sets $S_1, S_2, \ldots, S_6$, all measurable subsets of S, defined by

$$\begin{aligned}
S_1 &= \{x;\ \ f(x) \ge 0,\ g(x) \ge 0\},\\
S_2 &= \{x;\ \ f(x) < 0,\ g(x) < 0\},\\
S_3 &= \{x;\ \ f(x) \ge 0,\ g(x) < 0,\ f(x) + g(x) \ge 0\},\\
S_4 &= \{x;\ \ f(x) \ge 0,\ g(x) < 0,\ f(x) + g(x) < 0\},\\
S_5 &= \{x;\ \ f(x) < 0,\ g(x) \ge 0,\ f(x) + g(x) \ge 0\},\\
S_6 &= \{x;\ \ f(x) < 0,\ g(x) \ge 0,\ f(x) + g(x) < 0\}.
\end{aligned}$$

Clearly, $S_\nu \cap S_\mu = \varnothing$ for $\mu \ne \nu$ and $\bigcup_{\nu=1}^{6} S_\nu = S$. Because of Theorem 16-18 we shall have proved the present theorem when it has been shown that $\oint_{S_\nu} (f+g) = \oint_{S_\nu} f + \oint_{S_\nu} g$ for each S_ν. For S_1 it is true by virtue of Theorem 16-20, while that theorem, together with Theorem 16-19 implies the desired relation for S_2. A common argument leads to the proper conclusion for the remaining sets; it suffices to be explicit for S_3. Let $h = f+g$. Then on S_3 we have $f \ge 0$, $g < 0$ and $h \ge 0$. Since f, $-g$, and h are non-negative, and $f = h + (-g)$, we have $\oint_{S_3} f = \oint_{S_3} h + \oint_{S_3} (-g) = \oint_{S_3} h - \oint_{S_3} g$ or $\oint_{S_3} h = \oint_{S_3} f + \oint_{S_3} g$▲

Exercises

16-8 Prove Theorem 16-11.
16-9 Prove Theorem 16-12.
16-10 Prove Theorem 16-13.
16-11 Prove Theorem 16-18.

16-12 Show that the conclusion of Theorem 16-18 does not follow (a) if "summable" is replaced by "integrable" throughout the theorem, (b) if the finite collection of measurable sets is replaced by an infinite collection.

16-13 Prove that if g is non-negative and integrable on a measurable domain S and if $⨏_S g = 0$, then $g = 0$ almost everywhere on S.

16-14 Give an example of a function that is integrable but not summable.

16-15 Give an example of a function that is summable but unbounded.

16-16 Give an example of a function that is measurable but not integrable.

16-17 Give an example of a function g that is not integrable but such that $|g|$ is.

16-18 Cite an instance of summable functions f and g such that fg is not summable. Show that if either f or g is bounded and both are summable, then fg is necessarily summable.

16-19 Prove the *first mean value theorem* for Lebesgue integrals: If g is a measurable function on a domain S such that $m \leq g(x) \leq M$ for some m and M and for all x in S, and if h is a non-negative summable function on S, then a number η exists such that $m \leq \eta \leq M$ and $⨏_S gh = \eta ⨏_S h$.

16-20 (a) Use the function g defined on $\{x;\ x \geq 0\}$ by $g(x) = (-1)^n/n$ for $n-1 \leq x < n$, $n = 1, 2, \ldots$, to show that it is possible for a function to possess an improper Riemann integral without possessing a Lebesgue integral. (b) Show, however, that if the Lebesgue integral of a function exists, then it is equal to the improper Riemann integral if that, too, exists.

16-6 Sequences of Integrable Functions

We continue the analysis of convergence properties of sequences of integrable functions. The gulf between the Riemann and Lebesgue cases is seen to widen as bounded convergence (Theorem 16-9) is relaxed to monotone convergence of integrable but not necessarily summable functions, and then to domination by an integrable function. Theorem 16-22, immediately below, is in the nature of a lemma for use in the proof of Theorem 16-23.

Theorem 16-22 If $\{g_\nu\}$ is a sequence of non-negative functions converging to a limit g on a domain S, then, for each positive real number b, the sequence $\{g_\nu^{[b]}\}$ converges to $g^{[b]}$.

Proof If x is a point of S such that $g(x) > b$, then, for all sufficiently large ν it is necessarily the case that $g_\nu(x) > b$. Hence for such ν we have $g^{[b]}(x) = g_\nu^{[b]}(x) = b$ and the desired conclusion holds.

If $g(x) < b$, then $g_\nu(x) < b$ for large enough ν. Hence $g^{[b]}(x) = g_\nu^{[b]}(x)$, so that again the stated convergence is proved.

It remains to consider points x such that $g(x) = b$. In this event, for any $\varepsilon > 0$ and for all sufficiently large ν, we have $b - \varepsilon < g_\nu(x) < b + \varepsilon$, whence $b - \varepsilon < g_\nu^{[b]}(x) \leq b$. From this and the arbitrary nature of ε it follows that $g^{[b]}(x) = b = \lim_{\nu \to \infty} \{g_\nu^{[b]}(x)\}$ ▲

Theorem 16-23 (Monotone Convergence Theorem) If $\{g_\nu\}$ is an increasing sequence of non-negative integrable functions that converges to a limit on a domain S, then $\lim \{\int_S g_\nu\} = \int_S \lim \{g_\nu\}$.

Proof Since $\lim \{g_\nu{}^{[b]}\} = g^{[b]}$, where $g = \lim \{g_\nu\}$, we have, by Theorem 16-9,

$$\int_{S \cap [-b,b]} g^{[b]} = \lim_{\nu \to \infty} \left\{ \int_{S \cap [-b,b]} g_\nu{}^{[b]} \right\}.$$

From $\int_{S \cap [-b,b]} g_\nu{}^{[b]} \le \int_S g_\nu$ follows $\int_{S \cap [-b,b]} g^{[b]} \le \lim_{\nu \to \infty} \{\int_S g_\nu\}$, and since this is the case for all b,

$$\int_S g \le \lim_{\nu \to \infty} \left\{ \int_S g_\nu \right\}.$$

On the other hand, $g_\nu \le g$ for all ν. Therefore $\int_S g_\nu \le \int_S g$, whence

$$\lim_{\nu \to \infty} \left\{ \int_S g_\nu \right\} \le \int_S g$$

and the theorem follows▲

In Fatou's theorem which follows, the hypotheses that $\{g_\nu\}$ is increasing and convergent are dropped. A conclusion is drawn then, not for $\lim \{g_\nu\}$ but rather for $\underline{\lim} \{g_\nu\}$.

Theorem 16-24 (Fatou) If $\{g_\nu\}$ is a sequence of non-negative integrable functions on a domain S and if $\underline{\lim} \{g_\nu\}$ exists, then $\int_S \underline{\lim} \{g_\nu\} \le \underline{\lim} \{\int_S g_\nu\}$.

Proof Corresponding to each positive integer n define $h_n = \inf \{g_\nu; \quad \nu \ge n\}$. Then, by Theorem 15-48, we know that each h_n is measurable. Since $h_n \le g_\nu$ for $\nu \ge n$, we have $\int_S h_n \le \int_S g_\nu$ for $\nu \ge n$, and therefore

$$\int_S h_n \le \underline{\lim} \left\{ \int_S g_\nu \right\}$$

for all n.

The sequence $\{h_n\}$ is increasing and is convergent to the limit function $\underline{\lim} \{g_n\}$. (See Exercise 16-21.) Therefore the conclusion of the monotone convergence theorem may be stated for it. We obtain

$$\int_S \underline{\lim} \{g_n\} = \lim \left\{ \int_S h_n \right\} \le \underline{\lim} \left\{ \int_S g_\nu \right\} ▲$$

It is the restriction to non-negative functions in Fatou's theorem that insures integrability of $\lim \{g_\nu\}$. If it is desired to relax the non-negative condition, some other hypothesis must be called upon. One such concerns the notion of a *dominant function.*

Theorem 16-25 (Dominated Convergence Theorem) If $\{g_\nu\}$ is a sequence of summable functions on a domain S, if a function h exists which is summable on S

and such that $|g_\nu| < h$ almost everywhere on S for each ν, and if $\{g_\nu\}$ converges to a limit almost everywhere on S, then the limit function is summable on S and $\lim\{\int_S g_\nu\} = \int_S \lim\{g_\nu\}$.

Proof From $|g_\nu| \leq h$ almost everywhere in S follows $|g| \leq h$ almost everywhere in S, where $g = \lim\{g_\nu\}$. Hence g is summable (Theorem 16-17). Consider the sequence $\{h_\nu\}$ defined by $h_\nu = g_\nu + h$. Since h_ν is non-negative and summable for each ν, it follows, by Fatou's theorem, that

$$\int_S \underline{\lim}\,\{g_\nu + h\} \leq \underline{\lim}\left\{\int_S (g_\nu + h)\right\}$$

whence

$$\int_S g \leq \underline{\lim}\left\{\int_S g_\nu\right\}.$$

The function $h - g_\nu$ is non-negative and summable for each ν so that, again by Fatou's theorem,

$$\int_S \underline{\lim}\,\{h - g_\nu\} \leq \underline{\lim}\left\{\int_S (h - g_\nu)\right\}$$

from which

$$\int_S g \geq -\underline{\lim}\left\{-\int_S g_\nu\right\} = \overline{\lim}\left\{\int_S g_\nu\right\} \blacktriangle$$

Corollary Let $\{g_\nu\}$ be a sequence of summable functions on a domain S such that $\sum_{\nu=1}^{\infty} g_\nu$ converges almost everywhere in S. Suppose, too, that a function h exists which is summable on S and has the property that, for all n, $\left|\sum_{\nu=1}^{n} g_\nu\right| \leq h$ almost everywhere in S. Then $\sum_{\nu=1}^{\infty} g_\nu$ is summable on S and $\int_S \sum_{\nu=1}^{\infty} g_\nu = \sum_{\nu=1}^{\infty} \int_S g_\nu$.

The following theorem concerns not a sequence of functions, but a sequence of integrals based on a decomposition of the domain of a summable function.

Theorem 16-26 If g is summable on S and if $\{S_\nu\}$ is a sequence of disjoint measurable sets such that $\bigcup_{\nu=1}^{\infty} S_\nu = S$, then $\int_S g = \sum_{\nu=1}^{\infty} \int_{S_\nu} g$.

Proof For each ν let k_ν denote the characteristic function of the set S_ν. Then $g = \sum_{\nu=1}^{\infty} gk_\nu$. The functions gk_ν are summable on S since $|g|$ is summable on S and $|gk_\nu| \leq |g|$. Further, $\left|\sum_{\nu=1}^{n} gk_\nu\right| \leq |g|$. Thus the conditions in the corollary to Theorem 16-25 are satisfied and it follows that

$$\int_S g = \int_S \sum_{\nu=1}^{\infty} gk_\nu = \sum_{\nu=1}^{\infty} \int_{S_\nu} gk_\nu = \sum_{\nu=1}^{\infty} \int_{S_\nu} g \blacktriangle$$

It is essential that summability of g on S be included in the hypothesis of the above theorem. Summability of g on each S_ν is not sufficient, as the following example shows.

Example 16-5 Let $S_\nu = [\nu - 1/\nu, \nu]$, and consider the function defined on $S = \bigcup_{\nu=1}^{\infty} S_\nu$ by $g(x) = (-1)^\nu \nu$. Clearly $\int_{S_\nu} g = (-1)^\nu$ for all ν. However $\int_S |g| = \infty$, hence g is not summable on S.

Exercises

16-21 If $\{g_\nu\}$ is a sequence of non-negative functions, let $h_n = \inf\{g_\nu;\ \nu \geq n\}$. Prove that $\{h_n\}$ is an increasing sequence with limit $\underline{\lim}\ \{g_n\}$.

16-22 Suppose g to be a non-negative, bounded, and measurable function on a domain S of finite measure, and let M be an integer such that $g(x) < M$ for all x in S. Prove that $\int_S g$ and $\lim_{n\to\infty} \sum_{k=0}^{nM} (k/n)\mu(S \cap \{x;\ k/n \leq g(x) < (k+1)/n\})$ exist and have the same value, or neither exists.

16-23 Show that Theorem 16-26 remains valid if the condition that g is summable on S be replaced by the assumption that g is non-negative and integrable.

16-24 For positive integers n and k such that $1 \leq k \leq n$, let f_{nk} denote the characteristic function of the interval $[(k-1)/n, k/n]$. Let $\{g_n\}$ be the sequence $f_{11}, f_{21}, f_{22}, f_{31}, \ldots$. Show that $\int_{[0,1]} g_n$ has limit zero, yet $\{g_n(x)\}$ does not converge to zero almost everywhere in [0,1].

16-25 As an illustration of the case in which inequality holds in Fatou's theorem, consider the sequence $\{g_n\}$ defined on [0,1] by $g_n(x) = nxe^{-nx^2}$. Show that $\lim\{g_n(x)\} = 0$ for all x, while $\lim\{\int_{[0,1]} g_n\} = \frac{1}{2}$.

16-26 Use the sequence $\{g_n\}$ where $g_n(x) = 1/n$ if $|x| \leq n$ and 0 if $|x| > n$ to demonstrate the importance of dominance in the dominated convergence theorem.

16-7 Upper and Lower Derivatives

The remaining sections in the present chapter have to do with the interrelation between differentiation and Lebesgue integration. In this connection it will be necessary first to introduce a certain generalization of the derivative concept.

Suppose g to be a function defined on an interval I, bounded or unbounded, and of any type. We then define the *upper* and *lower derivatives of g at a point c of I* by

$$\bar{D}g(c) = \overline{\lim_{c}}\ \frac{g(x)-g(c)}{x-c} \quad \text{and} \quad \underline{D}g(c) = \underline{\lim_{c}}\ \frac{g(x)-g(c)}{x-c},$$

respectively. So that such derivatives may exist at each point of I, we shall admit ∞ and $-\infty$ as values. This is a departure, of course, from the convention adopted for derivatives in Chapter 9. Quite generally, for any g and c we have

$$\underline{D}g(c) \leq \bar{D}g(c).$$

Example 16-6 Consider the function g defined for all x in $\mathscr{R}$ by $g(0) = 0$ and $g(x) = x \sin(1/x)$ if $x \neq 0$. Here $g\bar{D}(0)$ and $\underline{D}g(0)$ are upper and lower limits, respectively, of the function defined by $\sin(1/x)$. Hence $\bar{D}g(0) = 1$ and $\underline{D}g(0) = 0$.

Example 16-7 Consider the function Ψ on [0,1]. If c is rational, then in any neighborhood of c, $\Psi(x)$ takes on the values 0 and 1; if c is irrational, then $\Psi(x) - 1$ assumes values 0 and -1. In either case $(\Psi(x) - \Psi(c))/(x - c)$ has no finite upper or lower bound. Thus $\bar{D}\Psi(c) = \infty$ and $\underline{D}\Psi(c) = -\infty$.

Certain properties of upper and lower derivatives are immediate consequences of our study of upper and lower limits of functions in Chapter 7. For one thing it is evident that a function has a derivative at a point if and only if its upper and lower derivatives at the point have a common (finite) value. Further, it is clear for any function g defined on an interval that at any point c of that interval $\underline{D}g(c) = -\overline{D}(-g(c))$. Also, if c is a point of an interval on which both f and g are defined, then $\overline{D}(f+g)(c) \le \overline{D}f(c) + \overline{D}g(c)$ and $\underline{D}(f+g)(c) \ge \underline{D}f(c) + \underline{D}g(c)$, provided the sums constituting the right members are meaningful.

Measurability of lower and upper derivatives is established next.

Theorem 16-27 If g is any function on an interval I, then $\overline{D}g$ and $\underline{D}g$ are measurable functions.

Proof Let k denote any real number and consider the set $S = \{x;\ \overline{D}g(x) > k\}$. We propose to show that S is measurable.

Corresponding to each pair of positive integers m and n, denote by S_{mn} the union of all intervals $[x_1,x_2]$ which are subsets of I and have the properties

$$|x_2 - x_1| < \frac{1}{m} \quad \text{and} \quad \frac{g(x_2) - g(x_1)}{x_2 - x_1} > k + \frac{1}{n}.$$

It is clear that each S_{mn} is a measurable set and that, as a consequence, so is the set $\bigcup_{n=1}^{\infty} \bigcap_{m=1}^{\infty} S_{mn}$.

Suppose $c \in S$. Then an integer n_0 exists such that $\overline{D}g(c) > k + 1/n_0$. Therefore, corresponding to each value of m, there exists a point x_0 of I with the properties

$$0 < |x_0 - c| < \frac{1}{m} \quad \text{and} \quad \frac{g(x_0) - g(c)}{x_0 - c} > k + \frac{1}{n_0}.$$

Thus the closed interval having c and x_0 as endpoints is a subset of the S_{mn_0}, whence $c \in S_{mn_0}$ for each m. Therefore $c \in \bigcup_{n=1}^{\infty} \bigcap_{m=1}^{\infty} S_{mn}$ and we conclude that

$$S \subset \bigcup_{n=1}^{\infty} \bigcap_{m=1}^{\infty} S_{mn}. \tag{7}$$

On the other hand, let c denote any point of $\bigcup_{n=1}^{\infty} \bigcap_{m=1}^{\infty} S_{mn}$. Then there is an n_0 such that $c \in \bigcap_{m=1}^{\infty} S_{mn_0}$, whence $c \in S_{mn_0}$ for all m. This implies the existence, corresponding to each m, of an interval $[x_1,x_2]$ such that $x_1 \le c \le x_2$,

$$|x_2 - x_1| < \frac{1}{m} \quad \text{and} \quad \frac{g(x_2) - g(x_1)}{x_2 - x_1} > k + \frac{1}{n_0}. \tag{8}$$

Suppose $x_1 < c < x_2$. Then at least one of the two statements

$$\frac{g(x_1) - g(c)}{x_1 - c} > k + \frac{1}{n_0} \tag{9}$$

and

$$\frac{g(x_2) - g(c)}{x_2 - c} > k + \frac{1}{n_0} \tag{10}$$

must be true; otherwise it leads directly to a contradiction of (8). Further, by (8) we see that (9) is a valid statement if $x_2 = c$ and that (10) is true if $x_1 = c$. It follows that, in all cases, a number x exists in I such that

$$0 < |x - c| < \frac{1}{m} \quad \text{and} \quad \frac{g(x) - g(c)}{x - c} > k + \frac{1}{n_0}.$$

We conclude that $\bar{D}g(c) \geq k + 1/n_0$ or $\bar{D}g(c) > k$. Thus the point c belongs to S. Therefore

$$\bigcup_{n=1}^{\infty} \bigcap_{m=1}^{\infty} S_{mn} \subset S. \tag{11}$$

Relations (7) and (11) imply that S is a measurable set and hence that $\bar{D}g$ is a measurable function. The argument for $\underline{D}g$ is similar▲

Corollary If g is a differentiable function on an interval I, then g' is measurable on I.

16-8 The Derivative of an Increasing Function

Suppose g to denote a function which is differentiable on an interval $[a,b]$ and whose derivative has the special property $g'(x) > k$ for some positive real number k and for all x in $[a,b]$. It is clear, of course, that g is continuous and strictly increasing. Let $\{x_0,x_1, \ldots ,x_n\}$ denote a Riemann partition of $[a,b]$. Then, by applying the mean value theorem to each segment of the partition, we find

$$g(x_\nu) - g(x_{\nu-1}) = g'(\xi_\nu)(x_\nu - x_{\nu-1}) \geq k(x_\nu - x_{\nu-1})$$

for $\nu = 1, 2, \ldots, n$, whence, by summing over ν,

$$g(b) - g(a) \geq k(b - a). \tag{12}$$

The image of $[a,b]$ by the function g is, of course, the interval $[g(a),g(b)]$; inequality (12) expresses a relationship between the length of domain and range of g. In the following theorem we see a generalization of this principle to the upper derivative of an increasing function. Frequent reference is made to the notion of an image of a subset of the domain of a function. For this one needs to recall that if S is a subset of the domain of g, then the image $g(S)$ of S is the set $\{g(x); \ x \in S\}$.

Theorem 16-28 is preliminary to the major results stated in Theorems 16-29 and 16-30.

Theorem 16-28 If g is strictly increasing on $[a,b]$ and k any real number, and if $S \subset \{x; \ \bar{D}g(x) \geq k\}$, then $\mu^*(g(S)) \geq k\mu^*(S)$.

Proof The theorem is trivial if $k \leq 0$. Assume, therefore, that k is positive.

By virtue of the increasing character of g it is continuous almost everywhere. Hence the set C of its continuity points is measurable, with measure $b - a$. As a consequence, if $S_1 = S \cap C$, then $\mu^*(S_1) = \mu^*(S)$. From $g(S_1) \subset g(S)$ it is then clear that it suffices for us to show that

$$\mu^*(g(S_1)) \geq k\mu^*(S_1). \tag{13}$$

Let $\varepsilon > 0$ be given. Then an open set G exists such that $g(S_1) \subset G$ and

$$\mu(G) < \mu^*(g(S_1)) + \varepsilon. \tag{14}$$

Let k_1 be a number such that $0 < k_1 < k$ and denote by $\mathscr{I}$ the collection of all closed intervals $[x',x'']$ with the properties $[g(x'),g(x'')] \subset G$ and

$$\frac{g(x'') - g(x')}{x'' - x'} \geq k_1. \tag{15}$$

(It will be observed that even if the numbers x' and x'' are members of S, the interval $[x',x'']$ itself need not be a subset of S.) Then $\mathscr{I}$ is a Vitali covering of S_1. This follows very simply by the following reasoning. If $c \in S_1$, then a neighborhood $\mathscr{N}(g(c),\varepsilon)$ of $g(c)$ is contained in G. Because of continuity of g at c, a neighborhood $\mathscr{N}(c,\delta)$ exists such that the image of $\mathscr{N}(c,\delta) \cap [a,b]$ by g is a subset of $\mathscr{N}(g(c),\varepsilon)$. Since $\bar{D}g(c) > k_1$, a point x of $\mathscr{N}(c,\delta)$ exists with the property $(g(x) - g(c))/(x - c) \geq k_1$. Because $g(c)$ and $g(x)$ are in $\mathscr{N}(g(c),\varepsilon)$, the interval with these numbers as endpoints is a subset of G. By a choice of x arbitrarily close to c it follows that $\mathscr{I}$ covers S_1 in the sense of Vitali.

By the Vitali theorem we conclude the existence of a countable collection of disjoint intervals $[x_\nu',x_\nu'']$ of $\mathscr{I}$ such that $\mu(S_1 - \bigcup_\nu [x_\nu',x_\nu'']) = 0$, whence

$$\mu^*(S_1) \leq \mu\left(\bigcup_\nu [x_\nu',x_\nu'']\right) = \sum_\nu \mu[x_\nu',x_\nu'']. \tag{16}$$

From (15) and (16) follows $k_1\mu^*(S_1) \leq \sum_\nu [g(x_\nu'') - g(x_\nu')]$.

Because g is strictly increasing, the intervals $[g(x_\nu'),g(x_\nu'')]$ are disjoint; and since they are all contained in G, we have

$$\sum_\nu [g(x_\nu'') - g(x_\nu')] = \sum_\nu \mu[g(x_\nu'),g(x_\nu'')]$$

$$= \mu\left(\bigcup_\nu [g(x_\nu'),g(x_\nu'')]\right) \leq \mu(G).$$

This and (14) now imply that $k_1\mu^*(S_1) < \mu^*(g(S_1)) + \varepsilon$, and since ε is arbitrary, $k_1\mu^*(S_1) \leq \mu^*(g(S_1))$ for all $k_1 < k$. This in turn implies (13)▲

Corollary If g is an increasing function on $[a,b]$, then the set $\{x;\ \bar{D}g(x) = \infty\}$ has measure zero.

Proof Let h be defined on $[a,b]$ by $h(x) = g(x) + x$. Then h is strictly increasing and, for any points x_1 and x_2 of $[a,b]$,

$$\frac{h(x_2) - h(x_1)}{x_2 - x_1} = \frac{g(x_2) - g(x_1)}{x_2 - x_1} + 1.$$

Therefore $\bar{D}g(c) = \infty$ if and only if $\bar{D}h(c) = \infty$. Thus it suffices to prove that $S = \{x;\ \bar{D}h(x) = \infty\}$ is a set having measure zero. This is easy to see, however, for from the fact that the range of h lies in $[g(a) + a, g(b) + b]$ it follows that $k\mu^*(S) \leq g(b) - g(a) + b - a$ for all real values of k. Hence $\mu^*(S) = 0$▲

We are now in a position to prove the following theorem, attributed to Lebesgue, concerning the existence of a derivative of an increasing function.

Theorem 16-29 An increasing function on an interval $[a,b]$ possesses a finite derivative almost everywhere on $[a,b]$.

Proof Corresponding to each pair of positive integers m and n, consider the set

$$S_{mn} = \left\{x;\ \underline{D}g(x) < \frac{m}{n} < \frac{m+1}{n} < \bar{D}g(x)\right\},$$

measurable by virtue of Theorem 16-27. To each $\varepsilon > 0$ there then corresponds an open set G_{mn} such that $S_{mn} \subset G_{mn}$ and $\mu(G_{mn}) < \mu(S_{mn}) + \varepsilon$. Let $\mathscr{I}$ be the collection of all closed intervals I with the properties $I \subset G_{mn}$ and $\mu^*(g(I)) \leq (m/n)\mu(I)$. Then $\mathscr{I}$ is a Vitali covering of S_{mn} so that a countable collection of disjoint intervals I_ν of $\mathscr{I}$ exists with the property $\mu(S_{mn} - \bigcup_\nu I_\nu) = 0$. Therefore $\mu(S_{mn}) = \mu(S_{mn} \cap \bigcup_\nu I_\nu)$.

From the definition of the I_ν we have

$$\mu^*\left(\bigcup_\nu g(I_\nu)\right) \leq \sum_\nu \mu^*(g(I_\nu)) \leq \frac{m}{n} \sum_\nu \mu(I_\nu).$$

Further, the fact that the I_ν are disjoint and are all subsets of G_{mn} implies $\sum_\nu \mu(I_\nu) \leq \mu(G_{mn})$. Therefore

$$\mu^*\left(\bigcup_\nu g(I_\nu)\right) \leq \frac{m}{n} \mu(G_{mn}) < \frac{m}{n} (\mu(S_{mn}) + \varepsilon).$$

Since, quite generally, $\bigcup_\nu g(I_\nu) = g(\bigcup_\nu I_\nu)$, we have, using Theorem 16-28,

$$\mu^*\left(\bigcup_\nu g(I_\nu)\right) = \mu^*\left(g\left(\bigcup_\nu I_\nu\right)\right) \geq \mu^*\left(g\left(S_{mn} \cap \left(\bigcup_\nu I_\nu\right)\right)\right)$$

$$\geq \frac{m+1}{n} \mu\left(S_{mn} \cap \bigcup_\nu I_\nu\right) = \frac{m+1}{n} \mu(S_{mn}).$$

By combining inequalities we have $[(m+1)/n]\mu(S_{mn}) \leq (m/n)(\mu(S_{mn}) + \varepsilon)$ or $\mu(S_{mn}) \leq m\varepsilon$. Because ε is arbitrary we conclude $\mu(S_{mn}) = 0$ for all m and n.

Finally, if $S = \{x;\ \bar{D}g(x) > \underline{D}g(x)\}$, it is clear that $S = \bigcup_{m=1}^{\infty} \bigcup_{n=1}^{\infty} S_{mn}$, whence $\mu(S) = 0$. Therefore $g'(x)$ exists for almost all x in $[a,b]$. By the corollary to the preceding theorem g' is finite almost everywhere▲

Corollary If g is a function of bounded variation on an interval $[a,b]$, then g' exists and is finite almost everywhere on $[a,b]$.

We consider next the problem of determining the nature of the Lebesgue integral of the derivative of an increasing function. Before doing so, however, a minor technicality must be disposed of. We have seen that if g is increasing on $[a,b]$, then g' is defined not necessarily on all of $[a,b]$, but only almost everywhere. To speak of the Lebesgue integral of g' on $[a,b]$ requires that g' have a value at each point. Thus it becomes necessary to provide g' with a definition that assigns values to it throughout the interval $[a,b]$—but, of course, this should be done in such a way as to not effectively distort the structure of g'. Actually, we may achieve the desired end by simply assigning to g' quite arbitrary values at points where it is not otherwise defined. By virtue of the vanishing of the Lebesgue integral of a function on a set of measure zero, such extension cannot effect significant values. To be definite, we shall adopt the convention of regarding g' to have the value zero at all points of $[a,b]$ where the definition of derivative itself fails to be meaningful. This should be understood without the statement having to be made each time a derivative is to be integrated.

Theorem 16-30 If g is an increasing function on $[a, b]$, then g' is Lebesgue integrable on $[a,b]$ and $\mathcal{L}\int_{[a,b]} g' \leq g(b) - g(a)$.

Proof Let the definition of g be extended to the interval $[a,b+1)$ by $g(x) = g(b)$ for all x in $(b, b+1)$. The definition of g' on $[a,b]$ (aside from points where the limit does not exist) may then be expressed in the form

$$g'(x) = \lim \left\{ n \left[g\left(x + \frac{1}{n}\right) - g(x) \right] \right\}.$$

By Fatou's theorem it follows that

$$\mathcal{L}\int_{[a,b]} g' \leq \underline{\lim} \left\{ n\, \mathcal{L}\int_{[a,b]} \left[g\left(x + \frac{1}{n}\right) - g(x) \right] dx \right\}.$$

However,

$$\begin{aligned}
\mathcal{L}\int_{[a,b]} \left[g\left(x + \frac{1}{n}\right) - g(x) \right] dx &= \mathcal{L}\int_{[a+1/n,b+1/n]} g(x)\, dx - \mathcal{L}\int_{[a,b]} g(x)\, dx \\
&= \mathcal{L}\int_{[b,b+1/n]} g(x)\, dx - \mathcal{L}\int_{[a,a+1/n]} g(x)\, dx \\
&= \frac{1}{n} g(b) - \mathcal{L}\int_{[a,a+1/n]} g(x)\, dx \\
&\leq \frac{1}{n} [g(b) - g(a)].
\end{aligned}$$

Therefore $\mathcal{L}\int_{[a,b]} g' \leq g(b) - g(a)$▲

If g is an absolutely continuous function on an interval (bounded or unbounded), then it is also of bounded variation and consequently possesses a finite derivative almost everywhere. We have seen it possible for a function to possess a zero derivative almost everywhere and yet the function not be constant on its domain. Because of the following theorem such a function does not possess the property of absolute continuity.

Theorem 16-31 An absolutely continuous function whose derivative vanishes almost everywhere is a constant.

Proof Suppose g defined on $[a,b]$ and denote by S the set $\{x;\ \ g'(x) = 0\}$. Let $\varepsilon > 0$ be given. Then $\delta > 0$ exists with the property

$$\sum_{\nu=1}^{n} |g(b_\nu) - g(a_\nu)| < \varepsilon \tag{17}$$

for any set of disjoint intervals (a_ν,b_ν), all contained in $[a,b]$, for which the sum of their lengths does not exceed δ.

Let $\mathscr{I}$ be the collection of all intervals $[x',x'']$ which are contained in $[a,b]$ and have the property $|g(x'') - g(x')| < \varepsilon |x'' - x'|$. Since $\mathscr{I}$ is a Vitali covering of S, there exists a countable collection of disjoint intervals $[x_\nu',x_\nu'']$ of $\mathscr{I}$ having a total measure equal to $\mu(S) = b - a$. However, because of the closed and disjoint characters of the intervals, a finite collection of them cannot cover $[a,b]$. It follows that they constitute an infinite set and, as a consequence, for any n not all can have length exceeding $1/n$. Effectively, intervals of arbitrarily small length exist in the collection. From this it follows that a finite subcollection $\{[x_1',x_1''],$ $[x_2',x_2''], \ldots, [x_n',x_n'']\}$ of members of $\mathscr{I}$ exists which has total measure exceeding $b - a - \delta$. Assuming the notation to be such that $x_\nu'' < x_{\nu+1}'$ for $\nu = 1, 2, \ldots, n-1$, it follows that the complementing intervals $[x_0'',x_1']$, $[x_1'',x_2'], \ldots, [x_\nu'',x_{\nu+1}'], \ldots, [x_n'',x_{n+1}']$, where $x_0'' = a$ and $x_{n+1}' = b$, are such that the sum of their lengths is less than δ. Hence for these intervals the relation (17) is satisfied and we conclude

$$|g(b) - g(a)| \leq \sum_{\nu=1}^{n+1} |g(x_\nu') - g(x_{\nu-1}'')| + \sum_{\nu=1}^{n} |g(x_\nu'') - g(x_\nu')| < \varepsilon(1 + b - a).$$

The arbitrary character of ε implies $g(b) = g(a)$. Upon applying the above reasoning to each interval of the type $[a,x]$, it then follows that $g(x) = g(a)$ for all x in $[a,b]$. Moreover, since a and b are arbitrary, it follows that if g is defined on an unbounded interval, then it is constant on that interval▲

Corollary If g and h are absolutely continuous and such that $g' = h'$ almost everywhere on their domain, then g and h differ by a constant.

16-9 The Indefinite Lebesgue Integral

In this section we establish the absolute continuity of the indefinite Lebesgue integral and determine the sense in which differentiation and Lebesgue integration may be regarded as inverse operations. We settle first, in Theorem 16-32, a minor point preliminary to the major result of Theorem 16-33.

Theorem 16-32 If g is a Lebesgue integrable function on a domain S and if $\varepsilon > 0$ be given, then a number $\delta > 0$ exists such that $|\oint_{S'} g| < \varepsilon$ for any subset S' of S with the property $\mu(S') < \delta$.

Proof Suppose $g \geq 0$ on S. Then N exists such that the function g_N defined on S by $g_N(x) = g(x)$ if $g(x) \leq N$ and $g_N(x) = N$ if $g(x) > N$ has the property $\oint_S |g - g_N| < \varepsilon$. Let $\delta = \varepsilon/N$ and suppose S' to be such that $S' \subset S$ and $\mu(S') < \delta$. From $g - g_N \geq 0$ on S follows $\oint_{S'}(g - g_N) \leq \oint_S (g - g_N) < \varepsilon$. Also $\oint_{S'} g_N \leq N\mu(S') < \varepsilon$. Hence

$$\left|\oint_{S'} g\right| = \oint_{S'} g = \oint_{S'} (g - g_N) + \oint_{S'} g_N < 2\varepsilon.$$

If g assumes negative values, the desired conclusion follows from $|\oint_{S'} g| \leq |\oint_{S'} g^+| + |\oint_{S'} g^-|$▲

Let g be a function which is Lebesgue integrable on each bounded interval. With respect to each real number a we define the *indefinite Lebesgue integral* of g on $\mathscr{R}$ to be the function G given by

$$G(x) = \begin{cases} \oint_{[x,a]} g & \text{if} \quad x < a, \\ 0 & \text{if} \quad x = a, \\ \oint_{[a,x]} g & \text{if} \quad a < x. \end{cases}$$

Theorem 16-33 An indefinite Lebesgue integral is absolutely continuous.

Proof Let $\varepsilon > 0$ be given. Then, by Theorem 16-32, $\delta > 0$ exists such that $|\oint_{S'} g| < \varepsilon$ if $\mu(S') < \delta$. Suppose $\{(a_1,b_1),(a_2,b_2), \ldots,(a_n,b_n)\}$ to be a finite collection of disjoint intervals such that $\sum_{\nu=1}^{n} |b_\nu - a_\nu| < \delta$. Then

$$\left|\sum_{\nu=1}^{n} (G(b_\nu) - G(a_\nu))\right| = \left|\oint_{\cup_{\nu=1}^{n} [a_\nu, b_\nu]} g\right| < \varepsilon$$

and absolute continuity of G follows from Exercise 13-23▲

The significant conclusion that an indefinite integral is differentiable almost everywhere leads to the following important theorem.

Theorem 16-34 The indefinite Lebesgue integral G of a function g possesses a derivative which is equal to g almost everywhere.

Proof That G is differentiable almost everywhere follows immediately. Denote by S_{mn} the measurable set $\{x; \quad g(x) < m/n < (m+1)/n < G'(x)\}$, where m and n are positive integers. Further, let $\varepsilon > 0$ be given. Then $\delta > 0$ exists with the property

$$\left|\oint_{S'} g\right| < \varepsilon \tag{18}$$

for any S' for which $\mu(S') < \delta$.

An open set G_{mn} exists for each m and n such that $S_{mn} \subset G_{mn}$ and

$$\mu(G_{mn}) < \mu(S_{mn}) + \delta. \tag{19}$$

Let $\mathscr{I}$ denote the collection of all intervals $[x',x'']$ with the properties $[x',x''] \subset G_{mn}$ and

$$\frac{G(x'') - G(x')}{x'' - x'} > \frac{m+1}{n}. \tag{20}$$

Then $\mathscr{I}$ is a Vitali covering of S_{mn}, so that a countable collection of disjoint intervals $[x_\nu',x_\nu'']$ belonging to $\mathscr{I}$ have the property

$$\mu(S_{mn} - B) = 0, \tag{21}$$

where $B = \bigcup_\nu [x_\nu',x_\nu'']$. As a consequence of (20) it follows that for each such interval $\oint_{[x_\nu',x_\nu'']} g > [(m+1)/n](x_\nu'' - x_\nu')$. Summing over ν then yields

$$\oint_B g > \frac{m+1}{n}\,\mu(B) \geq \frac{m+1}{n}\,\mu(S_{mn}). \tag{22}$$

Further, since $B \subset G_{mn}$, we have $B - S_{mn} \subset G_{mn} - S_{mn}$ whence, by (19), $\mu(B - S_{mn}) < \delta$. Therefore, by (18), $|\oint_{B-S_{mn}} g| < \varepsilon$ and we have $\oint_B g \leq \oint_{S_{mn}} g + \varepsilon$. Also, from $g(x) < m/n$ for any x in S_{mn} it follows that

$$\oint_{S_{mn}} g < \frac{m}{n}\,\mu(S_{mn}). \tag{23}$$

Combining the results exhibited in (21), (22), and (23), we obtain $[(m+1)/n]\mu(S_{mn}) < (m/n)\mu(S_{mn}) + \varepsilon$ whence $\mu(S_{mn}) < n\varepsilon$ for all ε. Therefore $\mu(S_{mn}) = 0$ for all m and n.

The set $S = \{x;\ \ g(x) < G'(x)\}$ is seen to be equal to $\bigcup_{n=1}^\infty \bigcup_{m=1}^\infty S_{mn}$. Hence $\mu(S) = 0$ and we conclude $G'(x) \leq g(x)$ almost everywhere.

Finally, let $h = -g$ and let H be an indefinite integral of h. Then $H = -G$ and $H'(x)$ exists if and only if $G'(x)$ does. By the argument above we have $H' \leq h$ almost everywhere. Hence $g \leq G'$ almost everywhere and we conclude $G' = g$ almost everywhere▲

The following assertion is essentially a restatement of the preceding theorem.

Theorem 16-35 If g is an absolutely continuous function, then $g(x) = g(a) + \oint_{[a,x]} g'$ for $x \geq a$. Similarly for $x < a$.

Proof Let $G(x) = g(a) + \oint_{[a,x]} g'$. Then $G' = g'$ almost everywhere, by Theorem 16-34. Therefore $G(x) = g(x) + c$ and, since $G(a) = g(a)$, it follows that $G(x) = g(x)$ for all x▲

We shall refer to a function as *singular* on an interval $[a,b]$ if it possesses a derivative which is equal to zero almost everywhere on $[a,b]$. An instance of such is the Cantor function. The following theorem shows how a singular function plays a significant role in the decomposition of an increasing function.

Theorem 16-36 If f is an increasing function, then $f = f_1 + f_2 + f_3$, where f_1 is a jump function, f_2 an absolutely continuous function, and f_3 a continuous singular function.

Proof We have already seen (Theorem 13-5) that if f_1 is the jump function corresponding to an increasing function f, then the function $g = f - f_1$ is increasing and continuous. Therefore g' exists almost everywhere and, by Theorem 16-30,

$$\int_{[x_1,x_2]} g' \leq g(x_2) - g(x_1) \tag{24}$$

for all x_1 and x_2 such that $x_1 < x_2$.

Let f_2 be an indefinite integral of g', and let $f_3 = g - f_2$. The function f_2 is increasing since $g' \geq 0$, and it is absolutely continuous since it is an indefinite integral. That f_3 is increasing follows from (24). Finally, we have $f_3' = g' - f_2' = g' - g'$ almost everywhere; that is, f_3 is singular. Further, it is continuous since g and f_2 are▲

Exercises

16-27 Let g be defined on $\mathscr{R}$ by $g(x) = 1/p^2q^2$ if $x = p/q$ where p and q are relatively prime and $p \neq 0$, and let $g(x) = 0$ otherwise. Show that g is an instance of a function that is differentiable almost everywhere but nondifferentiable on a dense set.

16-28 Prove that a continuous function g on an interval $[a,b]$ maps each measurable subset of its domain onto a measurable subset of its range if and only if the image by g of any set of measure zero is also a set of measure zero.

16-29 Prove that if g is an absolutely continuous function on $[a,b]$, then the image by g of any subset of $[a,b]$ having measure zero also has measure zero.

16-30 Prove that if g is absolutely continuous on $[a,b]$, then the image by g of any measurable subset of $[a,b]$ is measurable.

16-31 Use the function of Exercise 16-27 to demonstrate that a nonconstant function may have a zero derivative almost everywhere.

16-32 Prove that if f and g are Lebesgue integrable on $[a,b]$ and such that $\int_{[a,x]} f = \int_{[a,x]} g$ for all x in $[a,b]$, then $f = g$ almost everywhere.

16-33 Prove that if g is of bounded variation on $[a,b]$, then $\int_{[a,b]} |g'| \leq t_g(b)$. Moreover, equality holds if and only if g is absolutely continuous on $[a,b]$.

16-34 Prove the following formula for *integration by parts*: If f is absolutely continuous and h summable on $[a,b]$, and if $\int_{[a,x]} h = g(x) - g(a)$, then $\int_{[a,b]} fh = f(b)g(b) - f(a)g(a) - \int_{[a,b]} gf'$.

16-35 Prove the following *change of variable* formula: If h is absolutely continuous on $[p,q]$ with values in $[a,b]$ where $a = h(p)$ and $b = h(q)$, and if f is bounded and measurable on $[a,b]$, then $\int_{[a,b]} f = \int_{[p,q]} (f \circ h)h'$.

16-36 Prove the *second mean value theorem*: If f is increasing and h summable on $[a,b]$, then ξ exists such that $\int_{[a,b]} fh = f(a+) \int_{[a,\xi]} h + f(b-) \int_{[\xi,b]} h$ and $a < \xi < b$.

16-37 Establish the *Schwarz inequality*: If f and g are measurable and f^2 and g^2 summable on a domain S, then $|\int_S fg|^2 \leq (\int_S f^2)(\int_S g^2)$.

16-38 Establish the *Minkowski inequality*: If f and g are measurable and f^2 and g^2 summable on S, then $[\int_S (f+g)^2]^{1/2} \leq [\int_S f^2]^{1/2} + [\int_S g^2]^{1/2}$.

17

Generalized Measure and the Lebesgue-Stieltjes Integral

All animals are equal, but some animals are more equal than others.
GEORGE ORWELL (1903–1950), *Animal Farm*

17-1 Additive Classes of Sets

Let Ω denote an abstract set or *space* of entities of any character (not necessarily real numbers) and consider families or collections whose members are subsets of Ω. For example, $\{\varnothing,\Omega\}$ is such a collection, as is the family of all subsets of Ω and the set $\{A,B,\mathscr{C}A,\mathscr{C}B\}$, where A and B are subsets of Ω and $\mathscr{C}A$ and $\mathscr{C}B$ denote $\Omega - A$ and $\Omega - B$, respectively.

A collection $\mathscr{F}$ of subsets of Ω is called an *additive class*[1] if it satisfies the following requirements:

(i) $\Omega \in \mathscr{F}$;
(ii) $E \in \mathscr{F}$ implies $\mathscr{C}E \in \mathscr{F}$;
(iii) $E_\nu \in \mathscr{F}$ $(\nu = 1,2,\ldots)$ implies $\bigcup_{\nu=1}^{\infty} E_\nu \in \mathscr{F}$.

Thus, for $\mathscr{F}$ to be an additive class of subsets of Ω it must contain Ω itself. Also, for each subset E of Ω which belongs to $\mathscr{F}$, its complement, $\mathscr{C}E$, must belong as well. Further, for each countably infinite collection of subsets of Ω which belong to $\mathscr{F}$, their union must likewise be a member of $\mathscr{F}$.

One sometimes encounters a modification of the above definition in which condition (iii) is replaced by one requiring only that finite unions of members of $\mathscr{F}$ belong to $\mathscr{F}$. This being the case, $\mathscr{F}$ is described as *finitely additive* and

[1] The terms σ-*field* and *Borel field* are also used.

the term *completely additive* is used, for emphasis, to describe additivity of a set $\mathscr{F}$ as we have defined it. While finite additivity does not imply complete additivity, except if $\mathscr{F}$ is itself a finite set, in which case complete additivity is true vacuously, one consequence of Theorem 17-1 below is that a completely additive class is also finitely additive. We shall concern ourselves only with complete additivity and shall understand the term *additive class* to be employed in that sense.

Theorem 17-1 If $\mathscr{F}$ is an additive class of subsets of Ω, then

(a) $\varnothing \in \mathscr{F}$,

(b) $E_\nu \in \mathscr{F}$, $\nu = 1, 2, \ldots$, implies $\bigcap_{\nu=1}^{\infty} E_\nu \in \mathscr{F}$, and

(c) $E_\nu \in \mathscr{F}$, $\nu = 1, 2, \ldots, n$, implies $\bigcup_{\nu=1}^{n} E_\nu \in \mathscr{F}$ and $\bigcap_{\nu=1}^{n} E_\nu \in \mathscr{F}$, for any n.

Proof

(a) By (i) and (ii), $\varnothing = \mathscr{C}\Omega \in \mathscr{F}$.

(b) $E_\nu \in \mathscr{F}$, $\nu = 1, 2, \ldots$, implies $\mathscr{C}E_\nu \in \mathscr{F}$, $\nu = 1, 2, \ldots$. By (iii), this implies $\bigcup_{\nu=1}^{\infty} \mathscr{C}E_\nu \in \mathscr{F}$ or, what is the same, $\mathscr{C}\bigcap_{\nu=1}^{\infty} E_\nu \in \mathscr{F}$. From (ii) follows $\bigcap_{\nu=1}^{\infty} E_\nu \in \mathscr{F}$.

(c) $E_\nu \in \mathscr{F}$, $\nu = 1, 2, \ldots, n$ implies $E_\nu \in \mathscr{F}$, $\nu = 1, 2, \ldots$, where $E_\nu = \varnothing$ for $\nu > n$. Hence $\bigcup_{\nu=1}^{\infty} E_\nu \in \mathscr{F}$ or, what is the same, $\bigcup_{\nu=1}^{n} E_\nu \in \mathscr{F}$. If $E_\nu = \Omega$ for $\nu > n$, then $\bigcap_{\nu=1}^{n} E_\nu = \bigcap_{\nu=1}^{\infty} E_\nu \in \mathscr{F}$ ▲

The above theorem makes clear the structure of an additive class of subsets of a set Ω. The sets $\varnothing$ and Ω both belong, as do all complements and countable unions and intersections of members of the class. Further if E_1 and E_2 belong to an additive class, then their difference $E_1 - E_2$ does also.

Example 17-1

(a) Whatever the character of a set Ω, the collection $\{\varnothing,\Omega\}$ is an additive class of subsets of Ω.

(b) The collection $\mathscr{F}$ of all subsets of any set Ω is an additive class. If Ω is finite, then $\mathscr{F}$ is finite; if Ω is infinite, then $\mathscr{F}$ is noncountably infinite.

(c) The collection of all measurable sets of real numbers is an additive class of subsets of $\mathscr{R}$. The collection of all measurable sets having finite measure is *not* an additive class. The collection of all measurable subsets of an interval $\Omega = [a,b]$ is an additive class.

(d) The collection of all intervals of real numbers is not an additive class since the union of two intervals is not necessarily an interval. The family of all open sets is not an additive class since the complement of an open set need not be open.

(e) If A is any subset of Ω, then the collection $\{\varnothing,\Omega,A,\mathscr{C}A\}$ is an additive class.

(f) Suppose Ω to be any noncountable set and let $\mathscr{F}$ denote the collection of all subsets of Ω which are either countable or have countable complements. Then $\mathscr{F}$ is an additive class.

A possible alternate definition of the concept of additive class is apparent in the following. Proof is omitted.

Theorem 17-2 A collection $\mathscr{F}$ of subsets of Ω is an additive class if and only if the following conditions are met:

(a) $\Omega \in \mathscr{F}$,

(b) $E_\nu \in \mathscr{F}$, $\nu = 1, 2, \ldots$, implies $\bigcap_{\nu=1}^{\infty} E_\nu \in \mathscr{F}$,

(c) $E_1 \in \mathscr{F}$, $E_2 \in \mathscr{F}$ imply $E_1 - E_2 \in \mathscr{F}$.

17-2 Borel Sets

We consider now the construction of a certain family of subsets of the set $\mathscr{R}$ of all real numbers. Let $\mathscr{I}$ denote the set of all intervals of real numbers, bounded or unbounded, the empty set included. While this collection does not constitute an additive class, it is clear that additive classes of subsets of $\mathscr{R}$ exist which contain $\mathscr{I}$ as a subset—the set of all subsets of $\mathscr{R}$ has this property, as does the class of all measurable sets. Consider then the family $\mathscr{A}$ of all additive classes of subsets of $\mathscr{R}$ which have $\mathscr{I}$ as a subset, and denote by $\mathscr{B}$ the intersection of the members of $\mathscr{A}$. Thus, $\mathscr{F} \in \mathscr{A}$ if and only if $\mathscr{F}$ is an additive class of subsets of $\mathscr{R}$ such that $\mathscr{I} \subset \mathscr{F}$; and $\mathscr{B} = \cap\{\mathscr{F}; \quad \mathscr{F} \in \mathscr{A}\}$.

Concerning the set $\mathscr{B}$, it is clear that it contains $\mathscr{I}$ as a subset. Moreover, $\mathscr{B}$ is an additive class. To see this, note that $\Omega \in \mathscr{B}$ since $\Omega \in \mathscr{F}$ for all $\mathscr{F}$ in $\mathscr{A}$. Further, if E is any member of $\mathscr{B}$, then E belongs to each $\mathscr{F}$ in $\mathscr{A}$, and, since each $\mathscr{F}$ is an additive class, $\mathscr{C}E$ belongs to it. Hence $\mathscr{C}E \in \mathscr{B}$. Lastly, suppose $E_\nu \in \mathscr{B}$ for $\nu = 1, 2, \ldots$. Then $E_\nu \in \mathscr{F}$ for all ν and for each $\mathscr{F}$ in $\mathscr{A}$. Hence $\bigcup_{\nu=1}^{\infty} E_\nu \in \mathscr{F}$ for each $\mathscr{F}$ and, as a consequence, $\bigcup_{\nu=1}^{\infty} E_\nu \in \mathscr{B}$.

It is thus determined that $\mathscr{B}$ is an additive class of subsets of $\mathscr{R}$ with the properties

(i) $\mathscr{I} \subset \mathscr{B}$;

(ii) $\mathscr{B} \subset \mathscr{F}$ for any additive class $\mathscr{F}$ such that $\mathscr{I} \subset \mathscr{F}$.

The latter property establishes $\mathscr{B}$ as the *smallest* additive class (relative to set inclusion) which contains all intervals. We refer to each member of $\mathscr{B}$ as a *Borel set* and to $\mathscr{B}$ itself as the *additive class of Borel sets.*

Suppose $\mathscr{F}$ to denote an additive class of subsets of $\mathscr{R}$ which is known to contain all intervals of the form $(a,b]$. It is then possible to show that $\mathscr{F}$ necessarily contains *all* intervals and that, as a consequence, $\mathscr{B} \subset \mathscr{F}$. To this end, note that for any a and b ($a < b$) we have $(a,b) = \bigcup_{\nu=2}^{\infty} (a, b - (b - a)/\nu]$ whence $(a,b) \in \mathscr{F}$. Further, since $\{a\} = \bigcap_{\nu=1}^{\infty} (a - 1/\nu, a]$, it follows that sets consisting of single real numbers are in $\mathscr{F}$. This, in turn, implies that $\{a\} \cup (a,b) = [a,b)$ and $\{a\} \cup (a,b] = [a,b]$ are members of $\mathscr{F}$. Lastly, since any unbounded interval is a countable union of bounded intervals, we conclude that all such also belong to $\mathscr{F}$. Hence $\mathscr{I} \subset \mathscr{F}$ and $\mathscr{B} \subset \mathscr{F}$.

The reader may show, too, that if $\mathscr{F}$ is an additive class of subsets of $\mathscr{R}$ with the property that it contains all bounded open intervals (a,b), then $\mathscr{I}$ is properly contained in $\mathscr{F}$ and again $\mathscr{B} \subset \mathscr{F}$.

Insight into the character of Borel sets and the structure of the additive class $\mathscr{B}$ may be gained by pretending to construct that collection of sets in stages. As remarked above, it suffices to define $\mathscr{B}$ as the least additive class which contains all bounded open intervals. It then follows that the collection $\mathscr{G}$ of all open sets is properly contained in $\mathscr{B}$. With $\mathscr{G}$ as a starting point we construct $\mathscr{B}$ through a succession of union and intersection operations σ and δ, as defined in conjunction with Exercises 4-75 through 4-78. In these exercises it is established that $\mathscr{G} = \mathscr{G}_\sigma = \mathscr{G}_{\sigma\sigma} = \cdots$, while $\mathscr{G} \subset \mathscr{G}_\delta = \mathscr{G}_{\delta\delta} = \cdots$. Actually, $\mathscr{G}$ is a proper subset of $\mathscr{G}_\delta$ since the latter contains all closed sets. Since $\mathscr{B}$ is an additive class, we have $\mathscr{G}_\delta \subset \mathscr{B}$, whence it is determined that all closed sets, as well as all open sets are to be classified as Borel sets.

Because $\mathscr{G}_\delta$ contains all sets consisting of single points (closed sets), it follows that $\mathscr{G}_{\delta\sigma}$ contains, in particular, the set of all rational numbers. The latter set is not a member of $\mathscr{G}_\delta$; therefore $\mathscr{G}_\delta$ is properly contained in $\mathscr{G}_{\delta\sigma}$, and we have $\mathscr{G} \subset \mathscr{G}_\delta \subset \mathscr{G}_{\delta\sigma} \subset \mathscr{B}$, where each subset relation is a proper one. Continuing in this fashion we generate a sequence $\{\mathscr{G}^{(n)}\}$ of collections of sets of real numbers defined by

$$\mathscr{G}^{(1)} = \mathscr{G},\ \mathscr{G}^{(2)} = (\mathscr{G}^{(1)})_\delta,\ \mathscr{G}^{(3)} = (\mathscr{G}^{(2)})_\sigma,\ \mathscr{G}^{(4)} = (\mathscr{G}^{(3)})_\delta,\ \ldots$$

in which the δ and σ operations are seen to alternate. It is possible to demonstrate that for each n the collection $\mathscr{G}^{(n)}$ is a proper subset of $\mathscr{G}^{(n+1)}$. Further, since the successive collections are formed solely by appending unions and intersections, $\mathscr{G}^{(n)} \subset \mathscr{B}$ for each n. It may be shown, too, that none of the collections $\mathscr{G}^{(n)}$ is an additive class; indeed, neither is the limit set $\bigcup_{n=1}^{\infty} \mathscr{G}^{(n)}$. We have thus constructed a collection of sets $\bigcup_{n=1}^{\infty} \mathscr{G}^{(n)}$ which is a proper subset of $\mathscr{B}$ and which contains as a subset the set $\mathscr{I}$ of all intervals. (The reader should ascertain the stage at which all intervals appear as members of an element of the sequence $\{\mathscr{G}^{(n)}\}$.)

The above procedure may now be repeated with the collection $\bigcup_{n=1}^{\infty} \mathscr{G}^{(n)}$ as a starting point. A new sequence, $\{\mathscr{H}^{(n)}\}$, defined by

$$\mathscr{H}^{(1)} = \bigcup_{n=1}^{\infty} \mathscr{G}^{(n)},\ \mathscr{H}^{(2)} = (\mathscr{H}^{(1)})_\delta,\ \mathscr{H}^{(3)} = (\mathscr{H}^{(2)})_\sigma,\ \mathscr{H}^{(4)} = (\mathscr{H}^{(3)})_\delta,\ \ldots$$

is generated. It can be demonstrated that for all n the collection $\mathscr{H}^{(n)}$ is properly contained in $\mathscr{H}^{(n+1)}$ and that the limit $\bigcup_{n=1}^{\infty} \mathscr{H}^{(n)}$ is a proper subset of $\mathscr{B}$. Compatibly, $\bigcup_{n=1}^{\infty} \mathscr{H}^{(n)}$ is not an additive class.

One imagines the above procedure to be repeated again, and so on. However, the precise nature of "and so on" is difficult to explain within the limitations of the present text, since further analysis of the nature of infinite sets is required. Suffice it to say that by such construction one ultimately achieves the additive class $\mathscr{B}$. It is clear, at least, from the nature of the construction, that all sets generated are Borel sets and that after a certain quite early stage intervals of all types have been accounted for, as, indeed, have all of the sets one encounters in elementary analysis.

Exercises

17-1 Prove Theorem 17-2.

17-2 Let Ω be any noncountable set and $\mathscr{F}$ the collection of all those subsets of Ω that are either finite or have finite complements. Show that $\mathscr{F}$ is finitely additive but not completely additive.

17-3 Show that the collection of subsets of $\mathscr{R}$ that either have measure zero or have complements of measure zero is an additive class.

17-4 Prove that if E_1 is a member of an additive class $\mathscr{F}$ of subsets of Ω, then $\mathscr{F}_1 = \{E \cap E_1; \quad E \in \mathscr{F}\}$ is an additive class of subsets of E_1.

17-5 Prove that if an additive class of subsets of $\mathscr{R}$ contains all open intervals, then it contains all Borel sets.

17-3 Measurable Sets and Borel Sets

The additive class of Borel sets is a subset of every additive class containing $\mathscr{I}$. Since one of these is the class of all measurable sets, it follows that all Borel sets are measurable. It is not the case, however, that the class of measurable sets and the class of Borel sets are one and the same, since measurable sets exist which are not Borel sets. To the end of exhibiting such a set we first prove the following theorem concerning images of Borel sets by continuous increasing functions.

Theorem 17-3 If g is a continuous and strictly increasing function mapping the domain $\mathscr{R}$ onto the range $\mathscr{R}$, then the image by g of a Borel set is also a Borel set.

Proof Because of the one-to-one character of g, if A is any subset of $\mathscr{R}$, then $g(\mathscr{C}A) = \mathscr{C}g(A)$. Further, if $\{A_\nu; \quad \nu = 1, 2, \ldots\}$ is any countable collection of subsets of $\mathscr{R}$, then $g(\bigcup_{\nu=1}^{\infty} A_\nu) = \bigcup_{\nu=1}^{\infty} g(A_\nu)$.

Let $\mathscr{D}$ denote the collection of all inverse images of Borel sets, i.e., all subsets D of $\mathscr{R}$ with the property that $g(D)$ is a Borel set. Clearly, $\mathscr{D}$ is an additive class. Moreover, if I is any interval, then so is $g(I)$, and since this is a Borel set, I belongs to $\mathscr{D}$. Thus $\mathscr{D}$ is an additive class which contains all intervals. It therefore contains all Borel sets, and it follows that the image of a Borel set is a Borel set▲

Suppose f to denote the Cantor function on [0,1], and imagine the domain of f to be extended to $\mathscr{R}$ by defining $f(x) = 0$ if $x < 0$ and $f(x) = 1$ if $x > 1$. Consider the function g defined on $\mathscr{R}$ by $g(x) = x + f(x)$. It is continuous (since f is) and is strictly increasing with range $\mathscr{R}$. It satisfies, therefore, the hypotheses of Theorem 17-3.

Letting K denote the Cantor set, consider an interval (a,b) contained in $\mathscr{C}K$. Its image by g is also an interval, namely $(g(a),g(b)) = (a + f(a), b + f(a))$, with length $b - a$. As a consequence, the image of the set $\mathscr{C}K \cap [0,1]$ is a set of measure 1. Since $[0,1]$ itself maps into $[0,2]$, we conclude that $g(K)$, the image of the Cantor set, has measure 1.

Denote by A a nonmeasurable subset of $g(K)$. Then, because of the one-to-one character of g, the inverse image $g^{-1}(A)$ of A is a subset of K. Necessarily, $g^{-1}(A)$ is measurable, since K has measure zero. The set A, being nonmeasurable, is not a Borel set; hence $g^{-1}(A)$ is not a Borel set (by Theorem 17-3). Thus it is that measurable sets exist which are not Borel sets.

The essence of the difference between Borel sets and measurable sets involves sets having measure zero.

Theorem 17-4 Any measurable set is the union of a Borel set and a set of measure zero.

Proof If E is a measurable set, then, corresponding to each positive integer n, there exist Borel sets F_n and G_n, respectively closed and open, such that $F_n \subset E \subset G_n$ and $\mu(G_n - F_n) < 1/n$. Clearly, the sets $F = \bigcup_{n=1}^{\infty} F_n$ and $G = \bigcap_{n=1}^{\infty} G_n$ are Borel sets, and $\mu(G - F) = 0$. Let $Z = E - F$. Then $Z \subset G - F$, whence Z has measure zero. Further, $E = F \cup Z$▲

Theorem 17-5 If A is a set of measure zero, then a Borel set B exists such that $\mu(B) = 0$ and $A \subset B$.

Proof If $\mu(A) = 0$, then, corresponding to each positive integer n is an open set G_n such that $E \subset G_n$ and $\mu(G_n) < 1/n$. It follows that $B = \bigcap_{n=1}^{\infty} G_n$ is a Borel set with the properties $A \subset B$ and $\mu(B) = 0$▲

Corollary Every measurable set is the union of a Borel set and a subset of a Borel set of measure zero.

17-4 Measure—General Theory

Suppose $\mathscr{F}$ to be an additive class of subsets of $\mathscr{R}$, and let σ denote a set function with domain $\mathscr{F}$ and with range in the extended real number system $\mathscr{R}^*$. Thus σ assigns to each member E of $\mathscr{F}$ a number $\sigma(E)$, finite or infinite. We shall say that σ is *non-negative* if $\sigma(E) \geq 0$ for all E in $\mathscr{F}$, and refer to it as (completely) *additive* if

$$\sigma\left(\bigcup_{\nu=1}^{\infty} E_\nu\right) = \sum_{\nu=1}^{\infty} \sigma(E_\nu) \tag{1}$$

for any countably infinite collection $\{E_\nu;\quad \nu = 1,2,\ldots\}$ of disjoint members of $\mathscr{F}$.

A function σ which is both non-negative and additive and which, in addition, is such that $\sigma(E)$ is finite for each bounded set E in $\mathscr{F}$, is called a *measure function* on $\mathscr{F}$; the number $\sigma(E)$ is referred to as the *σ-measure* of E. (Many authors require of a measure function only that it be non-negative and additive.

We include the condition that it have finite values on bounded sets simply for convenience—we shall be concerned only with measures having this property.)

The Lebesgue measure μ, considered in Chapter 15, is an instance of a measure function defined on the additive class of measurable sets. In connection with this measure concept certain properties were established which tend also to describe the more general notion of measure defined above. Proofs of some of these were based on special characteristics of Lebesgue measure—others extend quite naturally to the general measure function.

Theorem 17-6 If σ is a measure function on an additive class $\mathscr{F}$ of subsets of $\mathscr{R}$, then

(a) $\sigma(\varnothing) = 0$;

(b) $\sigma(\bigcup_{\nu=1}^{n} E_\nu) = \sum_{\nu=1}^{n} \sigma(E_\nu)$ if $E_1, E_2, \ldots, E_n$ are disjoint members of $\mathscr{F}$;

(c) $E_1 \subset E_2$ implies $\sigma(E_1) \leq \sigma(E_2)$;

(d) $\sigma(E_1 \cup E_2) + \sigma(E_1 \cap E_2) = \sigma(E_1) + \sigma(E_2)$ for any two members E_1 and E_2 of $\mathscr{F}$;

(e) $\sigma(\bigcup_\nu E_\nu) \leq \sum_\nu \sigma(E_\nu)$ for any countable collection of members of $\mathscr{F}$;

(f) $E_\nu \subset E_{\nu+1}$, $\nu = 1, 2, \ldots$, implies $\lim \{\sigma(E_\nu)\} = \sigma(\bigcup_{\nu=1}^{\infty} E_\nu)$;

(g) $E_{\nu+1} \subset E_\nu$, $\nu = 1, 2, \ldots$, and $\sigma(E_1)$ finite, imply $\lim \{\sigma(E_\nu)\} = \sigma(\bigcap_{\nu=1}^{\infty} E_\nu)$.

Proof

(a) $\sigma(\varnothing)$ exists and is finite since $\varnothing$ is a bounded member of $\mathscr{F}$. From the additivity property (1) we have $\sigma(\varnothing) = \sum_{\nu=1}^{\infty} \sigma(\varnothing)$ from which follows $\sigma(\varnothing) = 0$.

(b) Let $E_\nu = \varnothing$ for $\nu = n+1, n+2, \ldots$. Then, by (1),

$$\sigma\left(\bigcup_{\nu=1}^{n} E_\nu\right) = \sigma\left(\bigcup_{\nu=1}^{\infty} E_\nu\right) = \sum_{\nu=1}^{\infty} \sigma(E_\nu) = \sum_{\nu=1}^{n} \sigma(E_\nu).$$

(c) Since $E_2 - E_1$ belongs to $\mathscr{F}$ we have, by (b), $\sigma(E_2) = \sigma(E_1) + \sigma(E_2 - E_1)$, from which follows $\sigma(E_1) \leq \sigma(E_2)$.

(d) See proof of Theorem 15-21.

(e) Let $B_1 = E_1$ and, for $\nu > 1$, $B_\nu = E_\nu - \bigcup_{k=1}^{\nu-1} E_k$. Then, for all ν, the set B_ν belongs to $\mathscr{F}$ and, by (c), $\sigma(B_\nu) \leq \sigma(E_\nu)$. Because the B_ν are disjoint and $\bigcup_{\nu=1}^{\infty} E_\nu = \bigcup_{\nu=1}^{\infty} B_\nu$, we have

$$\sigma\left(\bigcup_{\nu=1}^{\infty} E_\nu\right) = \sigma\left(\bigcup_{\nu=1}^{\infty} B_\nu\right) = \sum_{\nu=1}^{\infty} \sigma(B_\nu) \leq \sum_{\nu=1}^{\infty} \sigma(E_\nu).$$

A similar conclusion follows in the case of a finite union.

(f) See proof of Theorem 15-22.

(g) See proof of Theorem 15-23▲

Example 17-2 Suppose $\{a_\nu;\ \nu = 1, 2, \ldots\}$ to be a countably infinite set of real numbers, and let $\{p_\nu\}$ be a sequence of non-negative real numbers such that $\sum_{\nu=1}^{\infty} p_\nu$ converges. Denote by $\mathscr{F}$ the additive class of all subsets of $\mathscr{R}$, and for each member E of $\mathscr{F}$ define $\sigma(E) = \sum_{a_\nu \in E} p_\nu$, the p_ν's summed being those corresponding to the real numbers a_ν present in E. It is easily verified that σ is a measure function on $\mathscr{F}$; further,

σ is finite valued, that is, $\sigma(E) < \infty$ for all E in $\mathscr{F}$. If, in particular, $\sum_{\nu=1}^{\infty} p_\nu = 1$, the reader familiar with probability theory will recognize in σ a discrete probability measure.

Example 17-3 Consider the additive class $\mathscr{F} = \{\varnothing,\Omega,A,\mathscr{C}A\}$ where A is any non-empty proper subset of a space Ω. Let σ be defined by $\sigma(\varnothing) = 0$, $\sigma(A) = 2$, $\sigma(\mathscr{C}A) = 3$, and $\sigma(\Omega) = 5$. Then σ is a measure function on $\mathscr{F}$.

Example 17-4 Suppose g to be a non-negative function defined on $\mathscr{R}$ with the property that the Lebesgue integral $\mathfrak{L}_B g$ exists and is finite for any bounded Borel set B. Define σ on the additive class $\mathscr{B}$ of Borel sets by $\sigma(B) = \mathfrak{L}_B\, g$ if this exists and is finite and $\sigma(B) = \infty$ otherwise. The function σ is then a measure function on $\mathscr{B}$. It is easily verified that if $g(x) = 1$ for all x, then σ is precisely the Lebesgue measure on $\mathscr{B}$.

Exercises

17-6 Supply details of the proofs of parts (d), (f), and (g) of Theorem 17-6.

17-7 Let σ be a measure on an additive class $\mathscr{F}$ of subsets of $\mathscr{R}$. Let E_1 be a member of $\mathscr{F}$ and define a set function σ_1 by $\sigma_1(E) = \sigma(E \cap E_1)$ for each $E \in \mathscr{F}$. Prove that σ_1 is a measure on $\mathscr{F}$.

17-8 Suppose σ to be a finitely additive ((b) of Theorem 17-6) set function defined on an additive class $\mathscr{F}$ of subsets of $\mathscr{R}$. Prove that if σ also satisfies part (f) of Theorem 17-6, then it is completely additive on $\mathscr{F}$.

17-5 Complete Measure

We propose to demonstrate how a measure function may be extended to a larger domain. Suppose σ to be a measure function defined on an additive class $\mathscr{F}^\sigma$ of subsets of $\mathscr{R}$. Let $\mathscr{A}$ denote the collection of all subsets of $\mathscr{R}$ having the property that each is a subset of a member of $\mathscr{F}^\sigma$ having σ-measure zero. Explicitly

$$\mathscr{A} = \{A; \quad A \subset E,\ E \in \mathscr{F}^\sigma,\ \sigma(E) = 0\}.$$

Consider the collection formed of all possible unions of members of $\mathscr{F}^\sigma$ with those of $\mathscr{A}$. For convenience, let it be denoted by $\mathscr{F}^{\bar{\sigma}}$; thus

$$\mathscr{F}^{\bar{\sigma}} = \{A \cup E; \quad A \in \mathscr{A},\ E \in \mathscr{F}^\sigma\}.$$

It is easily verified that $\mathscr{F}^{\bar{\sigma}}$ is an additive class, and, of course, $\mathscr{F}^\sigma \subset \mathscr{F}^{\bar{\sigma}}$.

Let $\bar{\sigma}$ be defined on $\mathscr{F}^{\bar{\sigma}}$ by $\bar{\sigma}(A \cup E) = \sigma(E)$. That $\bar{\sigma}$ is a measure function on $\mathscr{F}^{\bar{\sigma}}$ is clear. We refer to it as the *completion* of the measure function σ.

If a measure σ on $\mathscr{F}^\sigma$ is completed to $\mathscr{F}^{\bar{\sigma}}$ and denoted by $\bar{\sigma}$, then, by the same construction, we may contemplate the completion of $\bar{\sigma}$ to a new measure function $\bar{\bar{\sigma}}$ on an additive class $\mathscr{F}^{\bar{\bar{\sigma}}}$. It is easy to see, however, that such a procedure is otiose, inasmuch as $\mathscr{F}^{\bar{\bar{\sigma}}}$ and $\mathscr{F}^{\bar{\sigma}}$ are one and the same collection of sets, and $\bar{\bar{\sigma}} = \bar{\sigma}$. We are prompted to define a measure σ as *complete* if $\mathscr{F}^{\sigma}$ and $\mathscr{F}^{\bar{\sigma}}$ are identical and $\bar{\sigma} = \sigma$. In effect, σ is complete if every subset of a set of σ-measure zero is σ-measurable.

Example 17-5 Consider ordinary Lebesgue measure defined on the class $\mathscr{B}$ of Borel sets. If one forms all possible unions of members of $\mathscr{B}$ with those subsets of Borel sets which have measure zero, the resultant collection is that of all Lebesgue measurable sets, and the completed measure is ordinary Lebesgue measure.

17-6 Distribution Functions

We consider next a correspondence between a rather wide category of measure functions, on the one hand, and a certain special class of functions of a real variable on the other. The advantage gained from such an association is clear. It enables us to make use of known facts about point functions to deduce corresponding properties of measure functions.

Let σ denote a measure function on an additive class $\mathscr{F}$ of subsets of $\mathscr{R}$ that contains all Borel sets. Let k be any real number and define a function F_k on $\mathscr{R}$ by

$$F_k(x) = \begin{cases} -\sigma((x,k]) & \text{if } x < k \\ 0 & \text{if } x = k \\ \sigma((k,x]) & \text{if } x > k. \end{cases}$$

We shall refer to F_k as a *distribution function* corresponding to the measure function σ. It is not unique, of course, since k is not committed. Variation in k is not significant, however, since, if $k_1 < k_2$, then $F_{k_1}(x) - F_{k_2}(x) = \sigma((k_1,k_2])$. Thus, distribution functions corresponding to different values of the parameter k differ only by a constant. This suggests the choice, once and for all, of a single value of k, say k_0. The function $F = F_{k_0}$ may then be regarded as *the* distribution function corresponding to σ.

The following theorems demonstrate essential features of distribution functions.

Theorem 17-7 The distribution function corresponding to a measure function is an increasing function which is continuous from the right at each point.

Proof For $x_1 < x_2$ we have

$$F(x_2) - F(x_1) = \sigma((x_1,x_2]) \tag{2}$$

and from the non-negative character of σ conclude that F increases. Further, with $a < x$,

$$\lim_{a+} [F(x) - F(a)] = \lim_{a+} \sigma((a,x]) = \sigma(\varnothing) = 0.$$

Therefore F is continuous from the right at a▲

Theorem 17-8 If F is a distribution function corresponding to σ, then
(a) $\sigma((a,b]) = F(b) - F(a)$,
(b) $\sigma([a,b]) = F(b) - F(a-)$,
(c) $\sigma([a,b)) = F(b-) - F(a-)$,
(d) $\sigma((a,b)) = F(b-) - F(a)$.

Proof With $x < a$ we have

$$\lim_{a-} [F(a) - F(x)] = \lim_{a-} \sigma((x,a]) = \sigma(\{a\}). \tag{3}$$

The various conclusions follow from (2) and (3)▲

One notes, from (3), that a distribution function is continuous from the left (and hence continuous) at a point a if and only if the σ-measure of the set consisting of a alone is 0.

17-7 Constructing a Measure

It is rather natural to query whether the development of the preceding section can be reversed. Suppose a function F to be defined on $\mathscr{R}$ with the properties that it is increasing and is continuous from the right at each point. Is it possible, then, to determine a measure function σ on an additive class of subsets of $\mathscr{R}$ of such a nature that F is its distribution function? The answer, very simply, is that such a set function does indeed exist and that its construction is quite straightforward. Actually, one has only to recall the definition of the Lebesgue measure μ of Chapter 15. With scarcely any modification, the development there may be applied to the more general measure concept.

Suppose F, defined on $\mathscr{R}$, to be an increasing function which is continuous from the right at each point. Let $I = (a,b)$ be any nonempty bounded open interval and associate with it the non-negative real number $\delta I = F(b-) - F(a)$. Further, define $\delta\varnothing = 0$. It is clear that for the special function F defined by $F(x) = x$ the number δI is the length of the interval I, denoted, in Chapter 15, by ΔI. It is apparent, then, that the construction of a general σ measure is to be based on a generalization of the concept of length of an interval.

Theorems 15-1 through 15-5 were fundamental to the development of the concept of measure. It is a simple matter to review statements and proofs of these theorems and to observe that all remains valid if ΔI is replaced by δI throughout.

Having thus established the basic properties of δI, we define the *outer σ-measure* σ^* of any bounded set E by

$$\sigma^*(E) = \inf\left\{\sum_\nu \delta I_\nu ;\quad E \subset \bigcup_\nu I_\nu\right\}.$$

Statements corresponding to Theorems 15-6 through 15-10 are valid for $\sigma^*(E)$. Proofs remain the same after replacement of symbols Δ and μ^* by δ and σ^*, respectively.

Next we define the notion of *inner σ-measure* $\sigma *$ in what should now be the obvious way; that is, if E is any bounded set, then

$$\sigma_*(E) = \sup \{\sigma^*(F); \quad F \text{ closed}, F \subset E\}.$$

It is plain that Theorems 15-11 through 15-15, with obvious modifications of notation, describe characteristic properties of inner σ-measure.

Finally, we define the *σ-measure* of a bounded set E to be the common value $\sigma(E)$ of its outer and inner σ-measures, if these be equal; and refer to E, when this is the case, as *σ-measurable.*

The extension of the definition of σ to unbounded sets follows the procedure in the special instance of measure μ.

The theorems of Section 15-6, which state essential properties of μ, may be restated and proved for σ. We conclude the following characteristics for the general σ-measure based on F which, by virtue of the definition of δI, is a distribution function corresponding to σ.

(i) The collection of all σ-measurable sets is an additive class.
(ii) Every interval is σ-measurable; hence each Borel set is σ-measurable.
(iii) σ is additive, that is, $\sigma(\bigcup_{\nu=1}^{\infty} E_\nu) = \sum_{\nu=1}^{\infty} \sigma(E_\nu)$ if $E_1, E_2, \ldots, E_\nu, \ldots$ are disjoint σ-measurable sets.
(iv) $\sigma(E) < \infty$ if E is bounded.
(v) σ is complete.

We conclude that σ is, indeed, a measure function and refer to it as the *Lebesgue–Stieltjes measure* induced or determined by F.

Example 17-6 It is well to note some properties of ordinary Lebesgue measure which do not hold for a general σ-measure. For one thing, the set $\mathscr{R}$ itself need not have infinite measure. Suppose, for example, that F is defined by $F(x) = 0$ if $x < 0$, $F(x) = x^2$ if $0 \leq x < 1$, and $F(x) = 2$ if $1 \leq x$. It follows that $\sigma(\mathscr{R}) = \lim_{\infty} F(x) - \lim_{-\infty} F(x) = 2$.

It is also the case for this distribution function that the corresponding σ-measure of the set of nonpositive real numbers is zero. As a consequence, all subsets of $(-\infty,0]$ also possess σ-measure zero. Since subsets of $(-\infty,0]$ exist which do not possess a Lebesgue measure, it is clear that sets which are nonmeasurable need not be non-σ-measurable when $\sigma \neq \mu$.

Remarks made in Section 15-8 do not apply to arbitrary σ-measure. It is easy to see, in fact, that the σ-measure induced by the function F of Example 17-6 is not invariant under translation.

The foregoing has been little more than a sketch showing how an increasing function which is everywhere continuous from the right may be used to induce a measure function on an additive class of subsets of $\mathscr{R}$. Details of proof have been omitted to avoid being repetitious—the reader should assure himself that the arguments of Chapter 15 do, indeed, apply in the general case. The next several sections will be devoted to the study of an even more comprehensive technique for constructing a measure function. Among other things we shall see that not all measure functions can be obtained by the procedure of the present section.

Exercises

17-9 (a) Prove that the collection of sets $\mathscr{F}^{\bar{\sigma}}$ as defined in Section 17-5 is an additive class. (b) Verify that the function $\bar{\sigma}$ is a measure function on $\mathscr{F}^{\bar{\sigma}}$.

17-10 Prove that if F_k is a distribution function corresponding to a measure function σ, then $F_{k_1}(x) - F_{k_2}(x) = \sigma((k_1, k_2])$ if $k_1 < k_2$.

17-11 Verify that Theorems 15-1 through 15-5 remain valid with ΔI replaced by δI.

17-12 State and prove facts for σ^* paralleling those for μ^* in Theorems 15-6 through 15-10.

17-13 State and prove facts for σ_* paralleling those for μ_* in Theorems 15-11 through 15-15.

17-14 Verify that statements (i) through (v) of Section 17-7 are valid for a general measure σ.

17-8 Outer Measure

A set function σ^* defined on the additive class consisting of all subsets of $\mathscr{R}$ is called an *outer measure* if it has the following properties.

(4)
(i) $\sigma^*(\varnothing) = 0$;
(ii) $E_1 \subset E_2$ implies $\sigma^*(E_1) \le \sigma^*(E_2)$;
(iii) $\sigma^*(\bigcup_{\nu=1}^{\infty} E_\nu) \le \sum_{\nu=1}^{\infty} \sigma^*(E_\nu)$.

Thus an outer measure is an increasing, subadditive set function which takes on the value zero on the empty set. That σ^* is finitely subadditive, i.e.,

$$\sigma^*\left(\bigcup_{\nu=1}^{n} E_\nu\right) \le \sum_{\nu=1}^{n} \sigma^*(E_\nu), \tag{5}$$

follows from (i) and (iii). It is *not* to be inferred that equality holds either in (iii) or in (5) should the sets E_ν be disjoint, as in the case of a measure function.

A set E is to be called *measurable with respect to* σ^*, or σ^**-measurable*, if for any subset A of $\mathscr{R}$ it is the case that

$$\sigma^*(A) = \sigma^*(A \cap E) + \sigma^*(A \cap \mathscr{C}E). \tag{6}$$

Note here that the outer measure of E itself is not involved in the definition but rather the manner in which E and its complement intersect all other subsets of $\mathscr{R}$. It is an immediate consequence of (5) and the simple set relation $A = (A \cap E) \cup (A \cap \mathscr{C}E)$ that a set is σ^*-measurable if and only if $\sigma^*(A) \ge \sigma^*(A \cap E) + \sigma^*(A \cap \mathscr{C}E)$ for all subsets A of $\mathscr{R}$.

The reader may have noticed a departure from a rather natural terminology. While each subset of $\mathscr{R}$ possesses a σ^**-measure*, only those sets are σ^**-measurable* which meet the requirement of the definition (6). The class of sets which are σ^*-measurable does not, in general, extend to the category of all subsets of $\mathscr{R}$, and, indeed, varies with the construction of σ^* itself. We shall show however, in the course of the present section, that for any σ^* the collection of σ^*-measurable sets is an additive class, and that σ^* is actually a measure function on that class if it has a finite value on each bounded interval. To this end, let σ denote a function defined on the class $\mathscr{F}^\sigma$ of σ^*-measurable sets by $\sigma(E) = \sigma^*(E)$. What we propose is to show that $\mathscr{F}^\sigma$ is an additive class and that σ is a measure function on it.

Theorem 17-9 If $\mathscr{F}^\sigma$ is the class of all σ^*-measurable sets, then

(a) $\mathscr{R} \in \mathscr{F}^\sigma$,

(b) $E \in \mathscr{F}^\sigma$ implies $\mathscr{C}E \in \mathscr{F}^\sigma$,

(c) $\sigma^*(E) = 0$ implies $E \in \mathscr{F}^\sigma$; in particular, $\varnothing \in \mathscr{F}^\sigma$.

Proof

(a) For any subset A of $\mathscr{R}$ we have $\sigma^*(A \cap \mathscr{R}) + \sigma^*(A \cap \mathscr{C}\mathscr{R}) = \sigma^*(A) + \sigma^*(\varnothing) = \sigma^*(A)$.

(b) This follows directly from the defining relation (6).

(c) $A \cap E \subset E$ and $A \cap \mathscr{C}E \subset A$. Therefore $\sigma^*(A \cap E) + \sigma^*(A \cap \mathscr{C}E) \leq 0 + \sigma^*(A)$, from which E is σ^*-measurable▲

In order to show, in Theorem 17-12, that $\mathscr{F}^\sigma$ is closed under countable unions, we require the preliminary results established in Theorems 17-10 and 17-11.

Theorem 17-10 If $\{E_\nu\}$ is a nonempty countable collection of disjoint members of $\mathscr{F}^\sigma$, then, for any subset A of $\mathscr{R}$, $\sigma^*(A \cap \bigcup_\nu E_\nu) = \sum_\nu \sigma^*(A \cap E_\nu)$.

Proof Consider first the case where $\{E_\nu\}$ is a finite collection, and let the proof be by induction. Applying the defining relation (6) to the set $A \cap \bigcup_{\nu=1}^n E_\nu$, we have

$$\sigma^*\left(A \cap \bigcup_{\nu=1}^n E_\nu\right) = \sigma^*\left(\left(A \cap \bigcup_{\nu=1}^n E_\nu\right) \cap E_n\right) + \sigma^*\left(\left(A \cap \bigcup_{\nu=1}^n E_\nu\right) \cap \mathscr{C}E_n\right)$$

$$= \sigma^*(A \cap E_n) + \sigma^*\left(A \cap \bigcup_{\nu=1}^{n-1} E_\nu\right).$$

Since the theorem is obviously true for $n = 1$, it follows that it is true for any n.

Next, let $E_1, E_2, \ldots, E_\nu, \ldots$ be an infinite collection of disjoint members of $\mathscr{F}^\sigma$. Then, for arbitrary n,

$$\sum_{\nu=1}^n \sigma^*(A \cap E_\nu) = \sigma^*\left(A \cap \bigcup_{\nu=1}^n E_\nu\right) \leq \sigma^*\left(A \cap \bigcup_{\nu=1}^\infty E_\nu\right),$$

whence

$$\sum_{\nu=1}^\infty \sigma^*(A \cap E_\nu) \leq \sigma^*\left(A \cap \bigcup_{\nu=1}^\infty E_\nu\right).$$

Since $A \cap \bigcup_{\nu=1}^\infty E_\nu = \bigcup_{\nu=1}^\infty (A \cap E_\nu)$, it follows, by (iii) of (4), that

$$\sigma^*\left(A \cap \bigcup_{\nu=1}^\infty E_\nu\right) \leq \sum_{\nu=1}^\infty \sigma^*(A \cap E_\nu),$$

and the theorem follows▲

Corollary If $\{E_\nu\}$ is any countable collection of disjoint members of $\mathscr{F}^\sigma$, then $\sigma^*(\bigcup_\nu E_\nu) = \sum_\nu \sigma^*(E_\nu)$.

Theorem 17-11 If $\{E_\nu;\ \ \nu = 1,2,\ldots,n\}$ is a collection of members of $\mathscr{F}^\sigma$, then $\bigcup_{\nu=1}^n E_\nu$ also belongs to $\mathscr{F}^\sigma$.

Proof Again we use induction. Let $\bigcup_{\nu=1}^{n-1} E_\nu = E$ and assume $E \in \mathscr{F}^\sigma$. By the defining relation (6) we have

$$\sigma^*(A \cap \mathscr{C}E) = \sigma^*(A \cap \mathscr{C}E \cap E_n) + \sigma^*(A \cap \mathscr{C}E \cap \mathscr{C}E_n),$$

while from $(A \cap E) \cup (A \cap \mathscr{C}E \cap E_n) = A \cap (E \cup E_n)$ and (iii) of (4) follows

$$\sigma^*(A \cap (E \cup E_n)) \le \sigma^*(A \cap E) + \sigma^*(A \cap \mathscr{C}E \cap E_n).$$

These relations, together with (6), then imply

$$\sigma^*(A) \ge \sigma^*(A \cap (E \cup E_n)) + \sigma^*(A \cap \mathscr{C}(E \cup E_n)),$$

which, in turn, proves the theorem▲

Theorem 17-12 If $\{E_\nu;\ \ \nu = 1,2,\ldots\}$ is an infinite collection of disjoint members of $\mathscr{F}^\sigma$, then $\bigcup_{\nu=1}^\infty E_\nu$ also belongs to $\mathscr{F}^\sigma$.

Proof For any n and for any set A we have, because of the disjoint character of the E_ν,

$$\begin{aligned}\sigma^*(A) &= \sigma^*\left(A \cap \bigcup_{\nu=1}^n E_\nu\right) + \sigma^*\left(A \cap \mathscr{C}\bigcup_{\nu=1}^n E_\nu\right)\\ &= \sum_{\nu=1}^n \sigma^*(A \cap E_\nu) + \sigma^*\left(A \cap \mathscr{C}\bigcup_{\nu=1}^n E_\nu\right).\end{aligned}$$

From $A \cap \mathscr{C}\bigcup_{\nu=1}^\infty E_\nu \subset A \cap \mathscr{C}\bigcup_{\nu=1}^n E_\nu$ we have

$$\sigma^*(A) \ge \sum_{\nu=1}^n \sigma^*(A \cap E_\nu) + \sigma^*\left(A \cap \mathscr{C}\bigcup_{\nu=1}^\infty E_\nu\right).$$

The arbitrary nature of n then implies

$$\begin{aligned}\sigma^*(A) &\ge \sum_{\nu=1}^\infty \sigma^*(A \cap E_\nu) + \sigma^*\left(A \cap \mathscr{C}\bigcup_{\nu=1}^\infty E_\nu\right)\\ &\ge \sigma^*\left(A \cap \bigcup_{\nu=1}^\infty E_\nu\right) + \sigma^*\left(A \cap \mathscr{C}\bigcup_{\nu=1}^\infty E_\nu\right)\blacktriangle\end{aligned}$$

Corollary If $\{E_\nu;\ \ \nu = 1,2,\ldots\}$ is an infinite collection of members of $\mathscr{F}^\sigma$, then $\bigcup_{\nu=1}^\infty E_\nu$ belongs to $\mathscr{F}^\sigma$ as well.

Proof Left as an exercise▲

The requisite properties for the collection $\mathscr{F}^\sigma$ of σ^*-measurable sets to be an additive class have thus been established. Further, if σ is defined on $\mathscr{F}^\sigma$ by

$\sigma(E) = \sigma^*(E)$ and if σ has a finite value on each bounded interval, then σ is a complete measure on $\mathscr{F}^\sigma$. We refer to σ as having been *induced* on $\mathscr{F}^\sigma$ by the outer measure σ^*.

17-9 Inducing Outer Measures

Having seen how an outer measure may be employed to induce a measure, we show next how an outer measure may be derived from a non-negative set function. A collection $\mathscr{M}$ of subsets of $\mathscr{R}$ is called a *countable covering class* if $\varnothing \in \mathscr{M}$ and if, corresponding to any subset A of $\mathscr{R}$, there exists a countable subset of $\mathscr{M}$ which covers A. Let λ denote a non-negative set function defined on a countable covering class $\mathscr{M}$ with the special property that $\lambda(\varnothing) = 0$. Consider then a set function σ^* defined on the collection of all subsets of $\mathscr{R}$ by

$$\sigma^*(A) = \inf\left\{\sum_\nu \lambda(M_\nu);\quad A \subset \bigcup_\nu M_\nu, M_\nu \in \mathscr{M}\right\}. \tag{7}$$

From $\lambda(\varnothing) = 0$ follows $\sigma^*(\varnothing) \le 0$. On the other hand, since $\lambda(M) \ge 0$ for all $M \in \mathscr{M}$, we conclude that σ^* is non-negative. Hence $\sigma^*(\varnothing) = 0$.

Let E_1 and E_2 denote subsets of $\mathscr{R}$ such that $E_1 \subset E_2$. If $E_2 \subset \bigcup_\nu M_\nu$ with $M_\nu \in \mathscr{M}$ for all ν, then certainly $E_1 \subset \bigcup_\nu M_\nu$ and it follows that $\sigma^*(E_1) \le \sigma^*(E_2)$.

Write $E = \bigcup_\nu E_\nu$, where $\{E_\nu;\ \nu = 1,2,\ldots\}$ is any countable collection of subsets of $\mathscr{R}$, and let $\varepsilon > 0$ be given. Corresponding to each ν there exists a countable collection $\{M_{\nu\omega};\ \omega = 1,2,\ldots\}$ of members of $\mathscr{M}$ with the properties $E_\nu \subset \bigcup_\omega M_{\nu\omega}$ and $\sum_\omega \lambda(M_{\nu\omega}) \le \sigma^*(E_\nu) + \varepsilon/2^\nu$. Since $\{M_{\nu\omega};\ \nu = 1,2,\ldots, \omega = 1,2,\ldots\}$ is countable and covers E, we have $\sigma^*(E) \le \sum_\nu \sum_\omega \lambda(M_{\nu\omega}) \le \sum_\nu \sigma^*(E_\nu) + \varepsilon$. Therefore $\sigma^*(\bigcup_\nu E_\nu) \le \sum_\nu \sigma^*(E_\nu)$.

It is thus established that σ^* is an outer measure defined on the collection of all subsets of $\mathscr{R}$. We refer to it as the outer measure *induced by the countable covering class* $\mathscr{M}$ *and the associated set function* λ.

Example 17-7 Suppose $\mathscr{M}$ to be the countable covering class consisting of all bounded open intervals, the empty set included. Define λ on $\mathscr{M}$ by $\lambda((0,1)) = 3$, $\lambda((0,2)) = 2$, and $\lambda(M) = 0$ for all members M of $\mathscr{M}$ other than (0,1) and (0,2). It is easy to see that if A is any subset of $\mathscr{R}$, then $\sigma^*(A) = 0$. The outer measure thus constructed is a trivial one. It serves to point out, however, that the value of $\sigma^*(M)$, where $M \in \mathscr{M}$, is not one and the same with the number $\lambda(M)$.

Example 17-8 Let $\mathscr{M}$ again denote the collection of all bounded open intervals, and define λ by $\lambda(\varnothing) = 0$ and $\lambda((a,b)) = b - a$. Then the outer measure σ^* induced by $\mathscr{M}$ and λ is precisely the outer measure function defined in Chapter 15 and referred to there as the Lebesgue outer measure. In turn, the measure function σ induced by σ^* is ordinary Lebesgue measure.

Example 17-9 For the collection $\mathscr{M}$ of the preceding examples, define λ by $\lambda(\varnothing) = 0$, $\lambda(I) = 1$ if $I \ne \varnothing$, $I \subset (0,1)$, and $\lambda(I) = 2$ if $I \not\subset (0,1)$. It follows for any nonempty set E that $\sigma^*(E) = 1$ if $E \subset (0,1)$, $\sigma^*(E) = 2$ if E is bounded but not a subset of $(0,1)$, and $\sigma^*(E) = \infty$ if E is unbounded.

Suppose $E \neq \varnothing$ and $E \neq \mathscr{R}$. Then $3 \in E$ or $3 \in \mathscr{C}E$; without loss of generality, suppose the former. Let a denote any point of $\mathscr{C}E$ and consider the set $A = \{a,3\}$. Clearly, $\sigma^*(A) = 2$, $\sigma^*(A \cap E) = \sigma^*(\{3\}) = 2$ and $\sigma^*(A \cap \mathscr{C}E) = \sigma^*(\{a\}) \neq 0$. As a consequence, E is not σ^*-measurable. It follows that $\varnothing$ and $\mathscr{R}$ are the only σ^*-measurable sets.

The last example makes it plain that if $\mathscr{M}$ and λ induce an outer measure σ^*, not all members of $\mathscr{M}$ need be σ^*-measurable. The following theorem cites a sufficient condition in this connection.

Theorem 17-13 If $\mathscr{M}$ is a countable covering class, λ a non-negative set function on $\mathscr{M}$ with $\lambda(\varnothing) = 0$, and if for any members M_1 and M_2 of $\mathscr{M}$ it is the case that $\sigma^*(M_1) = \sigma^*(M_1 \cap M_2) + \sigma^*(M_1 \cap \mathscr{C}M_2)$ (σ^* defined by (7)), then $\mathscr{M}$ is a subset of the additive class $\mathscr{F}^\sigma$ of σ^*-measurable sets.

Proof Let $M \in \mathscr{M}$ and $A \subset \mathscr{R}$. Suppose sets $M_1, M_2, \ldots$ of $\mathscr{M}$ to be such that $A \subset \bigcup_\nu M_\nu$. Then from $A \cap M \subset \bigcup_\nu (M_\nu \cap M)$ follows $\sigma^*(A \cap M) \leq \sigma^*(\bigcup_\nu (M_\nu \cap M)) \leq \sum_\nu \sigma^*(M_\nu \cap M)$. Also from $A \cap \mathscr{C}M \subset \bigcup_\nu (M_\nu \cap \mathscr{C}M)$ follows $\sigma^*(A \cap \mathscr{C}M) \leq \sigma^*(\bigcup_\nu (M_\nu \cap \mathscr{C}M)) \leq \sum_\nu \sigma^*(M_\nu \cap \mathscr{C}M)$. These relations imply

$$\begin{aligned}\sigma^*(A \cap M) + \sigma^*(A \cap \mathscr{C}M) &\leq \sum_\nu [\sigma^*(M_\nu \cap M) + \sigma^*(M_\nu \cap \mathscr{C}M)] \\ &\leq \sum_\nu \sigma^*(M_\nu).\end{aligned}$$

From the definition (7) of σ^* it follows that $\sigma^*(M_\nu) \leq \lambda(M_\nu)$. We conclude that $\sigma^*(A \cap M) + \sigma^*(A \cap \mathscr{C}M) \leq \sigma^*(A)$, whence M is σ^*-measurable▲

It should be noted that one cannot replace the hypothesis of the above theorem by a similar one with λ replacing σ^* since $M_1 \cap M_2$ need not be a member of $\mathscr{M}$.

Exercises

17-15 Prove the corollary to Theorem 17-12.

17-16 Prove that if σ^* is defined on subsets of $\mathscr{R}$ by $\sigma^*(\varnothing) = 0$ and $\sigma^*(E) = 1$ if $E \neq \varnothing$, then σ^* is an outer measure but not a measure. What is the class of measurable sets for σ^*?

17-17 With reference to the notations of Section 17-9, let $\mathscr{M}$ be the class of Borel sets, let $\lambda = \mu$, let σ^* be the outer measure induced by $\mathscr{M}$ and λ, and σ the corresponding measure induced on the additive class $\mathscr{F}^\sigma$ by σ^*. Prove that $\mathscr{F}^\sigma$ is exactly the class of Lebesgue measurable sets and that $\sigma = \mu$.

17-18 Discuss the outer measure induced by $\mathscr{M}$ and λ if $\mathscr{M}$ is the collection of sets consisting of $\varnothing$, $\mathscr{R}$, and all sets of one real number each, and if $\lambda(\varnothing) = 0$, $\lambda(\mathscr{R}) = \infty$, and $\lambda(E) = 1$ otherwise.

17-10 Lebesgue-Stieltjes Measure

Suppose F to denote a real valued function defined on $\mathscr{R}$. If F is increasing and if at each point it is continuous from the right, we shall refer to it as a *distribution function*. (The terminology here is borrowed from probability theory where a function having such properties describes the distribution of probabilities. In this connection, the further properties $\lim_{-\infty} F = 0$ and $\lim_{\infty} F = 1$ are essential, but in the context of our discussion we do not impose such restrictions. Continuity from the right is not a universal choice—continuity from the left would do as well.)

Let $\mathscr{M}$ denote the countable covering class comprising all bounded intervals of the type $(a,b]$, together with the empty set. (The use of half-closed intervals is a technical convenience; the advantage over open intervals is that the union of abutting intervals $(a,b]$ and $(b,c]$ is again a member of $\mathscr{M}$.) Define a function λ on $\mathscr{M}$ by $\lambda(\varnothing) = 0$ and $\lambda((a,b]) = F(b) - F(a)$. Then $\mathscr{M}$ and λ induce an outer measure, which we denote by $\sigma_F{}^*$, on the collection of all subsets of $\mathscr{R}$. In turn, $\sigma_F{}^*$ induces a measure σ_F on an additive class $\mathscr{F}^{\sigma_F}$ of subsets of $\mathscr{R}$. We call σ_F a *Lebesgue–Stieltjes measure*. By Theorem 17-9 it is clear that σ_F is complete.

Essential properties of the Lebesgue–Stieltjes measure corresponding to and determined by a distribution function F are established in the ensuing theorems.

Theorem 17-14 If F is a distribution function and $\sigma_F{}^*$ the corresponding outer measure, then $\sigma_F{}^*((a,b]) = F(b) - F(a)$.

Proof Suppose $\{(a_\nu,b_\nu];\ v = 1,2,\ldots\}$ to be such a collection of intervals that $(a,b] \subset \bigcup_\nu (a_\nu,b_\nu]$. For any $\varepsilon > 0$ a point a' exists such that $a < a' < b$ and $F(a') < F(a) + \varepsilon$. Since the collection $\{(a_\nu,b_\nu];\ \nu = 1,2,\ldots\}$ certainly covers $[a',b]$, it follows that a finite subset also has this property. Without prejudice to the argument, we may assume the finite covering to consist of the intervals $(a_1,b_1],\ldots,(a_n,b_n]$, where the arrangement is such that $a_\nu \le b_{\nu+1}$ for $v = 1, 2,\ldots, n-1$. If $b_\nu < a_{\nu+1}$ for some ν, the corresponding interval $(a_\nu,b_\nu]$ can be omitted—the remaining collection will cover $[a',b]$. It follows that we may further assume $a_{\nu+1} \le b_\nu$ for $\nu = 1, 2, \ldots, n-1$. As a consequence, we have

$$\begin{aligned}\sum_\nu \lambda((a_\nu,b_\nu]) &= \sum_{\nu=1}^{n} \lambda((a_\nu,b_\nu]) \\ &\ge F(b_n) - F(a_1) + \sum_{\nu=1}^{n-1} [F(b_\nu) - F(a_{\nu+1})] \\ &\ge F(b_n) - F(a_1) \ge F(b) - F(a) - \varepsilon\end{aligned}$$

and conclude that $\sigma_F{}^*((a,b]) \ge F(b) - F(a)$. Since it is a consequence of the definition of σ^* that $\sigma_F{}^*((a,b]) \le \lambda((a,b]) = F(b) - F(a)$, the theorem has been proved▲

Corollary If $\mathscr{M}$ and $\sigma_F{}^*$ are defined as above and if M is a member of $\mathscr{M}$, then $\sigma_F{}^*(M) = \lambda(M)$.

Consider a set E defined as the union of intervals $(a_1,b_1]$ and $(a_2,b_2]$, where $b_1 \le a_2$. From general properties established for outer measure we have

$$\begin{aligned}\sigma_F{}^*(E) &\le \sigma_F{}^*((a_1,b_1]) + \sigma_F{}^*((a_2,b_2]) \\ &\le F(b_1) - F(a_1) + F(b_2) - F(a_2).\end{aligned}$$

Further, from $(a_1,b_2] = E \cup (b_1,a_2]$ follows

$$\begin{aligned}F(b_2) - F(a_1) &= \sigma_F{}^*((a_1,b_2]) \\ &\le \sigma_F{}^*(E) + \sigma_F{}^*((b_1,a_2]) \\ &\le \sigma_F{}^*(E) + F(a_2) - F(b_1).\end{aligned}$$

These inequalities combine to yield $\sigma_F{}^*(E) = [F(b_1) - F(a_1)] + [F(b_2) - F(a_2)]$ or $\sigma_F{}^*((a_1,b_1] \cup (a_2,b_2]) = \sigma_F{}^*((a_1,b_1]) + \sigma_F{}^*((a_2,b_2])$.

The above arguments are easily extended to show that for any finite set of disjoint intervals $(a_\nu,b_\nu]$, $\nu = 1, 2, \ldots, n$, it is the case that

$$\sigma_F{}^*\left(\bigcup_{\nu=1}^{n} (a_\nu,b_\nu]\right) = \sum_{\nu=1}^{n} \sigma_F{}^*((a_\nu,b_\nu]).$$

The conclusions of the last theorem, together with the properties established immediately above for $\sigma_F{}^*$ may be seen to imply $\sigma_F{}^*(M_1) = \sigma_F{}^*(M_1 \cap M_2) + \sigma_F{}^*(M_1 \cap \mathscr{C}M_2)$ for any members M_1 and M_2 of the countable covering class $\mathscr{M}$. By Theorem 17-13 it follows that all members of $\mathscr{M}$ are $\sigma_F{}^*$ measurable. Further, the collection $\mathscr{F}^{\sigma_F}$ of $\sigma_F{}^*$-measurable sets is an additive class on which a measure σ_F is induced by $\sigma_F{}^*$. We conclude, in particular, that *all Borel sets are Lebesgue–Stieltjes measurable.*

The next two theorems point to the relation between Borel sets and Lebesgue–Stieltjes measurable sets.

Theorem 17-15 If E is a set possessing a Lebesgue–Stieltjes measure σ_F, then a Borel set B exists such that $E \subset B$ and $\sigma_F(E) = \sigma_F(B)$.

Proof Corresponding to each positive integer n, there is a sequence $\{(a_\nu,b_\nu]\}$ with the properties $E \subset \bigcup_\nu (a_\nu,b_\nu]$ and $\sum_\nu \sigma_F{}^*((a_\nu,b_\nu]) \le \sigma_F(E) + 1/n$. Let $B_n = \bigcup_\nu (a_\nu,b_\nu]$. Then $\sigma_F(B_n) \le \sum_\nu \sigma^*((a_\nu,b_\nu])$, whence $\sigma_F(B_n) \le \sigma_F(E) + 1/n$. The set $B = \bigcap_{n=1}^\infty B_n$ is a Borel set satisfying the proposals of the theorem▲

Theorem 17-16 If E is a Lebesgue–Stieltjes measurable set, then a Borel set B and a set Z having Lebesgue–Stieltjes measure zero exist with the property $E = B \cup Z$.

Proof Consider first the case in which E is bounded. Then, for some interval I we have $E \subset I$ and $I = E \cup E_1$, where $E_1 = I - E$ is σ_F-measurable. By the preceding theorem a Borel set B_1 exists such that $E_1 \subset B_1$ and $\sigma_F(B_1) = \sigma_F(E_1)$.

Without loss of generality we may assume, too, that $B_1 \subset I$ and write $I = B \cup B_1$ where $B = I - B_1$. It follows that $B \subset E$ and $\sigma_F(B) = \sigma_F(E)$. Hence the set $Z = E - B$ is of σ_F-measure zero. B and Z are the desired sets.

If E is unbounded, write $E = \bigcup_{\nu=1}^{\infty} D_\nu$, where $D_1 = E \cap [-1,1]$ and $D_\nu = E \cap ([-\nu,\nu] - [-\nu+1,\nu-1])$ for $\nu = 2, 3 \ldots$. Each D_ν is bounded and σ_F-measurable. Therefore $D_\nu = B_\nu \cup Z_\nu$, B_ν a Borel set and Z_ν having σ_F-measure zero. Then $B = \bigcup_{\nu=1}^{\infty} B_\nu$ and $Z = \bigcup_{\nu=1}^{\infty} Z_\nu$ have the desired properties▲

Corollary The measure function σ_F on the additive class $\mathscr{F}^{\sigma_F}$ is the completion of the measure σ_F on the class of Borel sets.

Measure functions and distribution functions may be resolved into components in a rather natural way. Suppose σ_1 and σ_2 to denote measure functions on additive classes $\mathscr{F}_1$ and $\mathscr{F}_2$, respectively, of subsets of $\mathscr{R}$. Let E denote any member of the additive class $\mathscr{F} = \mathscr{F}_1 \cap \mathscr{F}_2$, and define $\sigma(E) = \sigma_1(E) + \sigma_2(E)$. Then σ is easily seen to be a measure function on $\mathscr{F}$; we call it the *sum* of σ_1 and σ_2 and write $\sigma = \sigma_1 + \sigma_2$. If F_1 and F_2 are distribution functions and σ_{F_1} and σ_{F_2} corresponding measure functions, we are interested in the relation between the distribution function $F_1 + F_2$ and the measure $\sigma_{F_1} + \sigma_{F_2}$. Theorem 17-17 is preliminary to the main result in Theorem 17-18.

Theorem 17-17 If F_1 and F_2 are distribution functions, and if E is any subset of $\mathscr{R}$, then

$$\sigma^*_{F_1+F_2}(E) = \sigma_{F_1}{}^*(E) + \sigma_{F_2}{}^*(E). \tag{8}$$

Proof From Theorem 17-14 and its corollary it is clear that (8) is true if $E = (a,b]$. Consider $E \subset \bigcup_\nu (a_\nu,b_\nu]$. Then $\sigma_{F_1}{}^*(E) \le \sum_\nu \sigma_{F_1}{}^*((a_\nu,b_\nu])$ and $\sigma_{F_2}{}^*(E) \le \sum_\nu \sigma_{F_2}{}^*((a_\nu,b_\nu])$, so that

$$\begin{aligned}\sigma_{F_1}{}^*(E) + \sigma_{F_2}{}^*(E) &\le \sum_\nu \{\sigma_{F_1}{}^*((a_\nu,b_\nu]) + \sigma_{F_2}{}^*((a_\nu,b_\nu])\}\\ &= \sum_\nu \sigma^*_{F_1+F_2}((a_\nu,b_\nu]).\end{aligned}$$

This implies

$$\sigma_{F_1}{}^*(E) + \sigma_{F_2}{}^*(E) \le \sigma^*_{F_1+F_2}(E). \tag{9}$$

To prove the reverse inequality let $\varepsilon > 0$ be given. Then disjoint intervals I_ν', $\nu = 1, 2, \ldots$, exist such that $E \subset \bigcup_\nu I_\nu'$ and $\sum_\nu \sigma_{F_1}{}^*(I_\nu') \le \sigma_{F_1}{}^*(E) + \varepsilon$. Similarly, there are disjoint intervals I_λ'', $\lambda = 1, 2, \ldots$, with the properties $E \subset \bigcup_\lambda I_\lambda''$ and $\sum_\lambda \sigma_{F_2}{}^*(I_\lambda'') \le \sigma_{F_2}{}^*(E) + \varepsilon$.

Let $I_{\nu\lambda} = I_\nu' \cap I_\lambda''$, $\nu = 1, 2, \ldots$ and $\lambda = 1, 2, \ldots$. Then $I_{\nu\lambda}$ is either an interval of the type $(a,b]$ or is empty; and any two such intervals are disjoint. From $E \subset \bigcup_{\nu,\lambda} I_{\nu\lambda}$ follows

$$\sigma^*_{F_1+F_2}(E) \le \sum_{\nu,\lambda} \sigma^*_{F_1+F_2}(I_{\nu\lambda}) = \sum_{\nu,\lambda} \sigma_{F_1}{}^*(I_{\nu\lambda}) + \sum_{\nu,\lambda} \sigma_{F_2}{}^*(I_{\nu\lambda}).$$

Since, for each ν, $I_\nu' = \bigcup_\lambda I_{\nu\lambda}$, we have $\sigma_{F_1}{}^*(I_\nu') = \sum_\lambda \sigma_{F_1}{}^*(I_{\nu\lambda})$. Similarly, $\sigma_{F_2}{}^*(I_\lambda'') = \sum_\nu \sigma_{F_2}{}^*(I_{\nu\lambda})$. Therefore

$$\begin{aligned}\sigma^*_{F_1+F_2}(E) &\le \sum_\nu \sigma_{F_1}{}^*(I_\nu') + \sum_\lambda \sigma_{F_2}{}^*(I_\lambda'') \\ &\le \sigma_{F_1}{}^*(E) + \sigma_{F_2}{}^*(E) + 2\varepsilon.\end{aligned}$$

This, with (9), establishes the theorem▲

Corollary If F_1 and F_2 are distribution functions, and if E is a subset of $\mathscr{R}$ which is σ_{F_1}-measurable and σ_{F_2}-measurable, then E is $\sigma_{F_1+F_2}$-measurable, and $\sigma_{F_1+F_2}(E) = \sigma_{F_1}(E) + \sigma_{F_2}(E)$.

Theorem 17-18 If F_1 and F_2 are distribution functions, then the measure function $\sigma_{F_1} + \sigma_{F_2}$ is identical with the Lebesgue–Stieltjes measure $\sigma_{F_1+F_2}$.

Proof Write $\sigma_3 = \sigma_{F_1} + \sigma_{F_2}$. Then E is σ_3-measurable if and only if E is both σ_{F_1}-measurable and σ_{F_2}-measurable. By the corollary to the preceding theorem, if E is σ_3-measurable, then E is $\sigma_{F_1+F_2}$-measurable, and $\sigma_3(E) = \sigma_{F_1+F_2}(E)$.

On the other hand, suppose E to be $\sigma_{F_1+F_2}$-measurable. Then, if A is any set, $\sigma^*_{F_1+F_2}(A) = \sigma^*_{F_1+F_2}(A \cap E) + \sigma^*_{F_1+F_2}(A \cap \mathscr{C}E)$. From Theorem 17-17 we have

$$\begin{aligned}(10)\qquad \sigma_{F_1}{}^*(A) + \sigma_{F_2}{}^*(A) &= \sigma_{F_1}{}^*(A \cap E) + \sigma_{F_2}{}^*(A \cap E) \\ &\quad + \sigma_{F_1}{}^*(A \cap \mathscr{C}E) + \sigma_{F_2}{}^*(A \cap \mathscr{C}E).\end{aligned}$$

Also, for any sets A and E it is the case that

$$(11)\qquad \sigma_{F_1}{}^*(A) \le \sigma_{F_1}{}^*(A \cap E) + \sigma_{F_1}{}^*(A \cap \mathscr{C}E)$$

and

$$(12)\qquad \sigma_{F_2}{}^*(A) \le \sigma_{F_2}{}^*(A \cap E) + \sigma_{F_2}{}^*(A \cap \mathscr{C}E).$$

One observes that (10) is possible only if equality holds in (11) and (12). This implies that E is σ_{F_1}-measurable and σ_{F_2}-measurable. Therefore E is σ_3-measurable▲

It is thus seen that if F_1 and F_2 are given functions on $\mathscr{R}$, both increasing and right continuous, then Lebesgue–Stieltjes measures σ_{F_1} and σ_{F_2} are defined, and their sum $\sigma_{F_1} + \sigma_{F_2}$ is the Lebesgue–Stieltjes measure defined by $F_1 + F_2$.

An instance of the resolution of a measure $\sigma_{F_1+F_2}$ into a sum of measures σ_{F_1} and σ_{F_2} lies in the decomposition of a general increasing right continuous function. As we have seen, if F is such a function, then $F = F_1 + F_2 + F_3$, where F_1 is a step function, F_2 is absolutely continuous, and F_3 is continuous with a zero derivative almost everywhere. Each of the increasing functions F, F_1, F_2, and F_3 defines a measure, and, while the categories of sets which are measurable by the respective measures are not necessarily the same, the class $\mathscr{B}$

of Borel sets is common to all four categories. By Theorem 17-20, if σ_F, σ_{F_1}, σ_{F_2}, and σ_{F3} are all regarded on $\mathscr{B}$, then $\sigma_F = \sigma_{F_1} + \sigma_{F_2} + \sigma_{F_3}$.

A few remarks regarding the correspondence between F and σ_F are in order. Suppose F to be increasing and right continuous on $\mathscr{R}$. Then the measure σ_F defined by F is such that all Borel sets are measurable, and we may employ the construction of Section 17-6 to define an increasing right continuous function F_1 based on σ_F. The functions F and F_1 differ by a constant, whence for any $a < b$ we have $F(b) - F(a) = F_1(b) - F_1(a)$. From this it follows that the measure function σ_{F_1}, defined by F_1, is identical with σ_F.

17-11 Non-Lebesgue-Stieltjes Measures

The preceding remarks are meaningful only if one starts with a distribution function F, for if a measure function σ be given, it need not be the case that a corresponding distribution function exists. We have seen in Example 17-9 an instance of a measure defined on something less than the class of Borel sets; as such it is not a Lebesgue–Stieltjes measure. It is the purpose of the present section to present a nontrivial example of a measure function which is not a Lebesgue–Stieltjes measure.

Let C denote a non-Lebesgue-measurable set and define a countable covering class $\mathscr{M}$ by

$$\mathscr{M} = \{C \cap B; \quad B \text{ a Borel set}\} \cup \{\mathscr{C}C \cap B; \quad B \text{ a Borel set}\};$$

that is, $\mathscr{M}$ comprises all possible intersections of C and $\mathscr{C}C$ with Borel sets. Denoting outer Lebesgue measure by μ^*, let λ be a function defined on $\mathscr{M}$ by $\lambda(M) = \mu^*(M)$, $M \in \mathscr{M}$. Then $\mathscr{M}$ and λ induce an outer measure σ^* on the class of all subsets of $\mathscr{R}$ and, in turn, σ^* induces a measure σ on an additive class $\mathscr{F}^\sigma$.

To demonstrate that $B \subset \mathscr{F}^\sigma$, that is, the σ-measure thus constructed is defined for all Borel sets, we have only to show that if A is any set and B any Borel set, then $\sigma^*(A \cap B) + \sigma^*(A \cap \mathscr{C}B) \leq \sigma^*(A)$. To see that this is the case, let $\varepsilon > 0$ be given. Then a sequence $\{M_\nu\}$ of members of $\mathscr{M}$ exists such that $A \subset \bigcup_\nu M_\nu$ and $\sum_\nu \mu^*(M_\nu) \leq \sigma^*(A) + \varepsilon$. Since $A \cap B \subset \bigcup_\nu (M_\nu \cap B)$ and $M_\nu \cap B \in \mathscr{M}$, we have $\sigma^*(A \cap B) \leq \sum_\nu \mu^*(M_\nu \cap B)$. Similarly, $\sigma^*(A \cap \mathscr{C}B) \leq \sum_\nu \mu^*(M_\nu \cap \mathscr{C}B)$. Because B is Lebesgue measurable it follows that $\mu^*(M_\nu) = \mu^*(M_\nu \cap B) + \mu^*(M_\nu \cap \mathscr{C}B)$ for any set M_ν of $\mathscr{M}$. Therefore, $\sigma^*(A \cap B) + \sigma^*(A \cap \mathscr{C}B) \leq \sum_\nu \mu^*(M_\nu) \leq \sigma^*(A) + \varepsilon$ and we conclude that any Borel set B is measurable.

It is also the case that the set C, while non-Lebesgue-measurable, is σ-measurable. To see this let A and $\{M_\nu\}$ be as above. Consider subsets $\mathscr{M}_1$ and $\mathscr{M}_2$ of $\mathscr{M}$ such that the first consists of all sets M_ν of the form $C \cap B$ and the second all those of the form $\mathscr{C}C \cap B$. Then $C \cap A \subset \bigcup_{M_\nu \in \mathscr{M}_1} M_\nu$ and $\mathscr{C}C \cap A \subset \bigcup_{M_\nu \in \mathscr{M}_2} M_\nu$, whence $\sigma^*(C \cap A) \leq \sum_{M_\nu \in \mathscr{M}_1} \mu^*(M_\nu)$, and $\sigma^*(\mathscr{C}C \cap A) \leq \sum_{M_\nu \in \mathscr{M}_2} \mu^*(M_\nu)$. The σ-measurability of C then follows from $\sigma^*(C \cap A) + \sigma^*(\mathscr{C}C \cap A) \leq \sum_\nu \mu^*(M_\nu) \leq \sigma^*(A) + \varepsilon$.

A comparison of the outer measures μ^* and σ^* follows easily. If A is any set, then $A \subset \bigcup_\nu M_\nu$ where $\sum_\nu \mu^*(M_\nu) \leq \sigma^*(A) + \varepsilon$. Hence, from $\mu^*(A) \leq \mu^*(\bigcup_\nu M_\nu) \leq \sum_\nu \mu^*(M_\nu)$, we conclude $\mu^*(A) \leq \sigma^*(A)$.

Non-Lebesgue-measurability of C implies, of course, that $\mu^*(C) > 0$. As a consequence, it is also the case that $\sigma^*(C) > 0$. If I is a bounded interval, we have $I \subset (C \cap I) \cup (\mathscr{C}C \cap I)$. Therefore $\sigma^*(I) \leq \mu^*(C \cap I) + \mu^*(\mathscr{C}C \cap I) \leq 2\mu(I)$. We conclude that $\sigma^*(I)$ is finite; further, if A is any bounded set, then $\sigma^*(A) < \infty$.

If B is a Borel set such that $B \subset C$, consider $C - B = E$. If $\sigma^*(E) = 0$, then $\mu^*(E) = 0$, whence E is Lebesgue-measurable. It follows then from $C = B \cup E$ that C is Lebesgue-measurable, which is contradictory to the original hypothesis concerning C. We are forced to the conclusion that $\mu^*(C - B) > 0$, that is, no Borel subset of C has the same outer Lebesgue measure as that set. As a consequence of Theorem 17-16, σ is not a Lebesgue–Stieltjes measure.

Let F denote the distribution function corresponding to the measure σ as defined above on $\mathscr{F}^\sigma$; as we have seen, the function σ on $\mathscr{B}$ is sufficient to define F. From F we obtain a Lebesgue–Stieltjes measure σ_1 defined on an additive class $\mathscr{F}^{\sigma_1}$ consisting of sets of the form $B \cup E$, where B is a Borel set and $E \subset B_1$, where B_1 is a Borel set such that $\sigma^*(B_1) = 0$. It follows that C is not a member of the class $\mathscr{F}^{\sigma_1}$, whence σ and σ_1 are not the same.

Exercises

17-19 If F and σ_F are corresponding distribution function and measure, prove that $\sigma_F((a,b)) = F(b-) - F(a)$, $\sigma_F([a,b)) = F(b-) - F(a-)$, $\sigma_F([a,b]) = F(b) - F(a-)$, and $\sigma_F(\{a\}) = F(a) - F(a-)$.

17-20 Cite an example of a distribution function F and corresponding measure σ_F with the property $\sigma_F((a,b)) < F(b) - F(a) < \sigma_F([a,b])$.

17-21 Show that it is possible for a distribution function F to have the property that $F(b) - F(a)$ may exceed or be less than $\sigma_F([a,b))$ depending on the values of a and b.

17-12 The Lebesgue-Stieltjes Integral

It is possible to base an extension of the concept of Lebesgue integral on a general measure function. It is convenient, however, and more in keeping with applications, to restrict attention to Lebesgue–Stieltjes measures. Accordingly, we consider a distribution function F and the measure function σ induced by F on an additive class $\mathscr{F}^\sigma$ of subsets of $\mathscr{R}$.

Suppose E to be a member of $\mathscr{F}^\sigma$. Then by a *partition* (or $\mathscr{F}^\sigma$*-partition*) $\{E_\nu\}$ of E we mean a finite collection of disjoint members of $\mathscr{F}^\sigma$ whose union is E. We shall compare partitions and speak of refinements exactly as in the less general Lebesgue case of Chapter 16.

Consider a bounded function g defined on a domain E such that $E \in \mathscr{F}^\sigma$ and $\sigma(E) < \infty$. As usual, let m and M denote infimum and supremum, respectively, of g on E, and m_ν and M_ν similar quantities on E_ν. We define *lower* and *upper Lebesgue–Stieltjes sums of g with respect to F (or σ) and the partition* $\{E_\nu\}$ to be the numbers

$$\underline{LS}(g,\{E_\nu\}) = \sum_{\nu=1}^{n} m_\nu \sigma(E_\nu) \quad \text{and} \quad \overline{LS}(g,\{E_\nu\}) = \sum_{\nu=1}^{n} M_\nu \sigma(E_\nu),$$

respectively. It is immediate that

$$m\sigma(E) \le \underline{LS}(g,\{E_\nu\}) \le \overline{LS}(g,\{E_\nu\}) \le M\sigma(E).$$

Further, if $\{E_\nu\}$ and $\{E_\kappa'\}$ are any two partitions of E, then

$$\underline{LS}(g,\{E_\nu\}) \le \overline{LS}(g,\{E_\kappa'\}). \tag{13}$$

Next we define *lower* and *upper Lebesgue–Stieltjes integrals of g with respect to F (or σ) on E* to be $\sup\{\underline{LS}(g,\{E_\nu\})\}$ and $\inf\{\overline{LS}(g,\{E_\nu\})\}$, respectively, the $\{E_\nu\}$ varying over all partitions of E. Because of (13) it is the case that

$$\sup\{\underline{LS}(g,\{E_\nu\})\} \le \inf\{\overline{LS}(g,\{E_\nu\})\},$$

that is to say, the lower integral does not exceed the upper integral.

If the upper and lower Lebesgue–Stieltjes integrals of g on E are equal, we shall say that g is *Lebesgue–Stieltjes integrable with respect to F (or σ)* on E and that the common value of upper and lower integrals is the *Lebesgue–Stieltjes integral of g with respect to F (or σ) on E.* We denote the integral by either of two symbols $\mathcal{L}\!\!\int_E g(x)\,dF(x)$ and $\mathcal{L}\!\!\int_E g(x)\,d\sigma(E)$ or, for brevity, $\mathcal{L}\!\!\int_E g\,dF$ and $\mathcal{L}\!\!\int_E g\,d\sigma$.

Example 17-10 The case $F(x) = x$ is a natural one to consider. Clearly, in this instance $\sigma = \mu$ (Lebesgue measure), and the Lebesgue–Stieltjes integral $\mathcal{L}\!\!\int_E g(x)\,dF(x)$ reduces to the Lebesgue integral $\mathcal{L}\!\!\int_E g(x)\,dx$.

Example 17-11 Suppose g to be the familiar Ψ function on $E = [0,1]$ and let the distribution function F be defined by $F(x) = 0$ if $x < 0$, and $F(x) = x^2$ if $x \ge 0$. (That E is in the class $\mathscr{F}^\sigma$ on which the induced measure σ is defined is clear since $\mathscr{F}^\sigma$ contains all Borel sets.)

As for the existence of the Lebesgue–Stieltjes integral of Ψ with respect to F, we again consider the partition $\{E_\nu\}$ where $\nu = r$ and $\nu = i$, and E_r is the set of rationals in E and E_i the set of irrationals. Because F is continuous, the induced measure σ of a single point is zero. Hence $\sigma(E_r) = 0$. Further, from $\sigma(E) = F(1) - F(0) = 1$ we get $\sigma(E_i) = 1$. Both lower and upper integrals, therefore, have the value 1; hence $\mathcal{L}\!\!\int_{[0,1]} \Psi\,dF = \mathcal{L}\!\!\int_{[0,1]} \Psi\,dx^2 = 1$.

In the event g is defined on an interval $[a,b]$ it is instructive to compare the Lebesgue–Stieltjes integral $\mathcal{L}\!\!\int_{[a,b]} g\,dF$ with the Riemann–Stieltjes integral $\mathcal{R}\!\!\int_a^b g\,dF$ of a bounded function g with respect to an increasing integrator F. The difference between the two is simply one of generality in the notion of partition. In the Riemann–Stieltjes case partitions are collections of intervals, in the Lebesgue–Stieltjes case collections of σ-measurable sets. It follows that the lower and upper Lebesgue–Stieltjes integrals lie between the lower and upper Riemann–Stieltjes integrals. As a consequence, if $\int_a^b g\,dF$ exists, then so does $\int_{[a,b]} g\,dF$. Existence of $\int_{[a,b]} g\,dF$, however, does not imply that of $\int_a^b g\,dF$. The relationship between the two is reminiscent of that between the ordinary Riemann integral and the Lebesgue integral for a function on $[a,b]$. The reader may show, for example, that the Riemann–Stieltjes integral of Ψ with respect to the integrator F of Example 17-11 does not exist.

Extension of the definition of the Lebesgue–Stieltjes integral to cover the case of an unbounded function g and of a domain E having infinite measure follows precisely the pattern in the less general Lebesgue case. Suppose, first, that g is non-negative on E. Then for any real b we define $g^{[b]}$ on E by

$$g^{[b]}(x) = \begin{cases} g(x) & \text{if } x \in E \text{ and } g(x) \le b, \\ 0 & \text{if } x \in E \text{ and } g(x) > b. \end{cases}$$

We shall say that $\int_E g\,dF$ exists and that its value is $\lim_{b \to \infty} \int_{E \cap [-b,b]} g^{[b]}\,dF$ if $g^{[b]}$ is Lebesgue–Stieltjes integrable on $E \cap [-b,b]$ for each b. The indicated limit may be finite or infinite; if it is finite we describe g as *summable with respect to F on E.*

Let g be arbitrary, that is, relax the restriction to non-negativity assumed above. Consider that $g = g^+ - g^-$. If the two integrals $\int_E g^+\,dF$ and $\int_E g^-\,dF$ exist and if at least one is summable, we define $\int_E g\,dF = \int_E g^+\,dF - \int_E g^-\,dF$.

Because of the manner in which the Lebesgue–Stieltjes integral serves to generalize the Lebesgue integral, many properties of the latter are equally true of the former, with modifications reflecting the generality of the measure function σ (or distribution function F). Proofs essentially repeat earlier arguments and are left to the reader.

17-13 Measurable Functions

The definition of *measurable function* as introduced in Chapter 15 reflects an intimate connection with the notion of Lebesgue measure. Basically, however, the importance of the concept is derived from the fact that measurable sets constitute an additive class. Hence the following general definition.

Let g be a real valued function defined on a subset S of $\mathscr{R}$, and let $\mathscr{F}$ denote an additive class of subsets of $\mathscr{R}$. Then g is called *measurable relative to $\mathscr{F}$* if, corresponding to each real number k, the set $\{x;\ \ x \in S, g(x) \le k\}$ is a member of $\mathscr{F}$. It is clear that if $\mathscr{F}$ is the class of Lebesgue-measurable sets, then measurability of g relative to $\mathscr{F}$ means measurability of g in the earlier sense.

Example 17-12

(a) If $\mathscr{F} = \{\varnothing, \mathscr{R}\}$, only constant functions on $\mathscr{R}$ are measurable relative to $\mathscr{F}$.

(b) If $\mathscr{F} = \{\varnothing, \{x;\ \ x < 0\}, \{x;\ \ x \ge 0\}, \mathscr{R}\}$, then the function g defined on $\mathscr{R}$

by $g(x) = a$ if $x < 0$ and $g(x) = b$ if $x \geq 0$ (discontinuous if $a \neq b$) is measurable relative to $\mathscr{F}$. However, the continuous function $g(x) = x$ on $\mathscr{R}$ is not measurable relative to $\mathscr{F}$.

Because the present concept generalizes in a rather simple manner the earlier notion of measurable function, many properties established for the latter carry over, with or without modification. Without proofs, therefore, we reiterate some of the facts. The reader should recall properties enunciated in Sections 15-12 and 15-13.

If g is a measurable function relative to an additive class $\mathscr{F}$, then the domain of g is a member of $\mathscr{F}$. Further if g has domain S and if S_1 is a subset of S which is also a member of $\mathscr{F}$, then the function defined as the restriction of g to S_1 is also measurable relative to $\mathscr{F}$.

Each of the following statements is necessary and sufficient to measurability of g relative to $\mathscr{F}$: (i) for each k, $\{x;\ \ x \in S, g(x) < k\} \in \mathscr{F}$; (ii) for each k, $\{x;\ \ x \in S, g(x) \geq k\} \in \mathscr{F}$; (iii) for each k, $\{x;\ \ x \in S, g(x) > k\} \in \mathscr{F}$. It is also the case that if g is measurable relative to $\mathscr{F}$ and if I is any interval, then the set $\{x;\ \ x \in S, g(x) \in I\}$ is a member of the additive class $\mathscr{F}$.

We have seen (Theorem 15-42) that a continuous function defined on a measurable set is measurable (relative to class of Lebesgue-measurable sets). In the general sense, however, this is not the case. A continuous function is not automatically measurable relative to an additive class which contains its domain. Example 17-12 illustrates this point.

If f is a continuous function on $\mathscr{R}$ and if g is measurable relative to $\mathscr{F}$, then the function $f \circ g$ is measurable relative to $\mathscr{F}$. (See Theorem 15-13.) In particular, the functions $g + c$, cg (c a constant), g^2, and $|g|$ have this property. Moreover, if f and g are defined on a common domain and are measurable relative to one and the same additive class $\mathscr{F}$, then functions $f + g$ and fg are measurable relative to $\mathscr{F}$.

With respect to sequences of functions, the same statements valid for Lebesgue measurable functions are valid in the general sense. Thus, if $\{g_n\}$ is a uniformly bounded sequence of functions, each measurable relative to $\mathscr{F}$, then $\sup\{g_n\}$, $\inf\{g_n\}$, $\overline{\lim}\{g_n\}$, $\underline{\lim}\{g_n\}$, and $\lim\{g_n\}$, if it exists, are all measurable relative to $\mathscr{F}$. Further, if g is measurable relative to $\mathscr{F}$, then a sequence of simple functions, each measurable relative to $\mathscr{F}$, exists which converges to g.

The connection between Lebesgue–Stieltjes integrability and measurability relative to an additive class $\mathscr{F}^\sigma$ is essentially that between Lebesgue integrability and measurability with respect to μ. A bounded function g on a domain E of finite σ-measure is measurable relative to $\mathscr{F}^\sigma$ if and only if g is Lebesgue–Stieltjes integrable with respect to σ on E. If g is any function which possesses a Lebesgue–Stieltjes integral with respect to a measure function σ on a domain E of $\mathscr{F}^\sigma$, then g is measurable relative to $\mathscr{F}^\sigma$. Finally, if σ is a measure function on $\mathscr{F}^\sigma$ and if g is a measurable function relative to $\mathscr{F}^\sigma$ on a domain E belonging to $\mathscr{F}^\sigma$, then g is Lebesgue–Stieltjes integrable on E with respect to σ provided a function h exists which is Lebesgue–Stieltjes summable with respect to σ on E and is such that $|g| \leq h$ almost everywhere (σ).

17-14 Absolutely Continuous Distribution Functions

In the present section certain relationships are noted between general Lebesgue–Stieltjes measure and integration on the one hand, and the less general Lebesgue measure and integration on the other, in the event of an absolutely continuous distribution function. Theorems 17-19 through 17-21 are preliminary to the important result in Theorem 17-22.

Theorem 17-19 If F is an absolutely continuous distribution function and if $C = \bigcup_\nu I_\nu$ is a countable union of disjoint intervals I_ν, then $\oint_C dF = \oint_C F'$ (the integral on the left being, of course, of Lebesgue–Stieltjes type and that on the right Lebesgue).

Proof If I is an interval with endpoints a and b, then, since F is continuous, we have $\oint_I dF = F(b) - F(a)$. Further, by Theorem 16-35 on the Lebesgue integral, the absolute continuity of F implies $\oint_a^b F' = F(b) - F(a)$. As a consequence $\oint_I dF = \oint_I F'$. The same result is forthcoming if I is unbounded, simply by considering limits. Finally, $\oint_C dF = \sum_\nu \oint_{I_\nu} dF = \sum_\nu \oint_{I_\nu} F' = \oint_C F'$▲

Theorem 17-20 Let F denote an absolutely continuous distribution function, σ_F the Lebesgue–Stieltjes measure induced by F, and μ ordinary Lebesgue measure. If E is a Lebesgue measurable set with $\mu(E) = 0$, then E is σ_F-measurable, and $\sigma_F(E) = 0$. On the other hand, if E is σ_F-measurable with $\sigma_F(E) = 0$, then $F'(x) = 0$ almost everywhere (μ) on E.

Proof Let $C = \bigcup_\nu I_\nu$ be a countable union of disjoint intervals $I_\nu = (a_\nu, b_\nu)$, and let $\varepsilon > 0$ be given. Then, from the absolute continuity of F follows the existence of $\delta > 0$ such that $\sum_\nu |F(b_\nu) - F(a_\nu)| < \varepsilon$ if $\sum_\nu |b_\nu - a_\nu| < \delta$. Thus $\sigma_F(C) < \varepsilon$ if $\mu(C) < \delta$.

If $\mu(E) = 0$, then such a set as C above exists with the property $E \subset C$ and $\mu(C) < \delta$. Hence $\sigma_F(E) < \varepsilon$ for all $\varepsilon < 0$ and we conclude $\sigma_F(E) = 0$.

Suppose now that $\sigma_F(E) = 0$. Then corresponding to each integer n there exists a countable union C_n of disjoint intervals such that $E \subset C_n$ and $\sigma_F(C_n) < 1/n$. Then, by Theorem 17-19, $0 \leq \oint_{C_n} F' < 1/n$, and it follows that $\oint_C F' = 0$, where $C = \bigcap_n C_n$. Therefore $F'(x) = 0$ almost everywhere (μ) on C. We conclude that $F'(x) = 0$ almost everywhere (μ) on E, since $E \subset C$▲

Theorem 17-21 If F is an absolutely continuous distribution function and if E is a Lebesgue measurable set, then E is also σ_F-measurable. Further, if g is a Lebesgue measurable function, then it is also measurable relative to σ_F.

Proof If E is a Lebesgue measurable set, then $E = B \cup Z$, where B is a Borel set and Z a set such that $\mu(Z) = 0$. We have seen that all Borel sets are contained in the additive class $\mathscr{F}^{\sigma_F}$ of σ_F-measurable sets. Moreover, because of Theorem 17-20, Z also belongs to $\mathscr{F}^{\sigma_F}$ since $\sigma_F(Z) = 0$. Hence E is σ_F-measurable.

Further, if g is a Lebesgue measurable function and k any real number, then the set $E = \{x;\quad g(x) \leq k\}$ is Lebesgue measurable. Therefore g is measurable relative to σ_F since E is also σ_F-measurable▲

Theorem 17-22 If F is an absolutely continuous distribution function and E any Lebesgue measurable set, then $\fint_E dF = \fint_E F'$.

Proof By the preceding theorem $\fint_E dF$ exists since its value is $\sigma_F(E)$. Assume E to be bounded. Then $E \subset I$ for some bounded open interval I, and $\fint_E dF \leq \sigma_F(I) < \infty$. Also $\fint_E F' \leq \fint_I F' = \fint_I dF = \sigma_F(I) < \infty$.

Corresponding to each positive integer n there exists an open set G_n such that $E \subset G_n \subset I$ and $\mu(G_n - E) < 1/n$. Let $G = \bigcap_n G_n$. Then $E \subset G$ and $\mu(G - E) = 0$. Therefore $\sigma_F(G - E) = 0$, by Theorem 17-20.

Denote by K_n the characteristic function of the set G_n, for each n, and by K that of G. Then sequences $\{K_n\}$ and $\{K_n F'\}$ converge to K and KF', respectively. It follows that $\lim \{\fint_{G_n} F'\} = \lim \{\fint_I K_n F'\} = \fint_G F' = \fint_E F'$ and $\lim \{\fint_{G_n} dF\} = \lim \{\fint_I K_n\, dF\} = \fint_G dF = \fint_E dF$. Since $\fint_{G_n} F' = \fint_{G_n} dF$ by Theorem 17-19, we conclude the truth of Theorem 17-22 in the case E is bounded. The case in which E is unbounded is left to the reader▲

Theorems 17-19 through 17-22 have established the equality on a Lebesgue measurable set E of two integrals: (i) the Lebesgue–Stieltjes integral of the unit function on E with respect to an absolutely continuous distribution function F, and (ii) the Lebesgue integral of the derivative of the same distribution function F on E. We show in the following sequence of theorems how this may be generalized, again subject to absolute continuity of F, to the case of the Lebesgue–Stieltjes integral of an arbitrary function g relative to F and the Lebesgue integral of gF'. Theorems 17-23 and 17-24 are preliminary to the principal result in Theorem 17-25.

Theorem 17-23 If $Z = \{x;\quad F'(x) = 0\}$ where F is an absolutely continuous distribution function and if A is a σ_F-measurable set, then the set $A - Z$ is Lebesgue measurable.

Proof The set Z is Lebesgue measurable since F' is a Lebesgue measurable function. By Theorem 17-21, Z is therefore σ_F-measurable and, as a consequence, so is $A - Z$. It follows that $A - Z = B \cup C$, where B is a Borel set and $\sigma_F(C) = 0$. By Theorem 17-20, $\sigma_F(C) = 0$ implies $C = D \cup E$ where D is such that F' vanishes on it, and where $\mu(E) = 0$. Since $A - Z = B \cup D \cup E$ and since $F' \neq 0$ on $A - Z$, it follows that $D = \varnothing$. Therefore $A - Z = B \cup E$▲

Theorem 17-24 If F is an absolutely continuous distribution function, E a Lebesgue measurable set, and g a function defined on E, then g is σ_F-measurable if and only if gF' is Lebesgue measurable.

Proof Let $E^* = E - Z$, where $Z = \{x;\quad F'(x) = 0\}$. By Theorem 17-21 we see that the set E^* is σ_F-measurable. Consider the set $A = \{x;\quad x \in E^*, g(x) \leq k\}$

for any real number k, and suppose g to be a σ_F-measurable function. It follows that A is a σ_F-measurable set, whence $A - Z$ is a Lebesgue measurable set by Theorem 17-23. However, $A - Z = A$ since $A \subset E^*$. Therefore A is Lebesgue measurable and, as a consequence, g is a Lebesgue measurable function on E^*. From the absolute continuity of F then follows Lebesgue measurability of the function gF' on E^*. Finally, since $gF' = 0$ on $Z = E - E^*$, we conclude that gF' is a Lebesgue measurable function on E.

For the converse, assume gF' to be Lebesgue measurable on E. Then gF' is Lebesgue measurable on E^*. Note, too, that the function $1/F'$ is Lebesgue measurable on E^* because F is Lebesgue measurable and $F' \neq 0$ on E^*. It follows that g is a Lebesgue measurable function on E^* since $g = (gF')(1/F')$. By Theorem 17-21, g is then σ_F-measurable on E^*. Finally, since $\sigma_F(Z) = 0$, we conclude that g is σ_F-measurable on E▲

Theorem 17-25 Suppose F to be an absolutely continuous distribution function and g a function defined on a Lebesgue measurable set E. Then

$$\mathbf{L}\!\!\int_E g\,dF = \mathbf{L}\!\!\int_E gF' \tag{14}$$

if g is σ_F-summable on E or if gF' is Lebesgue summable on E.

Proof Since F' is finite except on a set of Lebesgue measure zero, it can be infinite only on a set of σ_F-measure zero (Theorem 17-20). Conveniently, where F' is infinite, redefine it to have the value zero; existence and values of integrals in (14) will not be affected.

It is clear that σ_F-summability of g implies that of g^+ and g^-. Further, if gF' is Lebesgue summable, then so are $(gF')^+ = g^+F'$ and $(gF')^- = g^-F'$. It suffices, therefore, to prove the theorem on the assumption $g \geq 0$.

Consider a sequence $\{g_n\}$ of σ_F-measurable simple functions such that $g_n \leq g$ for all n and all x in E, and converging almost everywhere (σ_F) on E to g. Then $\{g_n F'\}$ converges to gF' at all points of E except for a set A such that $\sigma_F(A) = 0$. It follows that F' and hence $g_n F'$ and gF' vanish on A except on a subset of A having Lebesgue measure zero. Therefore $\{g_n F'\}$ converges to gF' almost everywhere (μ) on A and hence almost everywhere (μ) on E.

For each simple function g_n we have $g_n = \sum_{\nu=1}^m c_\nu K_{E_\nu}$, where the c_ν are constants and K_{E_ν} are characteristic functions of disjoint σ_F-measurable sets E_ν such that $\bigcup_{\nu=1}^m E_\nu = E$. Let $Z = \{x;\ \ F'(x) = 0\}$. Then $\sigma_F(Z) = 0$ and we have

$$\mathbf{L}\!\!\int_E g_n\,dF = \mathbf{L}\!\!\int_{E-Z} g_n\,dF = \sum_{\nu=1}^m c_\nu\, \mathbf{L}\!\!\int_{E-Z} K_{E_\nu}\,dF = \sum_{\nu=1}^m c_\nu\, \mathbf{L}\!\!\int_{E_\nu - Z} dF.$$

Since $E_\nu - Z$ is Lebesgue measurable for each ν, it follows from Theorem 17-22 that $\mathbf{L}\!\!\int_{E_\nu - Z} dF = \mathbf{L}\!\!\int_{E_\nu - Z} F'$. Hence

$$\mathbf{L}\!\!\int_E g_n\,dF = \sum_{\nu=1}^m c_\nu\, \mathbf{L}\!\!\int_{E_\nu - Z} F' = \sum_{\nu=1}^m c_\nu\, \mathbf{L}\!\!\int_E K_{E_\nu} F' = \mathbf{L}\!\!\int_E g_n F'.$$

By the monotone convergence theorem, and its counterpart for Lebesgue–Stieltjes integrals (see Exercise 17-35), we have $\lim\{\fint_E g_n F'\} = \fint_E gF'$ and $\lim\{\fint_E g_n\, dF\} = \fint_E g\, dF$. The truth of the theorem follows▲

Example 17-13 The distribution function F given by $F(x) = 0$ if $x < 0$, $F(x) = x$ if $0 \le x < 1$, and $F(x) = 1$ if $1 \le x$ is clearly absolutely continuous with $F'(x) = 1$ if $0 < x < 1$ and $F'(x) = 0$ if $x < 0$ or $x > 1$. To find the Lebesgue–Stieltjes integral of the function x^2 with respect to F we have $\fint_{\mathscr{R}} x^2\, dF = \fint_{\mathscr{R}} x^2 F'(x)\, dx = \int_0^1 x^2\, dx = \frac{1}{3}$.

17-15 Sequences of Integrals

With respect to facts relating to sequences of increasing functions, the reader should recall Helly's compactness theorem. In the particular instance of a sequence of distribution functions, it affirms that if such a sequence is uniformly bounded, then a subsequence of it exists which converges weakly to a distribution function. Following are two theorems, also attributed to Helly, which relate Lebesgue–Stieltjes integrals with respect to elements and limit function of such subsequences. The first theorem concerns the restriction of distribution functions to a bounded interval, while the second generalizes the result to distribution functions proper, defined, as they are, on $\mathscr{R}$.

Theorem 17-26 Let $\{F_n\}$ be a uniformly bounded sequence of increasing functions, each continuous from the right at each point of an interval $[a,b]$. Suppose, too, that $\{F_n\}$ converges weakly to a limit function F which is increasing, continuous from the right at each point of $[a,b]$, and, in addition, is continuous at a and b. Then, if g is a continuous function on $[a,b]$,

$$\lim\left\{\fint_{[a,b]} g\, dF_n\right\} = \fint_{[a,b]} g\, dF.$$

Proof Let $\varepsilon > 0$ be given. Then, since g is uniformly continuous on $[a,b]$, a partition $P = \{x_0, x_1, \ldots, x_\kappa\}$ exists such that $|g(u) - g(v)| < \varepsilon$ if $|u - v| < \Delta P$. Further, the points $x_0, x_1, \ldots, x_\kappa$ may be chosen to be continuity points of F. Also, g is bounded; thus $|g(x)| \le M$ for some M and all $x \in [a,b]$.

Let a function g_ε be defined on $[a,b]$ by $g_\varepsilon(a) = g(x_1)$ and $g_\varepsilon(x) = g(x_\nu)$ if $x_{\nu-1} < x \le x_\nu$, $\nu = 1, 2, \ldots, \kappa$. We then have

$$\begin{aligned}\left|\fint_{[a,b]} g\, dF_n - \fint_{[a,b]} g\, dF\right| &\le \left|\fint_{[a,b]} g_\varepsilon\, dF_n - \fint_{[a,b]} g\, dF_n\right| \\ &\quad + \left|\fint_{[a,b]} g_\varepsilon\, dF_n - \fint_{[a,b]} g_\varepsilon\, dF\right| \qquad (15) \\ &\quad + \left|\fint_{[a,b]} g_\varepsilon\, dF - \fint_{[a,b]} g\, dF\right|.\end{aligned}$$

From the definition of g_ε it follows that $|g_\varepsilon(x) - g(x)| < \varepsilon$ for all $x \in [a,b]$. Further, because of the convergence of $\{F_n\}$ to F at continuity points of F, an integer N exists with the property $|F_n(x_\nu) - F(x_\nu)| < \varepsilon/\kappa M$ for all $n > N$. Lastly, since $\{F_n\}$ is uniformly bounded on $[a,b]$, we have $|F_n(x)| \le M'$ and $|F(x)| \le M'$ for $x \in [a,b]$ and $n = 1, 2, \dots$.

As a consequence of the above inequalities, the several terms in the right member of (15) may be estimated. We have

$$\left| \int_{[a,b]} g_\varepsilon \, dF_n - \int_{[a,b]} g \, dF_n \right| = \left| \int_{[a,b]} (g_\varepsilon - g) \, dF_n \right|$$

$$\le \varepsilon[F_n(b) - F_n(a)] \le 2M'\varepsilon.$$

Also

$$\left| \int_{[a,b]} g_\varepsilon \, dF_n - \int_{[a,b]} g_\varepsilon \, dF \right|$$

$$= \left| \sum_{\nu=1}^{\kappa} g(x_\nu)[F_n(x_\nu) - F_n(x_{\nu-1})] - \sum_{\nu=1}^{\kappa} g(x_\nu)[F(x_\nu) - F(x_{\nu-1})] \right|$$

$$= \left| \sum_{\nu=1}^{\kappa} g(x_\nu)[F_n(x_\nu) - F(x_\nu)] - \sum_{\nu=1}^{\kappa} g(x_\nu)[F_n(x_{\nu-1}) - F(x_{\nu-1})] \right|$$

$$\le \kappa M \cdot \frac{\varepsilon}{\kappa M} + \kappa M \cdot \frac{\varepsilon}{\kappa M} = 2\varepsilon.$$

Finally,

$$\left| \int_{[a,b]} g_\varepsilon \, dF - \int_{[a,b]} g \, dF \right| = \left| \int_{[a,b]} (g_\varepsilon - g) \, dF \right|$$

$$\le \varepsilon[F(b) - F(a)] \le 2M'\varepsilon.$$

From (15) we then have

$$\left| \int_{[a,b]} g \, dF_n - \int_{[a,b]} g \, dF \right| \le (4M' + 2)\varepsilon$$

for all $n > N$▲

Theorem 17-27 Let $\{F_n\}$ be a uniformly bounded sequence of distribution functions converging weakly to a distribution function F on $\mathscr{R}$. Suppose, too, that the limits $F(-\infty) = \lim \{F_n(-\infty)\}$ and $F(\infty) = \lim \{F_n(\infty)\}$ exist. Then if g is a continuous and bounded function on $\mathscr{R}$, $\lim \{\int_{\mathscr{R}} g \, dF_n\} = \int_{\mathscr{R}} g \, dF$.

Proof Let M be such that $|g(x)| \le M$ for all x, and let $\varepsilon > 0$ be given. Choose continuity points a and b $(a < b)$ of F such that $F(a) - F(-\infty) < \varepsilon$ and $F(\infty) - F(b) < \varepsilon$. Also, N_1 exists such that $|F_n(-\infty) - F(-\infty)| < \varepsilon$,

$|F_n(a) - F(a)| < \varepsilon$, $|F_n(b) - F(b)| < \varepsilon$, and $|F_n(\infty) - F(\infty)| < \varepsilon$ for all $n > N_1$. It follows that

$$\begin{aligned}\left|\oint_{(-\infty,a]} g\,dF_n + \oint_{(-\infty,a]} g\,dF\right| &\le M\,|F_n(a) - F_n(-\infty) + F(a) - F(-\infty)| \\ &\le M[|F_n(a) - F(a)| + 2\,|F(a) - F(-\infty)| \\ &\qquad + |F(-\infty) - F_n(-\infty)|] \\ &\le 4M\varepsilon\end{aligned}$$

for $n > N_1$. Similarly, $|\oint_{[b,\infty)} g\,dF_n - \oint_{[b,\infty)} g\,dF| \le 4M\varepsilon$ for $n > N_1$. Further, by the preceding theorem, N_2 exists such that $|\oint_{[a,b]} g\,dF_n - \oint_{[a,b]} g\,dF| < \varepsilon$ for $n > N_2$. It follows that, for $n > \max\{N_1,N_2\}$,

$$\left|\oint_{\mathscr{R}} g\,dF_n - \oint_{\mathscr{R}} g\,dF\right| < (8M + 1)\varepsilon\blacktriangle$$

Example 17-14 To observe the importance, in the hypothesis of the theorem, of the conditions $F(-\infty) = \lim\{F_n(-\infty)\}$ and $F(\infty) = \lim\{F_n(\infty)\}$, consider the sequence of functions $\{F_n\}$ defined by $F_n(x) = 0$ if $x < 0$, $F_n(x) = x/n$ if $0 \le x < n$, and $F_n(x) = 1$ if $n \le x$. Clearly, $\lim\{F_n(x)\} = F(x) = 0$ for each x, whence $F(\infty) = 0$. However, $\lim\{F_n(\infty)\} = \lim\{\lim_{\infty} F_n(x)\} = 1$. One sees, too, that $\lim\{\oint_{\mathscr{R}} dF_n\} = 1$, while $\oint_{\mathscr{R}} dF = 0$.

Exercises

17-22 Show that with $F(x) = 0$ if $x < 0$ and x^2 if $x \ge 0$, the Riemann–Stieltjes integral of Ψ with respect to F on $[0,1]$ does not exist. (See Example 17-11.)

17-23 Prove that if g is integrable on a domain E of σ-measure zero, then $\oint_E g\,d\sigma = 0$.

17-24 Prove that if g is non-negative and such that $\oint_E g\,d\sigma = 0$, then $g = 0$ almost everywhere (σ) on E.

17-25 Prove that if $\oint_E g\,d\sigma$ exists and if k is any constant, then kg is integrable with respect to σ and $\oint_E kg\,d\sigma = k\oint_E g\,d\sigma$.

17-26 Prove that if g is integrable with respect to σ on E, then so is $|g|$ and $|\oint_E g\,d\sigma| \le \oint_E |g|\,d\sigma$.

17-27 Prove that if f and g are summable with respect to σ on a common domain E, then fg and $f+g$ are summable on E; moreover, $\oint_E (f+g)\,d\sigma = \oint_E f\,d\sigma + \oint_E g\,d\sigma$.

17-28 Prove that if f and g are integrable with respect to σ on E, and if $f \le g$ almost everywhere (σ) on E, then $\oint_E f\,d\sigma \le \oint_E g\,d\sigma$.

17-29 Prove that if g is summable with respect to σ on E, and if $\{E_\nu\}$ is a countable collection of disjoint members of $\mathscr{F}^\sigma$, with $\bigcup_\nu E_\nu = E$, then g is summable on each domain E_ν, and $\oint_E g\,d\sigma = \sum_\nu \oint_{E_\nu} g\,d\sigma$.

17-30 State and prove the extension of Egoroff's theorem (Theorem 15-49) to the case of general measure.

17-31 Prove the following properties for measurability of a function relative to any additive class $\mathscr{F}$ of subsets of $\mathscr{R}$: (a) If g is measurable relative to $\mathscr{F}$, then the domain of g is a member of $\mathscr{F}$. (b) If f is continuous and g is measurable relative to $\mathscr{F}$, then $f \circ g$ is measurable relative to $\mathscr{F}$. (c) If f and g are measurable relative to $\mathscr{F}$, then so are $f + g$ and fg. (d) If $\{g_n\}$ is a uniformly bounded sequence of functions all measurable relative to $\mathscr{F}$, then $\sup\{g_n\}$, $\inf\{g_n\}$, $\underline{\lim}\,\{g_n\}$, and $\overline{\lim}\,\{g_n\}$ are measurable relative to $\mathscr{F}$. (e) A function g is measurable relative to $\mathscr{F}$ if and only if a sequence of simple functions, each measurable relative to $\mathscr{F}$, exists that converges to g.

17-32 Prove that a bounded function g on a domain E of finite σ-measure is measurable relative to $\mathscr{F}^\sigma$ if and only if g is integrable with respect to σ on E.

17-33 Prove that if g is integrable with respect to a measure function σ on a domain E belonging to $\mathscr{F}^\sigma$, then g is measurable relative to $\mathscr{F}^\sigma$.

17-34 If σ is a measure function on $\mathscr{F}^\sigma$ and if g is a measurable function relative to $\mathscr{F}^\sigma$ on a domain E belonging to $\mathscr{F}^\sigma$, then g is integrable on E with respect to σ provided that a function h exists that is summable with respect to σ on E and is such that $|g| \leq h$ almost everywhere (σ).

17-35 State and prove the extension of the monotone convergence theorem to the case of Lebesgue–Stieltjes integrals.

17-36 State and prove the extension of Fatou's theorem to the case of Lebesgue–Stieltjes integrals.

17-37 State and prove the extension of the dominated convergence theorem to the case of Lebesgue–Stieltjes integrals.

17-38 Suppose σ_1 and σ_2 to be measure functions on additive classes $\mathscr{F}^{\sigma_1}$ and $\mathscr{F}^{\sigma_2}$, respectively. Let g be a function defined on a domain E belonging to the additive class $\mathscr{F}^{\sigma_1} \cap \mathscr{F}^{\sigma_2}$ on which the measure function $\sigma_1 + \sigma_2$ is defined. Prove that if g is summable with respect to σ_1 and σ_2 on E, then it is summable with respect to $\sigma_1 + \sigma_2$. Moreover, $\oint_E g\,d\sigma_1 + \oint_E g\,d\sigma_2 = \oint_E g\,d(\sigma_1 + \sigma_2)$.

17-39 Prove Theorem 17-22 in the case where E is unbounded.

17-40 Evaluate $\oint_{\mathscr{R}} x\,dF$ if $F(x) = 0$ for $x < 0$ and $1 - e^{-x}$ for $0 \leq x$.

Bibliography

Birkhoff, G. and S. MacLane, *A Survey of Modern Algebra*, 3rd ed. New York: The Macmillan Company, 1965.

Burkill, J. C., *The Lebesgue Integral*. New York: Cambridge University Press, 1951.

Burrill, C. W., *Foundations of Real Numbers*. New York: McGraw-Hill Book Company, Inc., 1967.

Courant, R. and H. Robbins, *What is Mathematics?* New York: Oxford University Press, 1941.

Graves, L. M., *The Theory of Functions of Real Variables*, 2nd ed. New York: McGraw-Hill Book Company, Inc., 1956.

Halmos, P. R., *Measure Theory*. Princeton, N.J.: D. Van Nostrand Company, Inc., 1950.

Hobson, E. W., *The Theory of Functions of a Real Variable and The Theory of Fourier's Series* (2 Vols.). London: Cambridge University Press, Vol. I, 3rd ed., 1927, Vol. II, 2nd ed., 1926.

Jeffery, R. L., *The Theory of Functions of a Real Variable*, 2nd ed. Toronto: University of Toronto Press, 1953.

Kelley, J. L., *General Topology*. Princeton, N.J.: D. Van Nostrand Company, Inc., 1955.

Kestelman, H., *Modern Theories of Integration*. New York: Oxford University Press, 1937.

Knopp, K., *Theory and Applications of Infinite Series*, 2nd ed. New York: Hafner Publishing Company, 1948.

McShane, E. J., *Integration*. Princeton, N.J.: Princeton University Press, 1961.

McShane, E. J. and T. Botts, *Real Analysis*. Princeton, N.J.: D. Van Nostrand Company, Inc., 1959.

Munroe, M. E., *Introduction to Measure and Integration*. Reading, Mass.: Addison-Wesley Publishing Company, Inc., 1953.

Natanson, I. P., *Theory of Functions of a Real Variable* (2 Vols.). New York: Frederick Ungar Publishing Company, Vol. I, edited by L. F. Boron and E. Hewitt, 1955; Vol. II, edited by L. F. Boron, 1959.

Newman, M. H. A., *Elements of the Topology of Plane Sets of Points*, 2nd ed. London: Cambridge University Press, 1951.

Olmsted, J. M. H., *Real Variables*. New York: Appleton-Century-Crofts, 1959.

Royden, H. L., *Real Analysis*. New York: The Macmillan Company, 1963.

Rudin, W., *Principles of Mathematical Analysis*, 2nd ed. New York: McGraw-Hill Book Company, Inc., 1964.

Saks, S., *Theory of the Integral*, translated by L. C. Young, 2nd rev. ed. New York: Dover Publications, Inc., 1964.

Index